PRIMATE BIO-SOCIAL DEVELOPMENT

PRIMATE BIO-SOCIAL DEVELOPMENT:

Biological, Social, and Ecological Determinants

EDITED BY

Suzanne Chevalier-Skolnikoff
STANFORD UNIVERSITY

Frank E. Poirier
OHIO STATE UNIVERSITY

GARLAND PUBLISHING, INC.
NEW YORK & LONDON
1977

Copyright © 1977 by Garland Publishing, Inc.
All Rights Reserved

Library of Congress Cataloging in Publication Data

Main entry under title:
Primate bio-social development.

Includes bibliographies.
1. Primates—Behavior. 2. Social behavior in animals. I. Chevalier-
Skolnikoff, Suzanne. II. Poirier, Frank E. [DNLM: 1. Primates—
Congresses. 2. Socialization—Congresses. 3. Social environment—
Congresses. 4. Social behavior—Congresses. 5. Psychology, Compara-
tive—Congresses. 6. Models, Psychological—Congresses. QL737.P9
P952-23]
QL737.P9P6714 599'.05 76-25748

Hardcover: ISBN 0-8240-9900-1
Softcover: ISBN 0-8240-9854-4

Printed in the United States of America

Contributors

Stephanie Alpert, Primate Behavior Laboratory, Department of Psychiatry, State University of New York Downstate Medical Center, Brooklyn

Janice I. Baldwin, Department of Sociology, University of California at Santa Barbara

John D. Baldwin, Department of Sociology, University of California at Santa Barbara

Marc Bekoff, Ethology Group, University of Colorado at Boulder

Gershon Berkson, Illinois Institute for Developmental Disabilities, Chicago

Craig Bielert, Wisconsin Regional Primate Research Center, Madison

Suzanne Chevalier-Skolnikoff, G. W. Hooper Foundation, University of California Medical Center at San Francisco; Department of Anthropology, Stanford University

Cathleen B. Clark, Department of Anthropology, University of California at Davis

Christopher L. Coe, Primate Behavior Laboratory, Department of Psychiatry, State University of New York Downstate Medical Center, Brooklyn

William J. Demarest, Department of Anthropology, Stanford University

Laurence Fedigan, Department of Anthropology, University of Alberta

Linda Marie Fedigan, Department of Anthropology, University of Alberta

Daniel G. Freedman, Committee on Human Development, The University of Chicago

Kathleen R. Gibson, Department of Neurobiology and Anatomy, University of Texas Medical School at Houston

David Agee Horr, Peabody Museum, Harvard University

Joel N. Kaplan, Developmental Psychobiology Program, Stanford Research Institute, Menlo Park, California

W. C. McGrew, Department of Psychology, University of Stirling; Gombe Stream Research Centre, Kigoma, Tanzania

David Mack, Washington D.C. Zoo

Nancy A. Nicolson, Department of Anthropology, Harvard University

Sue Taylor Parker, Department of Anthropology, California State College at Sonoma

Frank E. Poirier, Department of Anthropology, The Ohio State University

Leonard A. Rosenblum, Primate Behavior Laboratory, Department of Psychiatry, State University of New York Downstate Medical Center, Brooklyn

Richard C. Savin-Williams, Committee on Human Development, The University of Chicago

Jefferson Slimp, Wisconsin Regional Primate Research Center, Madison; Department of Psychology, University of Wisconsin at Madison

Robert W. Sussman, Department of Anthropology, Washington University, St. Louis, Missouri

Thomas J. Testa, Department of Psychiatry, University of Pennsylvania at Philadelphia

Kim Wallen, Wisconsin Regional Primate Research Center, Madison; Neurosciences Training Program, University of Wisconsin at Madison

Contents

Preface

This work is part of a continuing effort to delineate some mechanisms and results of primate socialization and learning processes; our present hope is to disseminate information on the "current state of the art." The impetus for this work comes from two symposia on primate socialization (in New York City, 1972, and Dallas, 1973) of which most of the contributors were a part. The present articles attempt to isolate some of the early variables affecting behavioral development in the hope of eventually constructing an evolutionary framework, for until recently, potential contributions of biologically oriented explanations of social development and social behavior were largely ignored or misunderstood. However, as Simmel (1970) has noted, as an interest in the biological aspects of socialization has increased and its importance been recognized, there has been an acceleration of interdisciplinary research and communication. This book, with articles by authors in various disciplines, is witness to the fact that cooperation raises the level of sophistication in the use of biologically oriented concepts and explanations.

Even a cursory look at primate field studies shows that field investigators have spent considerable time viewing the end result of the socialization process instead of concentrating on the process itself. Some investigators have attempted to systematize the rapidly accumulating field data by relating ecology to social behavior and social organization; however, in order to understand the variables contributing to the diversity of nonhuman primate societies and social behavior, a wider framework is needed. Since most primates live in stable, complex, bisexual, year-around social groups, one fruitful avenue of research would attempt to discern how the socialization process affects behavior and, through this behavior, the social order.

This book contains articles derived from field and laboratory research. The field situation presents obvious problems and uncertainties for studying the socialization process, i.e., not finding a sufficient number of animals with which to work. Furthermore, field conditions present problems that make it very difficult to complete a detailed analysis of a restricted developmental stage. Traditional field studies of primate societies and ecology only partially answer the important questions about particular social behaviors and social organizations; however, we cannot fully understand nor can we explain primate behavior unless and until we have investigated the ontogeny of behavior. A closer examination of the socialization process enables us to concentrate on the definitive aspects of social behavior rather than the gross patterns normally investigated in current research.

Because of the problems inherent in the field situation, studies in laboratory situations whereby constant conditions can be maintained and elaborate testing procedures can occur are requisite. This work contains a number of articles compiled from laboratory studies. Close observational situations as at the Gombe National Park (as represented by Clark, McGrew, and Nicolson in this volume) perhaps provide the closest approximation to natural conditions from which good socialization data can be obtained.

The total web of social learning, sensorimotor coordination, and the reflex system can only be tentatively described from observations of free-ranging animals; thus if the complex factors determining the behavior of mature animals living in social groups are to be evaluated, detailed experimental analysis is needed. Laboratory and field studies must supplement each other in experiments concerned with behavioral development; wherever possible, hypotheses about social learning generated from field observations of socialization must be tested under controlled laboratory conditions. Laboratory observations may shed light on some factors of social interaction and learning not immediately obvious when seen within the context of the intricate structure of a free-ranging social group.

The organization of the book represents an attempt to examine some of the multiple determinants of the primate socialization process through a series of studies classified under the topics: Biological Determinants, Social Determinants, and Ecological Determinants. These chapters represent a series of exemplary studies rather than a systematic examination of the field. Some areas are more thoroughly covered than others, and some species are represented more thoroughly than others. To some extent, these strengths and shortcomings reflect the state of the field.

Part I, the biological determinants of socialization, focuses on the relationships between behavioral development (Parker), neural maturation (Gibson), and socialization potential (Chevalier-Skolnikoff), and the effects of various biological defects upon socialization (Berkson, and Fedigan and Fedigan). Genetics and hormones are not specifically represented in this section, although numerous chapters deal indirectly with genetics as they present data on sex and species differences. In addition, the chapters by Testa and Mack, and Wallen et al. present some data on hormonal influences, although their chapters focus mainly on social influences.

Part II considers social influences and socialization. The chapters by Kaplan, Clark, McGrew, Horr, and Demarest focus especially on the influence of the mother; those of Baldwin and Baldwin, Testa and Mack, and Wallen et al. focus especially on the effects of peers; and the studies of Rosenblum and Coe, Rosenblum and Alpert, and Kaplan examine the effects of multiple social influences. Kaplan's chapter also investigates the significance of direct stimuli, specifically, the effects of maternal perceptual properties on attachment. Examination of the effects of adult males on socialization are not represented here although they have been surveyed previously by Mitchell and Brandt (1972).

In Part III, the effects of indirect stimuli are examined in the chapters by Nicolson and Sussman, who focus upon ecological influences on socialization.

In Part IV, Savin-Williams and Freedman give a phylogenetic perspective to the book as they present material on human socialization, and Bekoff's chapter adds a general mammalian perspective as it examines some theories of mammalian socialization.

SUZANNE CHEVALIER-SKOLNIKOFF
FRANK E. POIRIER

1 Introduction

FRANK E. POIRIER

■ SOCIALIZATION DEFINED

One of the many problems associated with establishing a definition is that it is often harder to define socialization than it is to isolate its consequences. A key factor that concerns nonhuman primate studies is the lack of longitudinal behavioral studies. Most field studies last from twelve to twenty-four months which is hardly enough time to fully comprehend the socialization and learning processes. Therefore, much of the theoretical framework was originally drawn from studies of humans (i.e., Child 1954; Clausen 1968; Erickson 1950; Freud 1923, 1930), dogs (Scott 1945, 1950, 1950) and cats (Rosenblatt et al. 1961). This is a highly diverse group; however, that there are similarities in the mammalian socialization process is apparent (cf. Bekoff this volume). One result of the lack of data from primate behavior led to borrowing concepts from other works; early socialization studies were thus strongly influenced by the critical periods hypothesis as expounded by Lorenz (1937), Scott and Marston (1950), and Riesen (1961). However, this scheme cannot be completely superimposed upon the nonhuman primate process.

This volume presents a number of definitions for socialization and it is evident that each contributor refers to a wide range of complex phenomena. Each author presents his or her own definition of socialization thereby making an inclusive definition difficult to establish. Primates live in highly complex, bisexual, year-round social groups, and through an animal's learned social relationships with its fellow group members, individuals learn how to interact and behave within their social group. Socialization, as used here, refers to the sum total of an animal's past social experiences which, in turn, may be expected to shape future social behavior. "Socialization is that process linking an ongoing society to a new individual. Through socialization, a group passes its social traditions and life-ways to succeeding generations" (Poirier 1973:4). The socialization process ensures that adaptive behavior will not have to be discovered anew each generation (Poirier 1972a).

The socialization process simultaneously refers to the external stimuli received by an organism, the individual nature of the process, and to the end product or consequences of socialization (Poirier 1971, 1972a,c, 1973a,b). Each individual is the outcome, the result, of a given socialization process and we must look at the variables influencing this output both vertically (through time) and horizontally (in terms of social interactions). The results, or consequences, of socialization not only depend upon the original genetic material of the individual (see, for example, Fedigan and Fedigan; Berkson in this volume) and the degree to which climate, nurturance, and other factors permit realization of that potential; they are also influenced by the behavior of the adults and peers with whom the individual is or has been in regular contact (Burton 1972; Chalmers 1972; Lancaster 1972; Mitchell and Brandt 1972; Ransom and Rowell 1972; Sugiyama 1972). For a further discussion of these points, refer to the articles by Bekoff; Clark; Demarest; Fedigan and Fedigan; Horr; Kaplan; McGrew; Nicolson; Rosenblum and Alpert; and Rosenblum and Coe.

It is becoming clear that socialization variables have a differential effect depending on the time and level of the socialization process (Poirier 1973). We must investigate how an animal adapts to different developmental situations. Loy (1973) suggests that a nonhuman primate must learn to adapt to three conditions: (1) it must learn to become a member of its species, (2) it must learn to become a member of a particular social group—and we must remember that social groups exhibit great variability, even intraspecifically, and (3) it must adapt to its environment. These adaptations can be accomplished in many ways and we should, perhaps, be looking into more specific adaptive mechanisms to these different pressures. The fact that there are different adaptive pressures at different points in an animal's life, and the fact that at any one point in time animals may be unable to meet all these pressures (i.e., as discussed in this volume because of immature nervous systems, physical or emotional disabilities, and the lack of previously encountered social behaviors) is relevant to our research interests.

■ PRIMATE LEARNING AND SOCIALIZATION

Among other features, the nonhuman primate socialization process is influenced by the facts that these animals are social living, that they have a prolonged period of immaturity, and that the mother nurses the young. As pointed out by Bekoff (this volume) these traits are characteristic of mammals generally.[1] The fact that most primates are social animals residing in highly complex, bisexual, year-around social groups of varying size and composition is crucial. The social group has long been the primate niche, indeed, the social group has long been the mammalian niche. Group characteristics vary, and the degree of sociality, dominance, sexuality, and interanimal relationships varies; however, most primates spend part of their life in close association with conspecifics. (See Table 1.) Within the social group an animal learns to express its biology and adapt to its surroundings. Differences among primate societies depend upon the species' biology and, to a great extent, upon the circumstances in which animals live and learn. The composition of the social group, the

TABLE 1 / SOCIAL TRENDS IN PRIMATE EVOLUTION (Adapted from J. Pfieffer, 1969)

INCREASING SOCIAL COMPLEXITY		
Mother-infant groups	Mother-father-infant groups (family groups)	Groups of all ages and sexes (large social groups)
tree shrews	marmosets	baboons
lorises	lemurs	macaques
etc.	gibbons	colobines
		New World monkeys
		Old World monkeys
		chimpanzees
		gorillas
		hominids

particular balance of interanimal relationships, constitutes the social environment within which youngsters learn and mature. Because of the highly social nature of nonhuman primates we must view groups, as well as individuals, as the adaptive units of the species (Poirier 1973b).

Social living places a premium upon learning. Although many animals—i.e., birds and fish—have social behaviors, these are largely dependent upon fixed and innate cues. On the other hand, primates respond not only to fixed cues but to learned behaviors. Since there is considerable individual and behavioral variability within the social environment, primates must be flexible and discriminating in their social responses (as an example see McGrew this volume). In fact, one highly important adaptive trait common to nonhuman primates is the degree of behavioral flexibility and adaptability (Jay 1968; Poirier 1969c).[2] Since most primates live a rather complex social life, they must learn to adjust to one another, to get along; compared to most of the animal world, primate societies may have the greatest differentiation of *learned* social roles. Correspondingly, the primate brain has evolved into a complex and efficient learning mechanism.

Diamond and Hall (1969) specify the mammalian association cortex as the neocortex subdivision where prime evolutionary advancements have occurred. Although primate learning skills are not solely accountable by the volume of the neocortex relative to total brain volume, it is significant that the primate neocortex is proportionately larger than is true for carnivores and rodents (Harman 1957). The complex cognitive processes and advanced learning skills are accommodated by increased cortical fissuration, increased numbers of cortical units in the cortex fine structure, and the refinement of the subcortical structure interrelating the thalmus and cortex (Norback and Moskowitz 1963; Rumbaugh, 1970).

Social living is requisite for the younger primate to perform effectively as an adult of its species. Animals with restricted social experiences, for example, those raised in isolation or in unnatural conditions, exhibit some degree of social maladjustment, most especially in mothering, sexual, grooming, and aggressive behavioral patterns (Mason 1960, 1961a,b, 1963, and a number of authors in this volume). Laboratory studies suggest that the full development of an animal's biological potentialities requires stimulus and direction from social forces such as are usually supplied from the social group (Harlow 1963, 1965; Mason 1963, 1965).

While troop, or social life is important, it must be cautioned that not all primates have the "same degree" of social life (i.e., Horr, this volume). Among Nilgiri langurs, (*Presbytis johnii*) for example, social relations are not oriented toward individual protection by cooperative group action, but instead, toward self-protection by flight through the nearest trees (Poirier 1969a, 1973b).[3] Why then does the Nilgiri langur still live in a social group if the animals do not take full advantage of the opportunities of group life in the form of protection, grooming, and play? Washburn and Hamburg (1965) suggest that a primary reason for group existence is learning, the group being the center of knowledge and experience far exceeding that of its individual constituents. Within the group experience is pooled and generations linked—troop traditions are more advantageous than individual learning in many situations (Kummer 1971; McGrew, this volume; Poirier 1972c, 1973b). Tradition pools individual experiences and is superior to individual learning if the new behavior is difficult to acquire individually in direct interaction with the environment (Poirier 1973b). Troop tradition is based upon a long life expectancy (a primate biological trait) and a leading role for older animals. (In fact, primate societies may be loosely viewed as gerontocracies.)

Within the social context the animal is socialized, learns what foods to eat, who are existing predators, and the correct mode of behavioral interaction. Primates learn their mode of survival by living in a troop where they benefit from the shared knowledge and experience of the species (Poirier 1970a, 1971). The primary reinforcement for all normal primate learning is the social context, the group in which the infant is born and nurtured. Even independent sensorimotor activities like observing, manipulating, and exploring receive some facilitation, or inhibition, from the group setting (Hall 1968; Nicolson, this volume). Contrasting social structures impose differences in learning patterns leading to individualized behavior formation; this implies that group modification will alter the socialization process yielding individuals with different behaviors (Poirier 1970a; 1972b; 1973b; Rosenblum and Alpert, this volume; Sugiyama 1972).

Flexibility and adaptability, learning to exist and coexist within the social context, learning one's role in the social order, is at an optimum for primates. Primates inherit an ease of learning rather than fixed instinctive patterns; they easily, almost inevitably, learn behaviors essential for survival. Primates learn to be social, and under normal circumstances individual learning almost always occurs (Washburn and Hamburg 1965). Presumably, in most higher mammalian social systems, and particularly in primate social systems, individual behaviors are controlled by a continuous process of social learning arising from group interactional patterns. Learning to act according to social modes is extremely important, for animals whose behavioral traits do not conform sufficiently to group norms are less likely to reproduce and may be ejected. Social selection of this type apparently has a strong stabilizing influence upon the genetic basis of temperamental traits and motivational thresholds. Crook (1970) suggests that primate societies might determine the genetic basis of individual social responses.

Learning during socialization and the emergence of one's social role has a preponderant influence in shaping individual behavior. Social conformity and the maintenance of a group structure results from the adoption of traditional behaviors characteristic of the total social system. This is primarily accomplished by three interacting groups of factors (Crook 1970): (1) the species repertoire of biologically programmed neonate reflexes and social signals, plus innate factors affecting temperament and tendencies to learn some responses more readily than others, (2) the behavior of individuals comprising the social milieu, which partly controls the emergence of individual role playing, and (3) direct environmental effects, i.e., availability of need-reducing commodities and consequent behavioral learning that exploits the world in the manner ensuring greatest individual survival.

There are various social learning processes involved in these conformities and although learning processes vary interspecifically, as the contributions herein indicate, there are consistencies. Social facilitation and observational learning seem to be the most important vehicles for role assumption (Hall and Goswell 1964). However, when we study the complex interactional system of a primate group, we may suppose that any of the defined learning processes may be operating in the adjustments of individuals to their social and physical environment (Hall 1968).

Young primates are socialized; they learn by following their mother's actions, imitating their actions and the directions they take to the objects to which they relate. McGrew (this volume) provides an example of this modeling as it involves chimpanzee tool use and manufacture. Any animal's behavior may be set or facilitated by the perceived example of another group member (i.e., a peer, sibling, or an adult). Animals within a social group observe each other's behavior with or without awareness of the reference; later, they may behave similarly in a similar situation. Observational learning probably contributes largely to the broader problem of social learning (Hall 1963; Hall and Goswell 1964; Kawamura 1959; Menzel 1966; Tsumori 1967). An animal learns the consequences of another's behavior through direct visual observation within the group and adjusts its own behavior accordingly. As will be shown by later contributions, the frequency of events occasioning direct and indirect learning can be influenced by habitat (especially vegetational density which affects interanimal observation), seasonal food shortages occasioning increased travel, foraging and dispersal, and population density determining the extent and frequency of intra- and intergroup interactions.

Social conflict also seems to play a role in primate socialization and learning. Berlyne (1960:10) notes that conflict ". . . is an inescapable accompaniment of the existence of all higher animals, because of the endless diversity of the stimuli which act on them and of the responses that they have the ability to perform." Primate societies permit a continual interplay of friendly, sexual, aggressive, and fearful impulses, all of which must be balanced. The balancing of these helps maintain the social framework.

Primate socialization, through learning, may also occur during "instructed sessions," e.g., in some situations an animal's behavior may be directly molded by another animal. Nicolson (this volume) notes that two chimpanzee mothers directly detached their infants from their bodies, and when one of the infants was nine months old its mother attempted to "walk" him—she walked bipedally backwards, holding the infant's hands in hers. Such maternal manipulation probably encouraged the development of motor skills and independent behavior. Another chimpanzee mother repeatedly placed her infant's foot against the enclosure fence, bending the toes until they grasped the wire. She left the infant hanging under her watchful gaze. Furthermore, infants learn what to eat not only by watching the mother, but also by the mother's watching and limiting what the infants eat. Numerous instances of mothers taking nonedible items from their infants' mouths have been recorded. Youngsters also learn what animals and possible dangers must be avoided when a mother, or other adult, chases it from dangerous stimuli. Young females who handle infants roughly are often physically and vocally rebuked by the infant's mother; should they manage well, they receive a positive reward by being allowed to retain the infant (Jay 1962, 1965; Poirier 1969b, 1970a).

▪ BIOLOGICAL FACTORS INFLUENCING SOCIALIZATION

PROLONGED IMMATURITY

Primate socialization is influenced by certain primate biological characteristics. We mentioned previously the primate brain, especially the size of the neocortex. Washburn and Hamburg (1965) suggest that group life is the sociological response to the primate biological adaptation of prolonged immaturity. (It is, of course, possible that the situation is reversed, e.g., prolonged immaturity may be a biological response to group life.) Washburn and Hamburg argue that prolongation of preadult life is biologically expensive and that a major compensation is learning. Despite restraints imposed upon the social order, the long infancy period is advantageous; ". . . it provides the species with the capacity to learn the behavioral requirements for adapting to a wide variety of environmental conditions" (Washburn and Hamburg 1965:620).

Schultz (1956) has shown a progressive increase in preadult life as one moves from prosimian to human; one result of this prolongation of youth is an extension of time available for learning and socialization. With retardation of growth and a longer period of immaturity, there is a clear tendency for individual experiences to play a more subtle role in shaping behavior into effective patterns (i.e., Bekoff, this volume). An extended period of infant dependency enhances the amount and complexity of learning possible, while it increases opportunities to shape behavior to meet local environmental (social and natural) conditions. Flexibility of behavioral patterns may be one of the principal benefits of the longer dependency period (Poirier 1969a,c, 1970a, 1972a,c, 1973b). There appears to be a positive correlation between prolonged postnatal dependency and increasing complexities of adult behavior and social relationships. Prolonged youthfulness allows more time for contact with peers in the form of play (Baldwin and Baldwin, this volume) and more time for adult contact which probably promotes the socialization process and helps integrate youngsters into the social group.

A prolonged period of nursing is related to prolonged immaturity. The fact that a primate mother normally bears one infant per parturition, and the fact that this infant is relatively helpless for varying lengths of time, promotes strong mother-infant contact. The strength of this attachment is more fully realized when one studies the cessation of this close contact, i.e., during weaning (Clark, this volume). Furthermore, since the infant is relatively helpless, the mother is responsible for much of its early locomotion, carrying it about on her back or stomach (or in the mouth, as in some New World monkeys and prosimians). Many of the infant's early perceptions of the world are from the mother's back and stomach, serving to channel sensory input and setting the first stage in molding the infant's relations with other animals. Infants riding on the back or ventrum of dominant mothers witness different interanimal relations than infants riding on the back or ventrum of subordinate mothers.

POSTPARTUM ACTIVITIES

There is also the possibility that certain immediate postpartum activities help bind the mother and infant and vice versa. Perceptual data (i.e., Kaplan, this volume) suggest that infants relate to their mothers via visual and olfactory channels. Harlow's (1962, 1963) and Mason's (1960, 1965) work clearly shows that tactual senses are important in this early contact. However, there may be other factors "binding" the pair, i.e., the mother's cleaning of the infant. Among some mammals like sheep, rabbits, cats, and rats, a mother's postpartum licking and cleaning of her infant leads to proper maternal-infant attachment and to activities like proper bowel evacuation. Material on immediate postpartum primate activity is limited by the scant number of births witnessed; however, there are suggestions that a similar condition is found among nonhuman primates (Brandt and Mitchell 1971; Lindburgh and Hazell 1972).

Among prosimians and monkeys there is a rather consistent pattern in which the mother licks both herself and her infant clean soon after birth. Many primate mothers also consume the placenta and umbilical cord soon after expulsion. Loris (Doyle et al. 1967), lemur (Petter-Rousseaux 1964), and galago mothers generally consume the afterbirth before directly attending to the infant. Similar patterns have been observed in some New World monkeys, i.e., Takeshita (1961) notes that the third phase of squirrel monkey (*Saimiri sciureus*) delivery behavior involves consumption of the placenta. Hopf (1967) notes that squirrel monkeys also consume the umbilical cord and Williams (1967) reported that a woolly monkey may nibble the umbilical cord.

Similar patterns emerge in Old World monkeys. Rosenblum and Rosenblum (in Brandt and Mitchell 1971) report that a guenon (*Cercopithecus erythrotis*) mother licked the infant and afterwards consumed the placenta. A similar pattern has been noted for a mona monkey (*Cercopithecus mona*) (Takeshita 1961). A patas mother has been observed to lick her infant and eat the umbilical cord and placenta (Goswell and Gartlan 1965). Gillman and Gilbert (1946) reported that a baboon mother licked the infant clean and consumed the placenta. Rhesus (*Macaca mulatta*) mothers have been reported to consume both umbilical cord and placenta (Hartman 1928; Tinklepaugh and Hartman 1932) and presumably, stumptail macaques eat the placenta (Bertrand 1969). Fedigan and Fedigan (this volume) report that Japanese macaque mothers lick the infant clean.

Because few pongid births are recorded, the situation is less clear. However, gorilla infants born at the Columbus Zoo have been licked clean and mothers have been reported to consume the placenta and afterbirth. Chimpanzee mothers may consume the afterbirth, and an orangutan mother has been seen to suck the fetal membranes (Graham-Jones and Hill 1962).

Consumption of the afterbirth can serve a number of functions, i.e., the afterbirth might satisfy the mother's appetite for a few days which would allow her to give her full attention and energy to the infant. Consumption would also prevent an odorous source from possibly attracting predators. Interestingly, Meier (1965) reported that surgically delivered infants of some laboratory-reared rhesus females were not accepted by their mothers, whereas naturally delivered infants of laboratory mothers were cared for. The question arises as to whether a mother deprived of the natural birth process, including consuming the placenta, treats the infant differently.

PHYSICAL MATURATION

It has always been recognized that social influences will have varying effects depending upon the developmental state of the receptor, the receiving individual. The developmental state may be defined in a number of ways, but few will probably be more exact than studies of postnatal myelinization. There is little yet concretely known about the role of brain myelinization; the relationship between myelinization and onset of function in ontogeny is still incompletely understood, and there is considerable disagreement among the investigators. Although more is known about brain functions and development of social behavior among rhesus than in any other primate, little is known about the subject linking the two, e.g., the interaction of neurological and behavioral maturation. However, myelinization data appears to become more important as we learn more about the interrelationship between behavioral and physical maturation.

Myelinization studies are in a position to lend supporting information to neurophysiological studies of maturation, and their major advantage lies in the realm of comparative neuroanatomy. Furthermore, there is a wealth of data on the myelinization process, far exceeding that of any parameter of neurological maturation, available for many species, i.e., humans, baboons and other monkeys, dogs, cats, sheep, guinea pigs, sheep, rabbits, mice, hens, and fish such as salmon. Although the myelinization data is not yet well understood, especially concerning its relationship to the onset of behavioral patterns, it is conceivable that it has something to do with rates of development and the onset of behavioral patterns as these are noted by Gibson (this volume) and Parker (this volume) for example. The role of myelinization in species with short and prolonged periods of immaturity is speculative. Myelinization may be involved in the concept of heterochronous systemogenesis; that is, that the course of development depends upon demands being made on the infant at various times. As Bekoff (this volume) and Gibson (this volume) note, there are differences in the order of development of the sensory systems. Bekoff, citing Gottlieb (1971), notes that although sensory development is patterned according to differential needs within species, it has the same general patterning across species.

Gibson's (this volume) and Parker's (this volume) work (see also Chevalier-Skolnikoff 1973a,b) clearly shows that the development of social behavior is inextricably tied to CNS (central nervous system) maturation involving the ontogeny of sensory and motor patterns (Bekoff [this volume] also discusses this variable and its relationship to socialization). Parker's work clearly supports Kuo's (1967) contention that all behavior is a functional concomitant of the interrelationship of morphological, biophysical, and biochemical factors. Although the interaction of all variables is not clearly understood, Parker's application of Piaget's work suggests that such research would further clarify the interacting principles. Because we can understand the organization of behavior better by studying how it changes over time, Piaget's emphasis on developmental sequences is significant.

Piaget's approach can be valuable towards establishing an evolutionary perspective for the socialization process, for his methodology allows us to see how natural selection has evolved different behavioral programs in different lineages. Comparative longitudinal developmental studies such as Parker's facilitate behavioral analysis through contrast, without which we cannot fully comprehend the significance of behavioral organization. Although Piaget's model deals with the organization of human intelligent behavior, Parker proposes its utilitarian value as a framework for analyzing the organization of behavior generally. Piaget's emphasis is on the continuity of motor and mental operations making the model suitable for comparing the intelligence of human and nonhuman primates; however, the model has limited value as a cross-specific intelligence test. Parker's application of Piaget's scheme to the development of an infant macaque allows us to see differences in the behavioral organization in two species, and allows us to compare the contexts in which these behavioral differences occur. Parker's work is most interesting and surely will be followed by others.

Fedigan and Fedigan's work (this volume) on a severely handicapped Japanese macaque infant is germane to this discussion. The Fedigans note the development away from a self and self-and-mother stage towards a peer and then a group orientation. While normal infants showed evidence of this progression, the handicapped infant did not, and the infant never progressed beyond the self and mother-centered stage probably because of his disabilities. Using the readiness concept borrowed from human infant development, the Fedigans query whether the serious motor and sensory deficiencies of the handicapped infant arrested his development at an early stage of physical, social, and cognitive spheres. His inability to engage in normal play and exploratory behaviors and his lack of communicatory skills may have contributed to his poor perceptions of rank, affiliation, etc., manifest in his lack of appropriate responses to threats. The infant's responses to social situations were similar to those of a young infant still possessing a natal coat, and the infant remained socially naive until his disappearance. Berkson's (this volume) data on blind infants is also relevant to a discussion of physical, emotional, and social growth.

As the Fedigans (this volume) note, discussion of the stages of monkey development is still premature; however, the presence of an abnormal infant in a free-ranging group makes many otherwise assumed processes in social development apparent. The limited behaviors showed by Wania 6672 served to highlight the rapid and complex social development and social learning a normal infant monkey goes through in its first year of life.

ENDOCRINE DATA

Increasingly, endocrinology is playing a major role in helping interpret and define behavioral ontogeny. The work of Goy (1968), Goy and Phoenix (1962), and Goy and Resko (1972) has shown clearly the impact of hormonal bases upon behavioral expression. Much of the endocrine data is currently applied to the development of sexual behaviors, and this is the area of work which Wallen et al. and Testa and Mack address in this volume. As Wallen et al. note, little attention has been paid to the interaction between social experience and the prenatal endocrine environment, although both have a profound influence on behavioral development. Attempting to fill this void, Wallen et al. have examined how an individual's response within a social environment is determined by its hormonal history by testing the effects of prenatal androgen administration or testicular activity on the subsequent display of prepubescent sexual behavior. Goy and Goldfoot (1974) state that the basic psychosexual orientation is not determined by social experience as much as it is related to prenatal exposure to specific hormones.

The endocrine and maturation data add the needed dimension to help interpret behaviors normally witnessed by the naturalistic observer. The results currently appearing from endocrine and neurological studies strongly support the assumption that socialization can only be understood when field and laboratory workers cooperate. The naturalistic observer is at a grave disadvantage, being able only to observe outward manifestations (e.g., behavior) of inner processes.

CHRONOLOGICAL AGE

Certain key areas of social development are completed early in life. For example, Loy (1973) notes that over 85% of the communicative behaviors witnessed among adult rhesus (based on Altmann's 1965 ethogram for the Cayo Santiago population) are found in twenty-month-old juveniles. Some of the few behaviors not shown by the juveniles were attributed to physical immaturity. (This correlation between biological and behavioral maturation has been discussed by Chevalier-Skolnikoff [1973a,b] on the closely related stumptail macaque.) Sade (1971) has suggested that the rhesus behavioral repertoire is virtually complete by twelve months. Loy's (1973) data further suggests that young juveniles learn their roles and status within the group by an early age; the process normally termed "socialization" thus appears virtually completed by the end of the juvenile period. Adult socialization (which Loy [1973] termed modification) is characterized not so much by developing new behavioral patterns and relationships, as by modifying and specifying existing patterns.

Laboratory studies on nonhuman primates (cf. Harlow, 1960, 1962, 1966; Jensen et al. 1967, 1973; Mason, 1965), canids (Bekoff 1972; Scott 1958; Scott and Marston, 1950), and the work by Schneirla and Rosenblatt (1963), among others, suggest that the growth cycle can be conveniently divided into specific periods. Harlow and Harlow (1966), for example, divide the primate growth cycle into four periods, during each of which different developmental processes are operative. Since socialization is a continuous process occurring at all developmental stages, why specify age as a variable? It is becoming clear that the specific or average age of an individual may be a significant antecedent variable in the socialization process in the following ways: (1) There may be a simple causal relationship between age and some dependent variable; (2) there may be a more complex effect of age, not to be expressed in simple quantitative terms, e.g., the socialization process may have qualitatively different kinds of effects at various periods in the life cycle.

Harlow's four stages in the life cycle are commonly referred to as the neonatal period, period of transition, period of peer socialization, and the juvenile/subadult period. During the first period locomotor and ingestive patterns are suitable only for infantile life; during the neonatal period maternal-infant contact is continuous, close, and of long duration. The importance of the neonatal period is probably directly related to the amount and kind of maternal protection and solicitude afforded the infant. During the transitional period adult locomotor and ingestive patterns overlap, in unskilled manifestations, infantile forms. This period is gradual and extends over many months, terminating when the young leave the regular company of their mother. During peer socialization youngsters contact individuals other than their mother, primarily siblings, older females, and age mates. The juvenile/subadult period is characterized by the disappearance of infantile patterns and the appearance of such adult behaviors as copulatory activities.

While Harlow's suggestion that the primate growth cycle can be divided into four stages is most useful, Bekoff (this volume) and Rosenblum and Alpert (this volume) note some difficulties. Both agree that socialization is a lifelong process. Furthermore, Loy's (1973) suggestion that the primate socialization process is actually composed of two interacting processes—the developmental stage of socialization and the social modifiability stage—calls for some revision of Harlow's scheme. While we must be wary of establishing concrete stages in the socialization process, it is clear that at different ages qualitatively and quantitatively different things are happening. (Clark; Kaplan; McGrew; Nicolson; Parker; and Rosenblum and Alpert [this volume] all address this issue.) Perhaps dependent upon such variables as species, habitat, and phylogenetic position, the stages of primate socialization are sliding stages and not strictly cross-specifically comparable (Poirier 1973a). Such an interpretation is indicated by Parker's (this volume; 1973) data on stumptail macaques. Increasingly it is becoming incumbent upon researchers of primate socialization to define the adaptational process operative at different life stages, i.e., with studies of specific events like weaning (see Clark and Nicolson, this volume); they vary and play different roles. At different times in the life cycle qualitatively and quantitatively different things occur, and unless we isolate behaviors at an early age, we may be unable to understand later social development. Bramblett's (1973) studies on the social development of vervets strongly support this contention. By the same token, because we designate a behavior as important during one life stage, this doesn't necessarily imply that it will be important during another. Baldwin and Baldwin (1973), Rosenblum and Alpert (this volume), Testa and Mack (this volume), and Wallen et al. (this volume) all suggest that certain behaviors are more susceptible to alteration than others.

GENDER

Gender is a most important variable, for an infant's gender strongly influences its socialization. There is a direct relation between the socialization and learning processes of infant males and females and their subsequent adult roles (Poirier 1971, 1972a,c, 1973a,b, 1974, 1975b; Poirier and Smith 1974). Gender is rather immediately determined by conspecifics at birth by direct observation—many observers report that group members may pay close attention to the newborn's genitals by peering at, touching, sniffing, or mouthing them. This seems to be the first basic step towards classifying the gender of the new group member.

The manner in which an individual learns its role, including the male and female roles, may be heavily influenced by the immediate environment, as well as by its original genetic component (Lancaster 1972). Goy (1968, for example) and his associates (see also Scarf 1972) have studied hormonal influences on the development of sexual differences in rhesus; an up-date of this work is found in Testa and Mack and Wallen et al. in this volume. The high hormonal levels circulating in the blood of newborns suggest that during fetal or neonatal life, hormones act in an inductive way on the undifferentiated brain to organize certain circuits into male and female patterns. Early hormonal influences may affect the ease of learning and expression of behavior later in life, even though the hormonal level is low during the period of infancy and early adolescence.

Gender differences in behavior are also tied to the dynamics of group social interaction, i.e., by learning role patterns. Social roles are not strictly inherited in animal societies (Benedict 1969); laboratory studies indicate, for instance, that primates without social experiences lack marked sexual behavioral differences (Chamove et al. 1967). Studies of the mother-infant interactional dyad in pigtail (Jensen et al. 1968) and rhesus macaques (Mitchell and Brandt 1970) clearly show that there are sexual differences in the development of maternal independence. The effects of such early behavioral trends are apparent later in life (Poirier 1972a, 1973b).

Differential treatment occurs soon after birth due, perhaps, to the mother's reactions to dissimilar behavior in male and female young. Developmental studies of laboratory-reared rhesus (Mitchell 1969; Mitchell and Brandt 1970) and of the provisioned Cayo Santiago colony (Vessey 1971) show that mothers threaten and punish male infants at an earlier age, and more frequently, than female infants. Another early behavioral difference between males and females was described by Jensen et al. (1966) for pigtail macaques. They found that in a deprived laboratory environment, where an adult male was missing, the behavior of the male infant was more adversely affected than was the female's behavior. Nash (1965) reviewed much of the literature relevant to the role of the human father in early experience and indicated that the father's absence is more harmful to later behavior and role-playing of boys than girls.

Male and female differences are seen in such social activities as play (Baldwin and Baldwin this volume; Burton 1972; Goodall 1965; Hall 1968; Hall and DeVore 1965; Lindburgh 1971; Poirier 1970a, 1971, 1972a,c, 1973a,b; Poirier and Smith 1974; Smith 1972; Sorenson 1970), aggressiveness (i.e., Poirier 1970b, 1974), development of independence from the mother (Clark this volume; Jensen et al. 1968; Mitchell 1968; Mitchell and Brandt 1970; Ransom and Rowell 1972; Rosenblum and Alpert this volume; Vessey 1971), tool manipulation (McGrew, this volume), and demonstrativeness of role playing (Bekoff, this volume), among others. A major feature differentiating male and female primates is the expression of aggression, both by them, and directed at them. Males are more aggressive than females and mothers of male infants are more punishing than mothers of females. A rhesus mother, for example, threatens and punishes her male infant at an earlier age and at more frequent intervals than her female infant whom she restrains, protects, and retrieves (Mitchell 1969; Mitchell and Brandt 1970). Developmental studies of laboratory-reared infants elucidate some interesting points regarding male aggression; for example, even isolate-reared laboratory "motherless mothers" are more brutal towards male than female infants. Since exposure to excessive punishment has been correlated with later hostile behavior (Mead 1935; Mitchell and Mohler 1967), the infant male's predisposition towards rougher play and rougher infant-directed activity is subtly supported by its mother's behavior and through his observations of other mothers with their infants. Vessey (1971) reports that infant male macaques on Cayo Santiago began receiving aggression from their mothers in the second month, with a peak in months nine to eleven. Male infants received significantly more aggressive behavior than female infants. Similar data is generated by studies of feral baboons (Ransom and Rowell 1972).

The amount and frequency of affiliative and supportive mother-infant interaction is undoubtedly related to the amount of aggression. Studies of feral baboons (i.e. Nash and Ransom 1971; Ransom and Rowell 1972) show that there are consistent differences in the development of the mother-infant relationship and peer interactions as early as two months. Female infants are more consistently involved in close associative behaviors such as grooming, and are usually in closer proximity to other animals than are males. Males begin the process of peripheralization earlier (Nash and Ransom 1971) as the mother's earlier rejection of the male infant forces it into earlier contact with other male infants, usually in the form of peer play group interactions. Males are often found in age-graded play groups which range farther from the mothers as they mature. Young females, however, usually remain with the adults. Sociographic analyses show that male juveniles interact in larger groups than females, who mostly associate with only one partner. Preliminary data based on the same methods reveal a similar pattern in human children (Knudson 1971, 1973; Kummer 1971).

While young females maintain close ties with adult females, males also remain in proximity to one another. It may be important for the males (especially those destined to a subordinate position) to have a stable relationship with other males, especially older, more dominant males than with females who can be rather easily avoided. Juvenile females seem to develop their social relationships during long grooming sessions with other females and while holding or exchanging infants. For the males, peer group play seems more important and provides an opportunity for many diversified interactions with the animals with which they will eventually contend. Females are more likely to spend their total life within the group; males, on the other hand, leave or are driven from the group in the process of peripheralization (Fedigan 1972).

Rosenblum and Alpert (this volume) have shown that these gender differences vary among species. Among bonnet macaques neither males nor females showed a strong and significant preference for the mother until after twelve weeks. "Females however showed an earlier preference for the mother than did males throughout the entire first year" (468). Pigtail macaques, on the other hand, showed significantly less preference for their mother than did bonnet macaques, and pigtail males after twenty-four weeks showed no preference for their mothers.

Rosenblum and Alpert hypothesize that the strong and enduring bond of a bonnet macaque female infant to the mother may provide the basis for incorporating female infants into the adult female core of the group. Lancaster (1973) also argues for the importance of such female subgroups. The females' strong avoidance of strangers may reflect their hesitancy to establish new relationships, strengthening the cohesiveness of the female subgroup. Rosenblum and Alpert feel that this mechanism helps explain why females in the wild only rarely leave their natal groups.[4] Males, on the other hand, while showing a strong preference for the mother early in life are eventually forced from, and leave, the mother and show a greater readiness to move towards strange conspecifics. For males, the tendency towards establishing relatively less intense bonds with the mother, and a greater responsiveness to strangers, may interact to produce a greater inclination towards their peripheralization and facilitate a readiness to transfer troop allegiance.

Translating the above data into genetic possibilities, it is more likely that males will breed outside their natal group than will females. Peripheralization reduces the possibility of incest between males and their mothers, between siblings, and between a father and his daughter (Itani 1972). Instances of troop fission, i.e., among Japanese macaques (Furuya 1969), and infanticide among common langurs (Hrdy 1974; Mohnot 1971; Sugiyama 1967) may also be recalled. Perhaps such an explanation can be applied to the data which Demarest (this volume) discusses in his article, since it is clear that in the human situation individuals referring to themselves with kinship terms have a (cultural) bias against mating.

In her study of weaning behavior among chimpanzees, Clark (this volume) notes that male chimpanzees leave their mothers at an earlier age than the females. On the other hand, Nicolson (this volume) finds no gender differences in early infant development.

An interesting behavioral difference related to gender is discussed in McGrew's (this volume) article on object manipulation. Male chimpanzees apparently do more overall general manipulation than females; however, females are better in manipulatory skills, and at Gombe they exhibit more frequent tool use in obtaining insects by fishing and dipping. McGrew's findings are supported by other accounts suggesting female superiority in manipulatory skills, i.e., catching of thrown objects by Japanese macaques (Kawai 1967). Human female children are superior to males in manual dexterity from early childhood onwards (Garai and Scheinfeld 1968).

McGrew's data shows that males focus their object manipulation more often on growing vegetation, taking the form of vigorous locomotory play, either alone or socially. Females, however, show a greater tendency to manipulate movable detached objects rather than growing vegetation. How these activities relate to later adult behavior is unclear; however, the fact that females are more adept at termite fishing and dipping suggests that females are purveyors of this social tradition. Quite possibly females are more adept at these patterns than males because females spend more time with their mothers and thus have more opportunity for observational learning and actual experimentation.

Much has been written about primate play behavior. It is clear that males play harder, begin play at an earlier age and cease play at a later age, and play for longer periods of time than do females. There are qualitative and quantitative differences in the play of young male and female primates; however, Hinde and Spencer-Booth (1967) note that gender differences arise not so much in time of onset of play as in its expression. Harlow and Harlow (1966) distinguish the play of male and female laboratory rhesus at about two months. This dichotomy also appears in the field—by two years of age male and female baboons engage in differential play behavior. Male baboons play rougher and more frequently; female baboons groom more frequently and spend more time with newborn infants in the group (DeVore and Eimerl 1965). Kummer's (1968) data on cynocephalus and hamadryas baboons and Ransom and Rowell's (1972) data agree with DeVore. Similar data is found in studies of vervets (Fedigan 1972) and human children (Knudson 1971, 1973; McGrew 1972). Among humans, for example, the frequency of rough-and-tumble play for boys is significantly higher than for girls.

Dolhinow and Bishop (1970) suggest that a powerful endocrine effect influences sexual differentiation in play behavior. Females exposed to androgens during a critical period of early development become masculinized as pseudo-hermaphrodites (Goy and Resko 1972; Goy and Phoenix 1962; Phoenix et al. 1968). These masculinized animals tend to develop play patterns approximately halfway between typical male and typical female patterns.

In their extensive review of play behavior, Baldwin and Baldwin (this volume) suggest that a number of other factors affect the differential expression of play according to gender. They note that males are usually physically larger and stronger than females, a factor which alone may explain some differences between the frequency and duration of play. This, plus the endocrine influence, increases the likelihood that females experience less novelty and more aversive contact during exploration or social play than do males. This may be partly responsible for "... shaping the quiet, withdrawn, gentle activities of females and the tendency for females to orient to object manipulation play rather than the social contact play typical of males" (Baldwin and Baldwin p. 373). This agrees with McGrew's observations, and confirms Tsumori's (1967) observation that females learn to attend to lower but safer arousal levels afforded by manipulatory play and by exploration. As young females withdraw from rowdy play bouts, they may discover "play mothering," a pattern reported by Jay (1968) and Poirier (1969a, 1970a) for langurs, by van Lawick-Goodall (1967) for chimpanzees, by Lancaster (1971, 1972) for vervets, and by Baldwin (1969) for squirrel monkeys. Play mothering reinforces quiet, low arousal activities in adolescent females and accentuates gender differences in play behavior. A fourth variable that must be considered is the fact that females of many species reach social-sexual maturity prior to males, quickening the termination of female play behavior. Finally, social modeling may be operative, as is true for humans (Bandura and Walters 1963).

We might continue to list differences in the development of male and female primates, but the important point is that these differences are related to adult social roles and experiences (cf. Poirier 1970a, 1972a, 1973b, 1974, 1975b; Poirier and Smith 1974). In most species adult males and females play different roles, and their early social experiences seem to "condition" and prepare them for this. Since early in their development males are forced from their mothers out to the peer group where they mature and become less dependent upon the females, they learn to be assertive, aggressive animals. Females, on the other hand, remain with their mothers and other females—they learn to interact with other females, and most importantly they learn how to provide proper infant care. The major role which a female must learn is that of being a mother. The socialization and learning process of female nonhuman primates seems to be geared towards producing a healthy, effective mother. Although some basic patterns of mothering are relatively innate, learning plays an important role in the development of skill in performing them. The fact that young females soon drop out of play groups, that they have strong affiliative associations with adult females, and that they are continually in contact (visual or tactile) with infants is important. Many studies note that young juvenile females are inept at handling infants, but by adulthood they can carry and handle infants with ease and expertise (Jay 1962; Lancaster 1972; Struhsaker 1967). The dynamics of the maternal learning process occur under the mother's watchful eye; instances of carelessness, clumsiness, or real abuse are soon dissuaded. Through a simple conditioning process juvenile females learn appropriate behavior patterns with their reward being the continued presence of the infant.

Laboratory studies support the contention that early experience practicing the mothering role may be preparation for adult maternal behavior (Seay 1966). Field and laboratory studies clearly document that females are more closely attached to the mother, show more interest in young infants, and are gentler in their social relationships than males (Chamove et al. 1967; Spencer-Booth 1968). We know from labora-tory experience that a total lack of social experience leads to the development of infantile and aggressive mothers. Harlow et al. (1966) found that motherless mothers raised in semiisolation, or females deprived of peer relationships, responded to their firstborn with active hostility and rejection. However, social experience with an infant, no matter how minimal, affected maternal behavior; the same females who rejected their first infant often accepted the second.

■ OTHER VARIABLES INFLUENCING SOCIALIZATION

Among the many other variables affecting socialization, the following are discussed in this volume: the mother-infant dyad; the role of the mother; an infant's relationships with its siblings and consanguineal kin; an infant's relationships with its peers, espe-cially as these are expressed in play behavior; an infant's relationships with the adults of the group, excluding its mother; and the function of the particular type of social structure as it impinges upon an infant's social relationships. The following discussion will highlight some of these factors and their influence on the socialization process.

SOCIALIZATION AND THE SOCIAL STRUCTURE

Since nonhuman primates are group living, social animals, the socialization process must be viewed within the social structure to be fully understood. If, as we suppose, the development of nonhuman primate behaviors involves the interaction of a geneti-cally determined base with a set of environmental conditions, then we must re-member that the environmental context of most primate infants is largely social. Group structure reflects and influences individual behavior (Poirier 1969a, 1972b); not only the form of group organization, but group life itself, is dependent upon the early environment of individual animals.

The reciprocal relationship between socialization and the social structure is not necessarily one of discrete interactions, but may take the form of cycles or other sequences prolonged over substantial time periods. Primate social groups differ ac-cording to many variables, but with few exceptions animals learn to use their biology effectively and adapt to their habitat while living within the group (Washburn and Hamburg 1965). Differences in primate societies are due not only to biology, but to the circumstances in which an individual lives and learns. The character of a social group is related to the strength of interanimal affinities and to the degree to which relationships are tolerated by other group members. Group integration is a complex process relying on mutually reciprocal patterns of naturalistic behavior modified and made specific by learning and conditioning (Washburn and Hamburg 1965). Most social behavior enters into the determination of gregariousness and the qualities of a complex social organization.

Kummer (1967) suggests that differences in primate groups may be rooted in age-sex class affinities. The species-specific group structure determines with which animals an infant will interact. For example, Nilgiri langur subgroup assemblages allow an infant considerable contact with its mother, other females, and peers, and little contact with the adult male or males. (The consequences of this are discussed later). It is likely that some different behavioral and group structure characteristics of Indian leaf-eating monkeys are partially due to the amount of contact animals have with others.

Rosenblum and Alpert (this volume) discuss the relationship between the socialization process and social structure among bonnet and pigtail macaques. Bonnet mothers are more passive and pigtail mothers are more coercive in regulating their infants' behaviors. Bonnet macaque infants, therefore, exhibit more initiative than pigtail infants in promoting and maintaining proximity and contact. This is especially true early in life. "This different demand on infants of the two species correlates with the differences between bonnet and pigtail social structures within which the infants must assume their developing role" (page 475). The bonnet macaque group structure is less clearly hierarchical than the pigtail group structure, and the bonnet group allows a relatively fluid pattern of interanimal interactions. In the pigtail group, however, interactions are more restricted and individual roles more rigidly defined. Pigtail macaque individual behavior is thus more completely dictated by group structure; in bonnets, however, behavioral regulation appears to reside to a greater degree with the individual animal. Because of this difference, bonnet infants ultimately engage in a relatively looser social organization requiring greater social initiative on the part of its constituents. Pigtail macaques, in contrast, must be socialized to respond in a relatively fixed social structure requiring a greater adherence to predefined roles by the newly maturing animals.

KINSHIP TIES

Long-term studies on Cayo Santiago (cf. Missakian 1972; Sade 1965, 1967, 1968, 1972) and at the Japan Monkey Centre have clearly shown that a youngster's adult behavior is influenced by animals with whom it has consanguineal relationships, especially members of the matriline. Studies at the Japan Monkey Centre have shown that not only a mother's rank, but also the number of siblings on whom one can depend for support, influence one's development (cf. Itani et al., 1963; Kawai 1958a,b; Kawamura 1958; Kawamura and Kawai 1956; Koyama 1967, Yamada 1963). A similar situation is found among the Cayo Santiago rhesus (Sade 1969, 1972). Sade's 1972 data show that different mechanisms within the kinship network affect males and females differently.

Japanese macaque studies have shown that kinship ties affect the transmission and learning of new behavioral patterns (Itani 1958; Itani et al. 1963; Kawai 1965; Tsumori 1967; Tsumori et al. 1965). Pathways of habit formation follow preestablished networks of group affinities; in the Koshima, Ohiryama, and Takasakiyama troops, for example, habit propagation was strongly influenced by kinship. Animals of different kinship lines have different socialization and learning experiences; entire lineages consisting of a mother and her descendants tend to acquire or reject new behavioral components as a unit (Kummer 1971). Since animals of different groups exhibit different behavioral patterns (for an example of this in three different species see Poirier 1969c, 1975a), it seems reasonable to expect that animals within a kinship group will act more alike than animals in nonkin groups.

Fedigan and Fedigan (this volume) note an interesting case of kin support in their discussion of the severely handicapped infant, Wania 6672. The Wania matrilineage was not as tightly knit as some uterine groups in the Arashiyama West troop, but Wania 6672 spent considerable time with members of his matriline. On a few occasions when Wania could not find his mother, he located his grandmother or an aunt and attempted to follow and sit by them. Two of this handicapped infant's peers were his close relatives, and were among his most frequent companions and "retrievers." In this instance the matriline aided Wania's struggle with his environment and his attempts to function within the social group. Wania's severe handicap was somewhat alleviated by having a number of kin members within the social group to whom he could turn. (Likewise, Berkson, in this volume, suggests that the larger the social group the better a handicapped individual's chances of survival.)

Loy (1970) has demonstrated that the matriline remains stable during food shortages. Loy and Loy (1974) note that the effects of the matriline are much in evidence by the juvenile stage of development. Months after separation from the natal group, monkeys 20–30 months old could act in an orderly and predictable manner because of geneologically determined relationships forming the core of their interactions. "With the development of mother-infant specificity, the infant monkey is linked to its matriline and, to a large extent, its future behavior is determined. The infant develops close relationships with mother, siblings and other matrilineal kin, and its kin ties then serve as the basis for its integration into the remainder of the social group" (Loy and Loy 1974:94).

Loy and Loy (1974) discuss the relationship between kinship ties and the behavior of juvenile rhesus monkeys in grooming and play. Grooming between matrilineally related monkeys (comprising 6.6% of the study group) accounted for 34.4% of the total recorded. Related monkeys were involved in grooming over five times more frequently than expected from random selection of grooming partners. Sade (1965) reported that grooming among related animals (15% of the total possible grooming combinations) accounted for 62–64% of the observed grooming sequences. The Loys found that play behavior among related animals comprised 26.5% of the total sequences, a figure slightly more than four times greater than expected from random partner selection.

INFANT-ADULT MALE RELATIONSHIP

Most socialization studies focus on the mother-infant or peer relationship, and the possible socializing role of adult males is largely overlooked. One reason for this is perhaps the fact that the male's socializing role is likely to be the most variable relationship, in terms of the amount and kind of contact, which an infant has during its earlier years (an up-to-date review of this subject is found in Mitchell 1969; Mitchell and Brandt 1972; Poirier 1973b; Redican and Mitchell 1973). An adult male may kill a youngster, as among North and South Indian common langurs (Hrdy 1974; Itani 1972; Mohnot 1971; Sugiyama 1967); he may have little or no contact with it, as among Nilgiri langurs (Poirier 1969a, 1972a); or he may participate to a greater degree than the females in its care, as among Gibraltar macaques (Burton 1972). The adult male's contact with the infant is minimal in some species and extensive in others; in some species males play the role of group protector, in others they play an important role in socialization. In some species the male's role is active, in others it is passive. The adult male's role in the socialization process is influenced by such factors as social structure (Poirier 1969a, 1973b), habitat (Chalmers 1972), one male versus multiple male groups, and the phylogenetic position of the species (Mitchell and Brandt 1972; Poirier 1972a, 1973b). This diversity is hardly surprising given the amount of variability within the Order Primates. Generally, males of Old World species do not appear to care for infants as extensively as males of New World species.

Social structure influences the amount and type of adult male-infant contact; this may be illustrated by comparing Nilgiri langurs and Japanese macaques. Nilgiri langur groups are organized into fairly consistent subgroup assemblages, each of which is a social aggregate of individuals of similar age and/or sex. Most social interaction occurs within rather than between these subgroups; infants and juveniles rarely interact with adult males, especially when females in the subgroup give birth. All Nilgiri langur infants have minimal association with adult males who are largely physically and socially peripheral to the group. On the other hand, Japanese macaque troops are comprised of central and peripheral portions; infants born within the central part of the troop identify with troop leaders and eventually become leaders themselves. Whereas all Nilgiri langur infants go through a similar rearing process, some Japanese macaque infants receive special attention from leader males, passing through a different rearing process than infants on the troop's periphery.

The adult male's role varies according to age, dominance, specific social group, and individual idiosyncrasies. The role most often assumed is a protective response towards youngsters, but this is variable in pattern and extent (Chalmers 1967; DeVore 1963; Lahiri and Southwick 1966; Mitchell 1969). As Bekoff (this volume) notes, males of a number of mammalian forms perform various "maternal" tasks and paternal care is not uncommon in rodents and fissiped carnivores, as well as in primates (Spencer-Booth 1970). Among nonhuman primates, most paternal care is directed towards infants once they acquire the adult coat color. A male assuming a nurturing or protective role is likely to be a subadult or sometimes an adult of fairly high rank. Paternal roles have not been reported for young, low-ranking males except when a young male is protective of his young sibling (Kaufman 1966; Koford 1963; Sade 1965, 1967; van Lawick-Goodall 1967). Some instances of male "care" should not be considered paternal in nature. For example, Barbary macaque males may use a baby as a "buffer" in a situation where an approach without this infant "buffer" would increase the likelihood of aggression (Deag and Crook 1971).

Some interesting cases of paternal behavior are found among hamadryas baboons (Kummer 1968), where young males adopt young females who later become their consort, and among Gibraltar (Burton 1972; MacRoberts 1970) and Japanese macaques. Itani (1959) reports that in some troops dominant Japanese macaque males adopt yearlings when females are bearing new infants. Burton (1972) reports that from the fourteenth to the twentieth day of life, subadult Gibraltar macaque males become a predominating influence in an infant's life. Once an infant begins to walk, subadult males are allowed to carry them, sometimes as far as two hundred yards from the mother during progressions. The subadult male acts as a "babysitter" who conveys youngsters over any walking distance greater than five feet and over any jumping distance greater than a foot. The socialization of infant Gibraltar macaques occurs largely within the framework of the relationship to the subadult male, but is not confined to it.

The adult leader male also plays a major role in socializing the Gibraltar macaque infant; the mother acts primarily to reinforce learning. Consistently, the day after the leader male encourages the youngster to walk or climb dorsally, the mother does likewise. No female did these things before the male. The Gibraltar macaque male assumes four primary socializing roles (Burton 1972): (1) to encourage the infant to develop motor abilities permitting social interaction, (2) to reorient the infant away from himself to other troop members, (3) to reinforce socially acceptable behavior, and (4) to punish inappropriate behaviors.

The demeanor of individual adult males may also affect the infant's development. Simonds (1971) notes that the idiosyncrasies of the leader male in some bonnet macaque troops strongly influence the infant's social development. Similar observations are made among Japanese macaques. Rosenblum and Coe (this volume) suggest that adult squirrel monkey males may be an important stimulus for attracting infants into the larger social environment—the infants are strongly motivated to approach adult males as they become increasingly independent of their mother.

Poirier (1973a) has suggested that perhaps one reason we know so little about the male's socializing role, and therefore assume it is minimal, is because it may be subtle and we simply do not watch for it. Chalmers (1972) notes that the physical removal of an adult male Sykes monkey results in increased restrictiveness of the mother to her infant; whereas removal of a male patas does not. Two possibilities are suggested: (1) removal of the male disturbs the mother and she reacts more cautiously and protectively towards the infant than if relaxed, and (2) removal of the male upsets the group's social equilibrium, dissolving dominance dependent relationships under which the mother might have gained the adult male's protection (Imanishi 1960). The differences noted in Chalmers' study may also be due to species-specific behavioral patterns. Patas monkeys live in male groups in which the male is independent of other group members and the adult male rarely interacts with adult females. That the absence of a patas adult male might not upset the mother is therefore not surprising.

A male's presence may have other subtle ramifications; it has been noted, for example, that the presence of an adult male can affect the physical development of female mice (Fullerton and Cowley 1971). Females reared in the male's presence (though not necessarily in physical contact) gained weight more rapidly and showed earlier onset of eye and ear opening and eruption of lower incisors. Mugford and Nowell (1972) report that mice reared with males until weaning were more aggressive than those reared in the absence of a male.

INTERACTIONS WITH ONE'S PEERS AND PLAY BEHAVIOR

So much has been written about play behavior and its role in socialization, that it would be inappropriate to review all the material here. Baldwin and Baldwin's manuscript (in this volume) should be consulted as an up-to-date, comprehensive review of play behavior. That play behavior has an important role in socialization is without question; however, its precise function is debatable. Before discussing the role of play behavior, it is relevant to note the important role of peers in the socialization process. Fedigan and Fedigan's work reported here on the handicapped Japanese macaque youngster shows quite clearly what happens when an infant is unable to make the transition away from the mother out to the peer group. This data provides a good control for assessing the ontogeny of normal social behaviors.

The Fedigan and Fedigan article also notes an interesting situation in which peers apparently can help mediate severe defects. (A similar situation is found in Harlow's description of his "play therapists" where peers ameliorate the effects of mother deprivation.) In the instance which the Fedigans describe, the handicapped infant apparently makes a momentary reversal in its abnormal social life when a female age mate shows interest in him. This reversal occurred when the infant was ten to eleven months old, at which time the female age mate regularly sought the handicapped youngster and engaged him in gentle play. Sometimes other infants would join and the handicapped youngster would stay and attempt to play, contrary to his usual habit of avoiding infant groups. This period of increased social activity and peer interest lasted for a month and was probably provoked by the female's attention, for when the female no longer sought the handicapped infant, he returned to his pattern of mother-centered activity.

Washburn (1973) noted that if the field observer listed the kinds of daily behaviors witnessed among nonhuman primates according to the amount of time consumed, the usual order would be: sleeping, obtaining food, eating, playing, resting, and social contacts. Judging from the occurrences of play recorded in various studies, there is no question that play was a major adaptive behavioral trait during primate and mammalian phylogenetic histories. A multitude of variables affect the expression of play behavior (cf. Baldwin and Baldwin, this volume; Bekoff 1972; Bernstein 1962; De-Vore 1963; Poirier 1969a, 1970a, 1973b; Poirier and Smith 1974). Among these variables are age, gender, group structure and dynamics, and habitat.[5]

In a 1963 article Harlow demonstrated five developmental stages in rhesus play behavior; these stages may have some cross-specific applicability, for Poirier (1970a, 1973a) found similar stages among Nilgiri langurs. It is apparent that the more complicated aspects of play behavior appear with neuromuscular maturation as Sade (1973) has shown to be true for rhesus macaques. As each play pattern integrates into the next successive one during various phases of the life cycle, it undoubtedly assumes varying degrees of importance at any particular point in the life cycle. Not only do different patterns appear at different times, but they may be functionally different during the developmental sequence (Poirier and Smith, 1974).

The socializing functions of play behavior are debated; however, among the many possibilities the following are most often noted: play behavior functions as a mechanism for social development; for establishing the dominance hierarchy; as a means of achieving social integration; and as an experimental arena for learning the communication matrix. There is strong evidence that play behavior is essential for development of future skills requisite for survival, especially for meeting unexpected contingencies. Key elements of social life such as grooming, components of sexual behavior, and aggression are to some degree learned and rehearsed in the play group. Deprivation studies indicate that social peer group play interaction may be more important for the development of normal social behaviors than maternal interaction (Harlow 1969; Harlow et al. 1971; Tisza et al. 1970).

It is often suggested that the basis of the adult dominance hierarchy may be formed in the play group; e.g., that play behavior may help youngsters find their place in the existent social order (Carpenter 1934). During play, youngsters compete for many "valued items" such as food, sleeping places, and pathways. Through trial and error, through the constant repetition of play behaviors, infants learn the limits of their self-assertive capabilities and become familiar with both the dominant and subordinate positions (Dolhinow and Bishop 1970). Etkin (1967) suggests that play may be one means of reiterated stimulus exchange whereby social animals maintain familiarity with each other; in this way play may replace the kumpan relation found in some vertebrates. During play youngsters learn their place in the social group and perhaps develop appropriate ingroup feelings, e.g., play facilitates troop integration. During play, animals learn patterns of social cooperation without exceeding certain limits of aggression (Diamond 1970). The successful initiation of social or interactive play is critical to the appearance of the age mate affectional system (Harlow and Harlow 1965, 1966). Youngsters lacking the opportunity to play may be faced with the options of being maladjusted or of being excluded from the social group (Carpenter 1965). The play group is a major context for learning social and physical skills, and as such, is an important factor in social integration.

Play may also be an important vehicle whereby youngsters learn social communication skills (Poirier and Smith 1974; Smith 1973; Symons 1973). As an integral part of the practice of adult social roles, play serves to fully acquaint an animal with its species specific, and perhaps group specific, communication matrix. Socially deprived animals have problems with response integration and communication; although such animals exhibit most components of social behavior, these are not combined into an integrated pattern and effectively applied in social interaction. Mason (1963) believes this is due to a deficiency in sensory-motor learning or "shaping"; Berkson's (this volume) and Fedigan and Fedigan's (this volume) data included here suggest likewise. Although all basic postures, gestures, and vocalizations are probably unlearned, their effectiveness in social interaction is dependent upon experience, as social deprivation studies aptly demonstrate. This applies to the sending as well as the reception of messages, for messages can only be effective if individuals know their meanings.

A most useful approach to understanding the socializing role of play behavior is to consider play a kind of "grammatical structure" (Chomsky 1965). During ontogeny the players learn the behavioral syntax as a mathematical game (Kalmus 1969). Considering play as a mathematical game allows us to look at the two components of such a game: (1) a finite number of rules or positions, and (2) the rules governing the outcome. In play behavior there are undoubtedly precise rules which a youngster learns and the acquisition of an adequate performance and competence in rules of play (following Kalmus) is a developmental process. The rules a young primate learns are not lacking some sort of logical connection or structure; Altmann's (1965) stochastic approach to rhesus communication clearly shows that there is considerable predictability within the communication system of properly socialized animals. Again, Berkson's (this volume) and Fedigan and Fedigan's (this volume) data on physically and socially handicapped infants clearly demonstrate what happens when an animal is incapable of learning these rules.

Conceptualization of play as having a behavioral syntax may allow us to appreciate more fully the interrelationship of the various behavioral patterns comprising play. As a primate develops more elaborate and intense play behavior, it may order the rules of the game into the correct sequence for proper functioning in a social unit (Bekoff 1974, however, raises doubts about this). The key to the acquisition of these rules is in the sequencing of playful interactions and the association of relatively disjunctive units of behavior into larger functional categories. The behavioral syntax for certain adult behaviors may be learned through repetition in the play context, making an understanding of the complex repertoire of signals employed in the playful interaction an area of fruitful research (Goyer 1970; Poirier and Smith 1974).

Although we assume the primacy of play behavior for normal socialization, Baldwin and Baldwin's data (1974, and in this volume) question the relative importance of play for later social development. If their study of the Barqueta squirrel monkey population is cross-specifically applicable, then it suggests that the adaptive significance of play is more complex than ordinarily assumed. It may be that a behavior such as play functions more in simply bringing animals together than in developing certain adult behaviors. The Baldwins' Barqueta data show that, at least for squirrel monkeys, an adaptive modicum of competence develops even in the absence of play. However, play may be important in developing the full potential of an animal's behavioral repertoire beyond that point; e.g., play may be important in developing "complex" social behaviors, however they may be defined. Some of Baldwin and Baldwin's (1973) interpretations concur with my own observations on Nilgiri langurs. Nilgiri langurs, like the squirrel monkeys, who had limited play interactions, exhibited a full range of social behaviors but clearly the intensity of such behaviors differed from those behaviors of animals who frequently engaged in play and grooming. In contrast to the Baldwins' data, however, Nilgiri langur troops in which play and grooming were minimal were loosely structured and prone to fission (Poirier, 1969a, 1973a).

MOTHER-INFANT DYAD

Many primatologists, following the clinicians' lead, have suggested that all major social roles and classes of bonds (i.e., male-female, dominant-subordinate) may ultimately have their roots in the initial socialization of the infant by its mother (Jolly 1966; Poirier 1968b, 1970a, 1972a, 1973b; Tinklepaugh 1948). The maternal relationship is a youngster's first affectional bond (Harlow, 1962, 1963)[6] and is perhaps the prototype of all later such bonds (Jensen et al. 1968). With formation of a maternal attachment, favorable conditions exist for social learning; for social learning commences with the mother. Occasions for learning multiply as the growing youngster extends its social relationships beyond the mother; however, investigators have paid little attention to specifying how this is accomplished. One of the primary functions of contacts with other animals may be to sharpen, strengthen, or generalize the learned behavior originating in the mother-infant dyad. Hansen (1966), Jensen et al. (1967a,b, 1973) and Rosenblum (1971) discuss this point.

A newborn's ties to the mother are the earliest, become the strongest, and seem to last the longest. For varying lengths of time, depending upon the rates of neuromuscular development (see Berkson this volume and Fedigan and Fedigan this volume), a mother serves as the infant's locomotor organ and neocortex and thus determines the nature of the basic socialization environment. The neonatal infant, clinging to a mobile mother, forms an attachment to her and through her, to virtually her whole ecological-social setting. Physiological and morphological states influence the nature and extent of the early dyad, but psychological states and social habits formed during infancy influence the nature and extent of social relationships which persevere later in life (Tinklepaugh 1948). Later attachments may be differentiations and specializations of this early and relatively amorphous monolithic state (Poirier 1968b, 1972a, 1973b).

In mammalian neonates, behavior is typified by reciprocal stimulation between parent (especially mother) and offspring. The infant attracts the mother's attention; the mother then presents the newborn with a variety of tactile, thermal, and other stimuli, typically of low intensity and primarily approach-provoking. Socialization commences on this basis, and behavioral development is essentially social from the onset. The principal factors in this development involve the perceptual development of the mother, processes of individual perception of the infant, and the reciprocal stimulative relationships between the two. The dependent nature of the bond demands that participants arrive at a mutually satisfying interactional pattern whose consistency and flexibility allow the physical and emotional maturation of both parties. The interactional pattern derives its original form and later permutations from characteristics inherent to the pair itself and to social and physical surroundings of which they are a part. For example, the mother's age and parity (e.g., previous experience, or lack thereof) affect her behavior from the onset, including her degree of "restrictiveness" or "permissiveness," as well as the success with which she satisfies her own and her infant's needs with ease and economy of effort.

As is evident from this volume, there are notable differences in how nonhuman primate mothers handle their infants and in the amount of time they spend with them. Most field studies report that the mother and her newborn form the center of a cluster of interested group members, especially other females. Among baboons (De-Vore 1963) and Japanese macaques (Sugiyama 1965), this interest may be limited to peering at or trying to touch the infant. In other species, such as langurs (Jay 1962; Poirier 1968b; Sugiyama 1965), vervets (Gartlan 1969; Struhsaker 1967), some lemurs (Sussman, this volume), and chimpanzees (van Lawick-Goodall 1967), the mother permits other group members to hold and carry her youngster. Similar interest in infants has been reported in laboratory rhesus colonies (Hinde and Spencer-Booth 1967; Rowell et al. 1964; Spencer-Booth 1968).

Certain stimuli seem to elicit a female's solicitous reactions to her infant. Many newborns look different from the adult; for the first few months (depending upon the species) infants possess a natal coat which may be an essential element eliciting the female's maternal behavior (Booth 1962; Jay 1962).[7] The natal coat is generally present during the first few months of life when an infant most requires its mother's protection and nourishment. It is almost certainly more than mere coincidence that the duration of the natal coat coincides with a period of great dependency when it is essential that the infant be sheltered and protected (Poirier 1968b). Gartlan (1969) suggests that in species with a natal coat the infant has a special vulnerability to environmental dangers. Little is known of the behavior of species lacking a contrasting natal coat, but these infants may retain closer body contact with the mother for a longer period of their development.

Besides the natal coat, other factors affect the mother's initial attachment to the infant. Kaplan (this volume) notes a number of visual, tactile, and olfactory properties affecting the mother-infant attachment. Kaplan's data is interesting on two accounts: first, it provides clear evidence that early olfactory experience is involved in the development of emotional attachment, and second that the properties affecting attachment may change, or vary, with age. For example, his squirrel monkey data show that sight alone of familiar conditions (mother or environment) provided some degree of emotional comfort. Fedigan and Fedigan (this volume) provide evidence that infants (even their severely handicapped youngster) can recognize their mothers and vice versa on the basis of vocal patterns. Nevertheless, Rosenblum and Alpert (this volume) note that neither pigtail nor bonnet macaque infants showed an unequivocal preference for the mother until the third month of life. They conclude that the inability to discriminate the mother, as seen in their test situation, could be a potentially maladaptive social behavior under free-ranging conditions.

A multitude of variables affect the mothering experience. Some variability within the mother-infant relationship is phylogenetically related; the relationship tends to be more complex and longer lasting (probably as a function of maturation rates) in higher primates. This statement is demonstrated by comparing Sussman's (this volume) data on prosimians with Nicolson's (this volume) and Clark's (this volume) on chimpanzees; Bekoff's (this volume) article provides an added nonprimate dimension. Of the many variables affecting the mother-infant dyad, we will discuss habitat and social organization, mother's dominance status and mother's parity (i.e., her experience with previous infants and nipple length). The infant's gender was discussed previously. It must be cautioned that we are unable to set a standard for what is "good" and what is "bad" mothering, for the diversity is remarkable and ascription of good and bad or adequate and inadequate varies with many factors.

In his work with cercopithecines, Chalmers (1972) attempted to show how habitat can affect the mothering experience. Chalmers found that infants of the more terrestrial vervet and DeBrazza's monkey left their mothers at an earlier stage than infants of the more aboreal mangabey and Sykes monkey. A vervet or DeBrazza's infant which stumbles or is clumsy will come to little harm if it is near to, or on, the ground; therefore, it can leave the mother without facing this risk. A clumsy arboreal infant, however, is likely to be harmed if it falls. Even if it survived the fall, or did not sustain serious injury, it might be impossible for the mother or another adult to locate or retrieve it. This example suggests that it would be advantageous for the arboreal infant to stay close to the mother.

A contrary point, however, assumes that terrestrial animals are more prone to predation than arboreal animals, and that it would be selective for the terrestrial infant to stay by its mother. As an example of this proposition, Poirier (1968b) remarks that the arboreal Nilgiri langur mother has no hesitation about leaving her month-old infant screaming and clinging precariously to a branch as she descends to a lower level to feed.

In their article, Rosenblum and Alpert attempt to relate the pigtail and bonnet macaque mothering experience to differences in social structure and habitat. In her article, Nicolson notes that captive chimpanzee mothers broke contact with their infants proportionately more often than did Gombe Stream Reserve mothers, presumably because of differing environmental pressure. Sussman (this volume) notes that differences in how *Lemur catta* and *L. fulvus* attend to their infants may be due not only to locomotor and postural differences, but also to the large amount of time that *L. catta* spends terrestrially. Furthermore, *L. catta* are far more precocious than *L. fulvus* infants. Sussman also notes that when arboreal, not only the mother, but the group generally, slows its movement to accommodate the infant's lack of arboreal expertise. Similar observations are made among other arboreal and terrestrial species.

That previous mothering experience affects the degree and quality of infant care is obvious. Mitchell and Stevens' (1969) work with primiparous and multiparous mothers showed that at least for the first three months of an infant's life, a primiparous mother is consistently more anxious; she restrains her infant more and reacts more violently to a novel or slightly threatening situation. Harlow et al. (1966) found that social experience with an infant, no matter how minimal, affected maternal behavior. The same laboratory mothers who rejected their first infants often accepted the second. As noted for both human and nonhuman primates, there are consistent differences in the behavior of first-, second-, and last-born infants, due, in part, to differential maternal treatment. The number of instances of "aunting" or "babysitting" behavior reported for colobines, rhesus, and vervets, among others, suggests that maternal experience, even prior to giving birth, is important in developing the ability to care for youngsters.

One of the subtle variables associated with parity, and linked to socialization, is the length of the mother's nipple. In a feral study of baboons, Ransom and Rowell (1972) reported that nipple length affected the mother-infant relationship. Nipples of multiparous mothers are usually much longer than those of primiparous mothers. During the first days of life the baboon infant needs assistance in finding and maintaining nipple contact; its ability to cling in a flexed position against the mother's chest is limited and the mother frequently assists the infant. A first-born unable to maintain contact with the mother's short nipple soon begins to struggle and search for it. An infant born to a multiparous mother, however, can retain the longer nipple and often sleeps in a slumped position. During the first few days, young females frequently interrupt their own feeding and social activities to assist the first-born in nursing. Perhaps young mothers are more likely to experience nipple discomfort, due to stretching, and so continually set the infants down for short periods. These infants experience frequent frustration due to interruption of feeding and severance of body contact. The early interaction between a multiparous mother and her infant is more harmonious and secure, as these infants retain nipple contact when the mother is seated or walking without having to maintain tight flexion. Thus, they make fewer demands on the mother, and multiparous mothers are less disturbed by the infant's manipulation of the nipple and demands, and are less likely to displace them. A small difference in nipple length can produce variations in the quality of mother care and in the development of the mother-infant relationship.

There is a considerable amount of literature suggesting that a female's status affects her mothering behavior and eventually her infant's status. Data from feral baboon studies (i.e., DeVore 1963) suggest that the socialization process differs for youngsters (primarily males) at opposite ends of the dominance continuum. DeVore notes that infants of lower ranking mothers exhibited considerable insecurity in the form of a greater frequency of alarm cries and more demands on the mother, leading to an intensification of the mother-infant bond. Offspring of dominant females, however, acted more secure and exhibited more freedom from mother. A recent study by Poirier (unpub. ms.) on rhesus living in a corral generated similar data—higher ranking mothers allowed their infants more freedom sooner, and were less prone to retrieve the infants as they ventured from the mother's "protective shadow."

Behaviors expressing social ranking seem to be learned by the infant from the mother; Bekoff (this volume) notes a correlation in various mammals between a mother's rank and the social relationships which her infant develops. Loy and Loy's (1974) data on a group of thirty-three juvenile rhesus show that juvenile dominance rankings are 95% predictable from prior knowledge of the mother's rank. Longitudinal studies of macaques indicate that infants mimic their mothers' social interactions with other adults. Furthermore, infants of high status mothers (and infants of large matrilines) can count on their mother's support in case of trouble.

The concept of identification has been introduced into nonhuman primate studies to explain the fact that Japanese macaque infants with dominant mothers tend themselves towards dominance (Imanishi 1957). Infants of high-ranking mothers had substantial contact with troop leaders and identified with them; offspring of low-ranking mothers had minimal contact with troop leaders and were unable to identify with them. In the Takasakiyama troop, low-ranking infants were likely to become peripheral or desert the troop (Itani et al. 1963). Japanese primatologists speak of "acquired" or "derived" status to explain the fact that some infants "inherit" their mother's dominance status. Infants of high-ranking mothers have more contact with adult males and attain choice foods (Kawai 1958; Kawamura 1958). Because the Japanese macaque troop is comprised of a central and a peripheral part, infants born in the central part of the troop (infants of dominant mothers) associate with, and more importantly, identify with, troop leaders. These infants can "look to" dominant males in times of stress; in their turn they are more likely to become leaders.

Koford's (1963) and Sade's (1965, 1967, 1973) reports on the Cayo Santiago rhesus indicate that adolescent sons of the highest ranking females hold a high rank in the adult male hierarchy. Koford (1973) suggests this is apparently due to the protection given by the mother during adolescence. Sons of high-ranking females attain top status in the troop without becoming peripheralized or subdominant males. Ransom and Rowell's (1972) study of feral baboons suggest that one factor determining an infant baboon's rank is the relative rank of its mother.

WEANING

This discussion of the mother-infant relationship ends, appropriately, with the weaning process. Weaning entails the physical and emotional rejection of the infant by its mother, who, although she was once the major source of comfort, warmth, and food, is now hostile and denying (Kaufman and Rosenblum 1969). As Clark (this volume) notes in her study of chimpanzee weaning behavior, "Weaning from the nipple is the first serious break in the bond with the mother, and all the infant's behavior is affected when he is denied suckling and is supplanted by a younger sibling at his mother's breast. A period of depression sets in varying in intensity among individuals. Some infants come out of it within a few months whereas others remain depressed for a year or longer" (page 235). Clark's article on weaning among chimpanzees could profitably be consulted as an example of the factors which we will now discuss.

A number of variables affect the weaning process, its onset, tenor, and long-term results. Some mothers reject their infants more vigorously than others, and some infants are more persistent in their attempts to resist rejection (Jay 1962). Dominant mothers seem to have less trouble weaning their infants, who seem less reluctant to leave the mother, than do subordinate infants. Variations in weaning are also related to gender (Itani 1959; Jensen et al. 1968; Poirier 1968b, 1972b, 1974) and habitat (Chalmers 1972; Hinde and Spencer-Booth 1967; Nicolson this volume; Ransom and Rowell 1972; Rosenblum and Coe, this volume). Jensen et al. (1967) note, from a laboratory study of pigtailed macaques, that male infants left the mother more than female infants. Reports on feral Japanese macaques agree; male infants left the mother to form peer groups at an age when young females remained with the mother (Itani

1959). On the other hand, a feral study of Nilgiri langurs suggested no gender differences in weaning (Poirier 1968). This lack of gender differences in weaning, contrasted with the Japanese situation, may somehow be related to the greater behavioral variation noted for Japanese macaque males and females compared to Nilgiri langur males and females.

Investigation of the weaning process can reveal tantalizing leads; for example, Poirier (1971, 1972a,c, 1973b, 1974, 1975b) suggested a relationship between the method and duration of weaning and adult life. There may be a relationship between adult aggressiveness and the severity of rejection with which an infant is weaned (Mead 1935). Heath (quoted in Mead 1935) found a significantly higher degree of aggressiveness in nine early weaned rats compared with a similar sample which remained with the mother.

Anthoney (1968) suggests an ontogenetic development of grooming from nursing and weaning behavior. Grooming first becomes important for the infant when it is weaned; although the mother disallows nursing, she usually tolerates the infant's attempts to groom her. Thus, whenever the weaned infant is frightened or otherwise needs security, it comes to groom rather than nurse. There may be some link between the amount of grooming and the length of the nursing period.

▪ CONCLUSION

The task of discovering those factors, singly or in combination, that direct the form and development of social behavior is still in its infancy. While laboratory studies have made major strides towards understanding the socialization process, field studies are only now reaching a level of sophistication where such details can be gleaned from the material. A major outcome of primate socialization research has been the recognition that more longitudinal studies, experimental manipulation, and analysis of behavior and group structure will be rewarding. Crook (1970) set forth a number of areas which need to be investigated for a better understanding of primate socialization; a modified version of these still holds true: (1) The extent of mother-exclusive care of young in comparison to "aunting" and "babysitting" by peers, siblings, and others; (2) the rate and amount of maternal attitudinal change to the infant preceding, and subsequent to, weaning; (3) the extent, manner, frequency and importance of peer interactions; (4) the extent, manner and frequency of heterosexual contact; (5) the interactional patterns between juvenile males approaching puberty and adult males and females; (6) the manner and extent of exclusion of adult males from breeding units (i.e., Itani 1972); (7) the pattern of "friendship" relations; (8) the influence of consanguineal ties (i.e., Kawamura and Kawai 1965; Kummer 1971; Sade 1965, 1967); (9) the development of male alliances, leadership, and interactive "policing" of the group; and (10) the effects of parental roles and status on the behavioral development of youngsters (i.e., Kawai 1958a,b). Others can be added to this list, for example, recent unpublished research by Poirier suggests that old females without infants may play an important role in comforting yearlings as they are weaned and their mothers turn their attention to the new infant. Articles in this book suggest that other possibilities include hormonal influences and neurological and muscular rates of development.

■ NOTES

1. It is important to note, however, that primates are also characterized by a further extension of the period of immaturity. The importance of this trait as it concerns the socialization process can be realized by comparing Sussman's (this volume) prosimian data with that on chimpanzees, cf. Clark; McGrew; and Nicolson (this volume).

2. McGrew (this volume) would disagree on this point, however.

3. An analogous situation is found among patas (*Erythrocebus patas*) (Hall 1968) and their social organization shows many similarities to that of the Nilgiri langur.

4. Poirier (1968a, 1969a,b, 1970a,b) based on his Nilgiri langur study suggests other possibilities and argues that females may be behaviorally more conservative than males. Kummer (1971) discusses the importance, the selective advantage, of behavioral conservatism.

5. References to these variables are: age (Dolhinow and Bishop 1970; Frisch 1968; Itani 1958; Kawai 1965; Kummer 1971; Poirier 1969a, 1970a, 1973b; Sade 1973; Tsumori et al. 1965); gender (Burton 1972; DeVore 1965; Dohinow and Bishop 1970; Fedigan 1972; Goodall 1965; Hall 1968; Harlow and Harlow 1961; Knudson 1971; Kummer 1968; Lancaster 1972; Lindburg 1971; Poirier 1971, 1972a,b; Poirier and Smith 1974; Ransom and Rowell 1972; Smith 1972; Sorenson 1970); group structure and dynamics (Baldwin 1969; Ellefson 1967, 1968; Hinde 1971; Poirier 1972b); and habitat (Baldwin and Baldwin 1974; Jensen et al. 1968; Jolly 1972; Rowell 1966; Washburn and DeVore 1961).

6. Other bonds include the peer affectional system, the heterosexual bond, and the bond between older and younger animals (Harlow 1966).

7. Poirier (1973b) has argued that the soft, silky, artificially induced baby powdered smell of the human infant functions rather like an approach-provoking natal coat among nonhuman primates. So also do the human infant's neotonous features.

■ REFERENCES

Altmann, S. 1965. Sociobiology of rhesus monkeys, II. Stochastics of communication. *Journal of Theoretical Biology* 8:490–522.

Anthoney, T. 1968. The ontogeny of greeting, grooming, and sexual motor patterns in captive baboons (Superspecies *Papio cynocephalus*). *Behavior* 31:358–72.

Baldwin, J. 1969. The ontogeny of social behavior of squirrel monkeys (*Saimiri sciureus*) in a semi-natural environment. *Folia Primatologica* 11:35–79.

Baldwin, J., and Baldwin, J. 1973. The role of play in social organizations: Comparative observations of squirrel monkeys (*Saimiri*). Paper read at meeting of American Association of Physical Anthropologists, Dallas, Texas.

Bandura, A., and Walters, R. 1963. *Social learning and personality development*. New York: Holt, Rinehart and Winston.

Bekoff, M. 1972. The development of social interaction, play, and metacommunication in mammals: An ethological perspective. *Quarterly Review of Biology* 47:412–33.

————. 1974. Social play and play soliciting by infant canids. *American Zoologist* 14:323–40.

Benedict, B. 1969. Role analysis in animals and men. *Man* 4:203–14.

Berlyne, D. 1960. *Conflict, arousal and curiosity*. McGraw-Hill: New York.

Bertrand, M. 1969. The behavioral repertoire of the stumptail macaque. *Biblioteca Primatologica* 11:1–123.

Booth, C. 1962. Some observations of behavior of *Cercopithecus* monkeys. *Annals of the New York Academy of Science* 102:479–87.

Bramblatt, C. 1973. Acquisition of social behavior in male vervet monkeys: The first four and one-half years. Paper read at the American Association of Physical Anthropologists, Dallas, Texas.

Brandt, E., and Mitchell, G. 1971. Parturition in primates: Behavior related to birth. In *Primate behavior: developments in field and laboratory research*, vol. 2, ed. L. Rosenblum, pp. 170–223. New York: Academic Press.

Burton, F. 1972. The integration of biology and behavior in the socialization of *Macaca sylvana* of Gibraltar. In *Primate socialization*, ed. F. Poirier, pp. 29–62. New York: Random House.

Carpenter, C. 1934. A field study of the behavior and social relations of howling monkeys. *Comparative Psychology Monographs* 10:1–168.

———. 1965. *Naturalistic behavior of nonhuman primates*. University Park: Pennsylvania State University Press.

Chalmers, N. 1968. Group composition, ecology and daily activities of free-living mangabeys in Uganda. *Folia Primatologica* 8:247–62.

———. 1972. Comparative aspects of early infant development in captive cercopithecines. In *Primate socialization*, ed. F. Poirier, pp. 63–82. New York: Random House.

Chamove, A.; Harlow, H.; and Mitchell, G. 1967. Sex differences in the infant-directed behaviors of preadolescent monkeys. *Child Development* 38:329–35.

Chevalier-Skolnikoff, S. 1973a. The ontogenetic development of communication patterns in stumptail monkeys. Paper read at meeting of American Association of Physical Anthropology, Dallas, Texas.

———. 1973b. Visual and tactile communication in *Macaca arctoides* and its ontogenetic development. *American Journal Physical Anthropology* 38:515–19.

Child, I. 1954. Socialization. In *Handbook of social psychology*. vol. 2, ed. G. Lindzey, pp. 665–92. Cambridge: Addison-Wesley Press.

Chomsky, N. 1965. *Aspects of the theory of language*. Boston: Massachusetts Institute of Technology.

Clausen, A., ed. 1968. *Socialization and society*. Boston: Little Brown.

Crook, J. 1967. Gelada herd structure and movement: A comparative report. *Symposium Zoological Society of London* 18:237–58.

———. 1970. The socio-ecology of primates. In *Social behavior in birds and mammals*, ed. J. Crook, pp. 103–69. New York: Academic Press.

Crook, J., and Gartlan, S. 1966. Evolution of primate societies. *Nature* 216:1200–03.

Deag, J., and Crook, J. 1971. Social behavior and "agonistic buffering" in the wild barbary macaque *Macaca sylvana* L. *Folia Primatologica* 15:183–200.

DeVore, I. 1963. Mother-infant relations in free-ranging baboons. In *Maternal behavior in mammals*, ed. H. Rheingold, pp. 305–35. New York: John Wiley.

DeVore, I., and Eimerl, S. 1965. *The primates*. New York: Time-Life Books.

Diamond, S. 1970. *The social behavior of mammals*. New York: Harper.

Diamond, S., and Hall, W. 1969. Evolution of the neocortex. *Science* 164:251–62.

Dolhinow, P., and Bishop, N. 1970. The development of motor skills and social relationships among primates through play. *Minnesota Symposium Child Psychology* 4:141–98.

Doyle, G.; Pelletier, A.; and Bekker, T. 1967. Courtship, mating and parturition in the Lesser Bushbaby (*Galago senegalensis moholi*) under semi-natural conditions. *Folia Primatologica* 7:169–97.

Ellefson, J. 1967. A natural history of the gibbon of the Malay Peninsula. Ph.D. dissertation, University of California, Berkeley.

———. 1968. Territorial behavior in the common white-handed gibbon *Hylobates lar*. In *Primates: Studies in adaptation and variability*, ed. P. Jay, pp. 180–99. New York: Holt, Rinehart and Winston.

Erickson, E. 1950. *Childhood and society*. New York: Norton.

Etkin, W., ed. 1967. *Social behavior from fish to man*. Chicago: University of Chicago Press.

Fedigan, L. 1972. Social and solitary play in a colony of vervet monkeys (*Cercopithecus aethiops*). *Primates* 13:347–64.

Freud, S. 1923. *The ego and the id*. London: Hogarth.

_____. 1930. *An outline of psychoanalysis*. London: Hogarth.

Frisch, J. 1968. Individual behavior and intertroop variability in Japanese macaques. In *Primates: Studies in adaptation and variability*, ed. P. Jay, pp. 243–53. New York: Holt, Rinehart and Winston.

Fullerton, C., and Cowley, J. The differential effect of the presence of adult male and female mice on the growth and development of the young. *Journal Genetic Psychology* 119:89–98.

Gartlan, S. 1968. Structure and function in primate society. *Folia Primatologica* 8:89–120.

_____. 1969. Sexual and maternal behavior of the vervet monkey, *Cercopithecus aethiops*. *Journal of Reproduction and Fertility Supplement* 6:137–50.

Gerai, J., and Scheenfield, A. 1968. Sex differences in mental and behavioral traits. *Genetic Psychology Monographs* 77:169–299.

Gilman, J., and Gilbert, C. 1946. The reproductive cycle of the chacma baboon (*Papio ursinus*) with special reference to the problem of menstrual irregularities as assessed by the behavior of the sex skin. *South African Journal Medical Science, Biology Supplement* 11:1–54.

Goodall, J. van Lawick. 1965. Chimpanzees of the Gombe Stream. In *Primate behavior: Field studies of monkeys and apes*, ed. I. DeVore, pp. 425–73. New York: Holt, Rinehart and Winston.

_____. 1967. Mother-offspring relationships in free-ranging chimpanzees. In *Primate ethology*, ed. D. Morris, pp. 287–347. Chicago: Aldine.

Goswell, M., and Gartlan, S. 1965. Pregnancy, birth and early behavior in the captive patas monkey, *Erythrocebus patas*. *Folia Primatologica* 3:189–200.

Gottlieb, G. 1971. Ontogenesis of sensory function in birds and mammals. In *The biopsychology of development*, eds. E. Tobach, L. D. Aronson, and E. Shaw, pp. 67–128. New York: Academic Press.

Goy, R. 1968. Organizing effects of androgen on the behavior of rhesus monkeys. In *Endocrinology and human behavior*, ed. R. Michael, pp. 12–31. London: Oxford University Press.

Goy, R., and Phoenix, C. 1962. The effects of testosterone propionate administered before birth on the development of behavior in genetic female rhesus monkeys. In *Steroid hormones and brain function*, ed. C. Sawyer and R. Gorski, pp. 193–200. Berkeley: University of California Press.

Goy, R., and Resko, J. 1972. Gonadal hormones and behavior of normal and pseudohermaphroditic nonhuman female primates. In *Recent progress in hormone research*, vol. 28, ed. E. Astwood, pp. 707–33. New York: Academic Press.

Goyer, R. 1970. Communication, communication process, meaning: Toward a unified theory. *Journal Communication* 20:4–16.

Graham-Jones, O., and Hill, W. 1962. Pregnancy and parturition in a Bornean orang. *Proceedings Zoological Society London* 139:503–10.

Hall, K. 1968. The behavior and ecology of the wild patas monkey, *Erthyrocebus patas*, in Uganda. In *Primates: Studies in adaptation and variability*, ed. P. Jay, pp. 32–119. New York: Holt, Rinehart and Winston.

Hall, K., and DeVore, I. 1965. Baboon social order. In *Primate behavior: Field studies of monkeys and apes*, ed. I. DeVore, pp. 53–110. New York: Holt, Rinehart and Winston.

Hall, K., and Goswell, M. 1964. Aspects of social learning in captive patas monkeys. *Primates* 5:59–70.

Hansen, E. 1966. The development of maternal and infant behavior in the rhesus monkey. *Behaviour* 27:107–49.

Harlow, H. 1960. Primary affectional patterns in primates. *American Journal Orthopsychiatry* 30:676–84.

_____. 1962. The development of affectional patterns in infant monkeys. In *Determinants of infant behavior*, ed. B. Foss, pp. 75–97. New York: J. Wiley.

_____. 1963. Basic social capacity of primates. In *Primate social behavior*, ed. C. Southwick, pp. 153–61. Princeton: D. Van Nostrand.

_____. 1966. The primate socialization motives. *Transactions and Studies of the College of Physicians of Philadelphia* 33:224–37.

————. 1969. Age-mate or peer affectional systems. In *Advances in the study of behavior*, vol. 2, ed. D. Lehrman, R. Hinde and E. Shaw, pp. 333–83. New York: Academic Press.

Harlow, H., and Harlow, M. 1961. A study of animal affection. *Natural History* 70:48–55.

————. 1965. The affectional systems. In *Behavior of nonhuman primates*, vol. 2, eds. A. Schrier, H. Harlow and F. Stollnitz, pp. 287–334. New York: Academic Press.

————. 1966. Learning to love. *American Scientist* 54:244–72.

Harlow, H.; Harlow, M.; Dodsworth, R.; and Arling, A. 1966. Maternal behavior of rhesus monkeys deprived of mothering and peer associations in infancy. *Proceedings American Philosophical Society* 110:58–66.

Harlow, H.; Harlow, M.; and Suomi, S. 1971. From thought to therapy: Lessons from a primate laboratory. *American Scientist* 59:538–50.

Harman, P. 1957. Paleoneurologic, neoneurologic and ontogenetic aspects of brain phylogeny. James Arthur lecture on the Evolution of the Human Brain. American Museum of Natural History, New York.

Hartman, C. 1928. The period of gestation in the monkey, *Macaca rhesus*: First description of parturition in monkeys, size and behavior of the young. *Journal Mammalology* 9:181.

Hinde, R. 1971. Development of social behavior. In *Behavior of nonhuman primates*, vol. 2, eds. A. Schrier, H. Harlow and F. Stollnitz, pp. 1–68. New York: Academic Press.

Hinde, R., and Spencer-Booth, Y. 1967. The behavior of socially living monkeys in their first two and a half years. *Animal Behavior* 15:169–96.

Hopf, S. 1967. Notes on pregnancy, delivery and infant survival in captive squirrel monkeys. *Primates* 8:323–32.

Hrdy, S. 1974. Male-male competition and infanticide among the langurs (*Presbytis entellus*) of Cebu, Rajasthan. *Folia Primatologica* 22:19–58.

Imanishi, S. 1960. Social organization of subhuman primates in their natural habitat. *Current Anthropology* 1:393–407.

Itani, J. 1958. On the acquisition and propagation of a new food habit in the natural groups of the wild Japanese monkey at Takasakiyama. *Primates* 1:84–98.

————. 1959. Paternal care in the wild Japanese monkey, *Macaca fuscata fuscata*. *Primates* 2:61–93.

————. 1972. A preliminary essay on the relationship between social organization and incest avoidance in nonhuman primates. In *Primate Socialization*, ed. F. Poirier, pp. 165–71. New York: Random House.

Itani, J.; Tokuda, K.; Furuya, Y.; Kano, K.; and Shin, Y. 1963. Social construction of natural troops of Japanese monkeys in Takasakiyama. *Primates* 4:2–42.

Jay, P. 1962. Aspects of maternal behavior in langurs. *Annals New York Academy of Science* 102:468–76.

————. 1965. The common langur of North India. In *Primate behavior: Field studies of monkeys and apes*, ed. I. DeVore, pp. 197–250. New York: Holt, Rinehart and Winston.

————. ed. 1968. *Primates: Studies in adaptation and variability*. New York: Holt, Rinehart and Winston.

Jensen, G., Bobbit, R.; and Gordon, A. 1967a. Sex differences in social interaction between infant monkeys and their mothers. In *Recent advances in biological psychiatry*, ed. J. Wortis, pp. 283–98. New York: Plenum.

————. 1967b. The development of mutual avoidance in mother-infant pigtailed monkeys, *Macaca nemestrina*. In *Social communication among primates*, ed. S. Altmann, pp. 43–55. Chicago: University of Chicago Press.

————. 1968. Sex differences in the development of independence of infant monkeys. *Behavior* 30:1–14.

————. 1973. Mother's and infant's roles in the development of independence of *Macaca nemestrina*. *Primates* 14:79–88.

Jolly, A. 1966. *Lemur behavior*. Chicago: University of Chicago Press.

————. 1972. *The evolution of primate behavior*. New York: Macmillan.

Kalmus, H. 1969. Animal behavior and theories of games and language. *Animal Behavior* 17:607–17.

Kaufman, J. 1966. Behavior of infant rhesus monkeys and their mothers in a free-ranging band. *Zoologica* 51:17–28.

Kaufmann, I., and Rosenblum, L. 1969. The waning of the mother-infant bond in two species of macaque. In *Determinates of infant behavior*, IV, ed. B. Foss, pp. 41–59. London: Methuen.

Kawai, M. 1958a. On the rank system in a natural group of Japanese monkeys. *Primates* 1:84–98.

———. 1958b. On the system of social ranks in a natural troop of Japanese monkeys I: Basic and dependent rank. *Primates* 1:111–30.

———. 1965. Newly acquired precultural behavior of the natural troop of Japanese monkeys on Koshima Islet. *Primates* 6:1–30.

———. 1967. Catching behavior observed in the Koshima troop: A case of newly acquired behavior. *Primates* 8:181–86.

Kawamura, S. 1968. Matriarchal social ranks in the Minoo-B troop: A study of the rank system of Japanese monkeys. *Primates* 2:181–252.

Kawamura, S., and Kawai, M. 1956. Social organization of the natural group of Japanese macaque. *Animal Psychology* 6:120–35.

———. 1959. The process of subcultural propagation among Japanese macaques. *Primates* 2:43–60.

Knudson, M. 1971. Sex differences in dominance behavior of young human primates. Paper read at meeting of the American Anthropological Association, New York.

———. 1973. Sex differences in dominance behavior of young human primates. Ph.D. dissertation, University of Oregon, Eugene.

Koford, C. 1963. Rank of mothers and sons in bands of rhesus monkeys. *Science* 141:356–57.

Koyama, N. 1967. On dominance rank and kinship of a wild Japanese monkey troop in Arashiyama. *Primates* 8:189–216.

Kummer, H. 1967. Dimensions of a comparative biology of primate groups. *American Journal Physical Anthropology* 27:357–66.

———. 1968. Two variations in the social organization of baboons. In *Primates: Studies in adaptation and variability*, ed. P. Jay, pp. 293–312. New York: Holt, Rinehart and Winston.

———. 1971. *Primate societies*. Chicago: Aldine.

Kuo, Z. 1967. *The dynamics of behavior development*. New York: Random House.

Lancaster, J. 1971. Play-mothering: The relations between juvenile females and young infants among free-ranging vervet monkeys (*Cercopithecus aethiops*). *Folia Primatologica* 15:161–82.

———. 1972. Play-mothering: The relations between juvenile females and young infants among free-ranging vervet monkeys. In *Primate socialization*, ed. F. Poirier, pp. 83–104. New York: Random House.

———. 1973. In praise of the achieving female monkey. *Psychology Today* 7:30–37.

Lahiri, R., and Southwick, C. 1966. Parental care in *Macaca sylvana. Folia Primatologica* 4:257–64.

Lindburgh, D. 1971. The rhesus monkey in North India. In *Primate behavior: Developments in field and laboratory research*, vol. 1, ed. L. Rosenblum, pp. 83–104. New York: Academic Press.

Lindburgh, D., and Hazell, L. 1972. Licking of the neonate and duration of labor in great apes and man. *American Anthropologist* 74:318–25.

Lorenz, K. 1937. The companion in the bird's world. *Auk* 54:245–73.

Loy, J. 1970. Behavioral responses of free ranging rhesus monkeys to food shortage. *American Journal Physical Anthropology* 33:263–73.

———. 1973. Social development and modification among primates. Paper read at meeting of the American Association of Physical Anthropology, Dallas, Texas.

Loy, J., and Loy, K. 1974. Behavior of an all juvenile group of rhesus monkeys. *American Journal Physical Anthropology* 40:83–97.

MacRoberts, M. 1970. The social organization of barbary apes (*Macaca sylvana*) on Gibraltar. *American Journal Physical Anthropology* 33:83–101.

Mason, W. 1960. The effects of social restriction on the behavior of rhesus monkeys: I. Free social behavior. *Journal Comparative Physiological Psychology* 53:582–89.

————. 1961a. The effects of social restriction on the behavior of rhesus monkeys: II. Tests of gregariousness. *Journal Comparative Physiological Psychology* 54:287–96.

————. 1961b. The effects of social restriction on the behavior of rhesus monkeys: III. Dominance tests. *Journal Comparative Physiological Psychology* 54:694–99.

————. 1963. The effects of environmental restriction on the social development of rhesus monkeys. In *Primate social behavior*, ed. C. Southwick, pp. 161–174. Princeton: D. Van Nostrand.

————. 1965. The social development of monkeys and apes. In *Primate behavior: Field studies of monkeys and apes*, ed. I. DeVore, pp. 514–44. New York: Holt, Rinehart and Winston.

————. 1968. Naturalistic and experimental investigations of the social behavior of monkeys and apes. In *Primates: Studies in adaptation and variability*, ed. P. Jay, pp. 398–420. New York: Holt, Rinehart and Winston.

McGrew, W. 1972. *An ethological study of children's behavior.* New York: Academic Press.

Mead, M. 1935. *Sex and temperament in three primitive societies.* New York: Morrow.

Meier, G. 1965. Maternal behavior of feral and laboratory-reared monkeys following surgical delivery of their infants. *Nature* 206:492–93.

Menzel, E. 1966. Responsiveness to objects in free-ranging Japanese monkeys. *Behaviour* 26:130–50.

————. 1967. Naturalistic and experimental research on primates. *Human Development* 10:170–86.

————. 1968. Primate naturalistic research and problems of early experience. *Developmental Psychobiology* 1:175–84.

Missakian, E. 1972. Genealogical and cross-genealogical dominance relations in a group of free-ranging rhesus monkeys (*Macaca mulatta*) on Cayo Santiago. *Primates* 13:169–80.

Mitchell, G. 1969. Paternalistic behavior in primates. *Psychological Bulletin* 71:399–417.

Mitchell, G., and Brandt, E. 1970. Behavioral differences related to experience of mother and sex of infant in the rhesus monkey. *Developmental Psychology* 3:149.

Mitchell, G., and Brandt, E. 1972. Paternal behavior in primates. In *Primate socialization*, ed. F. Poirier, pp. 173–206. New York: Random House.

Mitchell, G., and Mohler, G. 1967. Long-term effects of maternal punishment on the behavior of monkeys. *Psychonomic Science* 8:209–10.

Mitchell, G., and Stevens, C. 1969. Primiparous and multiparous mothers in a mildly stressful social situation: First three months. *Developmental Psychology* 1:280–86.

Mohnot, S. 1971. Some aspects of social change and infant-killing in the hanuman langur *Presbytis entellus* (Primates: Cercopithecidae) in western India. *Mammalia* 35:175–98.

Mugford, R., and Nowell, N. 1972. Paternal stimulation during infancy: Effects on aggression and open-field performance of mice. *Journal Comparative Physiological Psychology* 79:30–36.

Nash, L., and Ransom, T. 1971. Socialization in baboons at the Gombe Stream National Park, Tanzania. Paper read at the meeting of the American Anthropological Association, New York, New York.

Norback, C., and Moskowitz, N. 1963. The primate nervous system: Functional and structural aspects of phylogeny. In *Evolutionary and genetic biology of primates*, vol. 1, ed. J. Buettner-Janusch, pp. 131–175. New York: Academic Press.

Parker, S. 1973. Intellectual development in an infant *Macaca arctoides* in terms of Piaget's paradigm. Paper read at meeting of the American Association Physical Anthropology, Dallas, Texas.

Petter-Rousseaux, A. 1964. Reproductive physiology and behavior of the Lemuroidea. In *Evolutionary and genetic biology of primates*, vol. 2, ed. J. Buettner-Janusch, pp. 91–132. New York: Academic Press.

Pfieffer, J. 1969. *The emergence of man.* New York: Harper and Row.

Phoenix, C.; Goy, R.; and Resko, J. 1968. Psychosexual differentiation as a function of androgenic stimulation. In *Perspectives in reproduction and sexual behavior*, ed. M. Diamond, pp. 33–49. Bloomington: Indiana University Press.

Poirier, F. 1968a. Analysis of a Nilgiri langur (*Presbytis johnii*) home range change. *Primates* 9:29–44.

———. 1968b. The Nilgiri langur (*Presbytis johnii*) mother-infant dyad. *Primates* 9:45–68.

———. 1969a. The Nilgiri langur troop: Its composition, structure, function and change. *Folia Primatologica* 19:20–47.

———. 1969b. Nilgiri langur (*Presbytis johnii*) territorial behavior. *Primates* 9:351–65.

———. 1969c. Behavioral flexibility and intertroop variability among Nilgiri langurs of South India. *Folia Primatologica* 11:119–33.

———. 1970a. Nilgiri langur ecology and social behavior. In *Primate behavior: Developments in field and laboratory research*, vol. I, ed. L. Rosenblum, pp. 251–383. New York: Academic Press.

———. 1970b. Characteristics of the Nilgiri langur dominance structure. *Folia Primatologica* 12:161–87.

———. 1971. Socialization variables. Paper read at the meeting of the American Anthropological Association, December, New York, New York.

———. 1972a. Introduction. In *Primate socialization*, ed. F. Poirier, pp. 2–29. New York: Random House.

———. 1972b. Nilgiri langur behavior and social organization. In *Essays to the chief*, ed. F. Voget and R. Stephenson, pp. 119–34. Eugene: University of Oregon Press.

———. 1972c. Primate socialization and learning. Paper read at the meeting of the American Anthropological Association, Toronto, Canada.

———. 1973a. Primate socialization—where do we go from here? Paper read at the meeting of the American Association of Physical Anthropology, Dallas, Texas.

———. 1973b. Primate socialization and learning. In *Culture and learning*, ed. S. Kimball and J. Burnett, pp. 3–41. Seattle: University of Washington Press.

———. 1974. Colobine aggression: A review. In *Primate aggression, territoriality, and xenophobia*, ed. R. Holloway, pp. 123–58. New York: Academic Press.

———. 1975a. Effects of ecological isolation on behavioral and morphological evolution in three primate species. Paper read at the meeting of the American Association of Physical Anthropology, Denver, Colorado.

———. 1975b. Socialization of nonhuman primate females: A brief overview. In *Being female: Reproduction, power and change*. ed. D. Raphael. pp. 13–19. The Hague: Mouton.

Poirier, F., and Smith, E. 1974. Socializing functions of primate play behavior. *American Zoologist* 14:275–87.

Ransom, R., and Rowell, T. 1972. Early social development of feral baboons. In *Primate socialization*, ed. F. Poirier, pp. 105–44. New York: Random House.

Redican, W., and Mitchell, G. 1973. The social behavior of adult male-infant pairs of rhesus macaques in a laboratory environment. *American Journal of Physical Anthropology* 38:523–27.

Riesen, H. 1961. Critical stimulation and optimum periods. Paper read at the meeting of the American Psychological Association, December, New York, New York.

Rosenblatt, J.; Turkewitz, G.; and Schneirla, T. 1961. Early socialization in domestic cats as based on feeding and other relations between female and young. In *Determinants of infant behaviour*, ed. B. Foss, pp. 51–74. London: Methuen and Co.

Rosenblum, L. 1971. The ontogeny of mother-infant relations in macaques. In *Ontogeny of vertebrate behavior*, pp. 315–67. New York: Academic Press.

Rowell, T. 1966. Forest-living baboons in Uganda. *Journal Zoological Society of London* 149:344–63.

Rowell, T.; Hinde, R.; and Spencer-Booth, Y. 1974. "Aunt"-infant interaction in captive rhesus monkeys. *Animal behavior* 12:219–26.

Rumbaugh, D. 1970. Learning skills of anthropoids. In *Primate behavior: Developments in field and laboratory research*, vol. 1, ed. L. Rosenblum, pp. 2–70. New York: Academic Press.

Sade, D. 1965. Some aspects of parent-offspring and sibling relationships in a group of rhesus monkeys with a discussion of grooming. *American Journal of Physical Anthropology* 23:1–17.

———. 1967. Ontogeny of social relations in a group of free-ranging rhesus monkeys (*Macaca mulatta* Zimmerman). Ph.D. dissertation, University of California, Berkeley.

———. 1968. Inhibition of son-mother mating among free-ranging rhesus monkeys. *Science and Psychoanalysis* 12:18–30.

———. 1971. Life cycle and social organization among free-ranging rhesus monkeys. Paper read at the meeting of the American Anthropological Association, New York, New York.

———. 1972. A longitudinal study of social behavior of rhesus monkeys. In *The functional and evolutionary biology of primates*, ed. R. Tuttle, pp. 378–99. Chicago: Aldine.

———. 1973. An ethogram for rhesus monkeys. I. Antithetical contrasts in posture and movement. *American Journal Physical Anthropology* 38:537–42.

Scarf, M. 1972. He and she: The sex hormones and behavior. *New York Times Magazine*, 7 May, pp. 30–31, 101–107.

Schnierla, T., and Rosenblatt, J. 1963. Critical periods in the development of behavior. *Science* 139: 110–15.

Schultz, A. 1956. Postembryonic age changes. In *Primatologia*, ed. H. Hofer, A. Schultz, and D. Starck, pp. 887–964. Basel: S. Karger.

Scott, J. 1945. Group formation determined by social behavior. *Sociometry* 8:42–52.

———. 1950. The social behavior of dogs and wolves. *Annals New York Academy Science* 55:1009–21.

———. 1958. Critical periods in the development of social behavior in puppies. *Psychological Medicine* 20:42–54.

Scott, J., and Marston, M. 1950. Critical periods affecting the development of normal and maladjustive social behavior of puppies. *Journal Genetic Psychology* 77:25–60.

Seay, G. 1966. Maternal behavior in primiparous and multiparous rhesus monkeys. *Folia Primatologica* 4:146–68.

Simmel, E. 1970. The biology of socialization. In *Early experience and the process of socialization*, ed. R. Hoppe, G. Milton, and E. Simmel, pp. 3–7. New York: Academic Press.

Simonds, P. 1971. Bonnet macaque infant socialization. Paper read at the meeting of the American Anthropological Association, New York, New York.

Smith, E. 1972. The interrelationship of age and status and selected behavioral categories in male pigtail macaques (*Macaca nemestrina*). Master's thesis, University of Georgia, Athens.

———. 1973. A model for the study of social play in non-human primates. Paper read at the meeting of the Midwest Animal Behavior Society, Oxford, Ohio.

Sorenson, M. 1970. Behavior of tree shrews. In *Primate behavior: Developments in field and laboratory research*, vol. II, ed. L. Rosenblum, pp. 141–94. New York: Academic Press.

Spencer-Booth, Y. 1968. The behavior of group companions towards rhesus monkey infants. *Animal Behavior* 16:541–57.

———. 1970. The relationship between mammalian young and conspecifics other than mother and peers: A review. *Adv. Study of Behavior* 3:119–194.

Struhsaker, T. 1967. Behavior of vervet monkeys, *Cercopithecus aethiops*. *University California Publications Zoology* 82:1–74.

Sugiyama, Y. 1965. Behavioral development and social structure in two troops of hanuman langurs (*Presbytis entellus*). *Primates* 6:213–48.

———. 1967. Social organization of hanuman langurs. In *Social communication among primates*, ed. S. Altmann, pp. 221–37. Chicago: University of Chicago Press.

———. 1972. Social characteristics and socialization among wild chimpanzees. In *Primate socialization*, ed. F. Poirier, pp. 145–63. New York: Random House.

Symons, D. 1973. Aggressive play in a free-ranging group of rhesus monkeys (*Macaca mulatta*). Ph.D. dissertation, University of California, Berkeley.

Takeshita, H. 1961. On the delivery behavior of squirrel monkeys and a mona monkey. *Primates* 3:59–72.

Tinklepaugh, O. 1948. Social behavior of animals. In *Comparative Psychology*, ed. F. Moss, pp. 366–94. New York: Prentice-Hall.

Tinklepaugh, O., and Hartman, G. 1932. Behavior and maternal care of the newborn monkey (*M. mulatta, M. rhesus*). *Journal Genetic Psychology* 40:257–86.

Tisza, V.; Hurwitz, I.; and Angoff, K. 1970. The use of a play program by hospitalized children. *Journal American Academy Child Psychiatry* 9:515–31.

Tsumori, A. 1967. Newly acquired behavior and social interaction of Japanese monkeys. In *Social communication among primates*, ed. C. Southwick, pp. 207–21. Chicago: University of Chicago Press.

Tsumori, A.; Kawai, M.; and Motoyishi, R. 1965. Delayed response of wild Japanese monkeys by the sand-digging test (I)—Case of the Koshima troop. *Primates* 6:195–212.

Vessey, S. 1971. Social behavior of free-ranging rhesus monkeys in the first year. Paper read at the meeting of the American Anthropological Association, New York, New York.

Washburn, S. 1973. Primate field studies and social science. In *Cultural illness and health*, ed. L. Nader and T. Maretzki, pp. 128–35. Washington: American Anthropological Association.

Washburn, S., and DeVore, I. 1961. The social life of baboons. *Scientific American* 204:2–11.

Washburn, S., and Hamburg, D. 1965. The implications of primate research. In *Primate behavior: Field studies of monkeys and apes*, ed. I. DeVore, pp. 607–23. New York: Holt, Rinehart and Winston.

Williams, L. 1967. Breeding Humbolt's wooly monkey (*Lagothrix lagothricha*) at Murryton Wooly Monkey Sanctuary. *International Zoo Yearbook* 7:86–89.

Yamada, M. 1963. A study of blood relationships in the natural society of the Japanese macaque. *Primates* 4:43–67.

Part One

BIOLOGICAL DETERMINANTS OF SOCIALIZATION

<table>
<tr><td>**2**</td></tr>
</table>

Piaget's Sensorimotor Period Series in an Infant Macaque: A Model for Comparing Unstereotyped Behavior and Intelligence in Human and Nonhuman Primates

SUE TAYLOR PARKER

■ INTRODUCTION

Very little is known about the organization of unstereotyped behavior and intelligence in nonhuman primates; even less is known about species' differences in the organization of unstereotyped behavior and intelligence. A great deal is known, however, about the organization and development of these patterns in our own species. Jean Piaget, the renowned Swiss psychologist, has developed a comprehensive model of human intellectual development from birth through adolescence. However well we study ourselves, we cannot fully understand the biological significance of our own pattern until we know what part of it is uniquely human and what part of it is shared with anthropoid apes, old world monkeys, and other primates and mammals; and we cannot know this until we compare the development and organization of unstereotyped behavior and intelligence in these groups.

Fortunately, Piaget's model provides the basis for a comprehensive framework for comparing the organization of unstereotyped behavior and intelligence in human and nonhuman primates. In fact, it is probably the only suitable basis for a framework for this purpose because no other model embodies equivalent concepts.

Ethological concepts such as the fixed-action-pattern and the innate-releasing-mechanism which have been so useful in analyzing and comparing the stereotyped behavior of insects, fish, birds, and even primates, are not appropriate for analyzing and comparing unstereotyped behavior and intelligence.

Although they have revealed important differences in species' abilities, psychological concepts such as insight learning and perceptual gestalts (Köhler 1927; Klüver 1933) are too limited in scope to provide a comprehensive picture of species' differences in the organization of unstereotyped behavior and intelligence. Although a developmental approach is vital to understanding the organization of behavior, most developmental models are too general or too limited in scope to reveal differences in the development of unstereotyped behavior and intelligence.

Whereas other conceptual frameworks offer insight into some aspects of behavioral organization, Piaget's model offers insight into virtually all aspects of behavioral organization.

Piaget's model is powerful because it encompasses the simplest and the most complex behavior of the most intelligent species in our order, and because it focuses on the meta-behavioral level, that is, on the organizational level of behavioral patterning. Piaget described parameters such as the nature and the locus of coordinations, their elicitors, their goals, their reinforcers, and their temporal patterning which usually remain implicit in behavioral descriptions. Moreover, he noted the longitudinal development of these parameters.

In the following pages I present the results of a longitudinal study of behavioral organization in an infant stumptail macaque monkey based on Piaget's model. This study demonstrates the value of a Piagetian approach to comparative studies of behavioral organization and intelligence.

Using a Piagetian approach requires some understanding of Piaget's model; before describing the results of my study and comparing development of an infant macaque with human infant development, I will summarize Piaget's sensorimotor period stages in human infants. Since his work on this period alone is encyclopedic, I will discuss only those concepts that are important in my study.

Because Piaget's specialized vocabulary of terms in his work on this period is difficult to master, I have substituted more common terms for his throughout this paper except in his stage names and the analogous stage names in the infant macaque. Whenever Piaget's terms appear they are placed in quotation marks.

In his three books on the sensorimotor period (1952, 1954, 1962), Piaget described the growth and development of intelligence from birth to two years of age in human infants beginning with neonatal reflexes and ending with the achievement of representation of new means. He described this period in terms of six separate series: the sensorimotor intelligence series; the object-concept series; the spatial, temporal, causal, and imitation series. Each series includes six stages; the stages in each series are approximately concurrent, e.g., the third stage occurs at the same time in each series.

Each stage is characterized by the emergence of new "schemata" (behavior patterns). The "schema" concept is a central element of Piaget's model: The "schema" is an organized unit of behavior of any level of complexity from reflex behaviors, such as neonatal sucking and grasping, to intelligent behavior patterns such as rotating a stick in order to get it through crib bars. (I will generally refer to behavior patterns as "schemata.")

The "schema" concept encompasses all the sensorimotor perceptual and experiential factors involved in any behavior pattern. Although behavioral organization can be inferred from particular classes of "schemata," the concept itself is theoretically neutral, implying nothing about the complexity or stereotypy of behavior. Unlike the innate-releasing mechanism and fixed-action-pattern, it can serve as a general descriptive term for all kinds of behavior.

Each stage is characterized by the emergence of new "schemata" which are transformations of old "schemata": "circular reactions," for example, are important classes of "schemata" which emerge and change form during the sensorimotor period. They are called "circular reactions" because they involve repeated motor acts. There are three classes of "circular reactions," the "primary circular reactions," the "secondary circular reactions," and the "tertiary circular reactions," which are characteristic of the second, third, and fifth stages in the sensorimotor intelligence series, respectively.

Each one of the "circular reactions" involves repetition of "schemata" but each one is quite different from the others in the aim, or goal, and the reinforcement of the repetition. (Each type of "circular reaction" is discussed in the following stage summaries.) According to Piaget, new "circular reactions" develop as the infant integrates experiences produced by the functioning of the preceding "circular reactions."

Piaget's model relies heavily on the concepts of "assimilation" and "accommodation" which he borrowed from biology: "assimilation" refers to the process of ingesting or absorbing nutrient or stimuli and then integrating them into the body structure and/or behavior; "accommodation" refers to the related process of organic adjustments to stimuli and objects during the process of "assimilation." "Assimilation" and "accommodation" are two reciprocal aspects of "adaptation" according to Piaget. The grasping "schema," for example, "assimilates" all kinds of objects and "accommodates" to their shapes, sizes, weights, etc. "Schemata" develop and change as they adapt to the environment through "assimilation" and "accommodation." Because the child can only integrate experiences which can be "assimilated" into his pre-existing "schemata" and to which he can "accommodate" his pre-existing "schemata," development is essentially conservative. The infant cannot imitate sounds or gestures, for example, until they occur in his own repertoire because he must "assimilate" the model's "schema" to his own and then "accommodate" his "schema" to the model's "schema."

Piaget described the emergence and developmental transformation of behavior patterns in human infants from birth to two years of age in his sensorimotor period series. Each new stage reflected an internal reorganization due in part to "adaptation" to the environment. Although the ages are approximately the same in the parallel stages of each series, they are not exactly the same because the acquisitions in some series are dependent on the development of acquisitions in parallel series. For example, imitation of new behavior patterns is dependent upon the "tertiary circular reactions" because the infant cannot imitate a novel "schema" until he can make trial-and-error "accommodations" of his own "schemata" to the behavior pattern of the person he is imitating thereby "assimilating" the model's "schema" to one of his own "schema."

■ PIAGET'S SENSORIMOTOR PERIOD SERIES IN HUMAN INFANTS

SERIES I: THE SENSORIMOTOR INTELLIGENCE SERIES

SERIES I, STAGE 1: *From birth through the first month—"The Reflex Stage"*

During this stage congenital reflexes operate in response to generalized stimuli; neonatal behavior patterns such as sucking, grasping, and looking "assimilate" (that is, take in or act upon) every accessible stimulus to themselves. There is little "accommodation" to the particular properties of objects during this stage; infants respond very generally to a large class of stimuli, sucking, for example, on whatever is presented to their mouth or cheek. The behavior patterns of this stage are largely involuntary.

SERIES I, STAGE 2: *From the second to the fourth month—"The Primary Circular Reactions and the First Acquired Adaptations to the Environment"*

During this stage infants begin to display some primitive voluntary behavior patterns extending and elaborating the action of reflexes of sucking, phonation, hearing, vision, and prehension. Piaget called the behavior patterns of this stage "primary schemata" because they are focused on the infant's own body or on the "assimilation" of objects to simple (consummatory) behavior patterns such as sucking, grasping, looking. Infants at this stage apparently respond to movement, touch, smell, taste, and proprioception primarily.

"Primary circular reactions" are characteristic of this stage; these are spontaneously repeated behavior patterns "reproducing interesting results produced by chance on his own body" (Piaget 1952:1954).

A. SOME EXAMPLES OF "PRIMARY CIRCULAR REACTIONS"

1. Small rhythmic tongue protrusions (mouthing movements).
2. Rhythmic "empty" grasping and letting go with hands.
3. Rhythmic arm, leg, trunk, and head movements.
4. Repeated hand-hand opposition.
5. Repeated thumb-sucking.

"Reciprocal coordinations" of behavior patterns with each other are also characteristic of this stage.

B. SOME EXAMPLES OF "RECIPROCAL COORDINATIONS" (WHICH ARE ALSO "PRIMARY CIRCULAR REACTIONS")

1. Thumb-sucking.
2. Bringing objects to the mouth and sucking.
3. Visual following own hand.
4. Reciprocal hand-hand grasping.
5. Touching own face and ears and hair.
6. Grasping objects he sees.

Primary adaptations to the environment are also characteristics of this stage; infants begin to display some "accommodation" to the environment.

C. SOME EXAMPLES OF PRIMARY ADAPTATIONS TO THE ENVIRONMENT

1. Moving the head around in order to "see" a sound.
2. Alternate glancing between two objects.
3. Visual following.

SERIES I, STAGE 3: *From the fourth to eight months—The "Secondary Circular Reactions"*

For the most part, "secondary circular reactions" develop out of the systematic coordinations of prehension and sight. The behavior patterns of this stage are called "secondary schemata" because they focus on the properties of objects and environmental results of acts. They are voluntary behavior patterns. "Secondary circular reactions" involve repetition of certain behaviors in order to recreate a (contingent) "interesting spectacle" such as the movement of a mobile or the sound of a rattle. "Secondary circular reactions" grow out of the "primary circular reactions" as the infant notices the movements and sounds of environmental objects associated with his own movements and repeats his actions in order to prolong or recreate the "interesting spectacle." "After reproducing the interesting results discovered by chance on his own body, the child tries sooner or later to conserve also those which he obtains when his action bears on the external environment. It is this very simple transition which determines the appearance of the 'secondary' reactions. . ." (Piaget 1952:1954). It is apparently the fact that the environmental effects were contingent upon his previous actions which makes the spectacles interesting. According to Piaget the behavior patterns of this stage are not truly intelligent because they lack foresight and a pre-existing goal; it is the happy accident which determines the "fortuitous" goal. Moreover they apply to only one object at a time.

A. SOME EXAMPLES OF SECONDARY CIRCULAR REACTIONS

(Shaking or swinging an object either directly with the hand or foot, or indirectly by moving an intervening object through movements of the whole body or forearm, head or feet; intentionally repeating actions.)

1. a. Swinging the bassinette ornament as legs thrash.
   b. Stopping.
   c. Repeating the leg thrashing while smiling and watching the bassinette ornament.
   d. Stopping.
   e. Repeating.
2. a. Waving the arm while grasping a rattle in the hand.
   b. Stopping.
   c. Repeating the arm shaking while listening to the rattle and smiling.

The third stage is also characterized by "differentiation of global behavior patterns;" for example, the infant first moves his whole body while watching the bassinette ornament and then thrashes his legs while watching it.

SERIES I, STAGE 4: *From eight months on—"The Coordination of the Secondary Schemata and Their Application to New Situations"*

During this stage infants coordinate "secondary schemata" (that is, behavior patterns focused on external objects) into new behavior patterns (i.e., new "schemata") which are fully intentional and goal-directed. These behavior patterns involve serial and simultaneous coordinations and applications of pre-existing behavior patterns (i.e., old "schemata") in new situations.

This stage is theoretically important because it marks the emergence of intelligent, i.e., intentional, behavior patterns. Behavior patterns are differentiated into final and auxiliary acts, with one act such as uncovering an object subsumed under the other, e.g., grasping the object, in pursuit of a goal, the final act being dependent upon the auxiliary act. Piaget called behavior patterns of this stage "mobile" because they are capable of inclusion and subsumption and exclusion.

A. SOME EXAMPLES OF SECONDARY COORDINATIONS (COORDINATION OF "SECONDARY SCHEMATA")

1. Shaking strings attached to a bassinette in order to free a toy entangled in them.
2. Taking mother's hand and placing it on an object she had just previously moved, apparently in order to make her repeat her action.
3. Putting down one object in order to grasp another object.
4. Uncovering an object which has been covered while the infant was watching.
5. Placing an object by the feet and kicking it.

The fourth stage is a critical stage in the sensorimotor intelligence series because it marks the beginning understanding of object-object relations. The beginning of object-object relations occurs when the child places one object in relation to another object acting on them both in one coordinated behavior pattern: Piaget says that when two primary "schemata" are reciprocally assimilated they apply to one and the same object, and when a secondary "schema" acts it applies to only one object, but in the fourth stage "on the contrary, the coordination of schemata bears upon two or several separate objects produced together (the objective and the obstacle or the objective and the intermediary, etc.) in such a way that the reciprocal assimilation of the schemata surpasses the simple fusion to construct a series of complicated relationships" (Piaget 1952:232).

During this stage the infant also displays another characteristic class of behavior patterns called the "derived secondary reactions" in which he applies each of his behavior patterns (especially prehensive behavior patterns) to a new object in serial order in a sort of primitive test of object properties.

B. SOME EXAMPLES OF THE "DERIVED SECONDARY REACTIONS" OF EXPLORATION OF OBJECTS THROUGH SERIAL APPLICATION OF SCHEMATA

1. The child explores a series of small toys one at a time in the following ways:
   a. Looking at it holding it still.
   b. Transferring it from hand-to-hand watching it.
   c. Studying it from different visual perspectives.
   d. Feeling it all over.
   e. Scratching it.

  f. Moving it slowly while watching it.
  g. Shaking it.
  h. Rubbing it against the bassinette.
  i. Sucking on it.

The fourth stage involves, then, a primitive sensorimotor distinction between means and ends, and an incipient capacity to explore object properties through serial application of different behavior patterns to the same object. During this stage the infant also becomes capable of "prevision" or anticipation of events on the basis of "signs"; for example, he recognizes sounds indicating that he will be fed soon.

SERIES I, STAGE 5: *From the twelfth to the eighteenth months—"The Tertiary Circular Reaction and the Discovery of New Means Through Active Experimentation"*

During this stage the infant begins to display trial-and-error manipulation of objects relative to other objects or relative to fields of force (i.e., friction, gravity, inertia). The behavior patterns of this stage are called tertiary because they all involve object-object or object-field interactions.

  The circular aspect of the "tertiary circular reactions" is the repetition of behavior patterns varying their intensity, frequency, orientation, etc., on objects relative to other objects and relative to fields and forces. The behavior patterns differ from those of the "derived secondary reactions" in their graded intensity and in the child's groping for new means and effects. He is apparently rewarded or reinforced by the experience of contingent novelty rather than simple contingency as in the "secondary circular reaction."

  Each "tertiary circular reaction" is a one-time event because after the child has discovered the new procedures he is no longer engaging in trial-and-error groping. The "tertiary circular reaction" is experimental by definition: "The subject will search on the spot for new means and will discover them precisely through tertiary reactions. It cannot be said that he will apply the tertiary schemata to these situations since, by definition, the tertiary circular reaction is vicarious and only exists during the elaboration of new schemata, but he will apply the method of the tertiary circular reaction" (Piaget 1952:267). The "tertiary circular reaction" occurs as a playful activity for its own sake and as directed groping in pursuit of an objective (i.e., trial-and-error groping).

A. SOME EXAMPLES OF PLAYFUL "TERTIARY CIRCULAR REACTIONS"

  1. Repeated dropping of objects and watching the results of varying their height, target, and position.
  2. Repeated dropping of objects into water from varying heights, submerging them, dumping water into water, and watching the results.
  3. Repeated sliding of various objects along surfaces at varying angles and velocities and watching the results.
  4. Repeated taking of objects in and out of containers and dumping them.

B. SOME EXAMPLES OF GOAL-DIRECTED TERTIARY CIRCULAR REACTIONS

1. Grasping and raising and tilting a stick through bars in order to bring it into the crib.
2. Rolling and dangling a chain in order to get it into a small container.
3. Using a stick in order to rake in a toy outside of crib bars.

SERIES I, STAGE 6: *From about eighteen months of age on—"Representation and the Invention of New Means Through Mental Combinations"*

The behavior patterns of this stage reveal the capacity for insight problem solving concerning object-object and object-field relations; this stage is an internalization of the "tertiary circular reaction."

This stage embodies one of the most important ideas in Piaget's model, that is the idea that representation develops from the internalization of the "tertiary circular reaction" through symbolic imitation of object-object relations and object behavior; he calls this process "object imitation." "Object imitation" allows the child to engage in trial-and-error groping for a solution to a problem without directly manipulating objects. "Object imitation" and representation allow the child to solve problems much more rapidly than he did during the fifth stage. Here is one of Piaget's examples of object-imitation leading to insight problem-solving:

In other words, instead of exploring the slit with his finger and groping until he discovered the procedure which consists in drawing to him the inner side (of the box) in order to enlarge the opening, the child is satisfied to look at the opening, except for experimenting no longer on it directly, but on its symbolic substitutes. Lucienne opens and closes her mouth while examining the slit in the box, proving that she is in the act of assimilating it and mentally trying out the enlargement of the slit. . . . Once the schemata have thus been spontaneously accommodated on the plane of simple mental assimilation, Lucienne proceeds to act and succeeds right away (Piaget 1952:344)

In other words, the infant performs the operations first through some sort of analogic mini-motor processes which inform him of the motor operations he needs to perform to solve the problem. (I call this "tertiary circular reaction" representation.)

SERIES II: THE OBJECT-CONCEPT SERIES

SERIES II, STAGES 1 AND 2: *From birth to third or sixth month—"The Absence of Special Behavior Relating to Vanished Objects"*

During this stage the infant shows very little interest in objects; he displays simple visual accommodations such as scanning stationary objects, incomplete following of slow-moving objects, and alternate glancing between two objects. When a slow-moving object disappears from his visual field the infant merely looks away or continues to stare at the same place.

SERIES II, STAGE 3: *From three to six months or nine months—"The Beginning of Object Permanence Extending the Movements of Accommodation"*

During this stage the child's object image is dependent upon his own actions of visual accommodation and prehension associated with the object; all the accomplishments of this stage follow from the prolongation of movements of accommodation to objects such as looking, reaching, etc., they all depend upon the beginning development of prehension and the coordination of vision and prehension. The object itself is subjective. "These images still remain at the disposal of the action itself" (Piaget 1954:20, 21). In other words, during the third stage the object does not exist in its own right, but only as an extension of the child's activities. The third stage is characterized by the following abilities and disabilities: visual accommodation to rapidly moving objects; "interrupted prehension"; the "deferred circular reaction"; the reconstruction of an invisible whole from a visible fraction, but the inability to uncover objects covered in view; and anticipation of the path of moving objects.

"Interrupted prehension" is Piaget's term for the act of reaching, without visual or tactile searching, for an object that has just escaped the hand. "Interrupted prehension" does not involve a concept of the object as separate from the action.

The "deferred circular reaction" is Piaget's term for a similar phenomenon of returning to a circular activity such as shaking, grasping, or sucking an object after a brief interruption, and without further stimulus from the object.

The "reconstruction of an invisible whole from a visible fraction" is Piaget's phrase for the behavior pattern of grasping an object which is partially covered by a screen, which he believes implies that the child recognizes the whole from the part. The child, during the third stage, is incapable, however, of searching for an object if it is completely hidden behind a screen as he watches: he will either give up, or search in the hand of the person who hid it. The child of this age does not lack the motor ability to uncover the object: "Everything occurs as though the child believed that the object is alternately made and unmade" (Piaget 1954:33). He limits his search to extending his own movements of "accommodation": he will give up if an object is covered as he is reaching for it.

> In short, so long as the search for the vanished object merely extends the accommodation movements in progress, the child reacts to the object's disappearance. On the other hand, as soon as it is a question of doing more, that is, of interrupting the movements of prehension, of visual accommodation, etc., in order to raise a screen conceived as such, the child abandons all active search. (Piaget 1954:44)

SERIES II, STAGE 4: *From about eight to ten months—"The Practical Object: The Active Search for the Vanished Object But Without Taking Account of the Sequence of Visible Displacements"*

The achievements of this stage are extensions of the behavior patterns of the previous stage; at about eight or nine months the child begins to uncover objects which have been covered in his view. He cannot, however, take into account sequential displacement of the object behind a series of screens; after he has recovered an object under one screen, he will continue to search for it again under that screen where he first discovered it, even when he sees the object hidden under an adjacent screen. Piaget calls this phenomenon the "residual reaction" and he interprets it to mean that the object is still at the disposal of the acts relating to it. In other words, the child has a concept of his own actions relative to the object rather than a concept of the object itself.

Correlated with this is the child's inability to conceive of objects as having a unique location in space at a given time (T. G. C. Bower 1971). He perceives them as belonging to particular contexts:

> Hence there would not be one chain, one doll, one watch, one ball, etc., individualized, permanent, and independent of the child's activity, that is, of the special positions in which the activity takes place or has taken place, but there would still exist only images such as "ball-under-the-armchair," "doll-attached-to-the-hammock," "watch-under-a-cushion," "papa-at-his-window," etc. (Piaget 1954:69)

(Papa-at-his-window refers to another illustration Piaget cites of the lack of identity characteristic of the fourth stage: Piaget waved to one of his children from his office window and then came outside and watched the child look at him, and then in response to the mother's query, "Where's papa?" look back at the window of the office from which his father [Piaget] had just come.)

SERIES II, STAGE 5: *From twelve months to eighteen months—"The Intermediate Object Permanence: The Child Takes Account of Sequential Displacements"*

During this stage the child is able to dissociate the object from the context in which it occurs, and, since it has identity as an object, he is able to follow its visible sequential displacements under adjacent screens; that is, to follow an object as it is moved from one screen to another as he watches, etc., but he is still unable to reconstruct invisible displacement, as, for example, placing an object inside a box and then dumping it out under a screen. When confronted with this situation he will search in the box and in the experimenter's hand.

SERIES II, STAGE 6: *From about eighteen or twenty-four months—"Object Permanence: The Representation of Object Displacements"*

During this stage the child becomes capable of directing his search for disappearing objects through representation. The object is objectified in the sense that it obeys its own laws of motion and is no longer dependent upon the child's actions. The child now perceives his own body as object by analogy with other objects.

These abilities are suggested by the child's ability to infer the itinerary of invisible displacements of an object sequentially moved under a series of screens while hidden in the experimenter's hand or another container, or dumped out under a screen from the experimenter's hand or another container.

SERIES III: THE SPATIAL FIELD STAGES

SERIES III, STAGES 1 AND 2: *From birth to three or four months—"The Practical and Heterogeneous Groups (of Displacement)"* [1]

During these stages the infant has no concept of an objective space containing himself and objects, but rather largely unconnected "practical" spaces associated with each sensory modality: tactile space, buccal space, auditory space, visual space, etc.:

> Before the prehension of visual objects the child is in the center of a sort of moving and colored sphere, whose images imprison him without his having any hold on them other than by making them reappear by movements of head and eyes." (Piaget 1954:163)

SERIES III, STAGE 3: *From three to six months or eight months—"The Coordination of the Practical Groups and the Formation of Subjective Groups (of Displacement)"*

The third stage is characterized by the coordination of the different "practical heterogeneous spaces" with one another; the development of prehension is central to this process. The child begins to use his hands on things and to bring things into relationship with each other, e.g., in the "secondary circular reaction," and he begins to watch himself doing these things. But despite these achievements the child still does not take into account displacements of objects relative to other objects or displacements of his own body; space is still determined by the child's actions.

"Interrupted prehension" (reaching for a dropped object without visual or tactile searching) characteristic of this stage stops at extending acts of accommodation to objects, and if it is unsuccessful, "the child behaves as if the object had reentered the void" (Piaget 1954:128).

Similarly, "anticipating the trajectory of a moving object" also stops at extending acts of accommodation: if he cannot find it, it ceases to exist as an independent object following an independent trajectory in space.

Similarly, the inadvertent rotation of objects, or rotation triggered by visible portions does not indicate an understanding of the other side and the space it inhabits.

During this stage the child begins to display a practical understanding of depth, specifically the difference between near space and far space as indicated by reaching for objects within grasp and failing to reach for objects beyond the grasp, but he is often uncertain about whether or not to reach.

Piaget calls these "subjective groups" because of the absence of a concept of object constancy in shape and dimension. They are not yet "objective groups of displacement" involving an "immobile" environment in which objects and bodies are displaced relative to each other.

SERIES III, STAGE 4: *From eight or nine months—"The Transition from Subjective to Objective Groups and the Discovery of Reversible Operations"*

The "application of familiar means to new situations" (fourth stage of the sensorimotor intelligence series) is the key to the achievements of this stage. All of the behaviors of this stage relate to the discovery of simple reversible operations (without understanding object-object interrelations): hiding and uncovering an object under a screen, moving an object near and far in front of the eyes, moving the head to study objects from different perspectives, persistent tactile searching for a lost object, and intentional rotation of objects looking for the other side.

During this stage the child also begins to understand the ordination of planes in space; he can reach behind objects and conceives of a space behind barriers.

The child displays a puzzling inability to realize that two objects, one on top of the other, are independent; he will not pick up matches on top of a book, for example, or draw objects to him by pulling a supporting object. This inability is part of the lack of understanding of object-object relationships.

SERIES III, STAGE 5: *From eight to twelve months—The "Objective Groups"*

This stage is marked by the understanding of the relative displacement of objects with respect to each other—that is, of object-object and object-field relationships. During this stage, space is perceived as having an objective existence; the child experiments with displacing objects "carrying objects from one place to another, moving them away and bringing them near, letting them drop or throwing them down to pick them up and begin again, making moving objects roll and slide along a slope, in short conducting every possible experiment with distant space as well as near space" (Piaget 1954:209). He perceives screens in relationship to himself and not in relationship to objects, and therefore he can search for an object behind a series of screens when he sees it hidden in sequence. He displays reversible operations involving object-object relations, e.g., putting solid objects into hollow containers and emptying them out. He forms spatial interrelations among objects by rotating objects relative to other objects, and stacking objects on top of other objects, and is concerned with the equilibrium of these objects; he consciously displaces himself relative to objects, but he cannot take into account displacements of himself and other objects outside of his perceptual field; he may try to lift himself up out of a ditch by pulling (bootstrapping) on his own foot (Piaget 1954:228), or pull an object up while standing on it.

SERIES III, STAGE 6: *From eighteen months on—"Representative Groups"*

During this stage the child finally perceives space as a "motionless environment in which the subject is himself located" (Piaget 1954:235). He represents his own position in space relative to the position of other objects including those he cannot see.

SERIES IV: THE CAUSAL SERIES

SERIES IV, STAGES 1 AND 2: *From birth to three months—"Making Contact Between Internal Activity and the External Environment, and the Causality Peculiar to the Primary Schemata"*

During these stages the infant does not inhabit a space (but, rather, many heterogeneous spaces; see spatial series), he does not experience time, he does not experience objects, and it is therefore impossible for him to conceive of a causality among these entities. What he experiences is a vague sense of "efficacy" or power associated with particular actions and desires.

SERIES IV, STAGE 3: *From three months to nine months—"Magico-Phenomenalistic Causality"*

The achievements of this stage are related to the "secondary circular reaction" of the third stage of the sensorimotor intelligence series which involves a primitive sense of causality; but this sense of causality is limited to the child's sense of the effect of his own actions in producing or reproducing "interesting spectacles." Whether he is focusing on actions of his own body, the action of objects in relation to his body, or the interaction of other objects, he attributes all these activities to his own actions: ". . . in these three cases it is to the dynamism of his own activity that the child attributes all causal efficacy, and the phenomenon perceived outside, however removed it may be from his own body, is conceived only as a simple result of his own action" (Piaget 1954:281). This explanation extends to the child's imitation of people; he is trying to prolong or recreate an interesting spectacle when he imitates them.

SERIES IV, STAGE 4: *From about nine to eleven months—"The Elementary Externalization and Objectification of Causality"*

During this stage the child begins to conceive of objects and people as centers of causality; he acts as if events are only partly caused by his actions, for example, he nudges toys as if to activate them, and he touches people's hands in order to make them repeat their actions. In other words he seems to attribute some power of causality to other people and objects, but that power is still limited to situations involving the child's own activities:

> . . . the child's action on persons during this fourth stage seems to reveal an intermediate causality, already partly objectified and spatialized (since persons already constitute external centers of particular activity) but not yet freed from the efficacy of the child's own movement (since these centers of activity are conceived as always depending on his personal procedures) (Piaget 1954:300)

SERIES IV, STAGE 5: *From about twelve months—"The Real Objectification and Spatialization of Causality"*

During this stage the child begins to recognize centers of causality entirely independent of his actions, and he conceives of causal links between objects external to himself. These realizations develop through the activities of the "tertiary circular reactions" and the "discovery of new means through active experimentation," for example, placing an object on an inclined plane and letting it slide down. "The first definite example of completely objectified causality seems to us to be the behavior patterns consisting of placing an object in such a position that it puts itself in motion" (Piaget 1954:308).

The "behavior pattern of the support" wherein the child draws an object to himself by pulling on an intermediary object such as a pillow on which the object, a toy, for example, is placed, involves the recognition of the cause and effect relationship between the movement of the toy on top of the pillow and the pillow.

The "behavior pattern of the stick" wherein the child uses the stick to manipulate an object, and no longer just as an extension of his own hand, also involves the recognition of the causal power of the stick in relation to the object.

During this period the child begins to conceive of himself as only one cause among other causes in the world. His understanding of causality is still limited to those phenomena he can perceive directly; he cannot place his whole body in an objective causality, and he still tries to lift up objects he's standing on or to lift himself up out of a ditch by pulling on his own leg (bootstrapping).

SERIES IV, STAGE 6: *From about eighteen months—"The Representative Causality and the Residue of the Causality of Preceding Types"*

During this stage the child displays the ability to reconstruct effects due to causes which are not perceptually apparent to him from the evidence of their effects. When an object won't yield to pushing, for example, he walks around it and searches for something blocking its movement. The child can also foresee the effects from a cause. "In a general way, therefore, at the sixth stage the child is now capable of causal deduction and is no longer restricted to perception of sensorimotor utilization of the relations of cause to effect" (Piaget 1954:337). But it is common for the child to regress to earlier stages of causality during this period if his deductions seem to fail or he meets other difficulties in explaining causality.

SERIES V: THE TEMPORAL SERIES

SERIES V, STAGES 1 AND 2: *Birth to four months—"Time Itself and the Practical Series"*

During these stages the infant's actions are coordinated without any perception of the duration or sequence of his actions in relation to a larger temporal field encompassing his actions and other events; time is not, as it will become, "an instrument of ordination interconnecting the external events with the subjective acts" (Piaget 1954:368). The infant displays a "memory of recognition" without a sense of past or present. He has only a subjective sense of duration of sensations.

SERIES V, STAGE 3: *From the fourth to eighth months—"The Subjective Series"*

During this stage the infant develops a sense of before and after, which is limited to events in which he has participated. (He cannot yet reconstruct or remember the sequence of events in which he did not participate. This subjective time sense is due to the "secondary circular reaction" which involves a primitive distinction between before and after.)

SERIES V, STAGE 4: *From the eighth month—"The Beginning of the Objectification of Time*

During this stage the child displays the ability to reconstruct or remember a series of events in which he did not participate; that is, he can arrange a sequence of events in time, but this perception is easily dominated by his "practical memory" of his own actions. The ability to reconstruct events is demonstrated by his ability to remember the position of an object after it is hidden behind a screen. The domination of practical memory is demonstrated by the "residual reaction" of continued searching for a hidden object under the screen where it was first discovered, even when it has been hidden under the second screen in clear view.

SERIES V, STAGE 5: *From about twelve months—"The Objective Series"*

During this stage the child displays the ability to arrange external events in a sequential order as demonstrated by the correct search for an object under a series of visible sequential displacements rather than being dominated by the memory of his own previous actions; he no longer displays the "residual reaction" in searching for objects sequentially hidden under a series of screens. He is still unable to reconstruct a series of events which he does not see; his sense of time is still limited by immediate perceptions.

SERIES V, STAGE 6: *From about eighteen months—"The Representative Series"*

During this stage in which the child displays the ability to reconstruct events outside his immediate perception, he can remember events from his past and use them to reconstruct invisible events. This capacity depends upon representation and object-imitation.

SERIES VI: THE IMITATION SERIES

SERIES VI, STAGE 1: *From birth to one month—"Contagious Crying"*

During this stage contagious crying (wherein the infant seems to be stimulated to cry more when he hears his own or another infant's crying) is the only evidence of imitation. This is a sort of "primary circular" crying.

SERIES VI, STAGE 2: *From one to three months—"Sporadic Self-Imitation"*

During this stage the infant repeats his own sounds spontaneously and also in response to hearing another person imitating his sounds. He shows no gestural imitation during this stage, but he does display visual accommodation and attention to the movements of people and objects.

SERIES VI, STAGE 3: *From fourth to eighth month—"Purposeful Self-Imitation"*

During this stage the infant purposefully imitates "schemata" from his own repertoire when he sees or hears them performed by another person. This activity is characterized by sustained attention to the person performing "schemata" from his repertoire. For example, he deliberately and systematically imitates his mother if she coos or makes some of his own sounds at him, and he deliberately and systematically imitates his mother when she hits the quilt, watching his mother's hands all the time.

During this stage the infant imitates people in order to get them to repeat their actions. It is therefore a specialized form of the "secondary circular reaction" characteristic of the third stage of the sensorimotor intelligence series. It is also interesting to note that "recognitory assimilation" during the third stage, that is, the brief motor outlining of a behavior associated with an object upon seeing the object, is a sort of early primitive object imitation.

SERIES VI, STAGE 4: *From eight or nine months—"Imitation of Movements Already Made by the Child but which are Not Visible to Him and Imitational Differentiation of Global Schema"*

During this stage the infant begins to differentiate his own "global schemata" (i.e., undifferentiated behavior patterns) by imitating another person who is differentiating a behavior pattern (in the infant's repertoire) into its components. He can, for example, differentiate his own "pah" vocalization and mouth opening into a silent "pah" movement of his mouth while watching his mother demonstrating this. He also begins to imitate actions which involve moving his face which is invisible to him, such as touching his nose and his eyes. He continues to display intense interest in the activities of the person he is imitating.

SERIES VI, STAGE 5: *From about twelve months to eighteen months—"Beginning Imitation of New Auditory and Visual Models"*

During this stage the child begins imitating new "schemata," which are analogous to his own, through trial-and-error groping to accommodate his behavior patterns to those of the person he's watching. This is a specialized form of the "tertiary circular reaction" in which he systematically varies and combines and differentiates his behavior patterns with the goal of matching those of the person he's imitating. During this process he carefully watches the actions of the person he's imitating and glances alternately between his own actions and the actions of this model when the actions are visible. It is also during this stage that children first attempt to imitate adult words.

SERIES VI, STAGE 6: *From twelve to eighteen months on—"Deferred Imitation"*

During this stage the infant displays the ability to imitate the actions of models for the first time when they are not present and, hence, reveals the capacity for mental representation. He also displays imitation of the actions and object-object relations of material objects (symbolic representation). Piaget gives the following example:

> T. was looking at a box of matches which I was holding on its end and alternately opening and closing. Showing great delight, he watched with great attention, and imitated the box in three ways: 1. He opened and closed his right hand, keeping his eyes on the box. 2. He said "tff, tff," to reproduce the sound the box made. 3. He reacted like L. (Lucienne) by opening and closing his mouth. (Piaget 1962:66)

During this stage the child also displays the ability to imitate complicated new procedures without overt trial-and-error groping as in the fifth stage. All of these abilities reveal mental representation of "schemata."

"Object imitation" is the basis of representation; in Piaget's terms, the image is the product of the internalization of object-object interactions through object imitation. In other words, stage six in the sensorimotor series depends on stage six in the imitation series.

Not only do the six stages in the six series run concurrently, but the achievements and disabilities of each stage in each series are reflected in the abilities and disabilities in the same stage in parallel series. For example, the trial-and-error groping and experimentation characteristic of the fifth stage (the "Tertiary Circular Reaction and the Discovery of New Means by Active Experimentation") in the sensorimotor series are necessary for the imitation of new behavior patterns through trial-and-error accommodation of the child's "schemata" to the model's "schema" characteristic of the fifth stage in the imitation series.

Summary

Similarly, the recognition of independent centers of causality in the fifth stage of the causality series depends upon the "tertiary circular reaction" and the "discovery of new means through active experimentation."

According to Piaget, the six parallel stages in all of the six series are reflections of a basic, integrated movement from "egocentric solipsism" (i.e., the belief that the world's existence is dependent on the self) to an objective awareness of the self as one object in a world of stable objects and many centers of causality operating independently of the self. This achievement, of course, is only on the sensorimotor level at this age and must be recapitulated later on subsequent levels of intellectual development.

Another point that Piaget emphasizes is that the behaviors characteristic of each stage may persist beyond that stage into following stages, ". . . behavior patterns characteristic of the different stages do not succeed each other in a linear way (those of a given stage disappearing at the time when those of the following one take form) but in the manner of the layers of a pyramid (upright or upside down), the new behavior patterns simply being added to the old ones to complete, correct or combine with them" (Piaget 1952:329).

Piaget's model is epigenetic, that is, each stage is conceived as emerging from the experiences of the previous stage. "Schemata" change through "assimilation" of and "accommodation" to experiences resulting from their own functioning. Development is essentially conservative because the child can only integrate experiences which can be "assimilated" into a pre-existing "schema" and "accommodated" by that "schema." According to Piaget, for example, the "secondary circular reaction" develops out of experiences produced by the "primary circular reaction"; "interesting spectacles" interrupt the "primary circular reaction" capturing the infant's attention and causing him to repeat his behavior pattern (e.g., kicking his feet) in order to recreate the "interesting spectacle" (e.g., the moving fringe on the bassinette). The interesting spectacle (the moving fringe) is "assimilated" to the "primary circular reaction" (repeated kicking) and the "primary circular reaction" is "accommodated" to the "interesting spectacle." Epigenesis, or progressive developmental transformation of behavior patterns, occurs then as a result of experience, but only experience which fits pre-existing "schemata."

Piaget's "schemata" are dynamic entities with a life of their own—*they* develop and differentiate. His reification of the "schema" concept makes behavioral development much more concrete and helps the observer see behavioral transformations more clearly than before. This is one reason it is valuable for comparative studies.

■ THE VALUE OF PIAGET'S MODEL FOR COMPARATIVE DEVELOPMENTAL STUDIES

The "schema" concept developed out of Piaget's longitudinal developmental studies of his own infants. It is ideally suited to longitudinal developmental studies because it illuminates behavioral organization by focusing attention on earlier and later versions of the same behavior patterns as connected entities. The contrast between the earlier and later versions of "schemata" reveals the organization of behavior. The gradual developmental unfolding of behavior patterns cannot be seen in cross-sectional studies, nor do longitudinal studies without the "schema" concept reveal the details of behavioral organization. They must be combined for maximum effect.

We can gain even greater insight into behavioral organization by comparing development of related primate species using a Piagetian framework. Just as comparing the behavior of an infant at one age with his own behavior at earlier ages illuminates behavioral organization, so comparing the behavioral development of infants of closely related species illuminates phylogenetic differences in behavior. In fact, the full power and significance of Piaget's model can only be appreciated in comparative studies which will reveal which aspects of development are unique to our species.

Piaget's stages provide a standard against which we can compare behavioral and intellectual development in closely related primate species.[2] Not only does he classify the most complex and intelligent behavior of our species,[3] but he traces the origins of these complex mental operations in adults back to early infant sensorimotor behavior patterns. This emphasis on the continuity of sensorimotor and mental operations makes his model particularly suitable for comparing human intelligence with the intelligence of nonhuman primates.

Piaget's model is also particularly well suited to comparative studies of primate development because he emphasizes the vital role of prehension (the hallmark of the primate order) in the development of intelligence—it is "an indispensable connecting link between organic adaptation and intellectual adaptation" (Piaget 1952:89–90).

Although some people may consider it anthropocentric, it is best to use the behavior of the most intelligent species as a standard so that we do not miss some complexity of behavioral organization. If we used a stumptail-based framework to analyze our behavior, for example, we might miss the "secondary and tertiary circular reactions" not to mention higher intellectual achievements of later developmental periods.

Using Piaget's approach as a basis for comparative studies implies that his categories can discriminate and classify all the behavior patterns in our order. It is too early to say for certain whether this is true, but a Piagetian approach worked well in the analysis of stumptail infant development. Comparative studies of great ape development will probably require applications of Piaget's model of later stages of intellectual development.

A Piaget-based framework is far more than a simple item-by-item comparative intelligence test (such a test would be of little value since it provides no information about the unique behavior patterns and behavioral organization of the species being studied). It is a framework for analyzing and classifying all behavior patterns in a species according to their behavioral organization. Since this framework can be applied to any primate species, it provides a consistent, comprehensive basis for comparing the organization of intelligence and other unstereotyped behavior in related taxa.

According to Piaget, intelligence is the purposeful invention of new procedures, or applications of old procedures, in a new context, to solve a problem. In other words, intelligence is a process of behavioral adaptation, involving the application of combinations or variations of behavior patterns occurring in other contexts. In order to understand intelligence we need to know the repertoire of motor patterns and the ways in which they are controlled, timed, organized, coordinated, stimulated, applied, and reinforced. We can only know these things by studying behavioral processes, particularly those early processes of behavioral development which are never repeated in later life. We cannot divorce the study of intelligence from the study of behavioral organization and its ontogeny.

■ CIRCUMSTANCES AND METHODOLOGY OF THE STUDY

This report is based primarily on a six-month longitudinal study of an infant stumptail macaque monkey undertaken at the Primate Research Station at University of California, Berkeley in 1970–71. I observed an infant male for approximately one hour a day for an average of five and one-half days a week from the day of his birth, October 8, 1970, until the day before his accidental death, March 31, 1971 (on the 174th day of his life), a total of approximately 137 hours. He was housed with his mother, and after he was two months old, with an older male infant, in a cyclone-fenced area of about 25′ × 15′ × 15′ which was partitioned off from a small colony of thirteen stumptail macaques including one adult male, several adult females and subadults, juveniles, and infants of both sexes. The infant's mother was removed from the larger group a day before she gave birth so that I could enter the cage to observe the infant at close quarters and introduce toys and objects. My observation method involved making continuous tape-recorded descriptions of all the behavior I could detect and describe including its context, what other animals were doing, its focus, its temporal patterning, its goals, results, etc. I made verbatim transcriptions of my tape-recorded notes and they, along with daily super 8mm films, formed the corpus for analysis.

I "tested" the stumptail infant for sensorimotor and object-concept "schema" in an informal way. From the first I kept a number of toys and "interesting" objects in a footlocker in the smaller cage and removed them daily during the observation period. These included the following objects: plastic rattles and plastic keys on a chain, cotton balls, a plastic troll-doll with long hair, rubber balls, a cloth monkey hand puppet, a small metal cup, toy cars with friction wheels, small receiving blankets and mirror. In addition, I left a number of items out in the cage permanently. These included a hard plastic clown with a weighted base, a large wooden box, a small metal pail, a stackpole toy with graduated plastic donut-shaped rings, a swing with a wooden bar attached to a low-hanging branch. A loose chain about three feet long hung on the door of the cage. I also introduced a low, large dish of water briefly, and I used an alarm clock to test the infant for visual searching for a sound source.

I tested the infant for object permanence "schemata" beginning on day 22 by repeatedly covering the small toys and pieces of oranges with the small blankets and later with the small metal cup and metal beach pail. Sequential displacements of the same object under two cloths was difficult because the older infant interfered and the younger infant was easily distracted.

Rattles and the weighted clown and hanging objects, such as a plastic donut ring on a string hanging under a branch, were provided to give the infant opportunities to display the "secondary circular reaction" and opportunities to pull up objects by strings; and the plastic stack toy, rubber balls, friction wheel cars, small blocks, and pail were provided to give him opportunities to display the "tertiary circular reaction." (In addition, I demonstrated these "schemata.")

These observations were supplemented by short-term observations of an Indian langur colony and a talapoin colony during the summer of 1972, study of a single Aotus female during 1972, and by an on-going weekly study of a human infant beginning September 26, 1974, two weeks after his birth, and of a gorilla infant beginning March 1975.

■ APPLYING PIAGET'S MODEL TO AN INFANT STUMPTAIL MACAQUE

The fact that some aspects of stumptail behavioral organization are strikingly different from human infant behavioral organization complicates comparisons of the two species. What series are pertinent to analyzing stumptail development? How many stages are there and what should they be called? A large element of subjective choice is involved in choosing stage names; the recognition of stages and the choice of series are more straightforward and should be subject to inter-observer agreement.

Piaget considers each successive stage in each parallel series (sensorimotor intelligence, object-concept, time, space causality, and imitation) in the human infant's sensorimotor period to be interrelated. In fact, he often includes the same behavioral patterns in the same stage in two or more series, emphasizing different aspects of its organization in each case. In the causality series, for example, Piaget discusses "tertiary circular reactions" from the sensorimotor intelligence series in terms of their implications about the child's understanding of causality.

The imitation series is also closely related to the sensorimotor intelligence series because it depends upon the "secondary and tertiary circular reactions," but it is logically separate because of the critical role of interactions with other people. The temporal series and the spatial series relate to all the other series, but especially to the object-concept series. I would argue that the object-concept series is related to the sensorimotor series in a different way than the causality and imitation series are since few, if any, of its achievements depend upon the "secondary and the tertiary circular reactions"; it depends upon other sensorimotor achievements, however. For example, stage four search for covered objects depends upon the hierarchical coordination of behavior patterns; stage five searching under several screens for successively displaced objects depends upon trial-and-error coordination of behavior patterns; and stage six search for invisibly displaced objects depends upon some kind of representation.

I have classified the stumptail infant's developmental stages under three series, the sensorimotor series, the object-concept series (including items from the spatial series), and the imitation series (except for a few items, such as bootstrapping, which I reclassified under the sensorimotor series). (I omitted the causal series because it is the most interpretive of all Piaget's series.)

I decided on these series because they reveal the complex patterns of similarities and differences between the stumptail and human infants. The stumptail infant completed all the stages of the object-concept series, only the third stage of the imitation series (this stage does not include gestural or vocal imitation) and displayed a unique set of parallel stages in the sensorimotor series. (See Table 1.)

The most striking difference between the stumptail infant and human infants is the brevity of "the primary circular reactions" and the absence of the "secondary and tertiary circular reactions" in the stumptail sensorimotor series. The differences between the two species are subtle; for example, although the stumptail infant lacked the "secondary and tertiary circular reactions," he displayed many of the coordinations characteristic of those behavior patterns.

Piaget's object-concept stage names were used for the monkey because the object-concept series were remarkably (though not completely) similar in the stumptail and human infants, except of course in timing: the stumptail infant completed this series by six months of age whereas human infants complete it between eighteen and twenty-four months of age.

TABLE 1 / SUMMARY OF DIFFERENCES BETWEEN A STUMPTAIL INFANT AND HUMAN INFANTS

SERIES I: SENSORIMOTOR INTELLIGENCE SERIES

Human Infants

Stage 1: Birth to one month—*"Reflex Stage"* (reflex rooting, sucking, grasping, looking, etc.)

Stage 2: Second to fourth month—*"The Secondary Circular Reactions and First Acquired Adaptations to the Environment"* (primitive voluntary prehensive and visual and oral motor patterns which are simultaneously "primary circular reactions," "reciprocal coordinations" and "adaptations to the environment").

Stage 3: From fourth to eighth month—*"The Secondary Circular Reactions"* (differentiation of behavior patterns).

Stage 4: From eighth month on—*"The Coordination of the Secondary Schemata and Their Application to New Situations"* (serial and simultaneous coordinations of "secondary schemata" in service of a goal and "derived secondary reactions" with serial application of "schemata" to an object).

Stage 5: From the twelfth to eighteenth months—*"The Tertiary Circular Reactions and Discovery of New Means Through Active Experimentation."*

Stage 6: From eighteen months—*"Representation and the Invention of New Means Through Mental Combinations"* (object imitation and representation of the "tertiary circular reaction").

SERIES II

Human Infants

Stages 1 and 2: From birth to third to sixth month—

Stumptail Infant

Stage 1: Birth to one week—*Reflex Primary Circular Reactions and Intercoordinations and Adaptations to the Environment* (reflex grasping and clinging, and rooting, and righting on mother: rapid decrement of "primary circular reactions").

Stage 2: One week to twelve weeks—*Primitive Voluntary Primary Intercoordinations and Adaptations to the Environment* (primitive voluntary visual and prehensive motor patterns without true "reciprocal coordinations"). Stage 2 is also characterized by *Primitive Voluntary Body Circular Reactions and Body Adaptations to the Environment* (primitive voluntary locomotion and climbing, often repeated in a circular way).

Stage 3: Fourth to eleventh weeks—*The Secondary Linear Schema* (with no "secondary circular reaction") (differentiation of visual and prehensive behavior patterns). Stage 3 is also characterized by *Voluntary Coordination and Differentiation of Body Schema* such as walking and running and climbing.

Stage 4: From ninth week on—*The Coordination of Secondary Linear Schema and Their Application in New Situations* (serial and simultaneous coordination of *secondary linear schemata* in service of a goal and *derived secondary linear reaction* with serial application of "schema" to an object). Stage 4 is also characterized by *Coordination of Body Schema* and *derived body reactions.*

Stage 5: From twenty-third week—*Trial-and-Error Secondary Linear Reaction and the Discovery of New Means Through Active Experimentation* and *Trial-and-Error Body Reactions* with no "tertiary circular reaction."

Stage 6: From week twenty-three—*(Secondary Linear Trial-and-Error) Representation and Planning Through Mental Combinations* (no object imitation or representation of the "tertiary circular reaction").

Stumptail Infant

Stages 1 and 2: From birth to fifth week—*Same*

continued

TABLE 1 / CONTINUED	
"Absence of Special Behavior Relating to Vanished Objects"	
Stage 3: From third to sixth to ninth month—*"The Beginning of Object Permanence Extending the Movements of Accommodation"*	Stage 3: Third to sixth week—*Same*
Stage 4: From ninth or tenth month—*The Practical Object: "The Active Search for the Vanished Object but Without Taking Account of the Sequence of Visible Displacements"*	Stage 4: Third to sixteenth week—*Same*
Stage 5: From twelve to eighteen months—*"The Intermediate Object Permanence: The Child Takes Account of Sequential Displacements"*	Stage 5: From twenty-third week—*Same*
Stage 6: From eighteen to twenty-four months—*"Object Permanence: The Representation of Object Displacements"*	Stage 6: From twenty-third week—*Same*
SERIES III: IMITATION SERIES	
Human Infants	*Stumptail Infant*
Stage 1: Birth to one month—*"Contagious Crying"* Stage 2: From one to three months—*"Sporadic Self Imitation"*	Stages 1 and 2: Birth to fourth week—*No Imitative Behavior*
Stage 3: From fourth to eighth month—*"Purposeful Self Imitation"*	Stage 3: Fourth week on—*Purposeful Self Imitation* (non-vocal and gestural; contagious performance of prehensive and body schema).
Stage 4: From eight or nine months—*"Imitation of Movements Already Made by the Child but Invisible to Him and Imitational Differentiation of Global Schema"*	Stages 4–6: No analogous stages.
Stage 5: From twelfth to eighteenth month—*"Beginning Imitation of New Auditory and Visual Models"*	
Stage 6: From twelfth to eighteenth month—*"Deferred Imitation"*	

Because the stumptail infant displayed no true imitation of vocal or gestural behavior, there was no problem deciding on analogous names for stages. (The stumptail infant did display a few contagious performances of behavior patterns from his own repertoire, but he displayed no purposeful matching of his behavior patterns to those of other animals, nor did he come to imitate behavior patterns outside his repertoire as human infants do.)

In the sensorimotor series stages, however, terminology was a real problem because of the complex and confusing mosaic of shared and unique features of behavioral organization. In this series I simply added and subtracted qualifying adjectives to Piaget's stage names in most cases.

During the brief first stage the stumptail infant displayed reflex "primary circular reactions," intercoordinations, and "adaptations to the environment," most of which do not occur in human infants.

During the second stage the stumptail infant displayed primitive, voluntary prehensive, visual, and body intercoordinations; "adaptations to the environment"; (these intercoordinations were different from the "reciprocal coordinations" of human infants) and no "primary circular reaction."

During the third stage the stumptail infant displayed no "secondary circular reaction" but a somewhat analogous set of behavior patterns which I called the *secondary linear schemata* (for example, twisting an object in two hands).[4] Piaget's "secondary circular reaction" is characterized by voluntary "reciprocal coordination" of prehensive or other acts with sounds or movements of one object, differentiation of these motor patterns, "fortuitous" goal direction, reinforcement by pleasure in exercise, accommodation, and perceived contingency. The stumptail's behavior patterns during this stage were voluntary, differentiated, had "fortuitous" goal direction, and were apparently reinforced and/or released by pleasure in exercise and non-contingent novelty. They were sometimes repeated, but they were not "reciprocally coordinated" with object sounds and motion (they did include object manipulations) and they were not reinforced/released by perceived contingency of the object's behavior on his preceding behavior (see Table 2 for definition of terms).

To say that the infant stumptail's acts were not released and/or reinforced by the pleasure in perceived contingency of an object's actions on his own preceding actions, does not mean that he failed to understand when the behavior of objects was contingent on his previous actions. He undoubtedly understood this relationship, but it apparently had no automatic reinforcement value for him as it does for human infants (Watson 1972). The stumptail infant did not repeat his actions idly; he used his understanding for practical purposes such as getting food or water, or opening doors. His actions apparently had a consummatory or exercise reinforcement.

Even in play the contingent actions of objects held no fascination for the young stumptail or his playmate. He knew perfectly well how to push the trigger on the faucet to take a drink, but neither he nor the other animals ever did so in play; similarly, he never repeated his actions on a rattle, or a swinging toy, or other objects I gave him, except in biting on them or climbing on them. The distinction is a subtle one, but it is clearly perceptible in terms of the organization of the animal's attention and the temporal organization of his behavior.

In stage four the stumptail infant displayed the *"derived secondary (linear) reaction"* (exploring objects through serial application of "schemata") and "hierarchical coordination" (performing an auxiliary act in order to perform a final act) of *secondary linear schemata* analogous to those of the fourth stage in human infants.

In stage five the stumptail infant displayed some trial-and-error coordination of "secondary schemata" which was somewhat analogous to the trial-and-error coordination of "secondary schemata" in the "tertiary circular reaction" in human infants which I called the *secondary trial-and-error reaction* (e.g., using different approach strategies to get the hanging chain away from his playmate). It differed in that it did not involve "reciprocal coordinations" between objects or objects and force fields, and was not reinforced/released by perceived variable contingency. Most of the trial-and-error coordinations involved locomotion and social/locomotor play rather than object manipulation.

In stage six the stumptail infant displayed evidence of *secondary trial-and-error representation* (this pattern had just begun to develop at the time of his death; it probably is more prominent in older animals), that is, representation of trial-and-error searching.

In addition to the "primary schemata" and the "secondary linear schemata," the stumptail infant displayed a class of behavior patterns which do not fit any of Piaget's categories. Piaget focused his attention on "schemata" growing out of the coordination of vision and prehension after the second stage because he was interested in the origins of intelligence. If we apply Piaget's ideas in a more general context of behavioral organization, we must add terms to refer to other classes of motor patterns whose development can also be traced. I have introduced the term *body schemata* to refer to voluntary locomotor, facial, and other behavioral patterns which undergo developmental changes parallel to those of "secondary schemata" during stages two, three, four, five, and six of the sensorimotor series.

■ PIAGET'S SENSORIMOTOR PERIOD IN AN INFANT STUMPTAIL MACAQUE

SERIES I: SENSORIMOTOR INTELLIGENCE SERIES

SERIES I, STAGE 1: *From birth to one or two weeks—"The Reflex Primary Circular Reaction and Intercoordinations and Adaptations to the Environment"*

During this stage the neonatal stumptail displayed a number of reflexes such as rooting, grasping, and mouthing which operated in a repeated way and, hence, had the character of "primary circular reactions." They were primary in the sense of being focused on "assimilation" of objects as consummatory stimuli. These oral and prehensive "primary circular reactions" rapidly diminished and disappeared as the infant's clinging and nursing abilities came under voluntary control; his grasping came under primitive voluntary control in the first week.

A. SOME EXAMPLES OF REFLEX PRIMARY CIRCULAR REACTIONS	AGE IN WEEKS
1. Random head bobbing.	1, 2
2. Repeated opening and closing mouth.	1, 2
3. Lipsmacking (reflex or primitive voluntary?)	1, 2
4. Rhythmic empty "chewing" movements of the mouth (reflex or primitive voluntary?)	1, 2

5. Tongue protrusion (reflex or primitive voluntary?)   1, 2
6. Rhythmic clinging and letting go.   1, 2
7. Random waving movements of hands and feet with flexion
   and extension of arms and legs.   1, 2
8. Rhythmic opening and closing of fingers.   1, 2

The infant stumptail also displayed reflex "adaptations to the environment" during the first stage including clinging and righting.

B. SOME EXAMPLES OF REFLEX ADAPTATIONS TO THE ENVIRONMENT	AGE IN WEEKS
1. Unfocused looking.	1
2. Clinging to mother or fence.	1, 2
3. Rooting on mother for nipple.	1, 2
4. Sucking on nipple.	1, 2
5. Righting self by grasping.	1, 2
6. Hoisting self on mother's ventrum.	1, 2
7. Undirected reaching and touching nearby objects.	1, 2
8. Alternate glancing.	1, 2
9. Alternate shoulder glancing.	1

The infant also displayed reflex intercoordinations of his motor patterns during the first stage, e.g., thumb sucking.

C. SOME EXAMPLES OF REFLEX INTERCOORDINATIONS	AGE IN WEEKS
1. Thumb sucking.	1, 2
2. Random scratching of self without contact.	1, 2
3. Touching own face.	1, 2
4. Visually triggered grasping own hand (rare).	1
5. Visually following own hand.	1

SERIES I, STAGE 2: *From one week to twelve weeks—Primitive Voluntary Adaptations to the Environment and Intercoordinations*

During this stage the stumptail infant displayed many of the same patterns as in the first stage, but they were now coming under voluntary control. He also displayed new patterns including visual and prehensive explorations. These patterns are still primary in that they are still focused on the consummatory aspects of stimuli and on intercoordinations of undifferentiated motor patterns. The "primary circular reaction" was absent during this stage; the primary voluntary adaptations and intercoordinations are noncircular, i.e., not repeated except for the *circular body schemata* and rhythmic mouth movements.

The stumptail infant did display persistent fingering of objects, but they were not "circular" in the sense of human infant object manipulation. This may be due partly to the early mobility of the stumptail infant who intersperses his object manipulation with walking and running bouts rushing from one object to another from the age of twelve weeks.

During this stage the stumptail infant increased his range of adaptations to the environment.

A. SOME EXAMPLES OF PRIMITIVE VOLUNTARY VISUAL AND PREHENSIVE
   ADAPTATIONS TO THE ENVIRONMENT   AGE IN WEEKS

   1. Undirected reaching and touching objects and surfaces with
      splayed fingers.   2, 3, 4, 5
   2. Incomplete visual following   2, 3
   3. Letting go of objects with difficulty.   2, 3
   4. Grasping objects and lifting them.   3, 4, 5 · · ·
   5. Ipsilateral clinging on mother with other side free (clinging
      with hand and foot of same side with other side free).   2 . . .
   6. Sidling off mother with ipsilateral grasping of fence.   2, 3, 4
   7. Alternate ipsilateral shifting (clinging with same side hand
      and foot with other side free, then reversing).   3, 4, 5, 11,
                                                                 12, 13
   8. Alternate glancing.   3, 4, 5, 8
   9. Alternate shoulder glancing.   3, 4, 5, 6, 8
  10. Sliding palm and raking open hand on surfaces (rare).   2, 4, 7, 12

During stage two the stumptail infant displayed primitive voluntary intercoordinations of simple motor patterns analogous to the "reciprocal coordinations" of motor patterns during stage two in human infants. I call these *intercoordinations* instead of "reciprocal coordinations" because they do not seem to involve the elements of "reciprocal assimilation" and "reciprocal accommodation" of the behavior patterns to each other which is so prominent in the primary coordinations of human infants. The absence of hand-mouth, hand-hand, and hand-eye "reciprocal coordinations" which are simultaneously "primary circular reactions" (repeated over and over) is very striking in the monkeys. The long period of hand-feet intercoordination in the stumptail is also striking.

B. SOME EXAMPLES OF PRIMITIVE VOLUNTARY INTERCOORDINATIONS   AGE IN WEEKS

   1. Thumb sucking (one hand at a time)
      hand to mouth   3, 4
      mouth to hand   3, 4
      reciprocal hand-mouth   2
   2. Visual following
      own hand (rare)   2, 4
      own foot (rare)   2, 3, 4, 11
   3. Foot-sucking
      foot to mouth   2, 5, 9, 10
      mouth to foot   3, 5, 7

4. Visually triggered grasping own hand on sight (rare) 3, 7, 9, 12, 15
5. Visually-triggered bringing hand to mouth 2, 3, 4
6. Clasping and unclasping hands (two times) 4, 15
7. Aimed scratching 2, 3 . . .
8. Tactile investigation of own body with hand
   face (rare) 3, 4, 5, 19, 20
   penis 3, 4, 7, 8, 9,
   12, 15, 18,
   22
   tail 12, 13, 15,
   18, 19
   cheekpouches 12, 13, 15,
   18, 19
9. Looking to sound source 6, 8, 11, 21,
   23
10. Disconcerted adaptations:
    a. hands letting go while feet continue to cling 1–5, 12, 14
    b. hands and feet both reaching for the same object 3, 11
    c. feet grasp object held in hands 6, 8, 11, 14,
    15, 19, 21,
    22, 23
11. Biting object while grasping 3, 4, 5, 7
12. Holding an object up in order to see it (rare) 3, 4, 9, 14,
    18, 23
13. Visually directed fingering of objects 3, 4, 5, 6,
    7 . . .

During the second stage the stumptail infant began to display primitive voluntary *body schemata* which also fell under the categories of the second stage: *body circular reactions, body adaptations to the environment.*

A. SOME EXAMPLES OF PRIMITIVE VOLUNTARY BODY CIRCULAR
   REACTIONS AGE IN WEEKS

1. Repeatedly climbing on and off mother and onto fences and
   back again. 2, 3, 4
2. Repeatedly climbing on and off branch or mother on branch
   onto fence and back again (an example of "temporal
   displacement," i.e., repeating same pattern later in a more
   difficult context). 7, 9, 10, 11
3. Leaving mother to look at an object and circling around
   behind her to climb on her other side. 2, 3, 4 . . .
4. Repeatedly climbing on and off objects using the same or
   functionally equivalent patterns (off mother, fence, tripod
   swing). 3, 4, 5, 6–11

B. SOME EXAMPLES OF PRIMITIVE VOLUNTARY BODY ADAPTATIONS TO
THE ENVIRONMENT   AGE IN WEEKS

1. Quadrapedal "baby" walking with flexed hips (and usually
   curled toes).   2, 3, 4
2. Sitting on floor.   2, 3 . . .
3. Pivoting around on back feet.   2, 3 . . .
4. Climbing the fence with alternate arm and leg movements   2, 3, 4 . . .

SERIES I, STAGE 3: *From fourth to eleventh weeks—The Secondary Linear
"Schemata"*

During this stage the stumptail displayed voluntary behavior patterns involving coordination of vision and prehension focused on the visual and tactile properties of
objects. This stage was characterized by the differentiation of "secondary schemata" as
it is in human infants, but there was no "secondary circular reaction," only the
analogous *secondary linear schemata*.

A. SOME EXAMPLES OF DIFFERENTIATION OF SECONDARY LINEAR
SCHEMATA DEVELOPING OUT OF THE COORDINATION OF VISION AND
PREHENSION   AGE IN WEEKS

1. Twisting objects in two hands.   5, 12
2. Pushing and pulling on grasped object on the floor.   4, 5, 6, 7,
   8 . . .
3. Poking with index finger.   4, 8, 10, 14,
   15, 23, 25
4. Pincer grasping small objects with thumb and index finger.   4, 5, 6, 11,
   13, 16
5. Involuntary dropping and voluntarily picking up objects.   4, 7, 10, 11,
   13, 17
6. Visual behavior patterns:
   a. Visual following fast-moving object in an arc.   6, 12
   b. Rotating eyes down to look at an object (rare).   11, 12, 13
   c. Looking at objects from different perspectives.   4, 8, 9, 12,
   13, 14, 15,
   16, 18, 19,
   20 . . .

Stage three in the stumptail infant was also characterized by the appearance and
differentiation of other behavior patterns which do not fit into Piaget's category of
"secondary schemata" because they did not involve the coordination of vision and
prehension, and do not fit into Piaget's category of "primary schemata" because they
are voluntary and often visually directed. These *body schemata* undergo a series of
transformations analogous to the "secondary schemata" in the third and subsequent
stages.

A. SOME EXAMPLES OF THE BODY SCHEMATA AND THEIR
DIFFERENTIATION

AGE IN WEEKS

1. Crouching:
   a. Fully flexed crouching
   b. Hips up, knees bent, arms extended crouching.
   c. Hips and knees flexed leaning on elbows.
2. Climbing down fence:
   a. Feet first.
   b. Head first after torso turning.
   c. Free fall (flipping over 180 degrees by letting go with
      hands).

   d. Quadrapedal jumping off.

3. Presenting rump:
   a. Foot back with leg twist "curtsey foot."

   b. One or two feet back (usually touching object).

   c. Plus hand back to touch object (rare).
4. Presenting ventrum (usually against partition to males):
   a. Presenting ventrum right side up.

   b. Presenting ventrum upside down.

5. Abducting leg (not in present).
6. Abducting arms: hanging under fences or chain.

7. Walking and running and jumping:
   a. Squat-walking

   b. Backward squat-walking
   c. "Scampering" running with hips out to the side.

   d. Hopping and springing with back legs together.

   e. Bipedal springing onto fence.
   f. Bipedal walking.

	AGE IN WEEKS
1a	3, 4, 5 . . .
1b	5, 19, 24, 25
1c	3
2a	4, 6
2b	5, 6, 7, 9, 10
2c	6, 15, 17, 19, 20, 21, 23, 25
2d	4, 5, 12, 13, 14, 15
3a	3, 6, 9, 12, 13, 15, 25
3b	6, 7, 8, 9, 10 . . .
3c	6, 14, 21, 23
4a	4, 5, 7, 10, 12, 15
4b	9, 11, 12, 15, 17, 18, 23, 25
5	10, 12, 13, 15
6	5, 12, 14, 18, 24
7a	3, 7, 9, 11, 12, 13, 14, 17, 18, 19, 25
7b	6
7c	5, 12, 14, 18, 19, 24
7d	3, 4, 6, 10, 19, 20, 21, 23, 25
7e	3, 4, 5, 6, 12
7f	13, 19, 24

g. Hip tossing; throwing legs out behind and standing on
   hands.   5, 12, 16, 18
h. Jumping onto objects.   6, 7, 12, 14,
       15, 17,
       23...

i. Jumping up under mother's belly with a torso twist.   4, 7, 14

8. Facial patterns:
   a. Grimace.   4, 20, 21, 22,
          23

   b. Play face.   10, 13, 21, 23
   c. Threat with intention biting.   6, 13, 14, 15,
          17, 18, 19,
          20, 21, 23

   d. Yawning (rare).   5, 8, 11
   e. Sniffing.   12, 13...
   f. (Lipsmacking continues).   1, 2...
   g. Square mouth pout; (round mouth pout, weeks 1 & 2).   14, 15, 19, 25
   h. Protruding small object from mouth.   7, 10, 11, 12,
          13

SERIES I, STAGE 4: *Week nine on—The Coordination of Secondary Linear Schemata and Their Application in New Situations*

During this stage the stumptail infant began coordinating a series of acts or a set of simultaneous acts in order to achieve some goal. He also displayed a *derived secondary (linear) reaction*, analogous to human infants, involving the serial application of several motor patterns to one object or to several objects. This stage differed from stage four in human infants in that the stumptail did not begin to place objects in relation to other objects and, hence, object-object interactions ("tertiary circular reactions") did not emerge from it.

A. SOME EXAMPLES OF THE COORDINATION OF SECONDARY (LINEAR) SCHEMATA   AGE IN WEEKS

1. Bimanual pulling objects apart.   10, 12, 14,
       15, 17, 19

2. Outstretched palm supporting an object.   12–17
3. Alternate "interrupted prehension" putting one object down and picking up another and manipulating it and putting it down and picking up the other.   12, 18, 23,
       24, 25

4. Hand-to-hand transferring an object.   13, 15, 19
5. Holding or pushing another animal away while holding an object in the other hand (one time only).   5

6. Biting a string or rope in the middle while pulling it apart with both hands.   20, 21
7. Rolling an object between two hands (one time only).   15
8. Holding object down with one hand in order to pull off a piece of object with other hand (rare).   22, 24
9. One hand involuntarily dropping object, the other hand catching it (rare).   22, 24
10. "Foot-storage schema"; placing an object in foot and later retrieving it (an adult pattern) (one time).   25

B. SOME EXAMPLES OF DERIVED SECONDARY (LINEAR) REACTIONS   AGE IN WEEKS

1. Serial application of several "schemata" to one object: e.g., hand-to-hand transferring, placing in mouth, extruding into hand, placing on flat palm, pincer grasping it with other hand.

13, 14, 17,
18, 19...

The infant stumptail displayed an early form of the *derived secondary linear reaction* which involved a pattern of rushing from one object to another in rapid succession, each new one catching his eye and distracting him as he was manipulating the former one. He would look at an object or an animal across the cage and start out toward it, and then spy a bright object, rush over to it, sit and pick it up, bite it, pull it out of his mouth and rub it on the floor, for example, that is, display serial application of *secondary linear schemata*; and then his eye would catch another object and he would rush over to it and go through the same procedure, describing a zig-zag path across the floor, interrupting these bouts with trips back to his mother. Earlier, he began these explorations one bout at a time, and often described a circle: stepping out of his mother's lap/ventrum and going around behind her back to explore an object, picking it up briefly before completing the arc back to her on the other side.

The stumptail infant also displayed serial and simultaneous coordinations of *body schemata* during the fourth stage that were quite analogous to his *secondary linear* (prehensive-object) coordinations, and serial applications which were analogous to the *derived secondary (linear) reactions*.

A. SOME EXAMPLES OF THE COORDINATION OF VOLUNTARY BODY
SCHEMATA   AGE IN WEEKS

1. "Penduluming"; rapid descent of the fence involving serial coordination of letting go with one hand, turning torso, grasping fence and bringing other hand to same position, and letting go with both feet so that they drop and repeating.   10, 11...

B. SOME EXAMPLES OF DERIVED BODY REACTIONS   AGE IN WEEKS

1. Repeated serial applications of functionally-equivalent motor patterns to same surfaces:

	AGE IN WEEKS
a. Alternate forelimb-hindlimb hanging under branches.	12, 14, 15, 17, 20, 23
b. Alternate foot bouncing one chain (a *circular body reaction*).	20
c. Alternate flat palm sliding on the ground (rare).	25
d. Repeatedly climbing on and off swing using different motor patterns: head first, hand over hand; feet first, hand over hand; (climbing up hand over hand); sliding down.	14, 15, 16, 17, 18

SERIES I, STAGE 5: *From twenty-third week on—The Trial-and-Error Secondary Linear Reaction and the Discovery of New Means Through Active Experimentation*

Trial-and-error secondary linear reactions were only beginning to develop in the infant stumptail at the time of his death; the only real evidence of this pattern was his search for a hidden object under two adjacent cloths (see Stage 5 of the object-concept series). (This class of behavior apparently continues to develop during the first four years of life [Harlow 1968].) This class of behavior differed profoundly from the "tertiary circular reaction" in that it involved no interest in potential object-object or object-force field interactions. Despite the introduction of objects which are used in the "tertiary circular reaction" in human infants, the infant stumptail and his older playmate failed to do any of the characteristic actions: they failed to roll or bounce balls, put objects into containers, stack objects, use sticks to rake in out-of-reach objects, etc. On one occasion the stumptail infant inadvertently dropped the ball which bounced, and on another, he inadvertently dropped an object into a container; he paid no attention to either event and did not repeat his actions.

During the fourth stage, the stumptail infant displayed *trial-and-error body reactions* which have some interesting parallels with the "tertiary circular reaction" in that they involve trial-and-error manipulation of body and locomotor actions and position in space relative to other animals during play bouts and during attempts to climb on his mother when she was rejecting him.

A. EXAMPLES OF TRIAL-AND-ERROR BODY REACTIONS   AGE IN WEEKS

	AGE IN WEEKS
1. Playing "keep away" with playmate on the loose chain: circling around under and above and grabbing chain in various places.	13, 19, 20 . . .

SERIES I, STAGE 6: *From week twenty-three—(Secondary Linear Trial-and-Error) Representation and Planning Through Mental Combinations*

Again, the only evidence of mental *representation of trial-and-error manipulation of secondary linear schemata* comes from stage six of the object-concept series: the stumptail's ability to uncover an invisibly displaced object suggests some such representation.

Long before he displayed this type of representation, however, he gave evidence of trial-and-error visual planning of pathways and detouring (beginning weeks 14, 18, 19, 20 . . .) which implies some sort of primitive representation. His ability to step off objects (rather than pulling on them while standing on them) which developed during the sixth stage suggests that he had some sort of body position image (stage 6, object-concept series); this is also suggested by "stealing" objects and maneuvering to protect them (beginning 18th week). On the other hand, he certainly did not have an image of his own face and obviously did not recognize himself in a mirror. It is interesting that the *body schemata* stages tend to develop more rapidly than the *secondary linear stages* and that the boundaries between stages are not clear-cut. The representation of *body schemata* may actually belong in stage four or five.

SERIES II: OBJECT-CONCEPT SERIES IN AN INFANT STUMPTAIL MACAQUE

SERIES II, STAGES 1 AND 2: *Birth to fifth week—The Absence of Special Behavior Relating to Vanished Objects*

During these stages the stumptail infant, like Piaget's infants, displayed no special behavior relating to vanished objects; he did display some visual accommodation to objects.

A. SOME EXAMPLES OF VISUAL ACCOMMODATION TO OBJECTS	AGE IN WEEKS
1. Alternate glancing between two objects.	1, 2, 3 . . .
2. Loss of interest in objects covered in his sight.	3
3. Continuing to look at the point of disappearance of a moving object	4, 5, 6
4. Slow following of moving objects through 180-degree arc.	5

SERIES II, STAGE 3: *From third to sixth week—Beginning Object Permanence Extending the Movements of Accommodation (with no "Deferred Circular Reaction")*

During this stage the stumptail infant displayed increasing visual accommodation to moving objects and increased prehensive accommodation to objects.

A. SOME EXAMPLES	AGE IN WEEKS
1. Visual accommodation to rapidly moving objects through a 180-degree arc.	6, 12
2. Glance returning to look at the point of disappearance of a moving object which had vanished.	4, 5
3. Visually anticipating the trajectory of moving objects.	5, 6, 7, 9 . . .
4. "Interrupted prehension" (reaching down for object previously dropped without looking).	4, 5, 7, 9
5. Grasping partially covered object.	3, 4, 5

SERIES II, STAGE 4: *From third to sixteenth week—The Practical Object—The Active Search for the Vanished Object*

During this stage the infant began to uncover objects that were hidden while he was watching. (Note that visual searching for a vanished object which Piaget classifies in this stage began earlier than glancing back to the point of appearance of a moving object which has vanished [stage 3].)

A. SOME EXAMPLES OF ACTIVE SEARCH FOR VANISHED OBJECTS	AGE IN WEEKS
1. Palpating covered objects.	5, 7, 8, 12 . . .
2. Visual searching for a vanished object at the point it disappears.	3, 4, 5, 7, 12, 13
3. Tactile searching for a dropped object.	5, 7, 9
4. Uncovering objects covered in sight by a single screen.	6, 12, 13, 17, 19
5. "Residual reaction" of searching under first screen where object was hidden and uncovered when object is sequentially hidden under two screens.	under 22
6. Intentional rotation of objects as indicated by looking toward desired end.	16, 17, 18

SERIES II, STAGE 5: *From twenty-third week—Intermediate Object Permanence*

During this stage the infant displayed the ability to correctly search for an object after visible displacements behind two screens (serially or simultaneously), and he displayed "interrupted prehension" *and* visual searching for a lost object.

A. SOME EXAMPLES OF INTERMEDIATE OBJECT PERMANENCES	AGE IN WEEKS
1. Correct search for object after two visible displacements (one time *secondary trial-and-error reaction*).	23
2. "Interrupted prehension" with visual searching.	24, 25
3. Pulling up an object by an attached string.	24, 25

During this stage the infant displayed certain disabilities such as the inability to reconstruct invisible displacements (dropping an orange or toy out of a cup or my hand while it is under a blanket); he continued to search in my hand or in the cup from the twelfth through the twenty-first weeks. He also displayed a behavior Piaget described during this stage which I call bootstrapping, i.e., he tried to pull up a blanket while standing on it (during week three, weeks ten to fifteen).

SERIES II, STAGE 6: *From week twenty-three on—Object Permanence: The Representation of Object Displacements*

During this stage the stumptail infant began searching behind a single screen for invisible displacements (week 23). During this stage he also stopped bootstrapping, or trying to lift things he was standing on. I was unable to test his ability to uncover a series of invisible displacements behind a series of screens before his unexpected death.

According to Piaget the analogous behavior in human infants means that the child is "directing his search by means of representation" (Piaget 1954: 91) (which means representation of the "tertiary circular reaction"). This interpretation does not seem to apply to the stumptail infant who did not display the "tertiary circular reaction." My interpretation of this behavior in the stumptail is that it signifies representation of the *secondary trial-and-error reaction*.

SERIES III: THE IMITATION SERIES IN AN INFANT STUMPTAIL

SERIES III, STAGES 1 AND 2: *Birth to fourth week—No Imitative Behavior*

This stumptail infant displayed no contagious crying or repetition of his own sounds.

SERIES III, STAGE 3: *From the fourth week and continued through the twenty-fifth week—Purposeful Self Imitation of "Schemata" from Own Repertoire*

During this stage the stumptail infant occasionally displayed contagious or imitative activities from his own repertoire when he saw them performed by another animal, especially his playmate, or other infants across the partition. Most of the "imitation" occurred during play; imitation of *body schemata* was more common than "imitation" of *secondary schemata* such as object manipulation. In addition, it was not true imitation in the sense that human infants and children imitate in that he did not focus on the other animals' motor patterns nor try to match his actions to those of the other animal; *social facilitation* or *contagious performance* might be a better name for these activities.

A. SOME EXAMPLES OF IMITATION OF <u>BODY SCHEMATA</u>   AGE IN WEEKS

1. Fence climbing upon seeing another infant on the fence.   4, 6, 10, 12, 13, 15, 19, 21, 24

2. Sitting on mother's head upon seeing another infant on his mother's head.   13, 17, 18
3. Presenting upon seeing another animal presenting.   18, 24
4. Lolling (lying and rolling on the side on the floor) upon seeing another animal lolling.   15, 20, 23
5. Bipedal running and leaping upon seeing a playmate do the same.   17, 18, 19
6. Climbing in and out of a box upon seeing a playmate do the same.   16
7. Bipedal bouncing upon seeing a playmate do the same.   19, 20, 21, 24
8. Somersaulting upon seeing a playmate somersaulting.   20

9. Hand sliding along the floor upon seeing a playmate do the
   same.   23
10. Flexing down in a crouch to drink water off the floor upon
    seeing mother do the same.   24, 25

B. SOME EXAMPLES OF IMITATION OF SECONDARY LINEAR SCHEMATA   AGE IN WEEKS

1. Eating upon seeing another animal eating.   11, 14
2. Coming over and looking at an object upon seeing other
   animals manipulating it.   9, 11, 12, 13,
                                                                                    15, 17, 18,
                                                                                    21, 24
3. Parallel object manipulation with another infant.   13, 17, 18, 19
4. Glancing from his mother's hand to his and placing his hand
   upon mother's hand after looking at mother's hand.   9, 13, 15, 24

In addition to these rudimentary forms of imitation, the stumptail infant displayed
"imitation" of a few vocal and facial "schemata":

                                                                                 AGE IN WEEKS

1. Contagious squeal (rare).   15
2. Contagious haaa noise (rare).   18
3. Contagious lipsmacking.   8, 9, 13, 15
4. Contagious threat face (rare).   18
5. Contagious play face.   10, 13, 21, 23

He displayed one puzzling behavior which might seem to imply a recognition of
his own face as the same as his mother's face though there was no other evidence of
this: he glanced at his mother's full cheek pouches and touched them and then
pushed food out of his own cheek pouches and ate it. I do not think it indicated such
recognition; it is more likely that he recognized the covered object as food and
remembered that he had some in his cheek pouches.

One of the most striking differences between the development of the stumptail
infant and human infants was the lack of mutual vocal and gestural imitation between
stumptail infant and mother: not only did the infant not repeat his sounds and
gestures, nor repeat the sounds and gestures of his mother, she did not repeat any of
his sounds and gestures as human mothers do. Needless to say the stumptail infant
displayed no trial-and-error imitation of new "schemata."

■ SYSTEMATIZING PIAGET'S MODEL FOR COMPARATIVE DEVELOPMENTAL STUDIES

Few investigators will use Piaget's model as a framework for an ethology of unstereotyped behavior unless it is simplified and systematized. Piaget's model is complicated because unstereotyped behavior, especially intelligent behavior, is complex;

TABLE 2 / META-BEHAVIORAL PARAMETERS ABSTRACTED FROM PIAGET'S SENSORIMOTOR
PERIOD SERIES IN AN INFANT MACAQUE AND IN HUMAN INFANTS

I. TYPE OF COORDINATIONS: nature of interaction between parts of the self, object and parts of the self, objects and
objects, objects, objects and forces, images and images, parts of the self and model's acts
  1. *reflex:* involuntary, preprogrammed, automatic, e.g., neonatal sucking and grasping and rooting (Stage 1,
  Human and Stumptail Sensorimotor Intelligence Series)
  2. *"reciprocal coordination":* simultaneous, mutual coordination of two acts (or two objects) with each other
  with mutual accommodation of the acts to each other, e.g., hand-hand twisting and grasping while watching
  (Stage 3, Human Sensorimotor Intelligence Series)
  3. *environmental adaptation:* accommodation to properties of objects, e.g., visual following of moving objects,
  turning toward sound source (Stage 2, Human Sensorimotor Intelligence Series)
  4. *intercoordination:* simultaneous coordination of acts to each other or in response to an object or a substrate
  occurring without mutual accommodation of acts to each other, e.g., coordination of limbs and hands in
  climbing, of two hands in picking up an object (Stage 2, Stumptail Sensorimotor Intelligence Series)
  5. *"serial coordination":* sequential coordination of a series of two or more acts, e.g., letting go of one object
  and grasping another (Stage 4, Human Object-Concept Series)
  6. *"hierarchical coordination":* sequential coordination of a series of two or more acts where one act is auxiliary
  to the other and is performed in order to achieve a goal, e.g., lifting a cover in order to grasp an object under
  it (Stage 4, Human Object-Concept Series)
  7. *"trial-and-error coordination":* voluntarily changing the position, sequence, orientation, intensity, or duration,
  etc. of acts, e.g., trial-and-error keep away chain from playmate (Body/Prehensive Schema, Stage 4, Stumptail
  Sensorimotor Intelligence Series)
  8. *reciprocal trial-and-error coordination:* voluntarily change the position, sequence, intensity, orientation, or
  duration, etc. of acts on one object relative to another object or of one object relative to a force field, e.g.,
  twisting a stick in order to bring it through bars, stacking a block on top of another block, or raking in an object
  with a stick (Stage 5, Human Sensorimotor Intelligence Series)
  9. *reversing operations:* repeating actions in reverse sequence, e.g., uncovering an object and covering it up
  again (Stage 4, Human Object-Concept Series)
II. LOCUS OF COORDINATIONS AND APPLICATIONS: the part or parts of the body acting, and/or the part or parts of the
physical or mental objects acted upon in the "schema"
  1. *object to self:* part of the self and the object being coordinated, e.g., nipple to mouth (Stage 2, Human
  Sensorimotor Intelligence Series)
  2. *self to self:* the parts of the body coordinated, e.g., thumb to mouth (Stage 2, Human Sensorimotor Intelligence
  Series)
  3. *self to substrates:* the parts of the body and the nature of the substrate coordinated, e.g., flat foot to floor
  (Stage 2, Stumptail Sensorimotor Intelligence Series)
  4. *series of objects to self:* objects and part of body coordinated, e.g., rattle, ball, cloth to mouth (Stage 3,
  Stumptail and Human Sensorimotor Intelligence Series)
  5. *objects in relation to objects or forces:* objects and forces, e.g., block in tub or ball rolling (inertia and friction)
  (Stage 5, Human Sensorimotor Intelligence Series)
  6. *images to images:* images, e.g., image of box with drawer coming open (Stage 6, Human Sensorimotor
  Intelligence Series)
  7. *own acts to model's acts:* own acts and model's acts, e.g., clapping hands after or while watching model
  clap hands (Stage 4, Human Imitation Series)

continued

TABLE 2 / CONTINUED

III. VOLITION OF MOTOR PATTERNS: degree of choice between alternative actions

1. *reflex or involuntary spontaneous*: automatic, unchosen, compulsive actions, e.g., neonatal sucking on objects in mouth, neonatal grasping (Stage 1, Stumptail and Human Sensorimotor Intelligence Series)

2. *primitive voluntary*: incipiently voluntary actions chosen among some options but clumsily and uncertainly performed and sometimes overwhelmed by reflex actions, e.g., early visually directed grasping, early thumb sucking (Stage 2, Stumptail and Human Sensorimotor Intelligence Series)

3. *voluntary*: chosen from among several alternatives (Stage 3 on, Stumptail and Human Sensorimotor Intelligence Series)

IV. DIFFERENTIATION OF BEHAVIOR PATTERNS: degree of separation of various components of a behavior pattern from other components

1. *undifferentiated "global"*: single behavior patterns involving many actions which cannot be broken down into components, typical of early neonatal behavior patterns, e.g., whole hand grasping (Stages 1 and 2, Stumptail and Human Sensorimotor Intelligence Series)

2. *primitively differentiated*: incipient, clumsy breakdown of "global" behavior patterns into component acts, e.g., index finger extension (Stage 3, Stumptail and Human Sensorimotor Intelligence Series)

3. *differentiated*: performance of individual acts which were previously part of a "global" behavior pattern, e.g., pincer grasping (Stage 3, Stumptail Sensorimotor Intelligence Series)

4. *consciously differentiated*: consciously purposely performed individual acts, e.g., demonstrating or imitating a motor pattern (Stage 4, Human Imitation Series)

V. NUMBER OF INCLUDED ACTS IN BEHAVIOR PATTERN: number of easily identified motor patterns (i.e., those motor patterns occurring singly in other contexts) occuring simultaneously or in sequence in a "schema", e.g., grasping, letting go, visually following and fixating, sliding hand, etc.

VI. GOAL DETERMINATION: degree of choice between alternative goals (i.e., stimulus control)

1. *reflexive goal*: no stimulus control, compulsive, automatic goal, e.g., neonatal sucking on object in mouth (Stages 1 and 2, Stumptail and Human Sensorimotor Intelligence Series)

2. *no goal*: no apparent goal

3. *accidentally discovered stimulus triggered goal ("fortuitous" goal)*: goal determined by some immediate stimulus in the environment that the actor happens to notice, e.g., movement of a mobile, presence of a toy, etc., creating an immediate response (Stage 3, Stumptail and Human Sensorimotor Intelligence Series)

4. *initial stimulus-triggered goal*: goal (determined by some environmental stimulus) which is maintained by the actor despite intervening stimuli, e.g., a toy across the room which the actor goes to without the actor being distracted by intervening toys (Stage 4, Stumptail and Human Sensorimotor Intelligence Series)

5. *mental-image goal*: goal image maintained in the actor's memory in the absence of the physical presence of the object, e.g., mental image of a covered object (Stage 4, Stumptail and Human Object-Concept Series)

6. *imitative goal*: image of another actor's behavior maintained in the actor's memory in the absence of the other actor and his actions (Stage 6, Human Imitation Series)

VII. CONTEXTS OF BEHAVIOR PATTERN APPLICATION: familiarity or novelty of setting in which "schema" occur

1. *reflexive*: no context in the ordinary sense, behavior stimulated by particular stimulus configurations (Stage 1, Stumptail and Human Sensorimotor Intelligence Series)

2. *old*: actor has performed "schema" in this context before, application of "schema" to familiar object, e.g., sucking nipple (Stage 2, Human Sensorimotor Intelligence Series)

3. *new*: actor has not performed "schema" in this context before, e.g., dropping objects from height into water for the first time (Stage 5, Human Sensorimotor Intelligence Series)

continued

TABLE 2 / CONTINUED

VIII. TEMPORAL/SPATIAL PATTERNING OF BEHAVIOR PATTERNS AND COORDINATIONS: position, orientation, sequence, frequency, repetition, etc., of acts

    1. *single occurrence:* acting only once

    2. *repetition of same pattern:* acting two or more times without intervening actions or sequences, e.g., repeatedly bringing hands together and apart (Stages 2 and 3 of Human Sensorimotor Intelligence Series)

    3. *delayed repetition of same pattern:* acting two or more times with intervening actions or sequences, e.g., clapping hand now, and clapping own hands later after intervening acts (Stage 4, Human Sensorimotor Series)

    4. *repetition of functionally-equivalent patterns:* doing the same thing two or more times but performing it in different ways, e.g., climbing on and off the fence different ways (feet first, head first, then jumping off) (Stage 4, Stumptail Sensorimotor Intelligence Series)

    5. *serial application of the same pattern:* repeating the same "schema" two or more times on different objects, e.g., picking up and shaking three objects; lifting three screens in a row (Stages 4 and 5, Human Sensorimotor Intelligence Series)

    6. *alternate application of same pattern with different body parts:* performing an act with one body part and then repeating it with another body part, e.g., banging object with one hand and then banging an object with the other hand; bouncing on fence with one foot and then bouncing on fence with other foot holding chain (Stage 4, Human Sensorimotor Intelligence Series, and Stage 4, Stumptail Body Schemata Stages)

    7. *serial application of different patterns:* applying a series of acts to a single object, e.g., biting, banging, rubbing, and pulling on a toy (Stage 4, Stumptail and Human Sensorimotor Intelligence Series)

    8. *variable repetition of pattern:* varying sequence, or intensity, or force, or orientation, etc. of acts on an object or a series of objects, e.g., dropping objects from different heights (Stage 5, Human Sensorimotor Intelligence Series)

    9. *repetition of model's acts or object's acts:* copying acts of a model or of a moving object (Stage 3, on Human Imitation Series)

IX. RELEASING AND/OR REINFORCING ASPECTS OF STIMULI OR ACTIONS: response to stimuli, or to feedback from actor's own actions, causing him to repeat and/or continue his actions

    1. *consummatory "assimilation":* pleasure in assimilation of taste, smell, or texture, etc. (Stages 1 and 2 Human and Stumptail Sensorimotor Intelligence Series)

    2. *simple "assimilation and accommodation" to prehension and vision:* pleasure in simple proprioceptive feedback from grasping, visual following, etc. (Stage 2, Human Sensorimotor Intelligence Series)

    3. *"reciprocal assimilation and accommodation":* pleasure in reciprocal "accommodation" and "assimilation" of acts to each other, e.g., moving hands up to eyes and visually following hands and touching and pulling two hands with each other while watching (Stage 2, Human Sensorimotor Intelligence Series)

    4. *exercise:* pleasure in feedback from exercise of long muscles, e.g., kicking, crawling, climbing (Stage 3, Human Sensorimotor Intelligence Series and Stumptail Sensorimotor Intelligence and Body Schemata Series)

    5. *noncontingent novelty:* pleasure in novelty unconnected to actor's actions of moderate stimulus contrast within a familiar context, e.g., picking up toys (Stage 3, Stumptail Sensorimotor Intelligence Series)

    6. *perceived "contingency" control of object action:* pleasure in perceived control over actions of objects, e.g., making a mobile move by kicking feet (Stage 3, Human Sensorimotor Intelligence Series)

    7. *perceived variable "contingency" control of variation in object action:* pleasure in perceived control of changes in object-object, object-field interaction, e.g., in causing objects to make a splash in hitting the water (Stage 5, Human Sensorimotor Intelligence Series)

    8. *reverse "contingency" control over self in response to object action (imitative matching):* pleasure in causing

continued

TABLE 2 / CONTINUED

own actions to match those of another actor or a moving object, e.g., imitating model's hand clapping (Stage 3, Human Imitation Series)

9. *reverse variable "contingency" control:* pleasure in causing own actions to match those of another actor or a machine, e.g., imitating model stacking blocks, imitating a model pushing a friction wheel car (Stage 5, Human Imitation Series)

X. SOURCES OF STIMULI ELICITING BEHAVIOR PATTERNS: the sources of stimuli associated with beginning, repeating, or continuing actions

    1. *internal state:* e.g., hunger

    2. *environmental objects*

    3. *other animal (allogenic)*

    4. *nonexercise feedback from own actions (internal autogenic):* sound of own vocalizations, sight of own hands, feel of own touch (Stage 2, Human Sensorimotor Intelligence Series)

    5. *self-induced environmental feedback (external autogenic):* sight or sound of objects whose actions are due to own behavior (Stage 3, Human Sensorimotor Intelligence Series)

    6. *feedback from exercise of long muscles (proprioception)*

    7. *memory image:* picture or image in the memory of any of the foregoing (Stage 4 on Human and Stumptail Object Concept Series)

    8. *transformed memory image:* picture of object-object or object-field interactions in memory image, e.g., image of opening a box (Stage 6, Human Sensorimotor Intelligence Series)

simplifying and systematizing is difficult and risky because of the danger of oversimplifying, misinterpreting, and excluding important formulations. Nevertheless it is worth trying.

I have developed a system for classifying Piaget's stages in terms of their *meta-behavioral parameters*, in other words, in terms of certain aspects of behavioral organization which usually remain implicit in behavioral descriptions. These include: type of coordination, locus of coordinations, volition, differentiation, goal direction, temporal/spatial patterning, contexts, number of included acts, sources of stimuli eliciting motor patterns, and releasing and/or reinforcing aspects of stimuli (see Table 2 for definitions). All of these parameters were taken as directly as possible from Piaget's behavior pattern and stage descriptions.

The behavior patterns of each stage are characterized by a unique set of *meta-behavioral parameters*; I am proposing that each set of *meta-behavioral parameters* represents a *behavioral program*, that is, an integrated neurological program, generating a particular class of behavior patterns. The "secondary circular reaction," for example, is characterized by the following *behavioral program* (unique set of *meta-behavioral parameters*): "reciprocal coordination" and incipient serial coordination of hands (or truncal or pedal or facial movements) with object movement and sounds; voluntary; differentiated motor patterns; "fortuitous" goal direction (determined by action of object); use of old motor patterns in new contexts; repetition of the same motor pattern or a functional equivalent; in response to autogenic (self-generated) external stimuli (e.g., noise of a rattle); and reinforced/released by perceived contingency, exercise, and/or accommodation.

See Tables 3 through 5 for a description of human infant *behavioral programs* characteristic of three of Piaget's sensorimotor series. Notice that each stage is characterized by the development of new *behavioral programs*, i.e., new sets of *meta-behavioral parameters.*

Each of Piaget's stages is characterized by a complex *behavioral program* with several new *meta-behavioral parameters* which did not occur during the previous stage. Obviously, a single name, such as the "secondary circular reaction," encompasses a large number of subtle distinctions; this fact makes it difficult to apply Piaget's concepts directly to another species with different patterns.

Just as in the case of human infants, each of the stumptail stages is characterized by a complex *behavioral program* with several new *meta-behavioral parameters* which did not occur during the previous stage.[5] See Tables 6 through 9 for a description of stumptail infant *behavioral programs* characteristic of the three sensorimotor series, the sensorimotor intelligence series, the object-concept series, and the imitation series.

The *meta-behavioral parameters* should describe *behavioral programs* in any primate species and thereby reveal detailed differences between species within a Piagetian framework without forcing inappropriate human categories on another species. It is important to realize the complexity and subtlety of Piaget's stage categories because other species, like the stumptail, may show a mosaic of *meta-behavioral parameters* which could be confused with the "secondary circular reaction" and other human stages.

Because Piaget's categories were developed to describe human infant development rather than nonhuman primate development, and because his stages are so complex and subtle, it is important to review the human stages and the analogous stumptail stages in terms of their *meta-behavioral parameters* (see Tables 3 through 9) so that we can make more systematic comparisons of the organization of behavior in the two species.

Each stage in each species is characterized by a *behavioral program*, that is a unique set of *meta-behavioral parameters* specifying the locus of behaviors, how they are coordinated and with what they are coordinated, and how many acts are coordinated, what their goal is, and their temporal patterning and the reinforcing and/or releasing aspects of stimuli.

Considering these *meta-behavioral parameters* (see Tables 2, 3–9) singly or in combination, we can compare the behavioral organization of the two species more systematically.

LOCUS AND TYPE OF COORDINATION

The stumptail infant showed virtually no interest in the behavior of his own hands, visually or tactually. He did not display the repeated (circular) hand-hand and hand-hand-mouth and hand-hand-eyes and hand-hand-object-eyes-mouth "reciprocal coordinations" which are so striking in human infants during the second stage of the sensorimotor intelligence series. Nor did he display the hand-eye-object movement or sound "reciprocal coordinations" characteristic of the human infant during the third stage (the "secondary circular reactions"). Nor did he display the object-object "reciprocal coordinations" nor *reciprocal trial-and-error coordinations* characteristic of the fifth stage (the "tertiary circular reaction").

In fact, although the stumptail infant showed intercoordinations between his hands and objects, and his hands and feet and objects or substrates and his hands and objects and eyes, he failed to show any true "reciprocal coordinations" (with mutual "assimilation" and "accommodation"). He did, of course, display other types of coordination characteristic of human infant development, i.e., serial and "hierarchical coordination" and trial-and-error coordination but not reciprocal trial-and-error coordination.

The face is one other area of the body which is the locus of strikingly different coordinations in the two species. It is not only that the stumptail displayed different facial expressions from those of human infants, but that his facial and mouth expressions did not participate in the "secondary circular reaction" and its specialized form, "the game" (Watson 1972).

In fact, there was no face-to-face interplay between the stumptail infant and his mother: although they engaged in face-to-face gazing, and mouth-to-mouth contact and sniffing (both mutual and unilateral) there was nothing even vaguely resembling the prolonged open-ended face-to-face interactions so characteristic of human infants and caretakers which Watson calls "the game" (Watson 1972). Nor, of course, did the stumptail infant smile as human infants do during "the game" and other "secondary circular reactions." Nor did the stumptail infant display vocal (or gestural) or facial imitation as human infants do.

VOLITION AND DIFFERENTIATION OF BEHAVIOR PATTERNS

Although the stumptail infant began life, as human infants do, with involuntary undifferentiated behavior patterns which gradually came under voluntary control and were differentiated, this occurred much earlier in the stumptail (i.e., during the first weeks of life) than in human infants; and furthermore, he never displayed voluntary control and conscious differentiation of vocal expressions, nor did he develop conscious differentiation of facial expressions or gestures. He did, however, develop differentiation and voluntary control of pincer grasping very early in life. It is clear from his motor control that the absence of the "tertiary circular reaction" from his repertoire was not due to a lack of motor coordination.

NUMBER OF ACTS INCLUDED IN A "SCHEMA"

The stumptail infant never displayed the complex sequences of hierarchically organized actions, or trial-and-error manipulations characteristic of human infants, though even at six months he could coordinate three acts in service of a goal as indicated by searching under two screens for a sequentially displaced object and then grasping it.

GOAL DETERMINATION

This is a rather subjective area, but it seems clear that the stumptail infant displayed a developmental series beginning with reflexive goals, going on to "fortuitous" stimulus-triggered goals (something catching his attention), and then to initial stimulus-determined goals which he could remember after the stimulus was gone; but he did not display evidence of mental image goals, or imitative goals involving object-object coordination.

CONTEXTS

Piaget uses this term to refer to the context in which "schemata" are used; during the "secondary circular reaction," for example, human infants use "primary schemata" in new contexts. The stumptail infant also used his behavior patterns in new contexts.

TEMPORAL/SPATIAL PATTERNING OF BEHAVIOR PATTERNS AND RELEASING AND/OR REINFORCING ASPECTS OF STIMULI

Repetition of behavior patterns plays an important part in human infant sensorimotor intellectual development: repetition of prehensive movements (hand-hand, hand-hand-mouth, hand-hand-eye) in the "primary circular reactions"; repetition of the hand and object movement, or foot and trunk and object movement, or face-to-face interactions in the "secondary circular reactions"; variable repetition of behavior patterns involving object-object and object-force field interactions in the "tertiary circular reactions"; and repetition of the behavior patterns of another person or an object in imitation. The reinforcements for these repetitions are quite different: the "primary circular reactions" are apparently reinforced by pleasure in "reciprocal assimilation and accommodation"; the "secondary circular reactions" are apparently reinforced by pleasure in contingency control; whereas the "tertiary circular reactions" are apparently reinforced by pleasure in "reciprocal assimilation and accommodation" of objects and by pleasure in variable contingency control (Watson 1966, 1967). Imitative repetitions are apparently reinforced by pleasure in reverse contingency control. All of these repetitions and reinforcements are characteristic of human infants during the sensorimotor period from birth to two years; all of them were absent from the stumptail's repertoire.

The stumptail infant repeated many behavior patterns but with rare exceptions they were *body schemata* such as jumping up and down on the fence holding the chain, or climbing on and off his mother; occasionally he repeated prehensive behavior patterns on objects, such as pushing and pulling on a stick on the floor. These repetitions were never aimed at reproducing some contingent action by the object and, hence, were not "circular" in the sense of the "secondary or tertiary circular reactions." Repetitions of *body schemata* in the stumptail were apparently reinforced by pleasure in proprioceptive feedback (exercise reinforcement); repetitions of prehensive-object manipulation schemata were also apparently reinforced by proprioceptive feedback; some object manipulations were apparently reinforced by pleasure in touch and visual patterns.

Serial applications of behavior patterns to two objects or to a single object is characteristic of the fourth stage in the sensorimotor intelligence series in human infants; according to Piaget this activity begins to form relationships among objects. It is interesting to note that the stumptail infant displayed these serial applications of behavior patterns to objects without forming relationships among objects. This difference may also be due to the absence of reinforcement from the pleasure of "reciprocal assimilation and accommodation" (of object with object in this case). The reinforcement for serial application of behavior patterns to an object, or a series of objects, in the stumptail may have been pleasure in touch and visual patterns and non-contingent novelty.

Reversing operations on objects begins in human infants during the fourth stage in the causality series; covering and uncovering an object during the fourth stage is an example of this, as is taking objects in and out of containers during the fifth stage. These activities are apparently reinforced by pleasure in reversing contingency. The stumptail infant displayed no evidence of reversing operations on objects. For example, when he uncovered an object, he never displayed the human infant pattern of covering it up again.

SOURCES OF STIMULI ELICITING BEHAVIOR PATTERNS

Like human infants, the stumptail infant responded to stimuli from his own internal states, from the environment, from other animals, from his own motor activities (proprioceptive stimuli), and, finally, from memory images. He did not respond to another class of stimuli which are important in human infant development: self-induced visual and tactile and auditory stimuli due to "reciprocal coordination" of hands, hands-eyes, etc. (internal autogenic stimuli) which play an important part in "primary circular reactions" and imitation are apparently insignificant in stumptail development; external self-induced positional visual and auditory stimuli (external autogenic stimuli) which play such an important part in the "secondary and circular reactions" were insignificant to the stumptail infant; finally, the mental image of a desired result (especially one involving object manipulations) which develops during the fifth or sixth stage in human infants was apparently absent in the stumptail infant.

One clear-cut difference that emerges from comparison of the stages in the two species is the early maturation of the stumptail: he almost completed all of his stages in each series by six months of age, whereas human infants reach an analogous stage of development at eighteen to twenty-four months of age.[7]

Another interesting difference between stumptail and human infant maturation is the overlapping, indeed, the almost simultaneous onset of some stages (especially stages two and three in the sensorimotor series, and stages three and four in the object-concept series) in the stumptail infant. For example, prehensive object manipulation began in stage two of the sensorimotor series in the stumptail.

Although he discusses various behavior patterns involving body displacements in the causality and spatial series, and imitation of facial and gestural and body movements in the imitation series, in the sensorimotor series and object-concept series, Piaget is primarily concerned with the behavior patterns growing out of the coordination of vision and prehension. These behavior patterns are extremely salient in human infants during the sensorimotor period because of their late locomotor development.

By contrast, prehensive and visual coordinations were less salient than locomotor and other *body schemata* coordinations in the infant stumptail macaque because of his early locomotor development. The contrast is startling.

Whereas the stumptail infant began walking and climbing at two weeks of age

TABLE 3 / HUMAN INFANT SENSORIMOTOR INTELLIGENCE BEHAVIORAL PROGRAMS: PIAGET'S SENSORIMOTOR INTELLIGENCE SERIES IN TERMS OF META-BEHAVIORAL PARAMETERS

META-BEHAVIORAL PATTERNS	STAGE 1 Reflex stage (birth through first month)	STAGE 2 "Primary circular reaction" (from two to four months)	STAGE 3 "Secondary circular reactions" (fourth to eighth months)	STAGE 4 Coordination of "secondary schemata" and their application to new situations (from eight months on)	STAGE 5 "Tertiary circular reactions" and active invention (from twelve to eighteen months)	STAGE 6 Representation (from about eighteen months on)
Type of Coordinations	reflex	"reciprocal coordinations" environmental adaptation	"reciprocal coordination," incipient "serial coordination"	serial and "hierarchical coordination"	reciprocal trial-and-error coordination	reciprocal trial-and-error coordination
Locus of Coordinations and Applications	head and mouth with mother's breast	mouth with tongue, phonation with hearing, hands with mouth, hand with hand, hands with eyes	body and prehensive movements with object movements (face with face movements of another person)	prehensive "schema" with one object or a series of objects	prehensive "schema" with object-object and object-force relations	mini-motor and image representation of object-object and object-force relations
Volition	reflex, involuntary	reflex and primitive voluntary	primitive voluntary	voluntary	voluntary	voluntary
Differentiation	undifferentiated	primitively differentiated	differentiated	differentiated	consciously differentiated	consciously differentiated
Number of Included Acts in Behavior Pattern	one	one or two	one	two or more	several	several

Goal Determination	reflex	"fortuitous"	"fortuitous"	incipient initial	initial	mental image
Contexts	reflex	old	new	new	new	new
Temporal Patterning of Behavior Patterns and Coordinations	repetition of same pattern and single occurrence	repetition of same pattern	delayed and immediate repetition of same or functionally equivalent patterns	serial application of same pattern, serial application of different patterns	variable repetition	variable repetition
Releasing and/or Reinforcing Aspects	consummatory "assimilation"	exercise and "reciprocal assimilation and accommodation"	exercise and perceived "contingency" control	perceived "contingency" control and novelty	novelty and perceived variable contingency control	novelty and perceived variable contingency control
Sources of Stimuli Eliciting Behavior Patterns	internal state and allogenic	internal autogenic and exercise	external autogenic	environmental object and external autogenic	environmental and external autogenic	goal image
EXAMPLES	rooting and sucking on mother's breast and other objects	sucking on own thumb, and touching own hands together	kicking feet repeatedly in order to make bassinette fringe move	shaking strings of bassinette in order to disentangle toy	repeated dropping of objects from different heights into water, watching results	insight problem-solving, opening and closing mouth and then opening box without trial-and-error groping

TABLE 4 / HUMAN INFANT OBJECT-CONCEPT BEHAVIORAL PROGRAMS: PIAGET'S OBJECT-CONCEPT SERIES IN TERMS OF META-BEHAVIORAL PARAMETERS

META-BEHAVIORAL PARAMETERS	STAGES 1 AND 2 The absence of special behavior relating to vanished objects (birth to three to six months)	STAGE 3 Beginning of object permanence extending accommodation (from three to six to nine months)	STAGE 4 Active search for vanished object (from eight to ten months on)	STAGE 5 Intermediate object concept (from twelve to eighteen months)	STAGE 6 Object permanence representation of invisible displacements (from about eighteen to twenty-four months)
Type of Coordinations	reflex and environmental adaptations and "reciprocal coordinations"	environmental adaptation and "reciprocal coordination"	"serial" and "hierarchical coordination" and reversals	trial-and-error coordination and reversal	trial-and-error coordination and reversal
Locus of Coordinations and Applications	beginning coordination of vision with object movement vision with prehension	hands with object position, vision with object movement, hand and eye to object and touch	hands and eyes with object position relative to own actions	hands and eyes with object position relative to own position	hands and eyes with image of object position relative to own position
Volition	reflex and primitive voluntary	primitive voluntary	voluntary	voluntary	voluntary

Differentiation	undifferentiated	primitively differentiated	differentiated	differentiated	differentiated
Number of Included Acts in Behavior Pattern	one	one	two	several	several
Goal Determination	reflex and "fortuitous"	"fortuitous"	"fortuitous"	initial	initial
Contexts	reflex and old	new	new	new	new
Temporal Patterning of Behavior Patterns and Coordinations	none	single occurrence and repetition and delayed repetition	serial application of same "schema"	serial application of equivalent "schemata"	serial application of equivalent "schemata"
Releasing and/or Reinforcing Aspects	assimilation and accommodation	"assimilation and accommodation"	"assimilation" and novelty and "accommodation"	"assimilation" and novelty and "accommodation"	"assimilation" and novelty and "accommodation"
Sources of Stimuli Eliciting Behavior Patterns	environmental and allogenic	environmental object and allogenic	environmental event and memory image	environmental event and memory image	environmental event and transformed image

continued

TABLE 4 / CONTINUED

META-BEHAVIORAL PARAMETERS	STAGES 1 AND 2 *The absence of special behavior relating to vanished objects*	STAGE 3 *Beginning of object permanence extending accommodation*	STAGE 4 *Active search for vanished object*	STAGE 5 *Intermediate object concept*	STAGE 6 *Object permanence representation of invisible displacements*
	(birth to three to six months)	(from three to six to nine months)	(from eight to ten months on)	(from twelve to eighteen months)	(from about eighteen to twenty-four months)
EXAMPLES	visual following slow-moving objects looking away when they disappear from view	"interrupted prehension": reaching without tactile or visual searching for object that has escaped hands; visual following of fast-moving object, continues to look at point of disappearance from view or will look back to point of origin	search under a screen for an object hidden there while he watches—he will continue to search under the first screen when he sees it hidden under second screen	correct search for an object which has been sequentially moved under several screens while he watches	correct search for an object hidden under one or more cloths while he could not see

TABLE 5 / HUMAN INFANT IMITATION BEHAVIORAL PROGRAMS: PIAGET'S IMITATION SERIES IN TERMS OF META-BEHAVIORAL PARAMETERS

META-BEHAVIORAL PARAMETERS	STAGE 1 Contagious crying	STAGE 2 Sporadic "self-imitation"	STAGE 3 Purposeful "self-imitation"	STAGE 4 Imitation of model displaying own "schema"	STAGE 5 Beginning imitation of novel and invisible "schema"	STAGE 6 Deferred imitation and object imitation
	(birth to one month)	(from one to three months)	(from four to eight months)	(from eight or nine months on)	(from twelve to eighteen months)	(from twelve to twenty-four months)
Type of Coordinations	reflexive	"reciprocal coordinations"	serial and "reciprocal coordination"	"hierarchical coordination"	reciprocal trial-and-error coordination	reciprocal trial-and-error coordination
Locus of Coordinations and Applications	vocal and auditory	vocal and auditory	vocal with model's vocal and prehensive with model's prehensive (from own schema)	facial, vocal and gestural and prehensive acts with model's vocal and gestural and prehensive acts	facial, vocal and gestural and prehensive acts with model's vocal and gestural and prehensive acts	analogic "schemata" such as body movement, mouth movement, hand movements to object movements and relations
Volition	involuntary	primitive voluntary	primitive voluntary	voluntary	voluntary	voluntary
Differentiation	undifferentiated	undifferentiated	primitively differentiated	differentiated	consciously differentiated	consciously differentiated
Number of Included Acts in Behavior Pattern	one	one	one	two or more	several	several

continued

TABLE 5 / CONTINUED

META-BEHAVIORAL PARAMETERS	STAGE 1 Contagious crying (birth to one month)	STAGE 2 Sporadic "self-imitation" (from one to three months)	STAGE 3 Purposeful "self-imitation" (from four to eight months)	STAGE 4 Imitation of model displaying own "schema" (from eight or nine months on)	STAGE 5 Beginning imitation of novel and invisible "schema" (from twelve to eighteen months)	STAGE 6 Deferred imitation and object imitation (from twelve to twenty-four months)
Goal Determination	reflex	"fortuitous"	"fortuitous" imitative	initial imitative	initial imitative	mental-image imitative
Contexts	reflex	old	new	new	new	new
Temporal Patterning of Behavior Patterns and Coordinations	repetition of same	repetition of same	repetition of same	repetition (of model's familiar acts)	variable repetition (of model's acts)	delayed repetition (of model's acts)
Releasing and/or Reinforcing Aspects	consummatory "assimilation"	"reciprocal assimilation and accommodation"	perceived "contingency" control	perceived reversed "contingency" control and novelty	reverse variable "contingency" control	reverse variable "contingency" control
Sources of Stimuli Eliciting Behavior Patterns	internal state internal autogenic	internal autogenic	allogenic autogenic	allogenic	allogenic	environmental and allogenic memory images

| EXAMPLES | contagious crying | repeats own sounds, no gestural imitation, visual accommodation | imitates own "schema" when they are performed by another person | imitates movements of face when performed by model, differentiates global vocal and facial "schemata" through imitation of model | imitates models when they are not present, imitates object actions and actions without groping | trial and error groping to imitate novel "schemata" of model |

TABLE 6 / STUMPTAIL INFANT SENSORIMOTOR INTELLIGENCE BEHAVIORAL PROGRAMS: STUMPTAIL SENSORIMOTOR INTELLIGENCE SERIES IN TERMS OF META-BEHAVIORAL PARAMETERS

META-BEHAVIORAL PARAMETERS	STAGE 1 Reflex "primary circular reactions," etc. (birth to two weeks)	STAGE 2 Primitive voluntary adaptations to environment (two to twelve weeks)	STAGE 3 The secondary linear schema (four to eleven weeks)	STAGE 4 Coordination of secondary linear schema (from nine weeks)	STAGE 5 Trial-and-error secondary linear reaction (from twenty-three weeks on)	STAGE 6 (Secondary linear) Representation (from twenty-three weeks on)
Type of Coordinations	reflex coordinations	environmental adaptations & coordinations	environmental adaptations & coordinations	"serial" and "hierarchical coordinations"	"trial-and-error coordinations"	"trial-and-error coordinations"
Locus of Coordinations & Applications	eyes to objects, mouth to nipple, prehension to fur, postural to upright and nipple, hand to mouth	mouth to nipple, prehension to fur, and objects, eyes to object, hands to own body, hands to feet, hand to mouth	eyes and hands to objects	hands to objects	hands to objects	hands to objects
Volition	reflex	primitive voluntary	voluntary	voluntary	voluntary	voluntary
Differentiation	undifferentiated	primitively differentiated	differentiated	differentiated	differentiated	differentiated
Number of Included Acts in Behavior Pattern	one	one	one or two	two or more	two or more	two or more

Goal Determination	reflex	reflex and fortuitous	fortuitous	initial	initial	initial
Contexts	reflexive	old	new	new	new	new
Temporal Patterning of Behavior Patterns and Coordinations	single occurrence and repetition of same patterns	single occurrences and repetition of same	repetition of functionally equivalent pattern	serial and alternate application of different patterns	variable repetition	variable repetition
Releasing and/or Reinforcing Aspects	consummatory "assimilation"	"assimilation and accommodation"	"assimilation and accommodation"	same plus noncontingent novelty	same plus noncontingent novelty	same plus noncontingent novelty
Sources of Stimuli Eliciting Behavior Patterns	internal state, allogenic	allogenic and proprioception and environmental	environmental	environmental	environmental	environmental
EXAMPLES	repeated rooting, grasping, mouthing, thumb-sucking, clinging	purposeful nipple location and nursing; alternate ipsilateral shifting, bimanual grasping objects and lifting them, biting them	picking up small objects with pincer grasp, pushing and pulling on objects on ground	bimanual pulling objects apart, hand-to-hand transferring of an object	searching under two screens for object hidden while he watches (see Object-Concept Series)	uncovering object hidden under screen while he could not see

TABLE 7 / SENSORIMOTOR BODY SCHEMATA STAGES

META-BEHAVIORAL PARAMETERS	STAGE 3 *Differentiation of body schemata* (from four weeks on)	STAGE 4 *Coordination of body schemata* (from nine weeks on)	STAGE 5 *Trial-and-error body schemata* (from thirteen weeks on)	STAGE 6 *Representation of body schemata* (fourteen to eighteen weeks on)
Type of Coordinations	environmental adaptation and inter-coordination	serial and "hierarchical coordinations"	"trial-and-error coordinations"	"trial-and-error coordinations"
Locus of Coordinations and Applications	hands and feet and arms and legs and head with torso and head with substrates and with each other	locomotor and suspensory "schemata" with each other	locomotor and suspensory and prehensive "schema" with actions of other animals	"visual templating" or tracing of alternative pathways before acting and strategic self-positioning
Volition	voluntary	voluntary	voluntary	voluntary
Differentiation	differentiated	differentiated	differentiated	differentiated
Number of Included Acts in Behavior Pattern	two or more	several	several	several
Goal Determination	fortuitous	initial	initial	initial
Contexts	new	old	new	old

Temporal Patterning of Behavior Patterns and Coordinations	repetition of same or functionally equivalent	serial and alternate application of different patterns	variable repetition	variable repetition
Releasing and/or Reinforcing Aspects	exercise and novelty	exercise and novelty	exercise and novelty	exercise and novelty
Sources of Stimuli Eliciting Behavior Patterns	environmental and proprioceptive	environmental and proprioceptive	environmental and proprioceptive	environmental and proprioceptive
EXAMPLES	climbing down fence, walking, jumping	"penduluming" down off fence	"keepaway" on chain	planning movements and positions

TABLE 8 / STUMPTAIL INFANT OBJECT-CONCEPT BEHAVIORAL PROGRAMS: STUMPTAIL OBJECT-CONCEPT SERIES IN TERMS OF META-BEHAVIORAL PARAMETERS

META-BEHAVIORAL PARAMETERS	STAGES 1 AND 2 Absence of special behavior relating to objects	STAGE 3 Beginning of object permanence extending movements of accommodation	STAGE 4 The practical object-search for vanished object	STAGE 5 Intermediate object permanence	STAGE 6 Representation of object displacement
	(birth to five weeks)	(from three to six weeks)	(from three to sixteen weeks)	(from twenty-third week on)	(from twenty-third week on)
Type of Coordinations	environmental adaptations	environmental adaptations	"serial" and "hierarchical coordination"	"trial-and-error coordination"	"trial-and-error coordination"
Locus of Coordinations and Applications	vision with objects	vision with objects, prehension with objects	hands and eyes with object position relative to own actions	hands and eyes with object position relative to own position	hands and eyes with image of object position relative to own position
Volition	reflex and primitive voluntary	voluntary	voluntary	voluntary	voluntary
Differentiation	primitively differentiated	primitively differentiated	differentiated	differentiated	differentiated
Number of Included Acts in Behavior Pattern	one and two acts	one or two	two	two or more	two or more

Goal Determination	reflex and fortuitous	"fortuitous"	"fortuitous"	initial	initial
Contexts	reflex and old	new	new	new	new
Temporal Patterning of Behavior Patterns and Coordinations	single occurrence and repetition of same	single occurrence or delayed repetition (single)	serial application	serial application	serial application
Releasing and/or Reinforcing Aspects	simple assimilation and accommodations	"assimilation and accommodation"	novelty and "assimilation and accommodation"	novelty and "assimilation and accommodation"	novelty and "assimilation and accommodation"
Sources of Stimuli Eliciting Behavior Patterns	environmental events	environmental events	environmental events and memory image	environmental events and memory image	environmental events and transformed image
EXAMPLES	alternate glancing between two objects, continuing to look at point of disappearance of moving object	visual "accommodation" to rapidly moving object—glance returning to point of appearance	search under screen for object hidden while he was watching	correct search under two cloths after object was hidden while he was watching	correct search under one cloth for object hidden while he could not see

TABLE 9 / STUMPTAIL INFANT IMITATION BEHAVIORAL PROGRAMS: STUMPTAIL IMITATION
SERIES IN TERMS OF META-BEHAVIORAL PARAMETERS

META-BEHAVIORAL PARAMETERS	STAGES 1 AND 2 No imitative behavior	STAGE 3 Purposeful imitation of "schemata" from own repertoire
	(birth to four weeks)	(from week four on)
Type of Coordinations		environmental adaptation
Locus of Coordinations and Applications		whole body to objects, self to substrates, object to self
Volition		voluntary
Differentiation		differentiated
Number of Included Acts in Behavior Pattern		one or two
Goal Determination		imitative goal
Contexts		new
Temporal Patterning of Behavior Patterns and Coordinations		repetition of model's act
Releasing and/or Reinforcing Aspects		"assimilation and accommodation" and exercise
Sources of Stimuli Eliciting Behavior Patterns		allogenic
EXAMPLES		sitting on mother's head after seeing another infant sitting on her mother's head; somersaulting after seeing another infant somersault

(during the second stage of the sensorimotor series) and was successfully clinging to his mother as she climbed during the first week, human infants usually begin crawling and walking at eight or nine months of age (during the fourth or fifth stage of the sensorimotor series). The net effect of this difference is that the stumptail infant began his visual and prehensive coordinations and explorations while he was clinging and walking and climbing, whereas human infants have virtually nothing to do with their hands and eyes for eight or nine months except coordinate them and explore with them.

In human evolution the priorities have been reversed, so to speak: visual and prehensive (and facial and vocal) development and coordination occur first and undergo the greatest elaboration, whereas locomotor behavior patterns develop later and undergo relatively little elaboration. In the stumptail infant, prehensive development underwent relatively little elaboration and was always competing with foot coordination and clinging and locomotor development which underwent early and extensive elaboration: each basic *body schema* such as walking and running and climbing and jumping underwent differentiation, serial and hierarchical and trial-and-error coordination. It is interesting to note that the stumptail displayed a *circular body reaction* involving repetition and trial-and-error coordination of his own movements which were analogous to the *tertiary circular reactions.*

These differences are all related to an obvious difference between monkeys and human beings: human beings are specialized bimanual manipulators and bipedal walkers whereas monkeys are quadrumanual manipulators and quadrupedal walkers. The quadrumanual aspect of monkey anatomy requires a long period of developing coordinations between the forehands and the hindhands: competing activities and miscoordinations of the two sets of hands were common in the infant stumptail and persisted even into the twenty-third week when the coordinated foot storage behavior pattern developed. Human infants, by contrast, have only to coordinate their two hands.

The stumptail infant also displayed a set of postural adaptations related to clinging which are absent in human infants, including righting, quadrumanual clinging and upright ipsilateral clinging and alternate ipsilateral clinging leaving one side free; unlike human infants he spent virtually no time lying supine when he was a neonate. He also, of course, displayed a set of characteristic monkey facial expressions and behavior patterns such as presenting his rump to his mother and other animals.

On a very general level we can say that infants of both species displayed many unstereotyped behavioral sequences involving free choice and coordination of motor patterns in service of various freely chosen goals. Moreover, some of their motor patterns, for example, prehensive patterns such as wholehand and pincer grasping, were similar, though many more, such as locomotor patterns, were not. The stumptail infant, however, was vastly less intelligent than human infants.

Intelligence is a peculiar class of unstereotyped behavior involving the recognition or definition of a problem, and the purposeful search for new procedures to solve that problem. Investigating and manipulating causality is the essence of intelligence. According to Piaget, truly intelligent behavior is characterized by directed groping for a solution to a problem; if we accept this definition, it is clear that very little of the stumptail infant's behavior falls into that category. In fact, only his uncovering of sequentially displaced objects and invisibly displaced objects falls into that category. I was unable to test him for trial-and-error searching behind a series of cloths for an invisibly displaced object before his death (but according to Harlow [1968], rhesus monkeys are capable of this kind of search by four years of age). Overall, I would characterize the stumptail infant's intelligence as *object-concept* and *secondary linear sensorimotor intelligence* (with representation of body and object position and "self imitation" of *body schemata* and representation of *secondary trial-and-error schemata*) as compared to the *"tertiary circular" sensorimotor intelligence* (and self representation and insight and imitative intelligence) of human infants at a comparable stage. The major difference between the species is the human focus on object-object and object-force field coordinations, i.e., the investigation and manipulation of causality.

It is interesting to speculate on the adaptive significance of these two species' patterns. The stumptail *behavioral program* produces a great diversity of unstereotyped locomotor and prehensive behavior patterns which must be very useful in traveling, foraging, and testing and preparing foods in the wild. These *behavioral programs* and representation of body and object position and self imitation are probably equally significant in social learning and social role playing.

It is important to note that stumptail macaque *behavioral programs* do not produce true tool use. By true tool use I mean directed trial-and-error or insight manipulation of one object by a secondary object.

Some investigators have used the term tool use to describe simpler manipulations of objects, for example, rubbing food on leaves (Chaing 1967), and potato-washing in the sea (Kawai 1965) by macaques, and face-cleaning with leaves by baboons (van Lawick-Goodall et al. 1973), which do not fit the usual definition of tool use.

There are, however, a few reports of individual baboons displaying true tool use in the sense of object-object manipulation (using a stick to rake in food [Bolwig reported in van Lawick-Goodall et al. 1973]). One case even involved cooperative tool use where a female hamadryas baboon brought a stick to her male (Beck 1973).

Tool use should not be confused with other types of object manipulation and searching abilities such as pulling on strings with objects attached, uncovering hidden objects, and trial-and-error manipulation of single objects. These object manipulations have been well documented (Kluver 1933; MacDougal 1929; Harlow 1968; Vaughter Smotherman and Ordy 1972; Wise et al. 1974). The fact that all these activities involve object permanence and *secondary linear trial-and-error reactions*, but not "secondary and tertiary circular reactions," is consistent with the hypothesis that with few exceptions the stumptail pattern is common to old world monkeys.[8]

Hall (1968) argues that tool use per se is not a significant measure of behavioral adaptability and intelligence since it must be evaluated in terms of its contexts and functions; for example, in many species of birds and mammals tool use involves the application of simple stereotyped motor patterns shaped by operant conditioning.

It is important to distinguish stereotyped tool use from unstereotyped tool use: unstereotyped tool use involves the application of the method of the "tertiary circular reaction" (i.e., a sensorimotor understanding of object-object and object-field interactions) and is, therefore, a significant measure of intelligence from a Piagetian perspective.

We must not conclude, however, that old world monkeys are unintelligent because they fail to display the "tertiary circular reaction." We must distinguish degrees of intelligence: old world monkeys are less intelligent than cebus monkeys, apes, and people, but they are more intelligent than lemurs and rats. They display unstereotyped *secondary linear* and object-concept intelligence.

The intelligence of old world monkeys is most clearly indicated by their long life span and their long period of socialization. Females are socially mature at about four years of age, and males are socially mature at about seven years of age. Despite the fact that young monkeys display many of their adult behavior patterns within the first six months of life, it takes them years to learn the proper contexts and combinations in which to deploy their behavioral repertoire (Bertrand 1971; Chevalier-Skolnikoff 1971).

During their relatively long period of socialization, monkeys learn many things about the biotic and abiotic elements of their home range and about their own changing status and capacities relative to other animals in their social group: they learn details of dispersion and seasonal variation of foods, water, sleeping sites, predators, and prey; they learn what foods to eat, techniques of food capture and preparation (Hall and DeVore 1965; Malmi 1973; Ransome 1971; Kawai 1965), and progression routes (Kummer and Kurt, 1968). Object manipulation in food preparation and capture in baboons, for example, includes digging for roots, lifting rocks and other objects to find insects (DeVore and Hall 1968), removing prickles from cactus, dropping oil nuts to break shells (Ransome 1971), and searching for and tracking small mammals in grass (Malmi, personal communication). All of these activities, except dropping nuts, require object permanence. Dropping nuts to break them is probably an example of operant conditioning of accidental dropping, but it could be due to trial-and-error or insight learning which would imply a limited "tertiary circular reaction." They learn the meaning of a complex set of social signals—postures, gestures and facial expressions—which allow them to predict more or less accurately the actions of other animals. They learn all these things through observation, "self imitation," and trial-and-error applications of *body schemata* and *secondary linear schemata*, especially during play (Dolhinow and Bishop 1970).

Social and sexual interactions often involve choosing an advantageous position in space relative to a dominant animal as, for example, in "tripartite" relations such as protected threat and enlisted threat where one animal threatens another animal in the troop when he is able to place himself strategically close to a dominant animal and induce that animal to back him up (Kummer 1971). Interactions of this sort apparently rely on the capacity for representation of body position relative to the position of other animals and objects (see stages five and six of the object-concept series).

Social success and dominance are probably based on skill in social learning (i.e., intelligence) as well as size, experience, mother's status, and temperament. In many species, survival and reproductive success and the probable survivorship of offspring are correlated with social status. Therefore, social selection (Crook 1972) must have favored some degree of intelligence in old world monkeys. The selective advantage of the capacity to discover new food resources and thereby increase the breadth of the dietary niche and possibly expand the species ranges would favor *secondary trial-and-error intelligence* in old world monkeys.

Very little is known about the selective advantages of human intelligence or its evolutionary history. When discussing the evolution of intelligence, anthropologists have focused primarily on the selective advantages of tool use and language. It seems likely that intelligence is equally important in human social life, especially in the acquisition of sex roles, moral codes (Kohlberg 1966), and life plans. Anthropologists have paid very little attention to the adaptive significance of infant and childhood intelligence. Comparative studies of child development in other cultures and in other primate species using a Piagetian framework, as well as studies in child language acquisition, should focus attention on this area. Because the early stages of intellectual development are comparable in monkeys, apes, and people, different species' patterns can suggest some hypotheses about the adaptive significance of certain aspects of the human patterns. For example, analysis of the contexts in which the "secondary circular reactions" and the "tertiary circular reactions" occur in human infants might shed some light on these classes of behavior patterns. The "secondary circular reactions" occur in both social and nonsocial contexts: it is the basis of "the game" interaction between infants and caretakers, and it is the basis of infant play with objects such as rattles and mobiles they can control. In "the game" the "secondary circular reactions" create and reinforce an affectional bond between infant and parents (Watson 1972). It may have been selected for this purpose with the progressive loss of infantile clinging and the evolution of bipedal locomotion.

The "tertiary circular reaction" (and representation) also occur in both social and nonsocial contexts. In social contexts it is the basis for understanding new events and for imitating new behavior patterns including words and intonations not in the infant's repertoire; it therefore has an important role in the early stages of language acquisition. It also forms the basis of a self-image and identification of the self as a boy or girl, and it is helpful in manipulating parents and siblings. In other words, it forms the early basis for learning language and social roles. Because children's early cognitive and imitative achievements are exciting and pleasing to parent and child they create and reinforce affectional bonds between parents and children.

The "primary circular reactions" and the "secondary circular reactions" and the "tertiary circular reactions" also operate to facilitate tool use: they form a series of stages beginning with "reciprocal coordination" of hands and eyes and ending with "reciprocal coordinations" of objects with objects. The behavioral programs producing these behaviors certainly must have been selected to facilitate tool use.

Moreover, a slow locomotor maturation may have been selectively advantageous in hominid infants because it allowed the prolonged development of hand-hand and hand-object manipulations. Bimanual prehensive and bipedal locomotor specialization certainly must have facilitated the evolution of these patterns.

Human intelligence only begins to develop during the sensorimotor period. This period is followed by five more periods of development: the symbolic and preconceptual period from two to four, the preoperational or intuitive operational period from four to seven, the concrete operations period from seven to eleven, and the formal operations period from eleven or twelve on (Piaget 1966). Human intelligence involves investigating, manipulating, and communicating about causality.

From a comparative perspective it is clear that the length of development and the number of discrete developmental periods and stages correlate positively with the degree of intelligence. It will be interesting to see how the great apes compare to the stumptail and human infants in these terms.

■ EVOLUTIONARY IMPLICATIONS OF COMPARATIVE STUDIES OF BEHAVIORAL ORGANIZATION AND INTELLIGENCE

We have just analyzed and compared the organization and development of unstereotyped behavior and intelligence in infants of two primate species in terms of *meta-behavioral parameters*. Using this framework we can understand differences between the two species in terms of the unique combinations of *meta-behavioral parameters* in their *behavioral program*, recalling that each *behavioral program* produces many related behavior patterns sharing common organizational features, such as the "secondary and tertiary circular reactions" in human infants (see Tables 3–9).

Comparative studies of living primate species can only reveal behavioral organization in terminal lineages and cannot tell us how our ancestors behaved. When most or all of the living species in a related group (i.e., a higher taxonomic category such as Cercopithecoidea or Hominoidea) share some of the same *behavioral programs*, however, we can conclude that these patterns were probably present in the common ancestral species giving rise to the group. (This is one reason that behavioral studies of our closest living relative, the great ape, are so important.)

It will be very interesting to see the taxonomic distribution of *behavioral programs* in various primate species. My reading and observation of some monkey species (stumptail macaques, Indian langurs, talapoins, and an Aotus monkey) lead me to believe that the stumptail infant pattern is common to all old world monkeys, but not to all new world monkeys and prosimians. Alison Jolly's (1964) investigations of lemur intelligence revealed that lemurs display little interest in object manipulation beyond determining the edibility of objects. My observations of an adult female Aotus (owl monkey) revealed the same lack of interest in objects and lack of searching for invisibly displaced objects. Cebus monkeys, on the other hand, have been seen to use tools in captivity, including cracking things open with a bone hammer (Ververs 1962), and raking objects in with a stick and positioning and climbing on a box to reach object (Klüver 1933), indicating the presence of the "tertiary circular reaction" in this genus. Vaughter et al. (1972) observed development of object permanence in captive squirrel monkeys by nine to twelve months of age. Greater diversity of behavioral organization among new world monkeys is not surprising since taxa in this group are more distantly related than are taxa of old world monkeys.

Studies of chimpanzee behavior, especially Köhler's (1927) work on insight problem-solving, and Gardner and Gardner (1969) and Premack's (1971) work on "language" learning suggests that chimpanzees display all of the stages in the sensorimotor intelligence and imitation (gestural but not vocal) series (probably with some interesting differences in context and frequency). It is also likely that they display some later stages of intellectual development characteristic of the symbolic and preconceptual period and perhaps even some patterns from the intuitive period.

Mosaic evolution must have occurred in the evolution of *behavioral programs* as it did in other behavioral-morphological complexes in primate evolution (Le Gros Clark

1964). It seems very likely, for example, that the similarity between the stumptail and human infants in the object-concept series as opposed to their differences in the sensorimotor intelligence and imitation series, is the result of mosaic evolution. The *behavioral programs* in the object-concept series may have been present in the common ancestral species giving rise to old world monkeys, apes, and people (or it may have evolved independently in the two lineages). The *behavioral programs* producing the "secondary and tertiary circular reactions" must have evolved much later and only in the hominoid lineages.

Even if it is possible to compare the intelligence of different species, is it meaningful? What sort of statements are we making about species' differences when we compare intelligence? In particular, are we measuring adaptive success? Washburn and Harding (1970) argue that psychological tests comparing the intelligence of different species, especially distantly related species such as rats, pigeons, and primates, are meaningless because of the radically different, but highly successful adaptations of different species. They believe that it is arbitrary to equate intelligence with success at learning sets based on visual cues.

Their discussion raises several important questions germane to this report: How do we define intelligence? Are our definitions necessarily arbitrary and anthropocentric? Is intelligence necessarily tied to visual and prehensive adaptations? Can intelligence be equated with adaptive success?

Piaget defines voluntary coordinations of action to achieve a goal as rudimentary acts of intelligence, and directed trial-and-error groping to solve a problem, as true acts of intelligence. Intelligence generally involves manipulating the self relative to objects and/or objects relative to other objects or forces. It seems clear from Piaget's work that investigating, manipulating, and communicating causality is the essence of intelligence.

Although investigating causality depends upon vision and prehension in human infants and children, and communicating causality also depends upon vocalizations and audition (language), in other species other sensorimotor modalities might allow analogous activities.

Among dolphins, for example, sound signaling and hearing are the major instruments of information gathering and communication, whereas the mouth and body muscles manipulate the self relative to objects and objects relative to other objects and forces. Certainly it is easier to compare species with perceptual abilities and motor patterns similar to our own. Studies of dolphin behavioral organization and intelligence, for example, require special equipment to compensate for human inabilities, but these difficulties do not render comparisons meaningless if we have some framework for comparing the differences.

Sonar, like vision and prehension, is apparently an instrument of intelligent problem-solving: it allows dolphins to investigate object location, distance, size, composition, shape, and movement (Kellogg 1961; Norris 1974). Dolphins can imitate vocalizations (Lilly 1967). They apparently display something like the "secondary and tertiary circular reactions," for example, repeatedly throwing an object and retrieving it, opening gates, and removing a hypodermic needle from another animal using their

FIGURE 2. *Infant climbing through hole in box—an apparent case of imitation of the older infant's body schema*

FIGURE 1. *Infant playing on swing rope—a common site for circular body schemata*

FIGURE 3. *Infant sitting by blanket in typical testing situation for object concept series*

FIGURE 4. *Infant watching mother lift blanket used in testing for object concept series*

(Photographs by Shirley Strum)

mouth (Norris 1974). Whether or not dolphins display true "secondary and tertiary circular reactions," it seems likely that dolphin behavioral organization and intelligence could be classified in terms of *meta-behavioral parameters* derived from Piaget despite the extremely different sensorimotor apparatus of that taxon. In other words, although vision and prehension are instruments of human intelligence, other sensorimotor modalities might act as instruments of intelligence in other species. Scent-marking and olfaction, however, are probably ill-suited for investigation of causality, because they do not convey information about object size, shape, movements, position, and interactions. Perhaps the evolution of higher intelligence always depends upon sensorimotor preadaptations for investigation and manipulation of causality, such as vision and prehension as in primates, sonar as in dolphins, or vision and prehensive trunks as in elephants.

Piaget's model and the *meta-behavioral parameters* derived from it provide a systematic and objective framework for comparing behavioral organization and intelligence in different species of primates, and possibly other mammals, without imposing a rigid set of anthropocentric criteria. This framework allows us to compare relative degrees and modalities of investigating and manipulating causality. Of course, species' patterns emerging from such comparative studies can only be understood in terms of the adaptive strategies of the taxa in question. In other words, each species pattern must be analyzed in terms of the naturalistic behavior of populations in the wild where the contexts and conditions which may have favored the evolution of that pattern may be apparent.

Finally, we must ask whether intelligence can be equated with adaptive success. Obviously it cannot; intelligence and adaptive success are two different phenomena reckoned by different criteria. Intelligence and adaptive success are often confused, however, because intelligence is an important human adaptation which is largely responsible for our adaptive success. Intelligence, however, plays no role in the adaptive success of many other taxa such as worms, insects, fish and reptiles.

Adaptive success is generally reckoned in terms of the number of individuals and/or species within a genus or higher taxonomic category, the extent of the geographic range of that group, its duration in geologic time, its degree of lineal transformation and adaptive radiation, and its replacement of competing groups.

Our lineage (Hominidae) is successful in terms of population density, geographic range, lineal transformation and replacement of competing forms; we are not successful in terms of our duration in geologic time. Our adaptive success is probably due largely to our intelligence. Many other species are highly successful on the basis of very different adaptive strategies.

■ ACKNOWLEDGMENTS

The research on which this paper is based was undertaken during 1970–71 at the Primate Research Station, University of California, Berkeley, under the supervision of Dr. Phyllis Jay Dolhinow in partial fulfillment of the requirements for a Ph.D. in anthropology. The U. C. Berkeley Primate Research Station was maintained by the U. S. Public Health Service Grant Number MH-08623 to Dr. Sherwood L. Washburn. My research was supported by Predoctoral U. S. Public Health Service Fellowship No. 1 Fol MH50116-01 from August 6, 1970, through August 5, 1971, and by small grants from the Department of Anthropology and Dr. S. L. Washburn. The help and support of my professors, Drs. Phyllis Dolhinow, John S. Watson, and S. L. Washburn, and the critical reading of this manuscript by Drs. Phyllis Dolhinow, Suzanne Chevalier-Skolnikoff, Frank Poirier, John S. Watson, David Gordon, and my husband, Peter G. Parker, is gratefully acknowledged, as is the manuscript preparation by Jeanne Ganas. I am also grateful to Dr. Thelma Rowell for giving me access to her talapoin colony during the summer of 1972.

■ NOTES

1. The "group" is a mathematical concept Piaget uses to describe certain sensorimotor and mental operations; on the sensorimotor plane it refers to "relations among things and relativity between one's movements and those of the object" (Piaget 1954:117). In particular, "there are groups in the sense that the child's activity is capable of turning back on itself and thus of constituting closed totalities which mathematically define the group" (Piaget 1954:110).

2. Alison Jolly (1972, personal communication, 1973) suggested the value of Piaget's sensorimotor period series and stages for comparing primate intelligence. Investigators have studied object-permanence in squirrel monkeys (Vaughter et al. 1972) and rhesus monkeys (Wise et al. 1974) and in cats (Gruber et al. 1971).

3. Piaget (1966) studied the development of intelligence in children through five periods: the sensorimotor period from birth to two years, the symbolic and preconceptual period from two to four years, the preoperational or intuitive period from four to seven years, the concrete operations period from seven to twelve years, and the formal operations period from twelve years through adolescence.

4. All the terms of my invention are in italics.

5. Robert Hinde (1971) has questioned the validity of stage formulations in primate developmental studies. He argues that they do violence to the essential continuity of primate infant behavioral development. I think that the stage specific *behavioral programs* described in terms of *meta-behavioral parameters* in Tables 3 through 9 demonstrate that Piaget's stages in human infants and my stages in the stumptail infant reflect real changes in behavioral organization. Stage formulations are attractive because they offer a nice solution to the problem of data presentations; they provide more information about behavioral organization than even a presentation of a full-behavioral repertoire would because they reveal parameters which are commonly ignored.

6. Piaget did describe "secondary schema" and "derived secondary reactions" which do not involve "secondary circular schema" (see stage four of the sensorimotor series in human infants), but he did not emphasize this category of behavior, nor did he emphasize noncircular "secondary schema" that do not pertain to interactions with objects or imitation.

7. Rhesus monkey infants tested for object-concept development by Wise et al. (1974) completed the sixth stage during their third month of life.

8. Caged talapoin monkeys in a small colony in Berkeley displayed an interesting behavior pattern verging on a "tertiary circular reaction": they repeatedly slid a metal jar lid open end in along the wall and sometimes "hung" it on a nail when it rested there (Parker, personal observation).

■ REFERENCES

Beck, B. B. 1973. Cooperative tool use by captive Hamadryas Baboons. *Science* 182:594–597.

Bertrand, M. 1969. The behavioral repertoire of the stumptail macaque. *Biblioteca Primatologica*, 11. Basel: Karger.

Bower, T. G. C. 1971. The object in the world of the infant. *Scientific American* 225:30–38

Chevalier-Skolnikoff, S. 1971. *The ontogeny of communication in Macaca speciosa.* Ph.D. dissertation, University of California, Berkeley.

Chevalier-Skolnikoff, S. 1974. The ontogeny of communication in the stumptail macaque (*Macaca arctoides*). *Contributions to Primatology*, 2. Basel: Karger.

Chiang, M. 1967. Use of tools by wild macaque monkeys in Singapore. *Nature* 214:1258–1259.

Le Gros Clark, W. 1964. *The fossil evidence for human evolution.* 2nd ed. Chicago: University of Chicago Press.

DeVore, I., and Hall, K. R. L. 1965. Baboon social behavior. In *Primate Behavior*, ed. I. DeVore, pp. 20–53. New York: Holt, Rinehart and Winston, Inc.

Dolhinow, P. J., and Bishop, N. 1970. The development of motor skills and social relationships among primates through play. In *Minnesota Symposia on Child Psychology* IV, ed. J. P. Hill, pp. 141–198. Minneapolis: University of Minnesota Press.

Gardner, R. A., and Gardner, B. T. 1969. Teaching sign-language to chimpanzees. *Science* 165:664.

Gruber, H. E.; Girgus, J. S.; and Banuazizi, A. 1971. The development of object permanence in the cat. *Developmental Psychology* 4:9–15.

Harlow, H. 1968. The development of learning in the rhesus monkey. In *Contemporary issues in developmental psychology*, ed. Endler, Boulter and Osser, pp. 226–243. New York: Holt, Rinehart and Winston, Inc.

Hinde, R. A. 1971. Development of social behavior. In *Behavior of nonhuman primates*, vol. 3, eds. A. Schrier and F. Stollnitz. pp. 1–68. New York: Academic Press.

Jolly, A. 1964. Prosimian's manipulation of simple object problems. *Animal Behavior* 12:560–570.

Jolly, A. 1972. *The evolution of primate behavior.* New York: The Macmillan Company.

Kawai, M. 1965. Newly-acquired pre-cultural behavior of the natural troop of Japanese Macaques on Koshima Isles. *Primates* 6:1–30.

Kellogg, W. 1961. *Porpoises and sonar.* Chicago: University of Chicago Press.

Klüver, H. 1933. *Behavior mechanisms in monkeys.* Chicago: University of Chicago Press.

Kohlberg, L. 1966. A cognitive-developmental analysis of children's sex-role concepts and attitudes. In *Development of sex differences*, ed. E. Maccoby, pp. 82–173. Stanford: Stanford University Press.

Köhler, W. 1927. *The mentality of apes.* New York: Vintage Books.

Kummer, H., and Kurt, F. 1968. *Social organization of Hamadryas Baboons.* Chicago: University of Chicago Press.

Lawick-Goodall, J. van. 1970. Tool using in primates and other vertebrates. In *Advances in the study of behavior*, vol. 3, ed. D. S. Lehrman, R. A. Hinde, and E. Shaw. New York: Academic Press.

Lawick-Goodall, J. van; Lawick, H. van; and Packer, C. 1973. Tool use in free living baboons in the Gombe National Park, Tanzania. *Nature* 241:212–213.

Lilly, J. 1967. *Mind of the dolphin.* Moonachie, N.J.: Pyramid Publishing.

McDougall, K. W. 1929. Insight and foresight in various animals—monkey, raccoon, rat, and wasp. *Journal of Comparative Psychology* 11:237–273.

Norris, K. S. 1974. *The porpoise watcher.* New York: W. W. Norton and Company, Inc.

Piaget, J. 1952. *The origins of intelligence in children.* New York: W. W. Norton and Company, Inc.

———. 1954. *The construction of reality in the child.* New York: Ballantine Books.

———. 1962. *Play, dreams, and imitation in childhood.* New York: W. W. Norton and Company, Inc.

———. 1966. *The psychology of intelligence.* Totowa: Littlefield, Adams, and Company.

Vaughter, R. M.; Smotherman, W.; and Ordy, J. M. 1972. Development of object permanence in the infant squirrel monkey: *Developmental Psychology* 7:34–38.

Washburn, S. L., and Harding, R. 1970. Evolution of primate behavior. In *The neurosciences: Second study program*, ed. F. S. Schmitt, pp. 39–47. New York: Rockefeller University Press.

Watson, J. S. 1966. The development of generalization of "contingency awareness" in early infancy: Some hypotheses. *Merrill-Palmer Quarterly of Behavior and Development* 12:2.

———. 1967. Memory and "contingency analysis" in infant learning. *Merrill-Palmer Quarterly of Behavior and Development* 13:1.

———. 1972. Smiling, cooing and "the game." *Merrill-Palmer Quarterly of Behavior and Development* 18:323–339.

Wise, K. L.; Wise, L. A.; and Zimmermann, R. R. 1974. Piagetian object permanence in the infant rhesus monkey. *Developmental Psychology* 10:429–437.

3	# Brain Structure and Intelligence in Macaques and Human Infants from a Piagetian Perspective

KATHLEEN R. GIBSON

■ INTRODUCTION

What is the relationship between brain structure and intelligence? Can intelligence be correlated with specific aspects of neural structure, or is the relationship which exists merely one of quantity: the larger the brain, the greater the intelligence? Attempts to answer this question usually meet with frustration comparable to that encountered in attempts to find neural correlates of other behavioral parameters such as learning, memory, or social behavior. This frustration results, in part, from failure to define behavioral parameters in terms amenable to neurological interpretation. Intelligence, per se, is not a function of the brain or any of its particular components. Rather, intelligence is the end result of a number of neurological capacities, some of which are mediated by particular neuroanatomical regions. The challenge to behavioralists and neurobiologists who seek neurological correlates of behavior is to define these capacities in terms that facilitate neuroanatomical correlation rather than hinder it.

According to Piaget (cited in Parker, this volume), "Intelligence is the purposeful invention of new procedures, or applications of old procedures, in a new context to solve a problem." The essence of intelligence is understanding and manipulating causality. Intelligence, thus defined, is basically an end product of the development of behavioral programs based on sets of metabehavioral parameters which Parker abstracts as age, locus of coordination, type of coordination, volition, differentiation, goal direction, temporal patterning of schema contexts, source of stimuli eliciting motor patterns, and releasing and/or reinforcing aspects of stimuli.

Examination of Parker's data indicates that certain basic neurological capacities form a necessary foundation for the development of many, if not all, of her metabehavioral parameters. These neurological capacities include the following:

1. The ability for fine differentiation of both motor actions and perceptual data.
2. The ability to coordinate two or more actions simultaneously and the corresponding ability to perceive simultaneous relationships among perceptual data.

3. The ability to coordinate two or more actions sequentially, and corresponding-ly, to perceive sequential relationships between sensory stimuli.
4. The ability to develop flexible and variable relationships between stimulus and response.
5. The ability to internalize perceptual and motor data.
6. The ability to anticipate, on the basis of this internalized data, actions and perceptions in the near or distant future.
7. The ability to compare the anticipated perceptions with the actual perceptions.

■ COMPONENTS OF INTELLIGENCE

1. DIFFERENTIATION

Differentiation, as used in this paper, implies the ability to break either a motor act or a sensory perception into its fine component parts. In the human neonate, this capacity is rudimentary, but gradually matures with age.

Progressive differentiation is the rule in the development of sensory perception (Bekoff, this volume; E. Gibson 1969; Kaplan, this volume). The ability to break perceptual data into its fine components lies at the basis of the ability to discriminate visual patterns, nonsense drawings, letters of the alphabet, phonemes, and other perceptual data. Gibson (1969) uses as an example the ability of a human to discriminate one goat in a herd from another. This discrimination is not made on the basis of generalization, but on the basis of noting very fine distinctions in patterns of coat color, facial form, etc. Similarly, in distinguishing one phoneme from another, one must note fine distinctions between the sounds produced by varying positions of the tongue, lips, and glottis. This ability to differentiate sensory stimuli is not intelligence, per se, but as an animal can only solve problems with respect to environmental stimuli, it can perceive (hence differentiate); the contexts in which intelligence is exhibited are limited in any given case by the animal's capacities of perceptual differentiation.

In the motor system, the progressive differentiation which occurs with age is illustrated by changing patterns of muscular use. Both prenatal reflexes and postnatal motor acts are initially global in nature, utilizing many muscles and body parts simultaneously. With maturation the capacity for fine individuated movements in-volving a few muscles and small areas of the body develops.

The prenatal precedence of global responses over individuated responses was first demonstrated by Coghill (1929) for the salamander, *Amblystoma*. The salamander's earliest reflexes are total pattern movements involving the neuromuscular apparatus of all body parts mature enough to function. As the earliest portion of the neuromuscu-lar apparatus to function is that in the region of the neck, the first reflexes include neck movements only. As soon as the neuromuscular apparatus of the extremities matures, movements of the limbs join those of the trunk and neck in total body reflexes. Later, the fetus develops the capacity for localized reflexes of individual extremities.

Early investigators of mammalian fetal behavior concluded that mammals differed from amphibians in that local reflexes preceded total body responses in development in the mammal (Windle 1944). More recent evidence from the study of the human fetus indicates that the human pattern, and probably that of other mammals, is, in actuality, similar to that found by Coghill for the amphibian; all body parts capable of neuromuscular activity function in the earliest movements (Humphrey 1970). The early controversies arose in part from a tendency to confuse the initial isolated neck movements which result from functional incapacity of the limbs with a local reflex. In addition, there is in the mammal a shorter time period between the development of the total reflex pattern and the local reflexes which partially obscures the sequential relationship between them.

Postnatally, the precedence of global responses over differential responses is clearly described by Parker (this volume) for the stumptail macaque, by Piaget (1952) for the human infant, and by others for the development of the grasp response in monkey and human (Chevalier-Skolnikoff 1974, this volume; Connolly and Elliott 1972; Halverson, 1943; Hines 1942; Jensen 1961). For instance, the young human infant, in shaking a bassinette, moves all four limbs; only after further maturation of the nervous system and experience with bassinettes does the infant develop the capacity to move a bassinette with only one arm, hand, or finger at a time. Similarly, the earliest grasping movements of both human and macaque infants result from a whole hand movement, all digits working together; the capacity to grasp with the thumb and forefinger alone is a late developing capacity in both rhesus monkey and humans. This behavioral sequence has been confirmed electrophysiologically for the rhesus monkey by Hines and Boynton (1940). The earliest movements elicited from stimulation of the motor cortex are global responses involving patterned actions of many muscles; isolated movements, particularly of separate digits, can be elicited only from the more mature cortex.

2. AND 3. SIMULTANEOUS AND SEQUENTIAL ORGANIZATION OF ACTS AND PERCEPTIONS

The neonatal repertoire of the macaque or human consists primarily of single acts: sucking, grasping, and visual following (Chevalier-Skolnikoff 1974, this volume; Parker, this volume). As the infant matures, it begins to coordinate actions with each other. Two or more motor patterns may be performed simultaneously as in sucking the thumb, visual following of the infant's own hand movements, or grasping of one hand with the other. Eventually, in humans at least, simultaneous coordination of actions may reach a very high level as in the simultaneous coordinations required for various mechanical and tool-using abilities or in some acrobatic maneuvers.

The most obvious example of the ability to perceive simultaneous relationships between sensory impressions is in the visual system. Sensory reception of any complex visual scene usually proceeds by means of a series of investigatory eye movements. Each visual fixation reports a slightly different picture to the brain, but the final perception is not one of a series of sequential visual impressions but, rather, of one simultaneous picture, including all the investigated visual parameters. All spatial perception is of a simultaneous nature and requires noting relationships between a number of visual stimuli at one point in time. Object recognition can also require the ability to perceive simultaneous relationships of many sensory stimuli present at one point in time, as in the simultaneous perception of color, texture, sheen, form, and pattern. Simultaneous sensory relationships can be perceived cross-modally as well as intramodally as the objects can be recognized, not only according to visual properties, but also according to combinations of sounds, smells, taste, and such tactual properties as temperature, roughness, and shape.

Of equal importance with simultaneous coordination of actions are sequential coordinations. Sequential coordinations are recorded for *Macaca arctoides* by Parker (this volume) in the following series of actions: "Simple serial applications of schema to one object, e.g., hand to hand transferring, placing in the mouth, extruding into hand, placing on flat palm, pincer grasping." Sequential coordination of actions is at the basis of any systematic investigation of object properties, of trial and error behavior, and any complex problem-solving behavior demanding a series of actions. In humans the ability to elicit motor patterns in series is essential for speaking, typing, writing, many athletic performances, and such simple actions as lighting a cigarette or cooking a meal.

The perception of sequential relationships between sensory stimuli is most obvious in the auditory mode, particularly in regard to the perception of speech. To comprehend a story presented verbally one must note the sequential relationships between the component words of sentences and, finally, sequential relationships between sentences and paragraphs. No understanding of spoken language is possible without the ability to retain long auditory sequences in order. Sequential perception is essential for non-auditory activities as well. In the visual mode, the ability to orient one's self geographically while traveling requires the ability to perceive the sequential relationships of separate spatial images.

Other behaviors also require the ability to perceive sequential relationships between sensory stimuli. Recognition of response-reward relationships requires the perception of the sequential order between the response and the reward. Conditioning only occurs if the reward follows the response, not if it precedes it. Causality is deciphered by the recognition of invariant sequential relationships between perceptions. In tool use or sports it is essential to note the sequential relationship between the motor action and the response of the object. Social behavior also demands the capacity to observe sequential relationships between the behavior of the actor and the probable responses of other animals.

4. FLEXIBILITY OF SENSORY-RESPONSE RELATIONSHIPS

Intelligence, as defined by Piaget, is problem-solving behavior. The abilities discussed above are perceptual and motor abilities. Although the complexity of perceptual and motor capacities limits the complexity of the problems that can be solved, perception itself is not the essence of intelligence, which is dependent on the creativity and flexibility of the animal's action in solving problems.

One behavioral capacity that is essential if problems are to be solved is a flexible relationship between stimulus and response. This depends on at least two subsidiary capacities: the ability to inhibit responses and sensations and the ability to determine actions on the basis of multiple sensory or conceptual schemes. Many actions of the infant macaque and human have an almost invariant relationship between stimulus and response. The infant's gaze is "captured" by moving stimuli or visual patterns; it acts as if it has no choice but to stare (E. Gibson 1969; Kagan 1972). Similarly, tactile contact of the palm always evokes a grasp reflex, while tactile contact of the cheek will elicit a rooting reflex. The only variations of these responses in the human infant correlate with sleep-wake cycles and hunger (Peiper 1963). In *Macaca arctoides* even these variations are inconsistent (Chevalier-Skolnikoff 1974).

Even non-reflexive actions may lack flexibility in young animals and humans. In some cases this inflexibility is demonstrated by a tendency to repeat the same action over and over again. For instance, a very young child asked to squeeze a ball once may do so many times in succession. Luria and Homskaya (1974) find that many actions are similarly repetitive because the young child has difficulty in inhibiting actions. In other cases, this inability to inhibit actions is evident in the inability of the young human to switch from one course of action to another without first completing the former, as in the case cited by Luria and Homskaya in which a young child, in the process of placing rings on a stick, was not able to remove them in response to a verbal command until he had first completed stacking them.

Possibly even more common and of great importance with regard to the creative aspects of intelligence is the situation in which previous perceptions and responses continue to dominate behavior despite changing conditions. For example, the human infant who has once found an object behind a particular screen and has then observed that object moved behind a second screen will continue to search for it behind the initial screen as the reality of its previous actions is more important than the more recent experience (Piaget, cited by Parker, this volume). In another instance, an infant observed his father (Piaget) at a window; Piaget then left the window and walked outside where he was observed by the infant. When asked where his father was, the infant again looked at the window. Young rhesus monkeys may also exhibit this inability to inhibit previous responses or perceptions, particularly on delayed alternation tests in which the monkey must alternate responses. Instead of doing so, the monkey continues to repeat the first response (Harlow 1959). Such invariant responses of necessity preclude creativity. With age, the infant monkey or human develops the ability to inhibit these invariant responses. Once this ability has developed, the macaque or human is then free to vary its responses with respect to the multiple perceptual and conceptual schemes at its command. (The total number of perceptual or conceptual schemes from which it may choose to act is based upon its capacities of differentiation, simultaneous and sequential synthesis, and internalization and upon its prior experience.)

Although the ability to inhibit past responses and past ideas and to elicit new ones on the basis of many varied stimuli forms a necessary basis of any creative act, these capacities are insufficient to permit goal-directed behavior or problem-solving behavior. Three additional capacities are needed: internalization of perceptions, anticipation of future perceptions, and the ability to compare anticipated perceptions with actual perceptions.

5. AND 6. INTERNALIZATION, ANTICIPATION, AND COMPARISON

Internalization is the ability to evoke images or ideas of perceptual events in the absence of the relevant environmental stimuli. It is an essential substrate of any behavior which demands the ability to act not only on the basis of concrete stimuli present in the environment at a given point in time, but also on the basis of past perceptions and anticipated future perceptions. Consequently, internalization is the basis of insight behavior which, according to Piaget (1952), results from mental combinations of internalized perceptions of past actions and results. It also is the basis of goal-oriented behavior when the goals are not present in the environment at the time, but only in the actor's mind.

In addition to internalization, goal-directed behavior demands the ability to plan in the mind a series of actions. If the goal is to be reached, these actions must be chosen intelligently in accordance with anticipated results. These results can be anticipated largely on the basis of internalized perceptions of sequential relationships between actions and results, derived from the actor's own experience or, in the case of the animal who possesses the ability to communicate, the experiences of its communicants. The number and complexity of serial actions and resulting perceptions which an individual can anticipate at one time result from internalized conceptions of time and sequential events, again derived from past actions of the actor or its communicants.

Finally, if the anticipated actions are to result in goal achievement, the individual must have the capacity to compare anticipated perceptions with actual results (perceptions) and if the two are not concurrent to change the course of its actions.

• CORTICAL FUNCTION AS RELATED TO INTELLIGENCE

Classically, intelligence has been considered to result from the functioning of the cerebral cortex. Certainly, the presence of the cortex is essential for nearly all aspects of intelligent behavior. The cortex plays an important role in precise sensory and motor discrimination, in the planning and execution of behavioral acts, and in creative and logical thought. Any discussion of intelligence and the brain must, of necessity, concentrate on the cerebral cortex.

The reader must realize, however, that the cortex does not function as an isolated entity. All cortical areas are intimately connected anatomically and physiologically with subcortical structures; cortical areas cannot function efficiently in the absence of input-output relationships with these structures, as intelligence depends on all parts of the brain functioning in concert. Unfortunately, however, the exact manner of interaction of cortical and subcortical areas in producing intelligent behavior is, in most cases, unknown, thus precluding much discussion of those interactions in the present context. Consequently, this discussion will be concerned only with the functioning of the cortex.

Classical neurologists divided the human cerebral cortex into primary sensory and motor areas; sensory association and premotor areas; and the parietal, temporal, and frontal association areas (Figure 1). In the classical scheme, there were three primary sensory areas: visual, auditory, and somatosensory. These were considered to mediate sensory reception; meaning was given to sensation within the sensory association areas which surrounded the primary sensory areas; motor acts were elicited by motor and premotor areas; and higher intelligence was considered a function of the association areas.

In general form, this scheme has stood the test of time; nevertheless, modern data necessitate some modifications. First and foremost, the sensory areas are not passive receptor areas, they are active areas of perception. As perception is a motor act involving active exploratory movements of the eyes, hands, etc., the sensory areas are also motor areas (E. Gibson 1969; Piaget 1952; Pribram 1971). The visual association areas, for instance, not only receive sensory stimuli, but they actively elicit eye and head-turning movements (Crosby 1962). These eye and head movements may enable the visual areas in infant macaques or humans to actively perceive visual stimuli of the animal's own choosing, not merely passively receive those stimuli which accidentally impinge upon its visual receptors. Similarly, electrical stimulation of the auditory association area elicits body and head-turning movement (Crosby 1962). These actions, if elicited by a sound stimulus in an infant animal, could enable the infant to actively orient its ears in the position most favorable for the perception of the auditory stimulus.

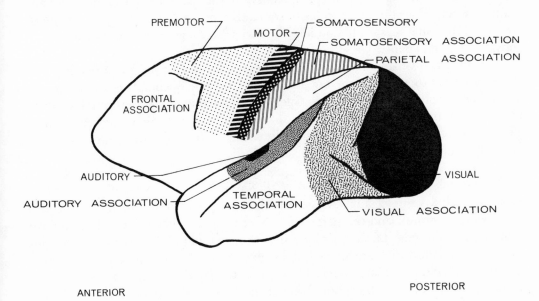

FIGURE 1. *Lateral surface of the cerebral cortex of the rhesus macaque,* Macaca mulatta.

The somatosensory areas also have motor functions. In part, they mediate active palpatory movements of the hands necessary to the perception of cutaneous and proprioceptive stimulation. The somatosensory areas may, however, be more than simple areas of somatosensory perception. They may also have an active motor role other than that involved in perception. In particular, they may provide the spatial coordinates for motion and thus aid in selection of which muscles will contract and to what degree (Luria 1966).

The concept of the motor area must also be revised in light of recent data. At one time it was thought that the motor cortex was the site of origin of voluntary motions; this is now considered false (Evarts 1973; Pribram 1971). Rather, actions are initiated elsewhere. The motor cortex is an area of sensory perception; it perceives stimuli of a somatosensory and kinesthetic nature and modifies, rather than elicits, motor actions on the basis of this data. In particular, the motor cortex aids in the differentiation of motor acts. In both human and monkey, the ability to perform fractionated movements of the fingers and, to a lesser extent, of other body parts is abolished by lesions of the cortex (Kuypers 1962). Muscle power is lost as is the speed of movement. It is considered that one role of the motor cortex is the modification of the force and speed of muscle contraction on the basis of somatosensory perception.

Because of the overlap of sensory and motor functions, the somatosensory and motor cortices are often considered together as the sensorimotor cortex. Nevertheless, they do differ in their apparent contribution to movement analysis as the contribution of the somatosensory cortex appears to be primarily of a spatial nature, while that of the motor cortex primarily relates to force and speed. In addition, they differ somewhat in their anatomical projections. Both project to the spinal cord and brain stem, but the motor cortex has more direct projections to the motor neurons in the cord and brain stem than does the somatosensory cortex (Kuypers 1956, 1958a, 1958b). These are valid functional reasons to consider the two as separate areas and they will be so considered in this paper.

This modified version of the classical scheme provides a convenient reference for discussion of cortical function, but it does not provide a complete model for the interaction of cortex and intelligence. Such a model requires a delineation not only of what sensory modalities are perceived, and where, but how they are perceived in terms of the components of intelligence listed above (i.e., discrimination, simultaneous vs. sequential perception, anticipation, etc.).

NEUROLOGICAL BASIS OF SEQUENTIAL AND SIMULTANEOUS PERCEPTION

Two major kinds of perceptual analyses are the perceptions of simultaneous relationships between stimuli and the perceptions of sequential relationships. The Russian neurologist, Alexander Luria (1966) has hypothesized that in humans, simultaneous and spatial relationships are perceived by the occipital and parietal lobes, while sequential relationships are perceived by the temporal and frontal lobes. This hypothesis is based, in part, on Luria's concepts of vision and touch (perceived by the occipital and parietal lobes, respectively) as spatial senses, and audition (perceived by the temporal lobe) as a sequential sense. In reality, visual and tactile impulses can be perceived sequentially as well as spatially, and auditory information can be analyzed spatially as well as sequentially, factors not accounted for in Luria's model of cortical function. In addition, data indicate that the visual association areas are found in the posterior temporal lobes as well as in the occipital and parietal lobes in both human and monkey (Milner 1967; Pribram 1971).

More recent investigators have hypothesized that in humans spatial information is perceived by the right hemisphere, while the left perceives sequential relationships (Butters et al. 1970; Carmon and Nachston 1971; Efron 1973; Gazzaniga 1970). Presumably, under this model the sequential aspects of all three modalities— somatosensation, vision, and audition—would be perceived by the left hemisphere, and the spatial aspects by the right in humans.

Luria's model of parietal and occipital lobe function in the perception of spatial aspects of somatosensation and vision still remains the most useful model of the functioning of these regions, even if future research should demonstrate that his concepts apply to the right cerebral hemispheres only, and must be extended to include the posterior temporal lobes as well as the occipital and parietal lobes. In brief, Luria considers that the primary somatosensory and visual areas perceive and analyze tactile and visual stimuli, respectively, which contain only one element. Tactile and visual materials containing several separate sensory impressions which must be synthesized into one simultaneous whole are perceived by the somatosensory and visual association areas, respectively. The parietal association area, which is located anatomically between the somatosensory, visual, and auditory areas, perceives material which requires the synthesis of a large number of separate tactile and visual elements into one whole. According to Geschwind (1965), this area also synthesizes auditory sensations with visual sensations.

Clinical data in humans support this interpretation. If the primary visual area is destroyed by a lesion, blindness over the corresponding area of the visual field results; but as long as some portion of the primary cortex remains functional, there is no deficit in spatial synthesis or visual discrimination capacity. The patient can compensate for partial visual field blindness by additional exploratory eye movements (Luria 1966; Pribram 1971). On the other hand, a lesion in the visual association areas may destroy either spatial perception or other visual capacities, depending on the location of the lesion. In both cases, these deficits may result from inability to perceive simultaneous relationships between stimuli. First, eye movements may be greatly diminished, leading to total lack of attention to some parts of the visual field, in which case, that part of the visual field will not be perceived or related to other visual stimuli. Even if eye movements appear to be intact, a visual agnosia, or inability to recognize objects, may result from lesions in visual association areas. According to Luria, the basic cause of visual agnosia is the inability for simultaneous synthesis of visual impressions. A human with visual agnosia, when presented with a visual stimulus of a component nature such as ⊕, will perceive only one portion of the stimulus—the triangle or the circle—he will not perceive both together. If several objects are present in a picture, he may see only one. One patient was shown a picture of a pair of glasses; he noticed only the rounded parts, not the whole, and guessed that they were bicycle wheels.

As a visual association cortex lesion approaches the parietal lobe, the defect becomes primarily spatial, an inability to relate the position of one object to others within the visual field. If the lesion is in the right temporal lobe, the difficulty will be in pattern discrimination and the discrimination of faces (Milner 1967). The discrimination difficulties are particularly severe in the case of complex stimuli that require the perception of many separate visual stimuli.

The somatosensory association area, like the visual association area, according to Luria's analysis, mediates simultaneous and spatial perception. While a lesion within the primary somatosensory area, in some instances, leads to an inability to localize tactile stimulation to one point on the skin, a lesion in the somatosensory association area leads to astereognosis, or an inability to recognize an object by touch. This inability is basically a defect of simultaneous and spatial synthesis. Recognition of object shape requires active tactile palpation; the separate perceptions from each palpatory act must then be synthesized into one simultaneous whole. In many cases, even this information will not itself suffice for recognition; the perceptions of shape must also be synthesized with perceptions of temperature, texture, and weight.

A second function of the somatosensory areas is proprioception, the perception of stimuli arising from muscles, joints, and tendons. The primary somatosensory areas perceive the position of single body parts or individual joints. The somatosensory association area perceives muscle and joint positions which require the synthesis of spatial impressions arising from two or more joints or body parts. As spatial perception, in many cases, demands both proprioceptive perception and visual perception, it is not mediated by somatosensory association areas alone; the parietal association areas, located anatomically between the visual and somatosensory association areas, cooperate in facilitating the perception of body position in space. With parietal lesions in the human, a defect in the ability to simultaneously coordinate visual images of space with proprioceptive body images results in such defects as inability to dress, to perceive right from left, to make a bed, or to recognize one's own fingers or other body parts.

Intellectual defects of a spatial nature, or requiring the simultaneous synthesis of several perceptions into one whole, also occur in parietal lesions in humans. The patient with such a lesion, particularly if it is on the right side, will be unable to draw three-dimensional figures or construct patterns from blocks, a defect known as constructional apraxia and considered to result from disrupted spatial perceptions. Other patients, because of spatial deficits, may be unable to read a clock or follow a map (Luria 1966). Parietal patients often have defective number concepts. They cannot perceive the simultaneous relationships between a written or spoken number and the actual numbers of objects presented; in other words, they cannot synthesize the perception of many separate objects into a perception of one simultaneous numerical whole. Arithmetic abilities will also be disturbed, particularly for problems written in a right-left orientation. Thus, they cannot distinguish $7 - 4$ from $4 - 7$ (Luria 1966).

In left parietal lesions, there may be language deficits related to difficulty in comprehending spatial concepts (under, on, in front, etc.), and difficulty in deciphering meanings based on the positional relationships between words. They cannot, for instance, distinguish "Kate is lighter than Sue" from "Sue is lighter than Kate" or "mother's brother" from "brother's mother" (Luria 1966).

The ability for simultaneous synthesis between auditory and visual or proprioceptive concepts may also reside in the parietal lobe as, according to Geschwind (1965), object-naming capacities are disrupted by left parietal lesions. The patient cannot perceive the simultaneous relationship between the sound of the word and the visual or proprioceptive image of the object (Geschwind 1965).

Luria considers sequential perception, as opposed to simultaneous perception, to be a function of the temporal lobes. His analysis is based primarily on auditory perception, although other data indicate that sequential perception of visual images is also mediated by the left temporal lobe (Carmon and Nachston 1971). In Luria's view, the primary auditory area perceives isolated aspects of auditory stimuli, such as pitch or tone. The synthesis of separate sounds into sequential wholes, as in the perception of words, is a function of the auditory association cortex, while the temporal association cortex is essential for the perception of longer sequences, as in the perception of the sequential relationships of lists of words, numbers, and nonsense syllables. An inability to perceive the sequential order of words and sentences prevents language comprehension, particularly, the comprehension of stories. Sequential deficits resulting from lesions in the temporal association areas may also result in difficulties with arithmetic. Such patients can perform written arithmetic problems, but they cannot solve problems requiring several sequential steps if the memory of one step must be kept in mind while performing the succeeding step. Luria's concepts of temporal lobe function are in accord with Milner's (1967), except that Milner considers them to be confined to the left temporal lobe while the right temporal lobe mediates visual discrimination.

While Luria's hypotheses of spatial and sequential perception are based on human clinical data, other evidence indicates that the visual and tactile areas may function similarly in spatial perception in macaques. Physiologically, the primary somatosensory and tactile areas in the macaque are organized so as to best perceive isolated stimuli, while the association areas are organized to perceive stimuli which contain a larger number of elements and represent a wider area of space.

The topographic projection in the sensory association cortices differs from that of the primary cortices in that each cell in the primary visual and somatosensory cortices has a very small receptive field from one portion of the retina, a small area of the skin, or one joint. Cells within the primary visual area respond to single isolated stimuli, such as brightness, edges, or angles, while those within temporal association areas respond to more complex shapes, such as objects of three-dimensional figures (Gross et al. 1972). The receptive fields in the association cortices are larger; neurons within the association cortices respond to stimuli applied to large areas of the retina or skin, or two or more joints (Duffy and Buschfiel 1971).

Visual discrimination defects, similar to those in humans, are noted in monkeys with lesions in the primary and visual association areas within occipital and posterior temporal cortical regions and, like humans, monkeys with lesions in the parietal association areas have difficulties in the perception of body position in space. They may be extremely disoriented in visual space and reject entire parts of their bodies in that they do not voluntarily move these body parts and may continually bump into them (Denny-Brown and Chambers 1958).

There is less evidence for temporal lobe mediation of sequential perception in the macaque than in humans; however, the significance of this meager evidence is uncertain as sequential perception has barely been investigated in the macaque. It is known that the auditory association cortex of the left temporal lobe of the macaque is essential for some types of auditory discrimination and for the ability to make an accurate sequential linkage between a particular sound and a red or green light (Dewson and Burlingame 1975). The anterior temporal lobes may disrupt the perception of reinforcement. Reinforcement is, of course, in part, a sequential perception, as reinforcement always follows—never precedes—a response. As the anterior temporal lobe in the macaque has anatomic connections with a subcortical nucleus (amygdala) which is thought to function to keep sequential events in memory (Pribram 1971), it is not unlikely that the anterior temporal lobe in the macaque also functions as a sequential perceptual or memory area.

Until recently, there was little evidence for laterality in the macaque brain. It was assumed that laterality of brain function distinguished the human brain from that of other primates and had evolved in conjunction with language. The work of Dewson and Burlingame (1975) on auditory recall indicates that there is laterality in the macaque brain in auditory recall functions (*Yerkes Newsletter* 1976). Further research is needed to ascertain if sequential perception in general is confined to the left hemisphere in the macaque and spatial perception to the right, as has been hypothesized for humans (Butters et al. 1970; Carmon and Nachston 1971; Efron 1973; Gazzaniga 1970).

DIFFERENTIATION

Perceptual differentiation can occur at all levels. Isolated stimuli such as colors, line orientations, or temperature can be differentiated from each other. Material that is highly synthesized in simultaneous and sequential modes, such as complex patterns, can also be differentiated from each other. No one area functions in differentiation—all areas do. Those areas which perceive simultaneity also differentiate simultaneity. The ability to distinguish phonemes and words occurs in the same temporal areas that perceive the sequence order of sounds, while the ability to differentiate spatial positions occurs in the same parietal areas that perceive spatial positions. For this reason, no special discussion of sensory differentiation will be given here.

FRONTAL LOBE FUNCTIONS IN GOAL DIRECTED ACTIONS

The frontal lobe is composed of three major functional areas: motor cortex, premotor area, and frontal association area. The motor cortex contributes to motor functions in several ways. In both human and macaque it facilitates the differentiation of motor acts, as the ability to perform fractionated movements of the fingers and, to a lesser extent, of other body parts, is abolished by lesions of the motor cortex (Kuypers 1962). Another major function of the motor cortex is the determination of the force of muscle contraction—in lesions of the motor cortex muscle power is lost. In extreme cases this loss of power may be manifested as paralysis. The motor cortex also contributes to an increase in speed of movement. In this role the motor cortex is an anticipator; it predicts which movements will be needed and sets a nervous system bias which facilitates the performance of these movements (Pribram 1971). Anticipation is one important function which the motor cortex shares with other frontal areas.

The premotor cortex plays a functional role in the performance of serial actions. Human patients with lesions in prefrontal areas are unable to perform coordinated serial movements such as those needed for playing a musical instrument or typing. The patients' actions do not flow; instead, the patient behaves as if every motor act requires a separate neural impulse (Luria 1966). Smooth-flowing serial actions require the ability to inhibit preceding actions, as well as the ability to anticipate succeeding actions. A loss of inhibitory capacity is well documented in premotor lesions. Patients with such lesions may be characterized by the re-emergence of neonatal reflexes such as the grasp reflex, and may demonstrate response preservation, a tendency to repeat the same motor response again and again. Anticipatory functions are less well documented for the premotor areas but, as other frontal areas have important anticipatory functions, it is not unreasonable to hypothesize similar functions for the premotor areas.

The frontal association area, more than any other area of the cortex, mediates the active aspects of intelligence: flexibility of relationship between sensation and response, internalization, anticipation of perceptions, and the comparison of expected perceptions to actual perceptions. The ability to inhibit responses and sensations is, as discussed above, a prerequisite to flexible perceptual/response relationships. The frontal lobes clearly have inhibitory functions, as patients with frontal lesions suffer from behavioral pathologies based on the inability to inhibit actions and/or perceptions, including the return of infantile reflexes (i.e., grasp, visual following, and urination reflexes). Patients suffering from response preservation may have difficulty in performing very simple acts such as lighting a cigarette or cooking dinner, because they cannot inhibit the first action in the sequence, but rather repeat it as in the continual striking of an already lighted match or repeated tossing of a salad (Luria 1966). At a more advanced intellectual level this inability to inhibit responses or ideas may be manifested in intellectual tasks such as the inability to understand metaphors which are interpreted literally or to solve different types of arithmetic problems, all which may be solved by the same method.

Rhesus monkeys with frontal damage may have similar problems caused by an ability to inhibit responses or sensations. This is particularly manifested by distractability and by inability to perform delayed alternation tasks (Stamm 1964; Grueninger and Pribram 1969).

Another important function of the frontal lobes, in addition to that of flexibility, is internalization. Patients with frontal lesions are sometimes described as stimulus bound; their behavior is determined by immediate environmental stimuli rather than by internalized concepts. Monkeys with frontal lesions who can solve a right-left discrimination task with ease as long as a visual stimulus is present to indicate the choice, cannot solve such a task if they must rely on internalized concepts of right and left (Pohl et al. 1972). Alternation tasks are generally difficult for monkeys with frontal lesions, but even these can be performed if external cues are provided and the monkey is not required to provide its own internal cues (Pribram 1971).

The most obvious and striking roles of the frontal cortex are in the realm of anticipation of the future and in the ability to match results to expectations. Patients with frontal lesions are characterized by an inability to plan for the future; they seem unable to anticipate goals and organize serial actions to meet goals. It is as if they cannot anticipate either the reinforcing aspects of goals or the results of their actions. All of their behavior is in the here and now; their behavior, as a result, is often described as lazy and shiftless and rarely do they have ambitions. This inability to anticipate and plan serial actions is not limited to the long-term future, but also to the immediate future. One patient, for instance, when asked to change a bedpan while in the process of taking a temperature, got very irritable for he could not, as he said, "do two things at once." The same patient, while working as a gas station attendant, would lose his temper if he was asked to wipe the windshield when he was in the process of changing the oil (Ackerly 1964).

Monkeys, if deprived of the frontal lobes, also have difficulty in serial organization of actions to meet a goal. This deficiency is displayed in their inability to solve the Hamilton Perseverance Test in which a food reward is placed in any of four boxes, its position varying at random in successive trials. Mature monkeys with intact frontal lobes develop systematic approaches to finding the food (such as opening each box once, in order, from right to left). Immature monkeys and those with frontal lobe lesions approach the problem haphazardly and may repeatedly open the same empty box (Harlow 1959).

Finally, when engaged in any problem-solving behavior, frontal patients do not match anticipated perceptions to results; they rarely correct errors. This defect shows up in other behaviors as well. Frontal patients often have social problems resulting from difficulties in anticipating reactions of other people to their behavior and correcting their behavior to match the response of others.

▪ ONTOGENY OF CORTEX AND INTELLIGENCE

The behavior of the newborn, be it stumptail macaque (Parker, this volume) or human (Piaget 1952), lacks most of the component capacities of intelligence discussed in this paper. Motor acts and sensory preceptions are basically undifferentiated; neither simultaneous nor sequential coordinations are evidenced in either the sensory or motor sphere; and actions show little goal direction or planning and are basically of an inflexible, nonvoluntary nature. This lack of intelligence does not mean lack of learning; the neonatal human and nonhuman infant can learn. Among human infants' demonstrated learning capacities are habituation to olfactory, auditory, and tactile stimuli; conditioned head-turning movements in response to auditory stimulation; avoidance of breast-feeding if the first experiences at the breast are unpleasant (Bridger 1961; Gunther 1961; Lipsett 1969; Robinson 1969); and adjustment of sleep-wake cycles. The neonatal macaque can learn to make visual discriminations and traverse a runway for food (Harlow 1959).

Whether the learning that occurs in the neonatal human or macaque is dependent on the functioning cerebral cortex is uncertain. Clearly, other animals can learn in the absence of cerebral cortex. Macaques and cats whose cortices have been removed surgically can learn basic brightness discrimination (Crosby et al. 1962). Planaria, who have no cortex at all, can learn (McConnell 1962). The fact that the neonate can learn is insufficient in itself to prove that its cortex functions. The learning that the human neonate can do is limited, requiring little sensory discrimination and no synthetic capacities. The macaque neonate is perhaps slightly more advanced in its ability to learn pattern discrimination, but its learning capacities, otherwise, are not of a sort that would require advanced cortical functions. Indeed, it has been argued in the past on the basis that the cortex is unmyelinated, that there is no function in the neonatal human cortex (Flechsig 1920, 1927), and that the behavior of the normal human neonate cannot be distinguished from that of the cortically damaged infant (Peiper 1963).

More recent evidence has indicated that neither of these arguments can be considered valid. Myelinization is now known to precede function in many, if not all, cases (Angulo y Gonzalez 1929; Hooker 1957; Langworthy 1928; Romanes 1947; Ulett et al. 1944). More important, the healthy neonate can be distinguished behaviorally from the brain-damaged. Hemiplegic infants, once considered to be undiagnosable at birth, now are known to differ from the healthy on the basis of muscle tone. Cortically damaged infants may have abnormal sleep-waking cycles and abnormal cries (Robinson 1969). In some instances, visual-following reactions or head-turning movements to visual stimuli are absent (Brazelton et al. 1966; Hill 1963). Neonatal epileptic fits of cortical origin can occur in the human (Caveness 1969). In addition, there is now positive evidence for cortical function in the human neonate; EEG activity can be obtained as early as 5–6 months of fetal life (Bergström 1969).

The rhesus macaque is more mature at birth in terms of its myelinization pattern than is the human, and most probably its cortical function is more mature as well. EEG activity is present in the monkey immediately after birth (Ellingson and Rose 1970); motor responses can be obtained subsequent to electrical stimulation of the motor cortex of the macaque as early as 112 days gestation age and evoked visual responses are present at birth (Hines and Boynton 1940). Epileptic fits can be induced in the neonatal rhesus monkey by application of penicillin to the cortex (Caveness et al. 1973).

The most thorough study of intellectual development in a monkey is that by Parker on the stumptail macaque, *Macaca arctoides* (Parker 1973; this volume). Studies of brain maturation of the stumptail macaque are lacking; the most detailed study of cortical maturation in monkeys is the study of Gibson (1970) on myelinization in the rhesus macaque, *Macaca mulatta*. Myelinization refers to the process by which a nerve fiber acquires a myelin sheath—a lipo-protein sheath which surrounds nerve fibers and increases the efficiency of fiber function. The acquisition of this sheath is one parameter of neuronal maturation. The stumptail and rhesus monkeys are very close phylogenetic relatives, as indicated by their mutual inclusion in the genus *Macaca*, and they bear close anatomical and behavioral relationship, although the stumptail may mature slightly in advance of the rhesus (Chevalier-Skolnikoff 1974). It is probable, therefore, that cortical function is also present in the neonatal stumptail macaque. It is also probable that the general order of maturation of cortical areas is similar in the stumptail to that found by Gibson (1970) for the closely related rhesus monkey, as cortical maturation is similar even in the more distantly related rhesus monkey, human, and laboratory rat.

The demonstration of some cortical function in neonatal humans and rhesus monkeys and of its probable function in the neonatal stumptail macaque does not mean that the cortex functions in a mature fashion—it does not. Function is less than mature in all cortical areas in the neonatal period, including the motor area which is histologically more mature at birth than any other cortical area. The prime hallmark of motor cortex function in the adult—the ability for fine fractionated movements—is absent from the neonatal behavioral repertoire. The infant's (both human and macaque) actions are global actions: sucking, grasping, and waving of several limbs at once.

Nevertheless, the motor cortex does function at birth in both humans and rhesus monkey, as evidenced by the fact that neonatal epileptic fits of motor cortex origin can be induced in both species (Caveness et al. 1973), but it functions differently from that of the adult. Electrical stimulation of the motor cortex of the newborn macaque yields primarily global movements such as sucking. With maturation, electrical stimulation of the motor cortex yields progressively more highly differentiated movements until approximately 6–10 months of age (Hines and Boynton 1940). Thus, the increasing differentiation of motor function, as demonstrated by electrical stimulation, parallels the increasing differentiation of motor function demonstrated by overt behavioral capacities.

Anatomical causes for this functional differentiation with age can be hypothesized. The neurons of all cortical areas are arranged in layers numbered I–VI from the cortical surface inward. All areas of the neocortex, except the motor area, contain all six cortical layers; the motor area has five layers, it lacks layer IV. Layers II–IV are receptive layers. Layer IV receives afferent fibers from subcortical areas, particularly the thalamus. Layers II and III receive afferent fibers from other cortical areas; layer III also sends fibers to other cortical areas. Layers V and VI, in contrast to the more superficial layers, are the primary source of efferent fibers which carry impulses downward to the brain stem and spinal cord.

The six layers do not mature at the same time. In the rhesus monkey, the layers myelinate in the order: VI, V, I, IV, III, II (Gibson 1970). A similar myelinization sequence is found in the human, and the layers of the human cortex mature in a similar sequence with respect to other parameters of maturation as well (Conel 1939,

1941, 1947, 1951, 1955, 1959, 1963). In the motor cortex of the rhesus macaque, only layers V and VI contain myelinated fibers at birth; the others myelinate subsequent to birth. In part, the changing function of the motor cortex with age is a result of the delayed maturation of the upper cortical layers.

The changing function of the motor cortex with age may also result from differential rates of maturation of the efferent fibers which leave the motor cortex; these descend to various parts of the brain stem and spinal cord. Those fibers which are known to be important in the control of finely differentiated muscular actions travel along with fibers from other cortical areas in the pyramidal tracts (Lawrence and Hopkins 1972). These tracts are still functionally immature at birth in the rhesus macaque (Kuypers 1962) and are incompletely myelinated at that time (Gibson 1970). As other cortical areas myelinate in a similar fashion to the motor cortex, they too would be expected to undergo changing functions with age, and their initial functioning would be at a primitive and nearly unrecognizable level.

This pattern of maturation of the cortical layers indicates that in each cortical area the efferent, or output, layers mature first, followed by the afferent layers. This prior development of efferent cortical layers over afferent is paralleled by developmental parameters of the spinal cord: motor cells in the ventral horn mature prior to sensory cells in the dorsal root ganglia (Hooker 1957, 1972) and motor fibers in the ventral roots myelinate in advance of sensory fibers in the dorsal roots. The precedence in maturation of motor fibers over sensory may seem unexpected to those accustomed to thinking in terms of reflexes and stimulus-response psychology. There is evidence, however, that in nearly all non-mammalian vertebrates the spontaneous movements precede reflex capacity at the spinal cord level (Hamburger 1963). Mammals may be an exception to this rule. In the rhesus monkey, spontaneous and reflexogenic activity appear to mature at the same time (Bodian 1966), while in humans, reflex activity may precede spontaneous activity (Humphrey 1970). At the cortical level, however, there is reason to believe that, at least in some instances, motor actions may precede perceptual capacity seen in monkey and human. In fact, perceptual learning as discussed above demands motor actions and, according to Piaget's scheme of behavioral development, motor actions must come first, as motor actions provide perceptual data (Piaget 1962).

All of the cortical areas mature in a similar fashion, but they do not all mature at the same time. In the rhesus macaque (Gibson 1970) and human (Conel 1939, 1941, 1947, 1951, 1955, 1959, 1963), myelinization begins in the motor and primary sensory areas, spreads to the sensory association and premotor areas, and lastly to the association areas. The motor area slightly precedes the other areas in its myelinization cycle, but essentially myelinates coincidentally in time with the primary sensory areas. The sensory association areas and premotor areas myelinate at the same time, with a slight precedence of the premotor area. The frontal, parietal, and temporal association areas all myelinate at the same time.

Myelinization begins in the lower layers of each cortical area and spreads upward. The whole myelinization process is gradual and occurs in such a manner that the upper layers of the motor and primary sensory areas are myelinating at the same time as the lower layers of some parts of the association regions. Thus, there is no point in time when one cortical area is capable of functioning at a mature level and others are incapable of functioning at all.

To the extent that behavioral maturation parallels brain maturation, behavior would be expected to mature in the following fashion. The earliest behaviors would have no cortical mediation, but would be determined by subcortical structures since subcortical structures, for the most part, mature prior to the cortex. The first cortically mediated behaviors to appear would reflect the functions at a very immature state of the motor and primary sensory areas. The second group of behaviors would reflect very immature functioning of premotor and sensory association areas accompanied by more definitive functioning of motor and primary sensory areas. Lastly, the association areas would begin to function. A gradually increasing functioning of all cortical areas would continue, at least until late childhood and perhaps longer, as myelinization and other histological parameters continue to mature to at least 8 years of age in humans (Conel 1939, 1941, 1947, 1951, 1955, 1959, 1963), and 3 ½ years of age in the rhesus macaque (Gibson 1970).

As each cortical layer and area develops, it interacts with, and modifies, the function of each preceding layer or area and, on this basis, one would expect a behavioral maturation pattern exactly like that described by Piaget, "behavior patterns characteristic of the different stages do not succeed each other in a linear way (those of a given stage disappearing at the time when those of the following one take form) but in the manner of the layers of a pyramid (upright and upside down), the new behavior patterns simply being added to the old ones to complete, correct or combine with them" (Piaget 1952: 329).

▪ PIAGET'S STAGES OF BEHAVIORAL DEVELOPMENT IN HUMAN INFANTS CORRELATED WITH CORTICAL DEVELOPMENT

The predicted sequences of behavioral maturation based on brain maturation coincide nicely with the stages of behavioral maturation in the human infant described by Piaget (1952, 1954, 1966) and in the stumptail macaque described by Parker (this volume). Piaget describes 6 stages of intellectual maturation in human infants in 6 parallel series: sensorimotor, object concept, spatial field, causal, temporal, and imitation. Since the behaviors characteristic of the same stage in different parallel series depend on similar cortical capacities and occur at similar ages, to discuss each stage separately in regard to cortical function would lead to needless redundance. Only the stages of the sensorimotor series will be named in this paper, behavioral characteristics of each series will be described as a unit under the parallel sensorimotor stage. Each stage of each series is summarized and discussed by Parker (this volume). Much of the discussion of Piaget in this article is derived from Parker.

The discussion of the behaviors is coupled with a discussion of probable cortical intermediates. At the earliest stages of development, the behaviors are quite primitive and do not always correspond closely with the behaviors of adult animals or humans from which most of the data on cortical function have been derived. Consequently, without further research, the exact cortical areas which mediate these behaviors can only be hypothesized and suggested; cortical correlations should not be considered proven. The hypotheses presented here are based on the correspondence of the intellectual parameters necessary for these early infantile behaviors, with adult intellectual parameters and adult cortical mediates as presented earlier in the paper.

In large part, early cortical function mediates the function of subcortical structures, such as the thalamus, which mature in advance of the cortex and have many interconnections with the cortex. It is possible that the examples of "initial" cortical functions listed below are really primarily thalamic functions, or are based primarily on early cortical influences upon the thalamus. Cortical mediates have been hypothesized here largely on the basis of the evidence cited above that the cortex influences behavior at an earlier age than would be expected simply on the basis of behavioral observations, and on the basis that the proposed cortical mediates correlate exactly with known sequences of cortical maturation.

STAGE 1. "THE REFLEX STAGE"—BIRTH TO ONE MONTH

Piaget's Stage 1 in the sensorimotor series is the *Reflex Stage*. This stage encompasses the first month of neonatal life in the human infant. The name is derived from the reflexive behaviors which predominate in this early postnatal period. The infant's actions are characterized by considerable sucking, grasping, and looking. Each of these behaviors is reflexive in that it is an almost invariant response to certain classes of stimuli. The infant sucks almost anything that touches its lips, grasps almost anything that touches the palmar surface of its hand, and looks at almost anything that moves slowly across its visual field. The infant at this age cannot inhibit these actions; they are automatic reactions to the presence of sensory stimuli.

These actions of sucking and grasping, as well as most other neonatal actions, are not only undifferentiated in terms of the sensory stimuli which elicit them, but are also undifferentiated in terms of their motor patterns: e.g., the grasping action is a whole-hand action; individual digits cannot be moved separately. The actions are single, isolated actions. Two actions do not occur together in a coordinated fashion, either serially or simultaneously, in that the infant does not focus two or more actions on the same object at the same time or in serial order. A final characteristic of the actions of the *Reflex Stage* is that they are not goal-directed.

Nothing in the nature of these actions indicates cortical function; quite the contrary, if the presence or absence of cortical function were to be deduced solely on the quality of these actions, the conclusion would be that the cortex does not function. No differentiation of motor actions implies no motor cortex function; lack of ability to inhibit reflexes, elicit variable responses, and direct actions towards goals indicates no premotor or frontal lobe function. The inability to coordinate two or more actions into one serial and simultaneous whole implies lack of premotor or parietotemporal association cortex functions, or both.

However, evidence cited above indicates that the motor cortex does function at this age, but it must function at a very immature level. As there is no evidence for premotor, frontal, parietal, or temporal association area function at this age, and as these areas myelinate later than the motor cortex, it is probable that they do not yet function at this age, although it is possible that they also function at a very immature level.

Each of these actions (sucking, grasping, and looking) generates sensory stimuli that can be perceived by the nervous system. The stimuli produced by these early actions are stimuli which are confined to one sensory modality at a time, i.e., looking produces only visual stimuli; grasping only tactile. Because the separate actions do not focus on the same objects in a coordinated fashion, two or more types of stimuli from the same object are not referred simultaneously or sequentially to the cortex and cannot be synthesized there. In addition, tactile stimuli which are produced by the infant's actions occur with reference to only one body part at a time; the cortex, therefore, does not receive data for the synthesis of tactile inputs from two or more body parts at a time. The results of these separate uncoordinated actions are that the infant at this age has no concept of a wholistic space, rather, it perceives many separate spaces: a visual space, a buccal space, a palmar space (Parker, this volume).

Within the visual system a similar pattern of focussing on isolated stimuli prevails. The newborn detects simple invariant visual stimuli: orientation of lines, angles, colors, movement, and binocular parallax (Bower 1969; Fantz 1965; Hubel and Wiesel 1963). Simple patterns and forms can also be detected. For the most part, however, the infant cannot synthesize these stimuli into one coherent whole; it can perceive pattern alone or shape alone, but it cannot perceive pattern and shape simultaneously (Bower 1969). Similarly, when presented with a compound figure, the human infant of less than 14 weeks will spend no more time staring at it than it would spend staring at each of the component parts alone.

The infant's inability to perceive simultaneous stimuli is parallelled by an inability to perceive sequential stimuli; the Stage 1 infant has no concept of time or sequential actions. The ability to perceive isolated stimuli within one sensory modality, coupled with an inability to perceive and synthesize simultaneous and sequential data from one or more sensory modalities, suggests that the primary sensory areas are functioning, but that the sensory association areas do not function at this stage. This is the expected situation on the basis of brain maturation data, since the primary areas mature in advance of the sensory association areas.

STAGE 2. "THE PRIMARY CIRCULAR REACTION AND THE FIRST ACQUIRED ADAPTATIONS TO THE ENVIRONMENT"—ONE TO FOUR MONTHS

The stage of the *Primary Circular Action* is very similar in most behavioral respects to the *Reflex Stage*, but it differs from it in some important characteristics of the motor repertoire. First, the infant develops some ability to inhibit reflexes and thus exert what Parker (this volume) terms "primitive voluntary control" over its actions. Second, much of the human infant's behavioral repertoire at this age is characterized by the primary circular reaction. "Circular" is a term that Piaget (1952) uses to refer to actions that are repetitive. "Primary" indicates that the actions are focussed on the infant's own body rather than on external objects. The repetitive actions of the primary circular reaction occur in a rhythmic form characterized by smooth starting and stopping of actions. Some characteristic primary circular reactions are: rhythmic tongue protrusions, rhythmic grasping actions in the absence of an object to grasp, and rhythmic waving of the arms. The ability to inhibit reflexes, as well as the ability to elicit primary circular reactions, indicates that at this age the premotor cortex is beginning to function since it is the premotor cortex which is responsible not only for the ability to inhibit reflexes, but also for the ability to inhibit one motor action and proceed to the next (Luria 1966).

Some of the primary circular reactions in the human infant are "reciprocal" in that the infant moves two or more body parts in relationship to each other in a simultaneously coordinated fashion: repeated hand-to-hand contact, repeated hand-mouth or hand-face contact, visual following of the infant's own hand. Piaget considers that the human infant who engages in these reciprocal actions attempts to mutually "assimilate" and "accommodate" the separate body parts to each other (see Parker, this volume). These reciprocal actions provide simultaneous sensory data from two different body parts and thus provide the cortex with its first opportunity for the simultaneous synthesis of sensory data. As elementary simultaneous synthesis is hypothesized by Luria (1966) to be a function of the sensory association cortices, these reciprocal actions are the first indication that these cortical areas may be beginning to function. A further indication that they may be functioning at this stage is the infant's visual behavior. Towards the end of the second stage—at about 14 weeks—the infant begins to devote more time to staring at compound stimuli than at isolated stimuli (Bower 1969).

In sum, the behavioral data for Piaget's Stage 2 indicate that the premotor and sensory association areas may be beginning to function at this stage. This is exactly the sequence which would be predicted by the brain maturation data which indicate that these four areas are the second group of cortical areas to mature after the motor and primary sensory areas.

STAGE 3. "THE SECONDARY CIRCULAR REACTIONS"—FOUR TO EIGHT MONTHS

The distinguishing characteristic of Stage 3 is the *Secondary Circular Reaction*. Like all circular reactions, the secondary circular reaction is characterized by repeated actions; it differs from the primary circular reaction in that the secondary circular reaction consists of repeated actions upon single objects. The secondary circular reaction arises when an infant's action fortuitously causes a reaction in an object. The resulting visual motion or noise captures the infant's attention and the action is repeated in order to prolong the visual or auditory "spectacle." Often, the motor component of the secondary circular reaction is a prehensive action of the hands, and the total behavioral act consists of repeated banging, shaking, or vibrating of an object such as a rattle or bassinette ornament. Occasionally, other body parts, such as the feet and legs, are used. A special form of the secondary circular reaction is the repeated use of vocal behavior and/or facial expressions in an attempt to elicit responses from the mother (Parker, this volume).

TABLE 1 / PIAGET'S STAGE OF SENSORIMOTOR INTELLIGENCE IN THE HUMAN INFANT AND HYPOTHESIZED CORTICAL CORRELATES*

SENSORIMOTOR STAGE	AGE	CHARACTERISTIC BEHAVIORS	IMPLICATIONS FOR CORTICAL FUNCTIONS	SUMMARY OF CORTICAL AREAS WHICH HAVE BEGUN FUNCTIONING BY THIS STAGE (HYPOTHESIZED)
1 The Reflex Stage	Birth to one month	Focal paroxysms (Caveness et al., 1973)	Motor cortex functioning	Initial functioning of motor and primary sensory areas
		Global actions	Motor cortex functioning at an immature level	
		No reflex inhibition	Little or no premotor cortex functioning	
		No goal direction	Little or no frontal association area functioning	
		No serially coordinated actions		
		Little or no flexibility of action patterns		
		Actions directed to unimodal sensory stimuli	Primary sensory areas functioning	
		Perception of isolated stimuli within one modality		
		No synthesis of separate sensory impressions in space or time	Little or no functioning of sensory association areas or association areas	
		Space exists as many separate entities		
		No concept of causality		

Stage	Age	Behavior		
2 The Primary Circular Reaction	1–4 months	Reflex inhibition Primary circular reaction	Initial premotor cortex functioning	Initial functioning of motor, premotor, and somatosensory association areas
		Reciprocal actions which provide stimuli for simultaneous synthesis More visual attention to compound stimuli than to isolated stimuli	Initial sensory association function	
3 The Secondary Circular Reaction	4–8 months	Beginning differentiation of motor patterns	Increasing motor cortex function	Initial functioning of motor, premotor sensory, sensory association, and frontal association areas (Initial functioning of temporal or parietal association areas?)
		Perception of sequential relationships between actions and visual or auditory data	Initial association area (temporal?) functioning	
		Fortuitous goal direction	Initial frontal association functioning	

continued

TABLE 1 / CONTINUED

SENSORIMOTOR STAGE	AGE	CHARACTERISTIC BEHAVIORS	IMPLICATIONS FOR CORTICAL FUNCTIONS	SUMMARY OF CORTICAL AREAS WHICH HAVE BEGUN FUNCTIONING BY THIS STAGE (HYPOTHESIZED)
4 The Coordination of the Secondary Schema and their Application to New Situations	8 months on	Goal directedness Serially and simultaneously coordinated actions Anticipation	Frontal association area functioning	All areas functioning; functioning of the association areas is not mature enough to permit the expression of all the parameters of intelligence discussed in this paper
		Simultaneous synthesis of unimodal and multimodal object properties Imitation of facial movements	Sensory association and parietal association area functioning	
		Beginning perception of causality outside of the infant's own actions Multimodal sequential perception	Temporal association area functioning	
		Incomplete body orientation in space	Immature parietal association area functioning	
		Inability to perceive sequential acts involving more than two acts Incomplete concepts of causality	Immature temporal association functioning	
		Inability to inhibit past sensations and responses Inability for internalization	Incomplete frontal association area functioning	

5 The Tertiary Circular Reaction and the Discovery of New Means through Active Experimentation	12–18 months	Trial-and-error search for new information and solutions to problems	Frontal association area functioning	All areas function, but the association areas do not function at a mature enough level to permit the expression of all the parameters of intelligence discussed in this paper
		Simultaneous and sequential synthesis of object-object and object-field relationships	Parietal and temporal association area functioning	
		Maturing concepts of causality	Immature frontal association area functioning	
		No internalization		
		Still does not completely perceive body as a separate object in space	Immature parietal association area functioning	
6 Representation and the Invention of New Means through Mental Combinations	18 months on	Internalized mental images of goals, and spatial and sequential concepts	Functioning of frontal, parietal and temporal association areas	All areas of the cortex function at a mature enough level to permit expression of all intellectual parameters discussed in this paper
		Perceives body as a separate object in space	Parietal association area functioning	

*Stage names and behavioral data from Piaget (1952) and Parker (this volume).

In essence, the secondary circular reaction is an action which demands an ability on the part of the infant to note sequential relationships between its own motor actions and the resulting behavior of objects. Developing synchronously are other behaviors which also demand these sequential abilities. One of these is a beginning perception of causality. The infant at this age recognizes cause and effect, as long as the cause is one of its own actions; in fact, this cause and effect capacity lies at the very basis of the secondary circular reaction. A primitive temporal sense based on the concept of before and after the infant's own actions and a beginning imitation based on the infant's attempts to cause a reaction in its mother by imitating her also develop at this time and are also based on the secondary circular reaction.

The Stage 3 infant displays the first rudimentary concepts of object permanence. This is an object permanence based on the concept that an infant's own actions of visual accommodation and prehension can sometimes maintain an object within the infant's sensory field when it would otherwise have disappeared and is signified by the appearance of more sophisticated visual accommodation abilities and "interrupted prehension" (see Parker, this volume), a behavioral action in which an infant who has just dropped an object tries to regain it by reinitiating prehensive actions. Thus, all of the behaviors in this stage indicate that the infant is object-oriented, but he perceives objects as they relate to his own motor actions.

The prime intellectual ability necessary for the infant's behavioral capacities during Stage 3 is the ability to perceive sequential relationships between its actions and resulting visual and auditory stimuli. Luria's model of cortical function relegated sequential perception to the temporal lobes, but it was based on sequential perception of auditory and visual, not somatosensory data.

It remains uncertain what cortical area actually mediates sequential perception of somatosensory-visual and somatosensory-auditory data; possibly it is the temporal association area. But as the parietal association area receives somatosensory data as well as visual and auditory data, the parietal association area is a prime candidate for this job. In any event, the ability for cross-sensory linkage of sequential data implies an initial function of at least one of the association areas, as it is the association areas which mediate multimodal perception.

None of the actions that an infant displays at this age are goal-directed in the true sense of the term. The infant initially discovers the relationship between its own action and the environmental results of that action by chance, although the action is later repeated with the goal in mind of reeliciting the environmental result. Parker (this volume) terms this chance discovery of the result "fortuitous goal direction." It is uncertain what area of the brain mediates "fortuitous goal direction," but as the frontal association area is known to be important in fully developed goal direction, this area is the most probable candidate. Possibly then, fortuitous goal direction indicates initial frontal association function. If so, this fits with the prediction of brain maturation data that the association areas mature in concert with each other and subsequent to the sensory association areas (which were hypothesized above to begin functioning in Stage 2), and with the findings which indicate definitive frontal association area function by Stage 4.

STAGE 4. "THE COORDINATION OF THE SECONDARY SCHEMA AND THEIR APPLICATION TO NEW SITUATIONS"—FROM EIGHT MONTHS ON

During Stage 3 the infant elaborates a number of behavioral schemata based on its own motor actions upon objects. During Stage 4, it begins to coordinate these schemata together both serially and simultaneously in pursuit of a common goal. The goal directedness emerges as a predominant characteristic of the Stage 4 behaviors and is a strong indication that the frontal association areas are beginning to function.

Other behaviors which are indicative of frontal association area function also emerge at this age. The infant develops rudimentary anticipation capacities as indicated by Parker's (this volume) statement that the infant can anticipate that he is about to be fed on the basis of sounds. The infant also displays the frontal association area capacity of inhibiting one action in order to move on to another, e.g., the ability to place an object by the feet and then kick it or the ability to drop one object in order to pick up another (Parker, this volume). Flexibility of behavioral patterns is another frontal lobe function which emerges at this age. The infant develops the capacity to combine and recombine his behavioral schemata in many diverse ways; each schema is capable of "inclusion and subsumption and exclusion" (Parker, this volume, page 48).

In addition to combining and recombining schemata in pursuit of goals, the infant begins to use each of the behavioral schemata developed during Stage 3 in a systematic examination of object properties, thus, each schema may be applied in serial order to a single object, i.e., the infant may shake, rub, mouth, and visually examine the object. These varied actions upon the same object provide diverse sensory stimuli from several modalities that can be synthesized in the sensory association and association areas into one coherent whole. The transition from the concept of the object as an entity which exists only in relationship to the infant's actions, to the concept of the object as a wholistic entity of many separate impressions which exists in its own right independently from the infant is considered by Piaget (1952) to be dependent upon this systematic examination of object properties. According to the neurological model presented above, it is also dependent upon the perceptual capacities of the parietal and temporal association areas.

Other behaviors also develop at this age which indicate a beginning capacity to perceive and synthesize simultaneous, spatial, and sequential information in visual and auditory modes and which are based, in part, upon this beginning understanding of object capacities developed through serial application of motor schemata to objects. The infant can find an object hidden behind a screen by someone other than itself, in other words, the infant has a concept of a space existing behind the screen and can perceive the behavioral sequences in placing the object in that space. It cannot, however, find the object if it did not see it being placed there or if it is moved behind several screens in sequential order. Possibly, it is unable to do this because its cortex is unable to perceive sequences based on multiple actions. Alternately, as Piaget (1952) suggests, this inability may be based on the fact that the infant's perceptions are still dominated by its own previous actions; thus, it continues to search behind a screen where it once found the object. If so, this would imply that although the frontal association areas are functioning, they are functioning at a still immature level as it is the frontal association area which mediates response inhibitions.

The Stage 4 infant still perceives its own actions as being the primary agents of causation, but at this stage it also begins to display behaviors which indicate that it has a concept that other people and objects may also be causal agents. For instance, the infant will touch its mother's hand in an effort to get her to move an object, indicating that it has an ability to understand that its mother's hand may act as a causal agent. This ability to perceive causation requires the ability to perceive sequential visual or auditory events and implies a definitive functioning of the temporal association areas which, according to the neurological model presented above, is a temporal lobe function.

Definitive functioning of the parietal association areas is displayed during Stage 4 as well; this is indicated by the infant's maturing abilities to imitate movements involving its own facial muscles which it cannot see. To do this, the infant requires well-developed abilities to synthesize simultaneous information from two or more sensory modalities as it must be able to equate the movements of its facial muscles, which it cannot see, to movements of the mother's facial muscles, which it cannot feel. Thus, the Stage 4 infant has definitive functioning of all association areas; none of these areas are yet functioning on a mature level, however, as indicated by the later emergence of new intellectual capacities in Stages 5 and 6.

STAGE 5. "THE TERTIARY CIRCULAR REACTION AND THE DISCOVERY OF NEW MEANS THROUGH ACTIVE EXPERIMENTATION"—TWELVE TO EIGHTEEN MONTHS

The tertiary circular reaction, like all circular reactions, is a repeated action. It is termed "tertiary" because it involves repeated actions on two or more objects at the same time, or repeated actions on objects relative to a field of force, i.e., gravity. The tertiary circular reactions are repeated in varying form in a manner which provides the infant with the opportunity to observe new properties of objects, i.e., an object may be dropped from varying heights and varying positions and directed toward varying targets (Parker, this volume). One special form of the tertiary circular reaction is trial-and-error problem-solving by means of object-object relationship, i.e., using a tool to rake in food, trial-and-error imitation of facial gestures and vocalizations.

The tertiary circular reaction is goal-directed in the sense that it is a purposeful search for new information or for solutions to problems. This, plus the flexibility of the reaction, indicates a maturing frontal association area.

The information provided by the tertiary circular reaction provides sequential and simultaneous sensory stimuli for synthesis in the parietal and temporal association areas. Through the synthesis of this information, the infant builds up concepts of object relationships, space and causality. By this age, the infant finally recognizes objects as independent entities, can follow a series of sequential object displacements behind screens, has a concept of space as a wholistic entity in which objects are located, and can recognize cause-and-effect relationships between objects and field forces outside of its own actions.

Thus, all association areas are functioning; however, they are not yet functioning in a completely mature fashion, as the infant still has no capacity for mental representation and still possesses immature spatial perception in that it does not yet perceive its own body as a separate object in space as demonstrated by its tendencies at this stage to attempt to lift objects it is standing on without first stepping off them.

STAGE 6. "REPRESENTATION AND THE INVENTION OF NEW MEANS THROUGH MENTAL COMBINATIONS"—EIGHTEEN MONTHS ON

Only one of the basic capacities of intelligence discussed in the introduction to this paper is still undeveloped at Stage 5: this is the capacity for internalization of mental images or representation which develops during Stage 6.

During Stage 5, the infant can recognize sequential displacements of an object behind a series of screens, but it can do this only if it directly observes the series of displacements; it has no internalized mental image of such displacements that allows it to find an object that is displaced at a time when the infant cannot see the displacement. Similarly, the infant can imitate its mother's expressions when she is present and it can observe them directly, but not when she is absent. By Stage 6 the infant has internalized mental images which allow it to find objects that have been hidden when it was not watching and to imitate its mother when she is not present.

Similarly, in Stage 5, the infant could solve problems only by means of overt trial-and-error; but by Stage 6, because of the infant's maturing capacities of internalization and representation, the infant can solve problems by means of mental combinations alone, i.e., insight. This capacity was hypothesized in the introduction to be primarily a frontal association area capacity, although it undoubtedly requires the frontal areas to function in conjunction with other cortical areas which mediate the simultaneous and sequential perceptions which have been internalized.

This capacity for internalized mental images implies that, by Stage 6, the cortex is functioning at a sufficiently mature level to permit the functioning of each of the parameters of intelligence listed in the introduction. Thus, each cortical area is functioning at this stage, even though the upper layers of many of these areas are still immature. The later maturation of intellectual properties and of the brain may be expected to be based on a more sophisticated functioning of each of these cortical areas rather than on the emergence of new neurological capacities based on the functioning of new cortical areas.

The intellectual parameters discussed are, of course, not the only factors which affect behavioral development—emotions and motivation are equally important. An important concept, however, emerges when the reinforcements that Parker (this volume) hypothesizes are functioning during each Piagetian stage are compared to the behaviors of that stage. There is a progressive change in reinforcers with neurological maturation. At each stage the infant behaves as if the use of the currently developing neurological capacity is reinforcing. In Stage 1, when only the motor and primary sensory areas are functioning, the infant is reinforced by performing all of the actions it is capable of performing—i.e., sucking, grasping, visual following, and other reflexes—and seeking only unimodal perceptions. As the sensory association areas which provide the capacity for simultaneous and sequential synthesis begin functioning, the infant seeks multimodal stimulation. During Stage 3, the secondary circular reaction is a predominant behavior. Parker hypothesizes, based on Watson's work (1972), that the secondary circular reaction is reinforced by perceived contingency, a new reinforcer that also emerges during Stage 3. Contingency reinforcement means that the infant is reinforced by the knowledge that results, i.e., sensory stimuli, are contingent on its own actions.

When the abilities for planned actions and for complicated simultaneous and sequential synthesis of perceptions mature, the infant finds exploration, novelty, and imitation reinforcing. At the same time, some of the early reinforcers become less important. While they probably never completely disappear, non-goal directed simple looking, touching, exploration, and imitation seem less important to adult humans or monkeys than to infants.

This pattern of behavioral maturation is undoubtedly biologically adaptive, for it assures that each portion of the cerebral cortex is subject to maximum sensory stimulation during what may be very critical periods of development. It is known, for instance, that the visual cells of the monkey cortex must be stimulated during the first 6–8 weeks of life or they become degenerate and the monkey can never develop normal visual perception (Hubel and Wiesel 1970). During this period, the visual cortex is just beginning to myelinate and the lateral geniculate body is exhibiting very rapid myelinization (K. R. Gibson 1970). If, in kittens, the visual stimulation is limited to horizontal or vertical lines during this early period, the neurons become sensitive only to horizontal or vertical lines and do not respond to other orientations (Blakemore 1974; Hirsch and Spinelli 1970). There is less evidence for this type of environmental effect for other cortical cells but it is possible, if not probable, that it exists.

■ **PARKER'S STAGES OF BEHAVIORAL DEVELOPMENT IN THE STUMPTAIL MACAQUE INFANT CORRELATED WITH CORTICAL DEVELOPMENT**

Parker (this volume) discerns six stages of the sensorimotor intelligence series in the stumptail macaque, *Macaca arctoides*:

Stage 1. *"The Reflex Primary Circular Reaction and Intercoordinations and Adaptations to the Environment"—birth to one or two weeks*

Stage 2. *"Primitive Voluntary Adaptations to the Environment and Intercoordinations"—one to twelve weeks*

Stage 3. *"The Secondary Linear Schema"—four to eleven weeks*

Stage 4. *"The Coordination of Secondary Linear Schema and their Application in New Situations"—nine weeks on*

Stage 5. *"The Trial-and-Error Secondary Linear Reaction and the Discovery of New Means through Active Experimentation"—twenty-three weeks on*

Stage 6. *"(Secondary Linear Trial-and-Error) Representation through Mental Combinations"—twenty-three weeks on*

Like the Stage 1 human infant, the Stage 1 stumptail macaque exhibits behavior patterns which possess few of the characteristics of intelligence; much of its behavior is reflexive. Its motor actions (sucking, clinging, looking) are global actions; these actions are characterized by little muscular differentiation and produce sensory stimuli in one modality only. Thus, the Stage 1 macaque resembles the human infant in that only the primary sensory areas and the motor cortex appear to be functioning. Even the motor cortex exhibits little function that can be discerned by observations of overt behavior. The evidence for motor cortex function in the neonatal macaque comes primarily from the work of Caveness et al. (1973) which indicates the existence of focal seizures of motor cortex origin.

By Stage 2, the infant macaque exhibits some premotor function in that it can inhibit reflexes. It also engages in "intercoordinations" (actions which result from the interaction of two or more body parts), i.e., "foot sucking" and "visually triggered bringing hand to mouth" (Parker, this volume), and which provide the first opportunity for simultaneous perceptions within and between sensory modalities. As in the human infant, this opportunity for simultaneous perceptions may indicate initial functioning of the sensory association areas.

The Stage 3 infant macaque displays some behaviors which are analogous to Stage 3 human infants: secondary schemata which consist of applying motor actions to single objects, "fortuitous goal direction," "interrupted prehension," and increased powers of visual accommodation. These actions demand rudimentary synthesis of stimuli from two or more sensory modalities and rudimentary goal direction; consequently, it is possible that they indicate initial association area functioning. However, the cross-modal perceptions of this stage are still too primitive to be certain of this interpretation.

By Stage 4, however, more definitive evidence of the functioning of the association areas exists. Frontal association area function in Stage 4 is indicated by the infant's ability to coordinate two or more actions in the pursuit of a common goal. Definitive parietal lobe function is indicated by the infant's ability to orient its body in space, i.e., the infant can engage in motor actions which require serial and simultaneous coordination and differentiation of locomotor patterns in respect to objects such as swings and fences. Parietal and temporal association area functions are implied during this stage by the infant's capacity to apply many separate schema to a single object, thus providing opportunities for simultaneous and sequential synthesis of sensory stimuli in several modalities. Temporal association area function is also indicated by the infant's ability to uncover an object after it observes it being hidden behind a screen, which implies the existence of the temporal association area function of sequential perception of auditory and visual stimuli.

Although all of these areas are functioning in the stumptail macaque during this stage, they are not functioning at a mature level as exhibited by the subsequent maturation during Stages 5 and 6 of certain types of trial-and-error behaviors, representation, and more definitive concepts of body orientation in space (Parker, this volume).

Thus, the pattern of behavioral maturation in the infant macaque is similar to that

TABLE 2 / PARKER'S STAGES OF SENSORIMOTOR INTELLIGENCE IN THE STUMPTAIL MACAQUE, MACACA ARCTOIDES, AND HYPOTHESIZED CORTICAL CORRELATES

SENSORIMOTOR STAGE	AGE	CHARACTERISTIC BEHAVIORS	IMPLICATIONS FOR CORTICAL FUNCTIONS	SUMMARY OF CORTICAL AREAS WHICH HAVE BEGUN FUNCTIONING BY THIS STAGE (HYPOTHESIZED)
1 The Reflex Primary Circular Reaction	Birth to one month	Focal paroxysms (Caveness et al., 1973)	Motor cortex functioning	
		Global actions	Motor cortex functions at an immature level	
		Little or no reflex inhibition; No goal direction; No serially coordinated actions	Little or no premotor or frontal association cortex functioning	Initial functioning of motor and primary sensory areas
		Actions directed toward unimodal sensory stimuli, only	Functioning of primary sensory areas; little or no functioning of sensory association areas	
2 Voluntary Adaptations to the Environment and Intercoordinations	1–12 weeks	Reflex inhibition	Initial premotor functioning	Initial functioning of motor, premotor, sensory and sensory association areas
		Intercoordinations which provide sensory stimuli for simultaneous synthesis	Initial sensory association area functioning	

3 The Secondary Linear Schema	4–11 weeks	Fortuitous goal direction	Initial frontal association area functioning	Initial functioning of motor, premotor, and sensory association areas
		Rudimentary sequential and simultaneous cross-modal perceptions	Initial parietal and temporal association area functioning?	Initial functioning of frontal, parietal, and temporal association areas?
4 The Coordination of Secondary Linear Schema and Their Application in New Situations	9 weeks on	Actions directed towards goals Serially and simultaneously coordinated actions	Frontal association area functioning	All areas function but functioning of the association areas is still not mature enough to permit the expression of all the parameters of intelligence discussed in this paper
		Rudimentary simultaneous and sequential synthesis of multi-modal object properties	Functioning of parietal and temporal association areas	
		Ability to orient body in space and differentiate actions according to spatial patterns	Parietal association area functioning	
		Ability to perceive sequential actions in visual modes	Temporal association area functioning	
		Little or no internalization No trial-and-error behavior	Incomplete frontal association area function	
		Incomplete concept of the body as a separate object in space	Incomplete parietal association area functioning	

continued

TABLE 2 / CONTINUED

SENSORIMOTOR STAGE	AGE	CHARACTERISTIC BEHAVIORS	IMPLICATIONS FOR CORTICAL FUNCTIONS	SUMMARY OF CORTICAL AREAS WHICH HAVE BEGUN FUNCTIONING BY THIS STAGE (HYPOTHESIZED)
5 and 6 The Trial-and-Error Secondary Linear Reaction and the Discovery of New Means through Active Experimentation (Secondary Linear Trial-and-Error) Representation through Mental Combinations	23 weeks on	Trial-and-error behavior Rudimentary internalization Awareness of body in space	Functioning of frontal, parietal and temporal association areas	All areas function at a mature enough level to permit the expression of all the parameters of intelligence discussed in this paper

*Stage names and data from Parker (this volume).

in the human infant in that the first cortically mediated behaviors to appear are those based on the functioning of the primary sensory and motor areas, while behaviors based on sensory association cortex and frontal, temporal, and parietal association cortex mature later. There is also a gradual increase with age in each of the cortically mediated aspects of intelligence: differentiation, simultaneous and sequential perception, goal directedness, flexibility of stimulus-response relationships, and anticipation.

■ DIFFERENCES BETWEEN STUMPTAIL MACAQUE AND HUMAN

Despite these similarities, there are important differences between the stumptail macaque and the human. A major and obvious difference is in the age of maturation of the various capacities. The human infant achieves all of the capacities of the sensorimotor series by two years, the macaque does so in about 6 months (Parker, this volume). In addition, the time frames for each of the 6 stages in the macaque overlap much more extensively than in humans. This difference in timing relates in a direct and obvious fashion to differences in rates of cortical maturation—the macaque cortex matures at a much faster rate and there is less time differential between the maturation of individual cortical areas (K. R. Gibson 1970).

A second obvious difference is the rate of locomotor maturation which is considerably faster in the macaque, in comparison to intellectual maturation, than in humans. An obvious neurological correlate of this increased speed of locomotor maturation, in comparison to intellectual maturation, does not exist in terms of cortical maturation alone. In the macaque, the sequence of maturation of cortical areas is similar to that in humans; in both, the motor cortex matures earliest (Gibson 1970). Locomotion is not determined primarily by the cortex, however. Non-cortical structures, such as the cerebellum and basal ganglia, play an important role in locomotion and it may be that the early locomotor maturation in the macaque is related to an earlier maturation of these subcortical areas relative to the cortex in the monkey as compared to the human. Further research is needed to answer this question.

Other differences between the stumptail macaque and the human lie in the forms of intelligence that each displays in adulthood and during ontogeny. The human has a greater overall intelligence which is superior to that of the macaque in many ways, including tool-using, imitative, and language capacities. Each of these skills requires a greater development of each of the parameters of intelligence listed at the beginning of this paper: differentiation, serial and simultaneous perception and coordination of actions, flexibility of stimulus-response relationships, anticipation of future perceptions, and internalization.

These increased intellectual capacities result, in part, from the greatly increased size of the human cerebral cortex, particularly in the frontal, parietal, and temporal association areas. Some neurological reorganization has also accompanied these increased capacities, although the total extent of this reorganization and its exact contribution to the behavioral differences between humans and macaques is uncertain.

One form this reorganization has taken is a decrease in neuronal density and an accompanying increase in the density of dendritic branching in the human cortex as compared to the macaque. Thus, although humans have more cortical nerve cells than macaques, they do not have as many additional neurons as they have increased cortical mass. The neurons they do have, however, have more interconnections; thus, the human cerebral cortex has a greater capacity for neuronal interactions than does the macaque cerebral cortex. This suggests greater information processing capacity in the human cortex (Jerison 1973).

A second form that neurological reorganization has taken is a change in the patterns of axonal connectivity between cortical and subcortical regions. The best example of this is in the pyramidal tracts. The fibers of the pyramidal tracts descend from the cortex (primarily motor and somatosensory cortex, although other cortical areas contribute fibers to this tract as well) to the spinal cord where they exert a modulating action on muscle function. In most mammals the majority of pyramidal fibers project to the dorsal horn of the spinal cord which contains sensory neurons. In the primate order there is an increase in the number of pyramidal fibers which project directly to the motor neurons of the cord (Kuypers 1958a, 1958b). These direct connections permit greater cortical control of individual muscles and correlate with the ability for fine control of muscular actions, particularly of the hands. A higher percentage of these direct motor projections are found in the human than in the monkey, thus permitting a greater differentiation of muscle actions, at least in some body parts. The extent to which similar differential projections in other tracts may account for other species' differences in behavior is as yet unknown.

These differential action capacities based on the greater human abilities for discrete muscular control are an important factor distinguishing many human skills from the skills of macaques. Tool use, vocalization, and imitation of facial gestures are all dependent on the ability to differentiate and control discrete muscular actions.

These capacities, of course, demand more than the ability to produce discrete muscle actions; the actions must be coordinated both in space and time. Many human activities such as typing, writing, and speaking require the ability to make smooth, quick, and precise transitions between the discrete muscle actions of the fingers or vocal apparatus and to serially organize and reorganize these actions. According to Luria (1966), such smooth serial transitions are a function of the premotor cortex. The macaque can produce smooth, serially organized and reorganized transitions between some motor actions, especially locomotor patterns (Parker, this volume), but apparently does not do so for the vocal muscles and the facial muscles, or at least does not do so voluntarily. These differences between macaque and human imply differential function of the premotor region. Perhaps this differential function is based on the anatomical projections of the pyramidal tract which provide for discrete muscular control in general, or possibly it requires some reorganization of the premotor cortex itself. Further research is necessary to solve this problem.

Spatial organization of muscle actions is equally important to serial organization in tool-use, vocalization, and imitation; e.g., in vocalization a very small difference in tongue position can make a difference between a *t* sound and a *d* sound. In order to learn a spoken language, the infant not only needs fine spatial control of the muscles of the vocal apparatus, but it must also be able to perceive the simultaneous relationships of the muscular actions and the sound, a capacity the macaque may be lacking. This ability for simultaneous perception of cross-modal stimuli is essential for object-naming capacities which require the ability for cross-modal perception in the auditory-visual channels (Geschwind 1965).

Simultaneous perception and synthesis also play a role in imitation. Even a relatively simple imitative act, such as the imitation of the hand position of another animal, demands simultaneous coordination of finely distinguished proprioceptive sensations of the imitator's hand with differential visual perceptions of both its hand and that of another. A more complex imitative act, such as that of an infant imitating its mother sticking her tongue out, requires very well-developed simultaneous synthetic capacities. The infant must somehow relate the proprioceptive sensations of its own tongue, which it has never seen, to the visual perceptions of its mother's tongue, which it has never felt. Piaget (1952) hypothesizes that this occurs via an auditory intermediate. According to Piaget, the infant notes the sounds produced by its own tongue movements and relates those to its proprioceptive sensations. The infant also notes that the sounds it produces are identical to those of its mother, and that the sounds its mother makes occur simultaneously with visual movements. The infant may then hypothesize that since proprioceptive stimuli from its own tongue and visual impulses from its mother's tongue occur simultaneously with the same sounds, that they may, in fact, be the same organ. Rarely, if ever, do monkeys engage in actions requiring this much simultaneous perception.

The ability to perceive and anticipate sequential relations is also expanded in humans as compared to macaques. Many human behaviors require the ability to note relationships between many successive stimuli, not just two or three as in most monkey behaviors. It is doubtful, for instance, that any behavior of the monkey demands the ability to note relationships between many successive stimuli, not just two or three as in most monkey behaviors. It is doubtful, for instance, that any behavior of the monkey demands. the ability to perceive and anticipate as many sequential relationships as is essential for a skilled chess player.

Moreover, many human behaviors require the ability to perceive sequential relationships not only between relatively simple perceptions but also between perceptions which are already highly synthesized spatially in the simultaneous mode. This is true of a human operating a complicated machine and also in seemingly simple mechanical actions. An individual who attempts to pry open the lid of a box with a lever must perceive the spatial configuration of the proprioceptive impulses of its arm, hand and body; the tactile impressions of the lever in the hand; the weight and force which he exerted on the lever; the resistance of the lid; the spatial configuration of the lid and box; and all of the visual and proprioceptive responses which follow its actions. If, instead of a lever and a box, the individual has a ping-pong paddle and ball, most of the above factors must be perceived, but the player must also note how the ball moves spatially and temporally through the air, where and with how much force it lands, and the relationship between all of these factors and how the ball will be returned. When it is returned he must then synthesize many other separate impressions of the relationship between where, when, and how he attempts to hit the ball and where and how it lands.

Neither simultaneous nor sequential perception, of course, is new in humans; both are present in less developed form in the other primates. The neural mechanisms which provide the human with expanded simultaneous and sequential capacities lie within the parietal and temporal association areas. These are very much larger in humans than in most other primates (Geschwind 1965) and the increased capacities almost certainly relate, in part, to size increase. In part, the increased capacities may also result from an anatomical and functional reorganization of these areas; there is little evidence for such reorganization, however.

Another neuroanatomical area which has demonstrated some enlargement in humans, as compared to most other primates, is the frontal lobe. This area mediates behaviors which, perhaps more than any other, characterize the human's greater intellectual capacity. Prime functions of the frontal lobe include internalization, flexibility, and the ability to anticipate future perceptions. The human's expanded capacities for flexibility are the basis of all creative endeavors, while the human's expanded anticipatory capacities are at the basis of the concept of future and of the ability to anticipate and plan many serially organized acts extending far into the future. This capacity, of course, is not based on the frontal lobe function alone, but also on internalized concepts of sequential relationships between actions and results derived from sequential perceptions of the left temporal lobe. In addition, frontal lobe functions are essential to speech. Reproduction of speech requires the ability to sequentially order fine motor movements. Discussion of objects not present in the environment depends on the capacity of internalization, and the utterance of original observations requires flexibility of actions and ideas.

Less obvious to the uninitiated observer than the differences in intelligence between adult stumptail macaques and humans are differences in the intellectual performances of the infants. The recent work by Parker (this volume) elucidates some of these differences.

The distinctions begin in infancy with the tendencies of the human infant, but not the macaque infant, to perform primary circular reactions of a voluntary nature, voluntary reciprocal coordinations, secondary circular reactions, and tertiary circular reactions. The primary circular reaction consists of repetitive actions involving single organs—tongue, hand, etc. Reciprocal coordinations are actions in which two or more body parts are coordinated; they are usually repetitive. Secondary circular reactions are repetitive actions upon a single object; tertiary circular reactions are repetitive actions upon two objects at a time or upon one object relative to a field of force. Primary circular reactions and secondary circular reactions, as performed by the human infant, do not differ from the actions of the stumptail macaque so much in the form of the motor actions as in their repetitive nature. The neurological correlate of these repetitive actions is unknown; the actions may relate to the delayed human locomotor capacity. The young macaque spends much of its time practicing new locomotor patterns; the young human infant is incapable of locomotor behavior and thus has more time to devote to the practice of motor patterns involving the hands and mouth. The primary and secondary circular reactions may result primarily from this increased time for practice.

Alternately, the primary circular reaction may be based on differences in cortical function between macaques and humans. The smooth starting and stopping components of the human primary circular reaction suggest premotor control, as it is the premotor areas which are responsible, in part, for smooth transitions between "intelligent" motor actions in the adult (Luria 1966). Many behavioral capacities in the human adult, such as typing and writing, suggest increased premotor function in the human as compared to the macaque. Thus, the primary circular reaction may indicate differential premotor function in the macaque and human infant.

The secondary circular reaction also is repetitive and its repetitive nature may also be based on premotor differentiation. In addition, the secondary circular reaction is characterized by the ability to perceive sequential relationships between actions and results, i.e., to perceive causality. The stumptail macaque has some ability to perceive causality and apparently can perceive causality between the types of actions and results which form a basis for many of the secondary circular reactions displayed by the human infant (Parker, this volume), but the adult macaque's capacities for perception of sequential actions and causality are not as fully developed as the adult human's. This greater development of sequential and causal perception in the human most likely results from increased temporal association area functioning as the temporal association area is important in sequential perception (Luria 1966). One possible explanation of the secondary circular reaction is that the temporal association or other association areas are already functioning in an increased capacity in the human infant as compared to the macaque infant.

Typical secondary circular reactions are the repetitive shaking of a rattle and the repetitive movement of a bassinette ornament. According to Piaget, the infant performs the secondary circular reaction for the purpose of "recreating the interesting spectacle he realizes he has engendered" (Parker, this volume). An alternate reason for the lack of a secondary circular reaction in the macaque is that the macaque may not consider the sounds of the rattle or the movement of the bassinette interesting. In the case of rattles and bassinette ornaments, the resulting perceptions are simple perceptions demanding little simultaneous or sequential synthesis. One might expect them to be most reinforcing to an immature brain in which the association areas are not well developed. In the monkey the association areas develop more quickly than in humans and it is possible, for this reason, that simple perceptions are less reinforcing to the young monkey than to the young human. If so, the monkey should engage in circular reactions if the resulting perception is "interesting."

"Reciprocal coordinations" of the human infant resemble intercoordinations of the macaque infant in that both are based on coordinated actions between two separate body parts. Voluntary (as opposed to reflex) reciprocal coordinations of the human infant are repetitive actions, while the macaque's intercoordinated actions are not. The reciprocal coordinations of the human infant also differ from the intercoordinations of the macaque in that the human attempts to "assimilate" and "accommodate" the two organs to each other (Parker, this volume). This "assimilation" and "accommodation" provide data for a synthesis of separate sensory impressions into one simultaneous whole. The repetitive aspect of the reciprocal coordinations may be based on premotor functioning, just as the repetitive aspects of the primary and secondary circular reactions may be based on the premotor cortex. The mutual assimilation and accommodation, with its accompanying synthesis of sensory impressions, is most likely to result from sensory association area and parietal association area functioning, as it is these areas which are responsible for simultaneous synthesis of separate sensory impressions. If so, this would indicate differential functioning of the association areas in the human infant as compared to the macaque infant.

The tertiary circular reaction consists of trial-and-error search for new object properties. The actions of the tertiary circular reaction are repeated actions, but they are repeated in variable form and are directed towards two objects at a time or towards object-field interrelationships. The variable form of the repeated actions implies frontal association cortex functioning. The ability to focus on object-object interactions is based on the capacity to note simultaneous and sequential relationships between objects; these are parietal and temporal association area capacities. The existence of the tertiary circular reaction in the human infant as compared to the macaque seems, therefore, to denote differential frontal, parietal, and temporal association area functioning in the infants of the two species; differential functioning, as well as differential size in parietal and temporal areas, is known to exist between the adults of the two species (see discussion above).

In conclusion, the neurological capacities which underlie the human infant's capacities for the primary, secondary, and tertiary circular reactions and reciprocal coordinations are not known with any certainty. One hypothesis that would fit most of the data is that cortical areas which function differentially in the human adult as compared to the stumptail macaque adult are reorganized in such a manner that they already function differentially in human and macaque infants when the cortical areas themselves are largely immature, that is, differential functioning of the immature premotor area in human and stumptail macaque infants may account for the existence of the primary and secondary circular reactions in the former species but not the latter; and differential functioning of immature frontal, parietal, and temporal association areas may account for the tertiary circular reaction in the human infant but not the stumptail macaque infant.

If so, this implies that the anatomical reorganization of the cortical area which permits these differential action capacities in the infants is a reorganization which involves the earliest maturing portions of the cortical areas, i.e., layers V and VI, rather than layers II–IV. Layers V and VI give rise to long efferent projection fibers which descend from the cortex to subcortical regions. Some investigators consider that these long fibers are "prewired" genetically (Jacobsen 1974; Sperry 1973); in other words, the anatomical connections are genetically inflexible and not usually subject to environmental modifications. In contrast, layers II–IV, which are the last layers to mature, are populated by small cells with short axons and are primarily receptor areas which receive sensory stimuli from subcortical or other cortical regions. Small cells and the upper cortical layers are subject to environmental modification of all types (Jacobsen 1974).

Anatomically, then, it appears that certain genetically inflexible structures of each cortical area mature first and are followed by structures whose anatomy is more subject to environmental modification. This may correlate with behavioral maturation—new capacities of the human infant, such as the capacity for the secondary and tertiary circular reactions, may be provided by the earliest maturing fibers and layers of the cerebral cortex, i.e., they may be based on these long genetically inflexible fibers and the lower cortical layers. On the other hand, the information gained from these actions, and the overall cognitive level reached by any given infant on the basis of these actions, may depend in large part on cortical and environmental interactions involving the more genetically flexible small cells and upper cortical layers.

▪ CONCLUSION

A number of hypotheses were presented relevant to the organization and maturation of intelligence and the brain in the stumptail macaque and human infant. It is suggested that intelligence is composed of the following general factors which form a basis of most of the Piagetian stages of intelligence: differentiation, serial and simultaneous perception and coordination of motor acts; flexibility of stimulus-response relationships; anticipation; comparison of anticipated perceptions to actual perceptions; and internalization. Each of these parameters of intelligence may be mediated by particular cortical regions: differentiation by all sensory and motor areas; serial and simultaneous coordination of motor acts by the premotor areas; perception of simultaneous and spatial relationships between sensory stimuli by the right cerebral cortex, particularly the parietal lobe; perception of sequential relationships between stimuli by the left cortex, particularly the left temporal lobe; flexibility of stimulus-response relationships; comparison of anticipated perceptions to actual perceptions; and internalization by the frontal lobes.

The cortical regions mature in the sequence: motor and primary sensory areas; premotor and somatosensory association areas; frontal and parietal association areas. The parameters of intelligence mature in the same sequence as the cortical areas which are hypothesized to mediate them. Differences in intelligence between stumptail macaque and human adults are based in large part on differential development of each of the factors of intelligence suggested in this paper. Differences in intelligence between stumptail macaque and human infants indicate that the cortical areas of the human and macaque infants function differently even at a stage when their functioning is still immature. This implies that differential functioning of each cortical area is based in part on the anatomical properties of the earliest maturing portions of each cortical layer, i.e., layers V and VI.

▪ REFERENCES

Ackerly, S. S. 1964. A case of paranatal bilateral frontal lobe defect observed for thirty years. In *The frontal granular cortex and behavior*, eds. J. M. Warren and K. Akert, pp. 192–218. New York: McGraw-Hill.

Angulo y Gonzalez, A. W. 1929. Is myelinogeny an absolute index of behavioral capability? *J. of Comp. Neurol.* 48:439–464.

"Assymmetry found in monkey brain." *Yerkes Newsletter* (February, 1976), p. 22.

Bergström, R. M. 1969. Electrical parameters of the brain during ontogeny. In *Brain and early behavior*, ed. R. J. Robinson, pp. 15–84. New York: Academic Press.

Blakemore, C. 1974. Developmental factors in the formation of feature extracting neurons. In *The neurosciences*, eds. F. O. Schmitt and F. G. Worden, pp. 105–113. Third Study Program, M.I.T.

Bodian, D. 1966. Development of fine structure of spinal cord in monkey fetuses. *Bulletin of the Johns Hopkins Hospital* 119:129–149.

Bower, T. G. R. 1969. Perceptual functioning in early infancy. In *Brain and early behavior*, ed. R. J. Robinson, pp. 211–223. New York: Academic Press.

Brazelton, T.; Scholl, M. D.; and Robez, J. S. 1966. Visual responses in the newborn. *Pediatrics* 37:284–290.

Bridger, W. H. 1961. Habituation and discrimination in the human neonate. *Amer. J. Psychiatry* 117:991–996.

Butler, R. A. 1960. Acquired drives and the curiosity-investigative motives. In *Principles of comparative psychology*, eds. R. H. Waters, D. A. Rethlingshafer, and W. E. Caldwell, pp. 154–173. New York: McGraw-Hill.

Butters, N.; Barton, M.; and Brody, B. A. 1970. Role of the right parietal lobe in the mediation of cross-modal association and reversible operations in space. *Cortex* 6:174–190.

Carmon, A., and Nachston, I. 1971. Effect of unilateral brain damage on perception of temporal order. *Cortex* 7:410–418.

Caveness, W. F. 1969. Ontogeny of focal seizures. In *Basic mechanisms of the epilepsies*, eds. H. H. Jasper, A. A. Ward, Jr., and A. Pope, pp. 517–534. Boston: Little, Brown Co.

Caveness, W. F.; Echlin, F. A.; Kemper, T. L.; and Kato, M. 1973. The propagation of focal paroxymal activity in the *Macaca mulatta* at birth and 24 months. *Brain* 96:757–764.

Chevalier-Skolnikoff, S. 1974. *The ontogeny of communication in the stumptail macaque*. Basel: S. Karger.

Coghill, G. E. 1929. *Anatomy and the problem of behavior*. London: Cambridge University Press.

Conel, J. S. 1939. *The postnatal development of the human cerebral cortex*. Harvard University Press, Cambridge: Vol. 1, *The cortex of the new born child*; Vol. 2, *The cortex of the one month child* (1941); Vol. 3, *The cortex of the three month child* (1947); Vol. 4, *The cortex of the six month child* (1951); Vol. 5, *The cortex of the 15 month child* (1955); Vol. 6, *The cortex of the 24 month child* (1959); Vol. 7, *The cortex of the six year old child* (1963).

Connolly, K., and Elliott, J. 1972. The evolution and ontogeny of hand function. In *Ethological studies of child behavior*, ed. N. B. Jones, pp. 329–383. New York: Cambridge University Press.

Crosby, E. C.; Humphrey, T.; and Lauer, E. W. 1962. *Correlative anatomy of the nervous system*. New York: The Macmillan Company.

Debecker, J.; Desmedt, J. E.; and Corazyna, H. 1973. *Société belge de pharmacologie reunion de Bruxelles*, pp. 554–556.

Denny Brown, D., and Chambers, R. A. 1958. The parietal lobe and behavior. In "The Brain and Human Behavior," *Research Pub. Assoc. for Research in Nervous and Mental Disease* 36:35–117.

Dewson, J. H., and Burlingame, A. C. 1975. Auditory discrimination and recall in monkeys. *Science* 187:267–268.

Duffy, F. H., and Buschfiel, J. L. 1971. Somatosensory system: Organizational hierarchy from single units in monkey Area 5. *Science* 172:273–275.

Efron, R. 1963. The effect of handedness on the perception of simultaneity and temporal order. *Brain* 86:261–284.

Ellingson, R. J., and Rose, G. H. 1970. Ontogenesis of the electroencephalogram. In *Developmental neurobiology*, ed. W. Himwich, pp. 441–474. Springfield: Charles Thomas.

Evarts, E. 1973. Brain mechanisms in movement. *Scientific American* 229:92–103.

Fantz, R. 1965. Visual perception from birth as shown. by pattern selectivity. *Annals of New York Academy of Science* 118:793–814.

———. 1965. Ontogeny of perception. In *Behavior of non-human primates*, eds. A. M. Schrier, H. F. Harlow and F. Stollnitz. Vol. 11, pp. 365–404.

Flechsig, P. 1920. *Anatomie des menschlichen Gehirns und Rückenmarks auf Grundlage*. Leipzig: George Thomas.

———. 1927. *Meine myelogenetische Hirnlehse*. Berlin: Springer.

Gazzinaga, M. S. 1970. *The bisected brain*. New York: Appleton-Century Crofts.

Geschwind, N. 1965. Disconnexion syndromes in animals and man. *Brain* 88:237–294, 585–652.

Gibson, E. J. 1969. *Principles of perceptual learning and development*. New York: Appleton-Century Crofts.

———. 1970. The development of perception as an adaptive process. *American Scientist* 58:98–107.

Gibson, K. R. 1970. Sequence of myelinization in the brain of *Macaca mulatta*. Ph.D. dissertation, University of California, Berkeley.

Gross, C. G.; Rocha-Miranda, C. E.; and Bender, D. B. 1972. Visual properties of neurons in inferotemporal cortex in the macaque. *J. Neurophysiol.* 35:96–111.

Grueninger, W. E., and Pribram, K. H. 1969. Effects of spatial and nonspatial distractors on performance latency of monkeys with frontal lesions. *J. Comp. Physiol. Psych.* 68:203–209.

Gunther, M. 1961. Behavior at the breast. In *Determinants of infant behaviour*, ed. B. M. Foss, pp. 37–40. New York: John Wiley and Sons.

Halverson, H. M. 1943. The development of prehension in infants. In *Child Development and behavior*, eds. R. G. Barker, J. S. Kounin and H. F. Wright. London: McGraw-Hill.

Hamburger, U. 1963. Some aspects of the embryology of behavior. *Quart. Rev. of Biol.* 38:342–365.

Harlow, H. F. 1959. The development of learning in the rhesus monkey. *American Scientist* 47: 459–479.

Hepp-Reymond, M. C., and Wiesendanger, M. 1972. Unilateral pyramidotomy in monkeys: Effect on force and speed of a conditioned grip. *Brain Research* 36:117–131.

Hill, K.; Cogan, D.; and Dodge, P. 1963. Ocular signs associated with hydranencephaly. *Am. J. Ophthalmology* 51:267–275.

Hines, M. 1942. The development and regression of reflexes and progression in the young macaque. *Carnegie Inst. of Washington Contr. to embryol.* 30:153–209.

Hines, M., and Boynton, E. P. 1940. The maturation of "excitability" in the precentral gyrus of the young monkey (*Macaca mulatta*). Carnegie Inst. of Washington, Pub. 518, *Contr. to embryol.* 28: 309–451.

Hirsch, H., and Spinelli, D. N. 1970. Visual experience modifies distribution of vertically oriented receptive fields in cats. *Science* 168:869.

Hooker, D. 1957. *Evidence of prenatal function in the central nervous system in man*. James Arthur Lecture on the Evolution of the Human Brain, American Museum of Natural History, New York.

———. 1972. *The prenatal origins of behavior*. University of Kansas Press.

Hubel, D. H., and Wiesel, T. N. 1963. Receptive fields of cells in striate cortex of very young visually inexperienced kittens. *J. of Neurophysiology* 26:994–1002.

———. 1970. The period of susceptivity to the physiological effects of unilateral eye closure in kittens. *J. Physiol.* 206:419–436.

Humphrey, T. 1970. The development of human fetal activity and its relation to postnatal behavior. In *Advances in Child Development and Behavior*, eds. H. W. Reese and L. P. Lipsitt. 5:1–56. New York: Academic Press.

Jacobsen, M. 1974. A plentitude of neurons. In *Aspects of Neurogenesis*, ed. G. Gottlieb. 2:151–166. New York: Academic Press.

Jensen, G. D. 1961. The development of prehension in a macaque. *J. Comp. Physiol. Psychol.* 54:11–12.

Jerison, H. J. 1973. *Evolution of the brain and intelligence.* New York: Academic Press.

Kagan, J. 1972. Do infants think? *Scientific American* 226:74–83.

Kuypers, H. G. J. M. 1956. Cortical projections to the pons and the medulla oblongata in cat and man. *Anat. Rec.* 124:322–323.

———. 1958a. Some projections from the pericentral cortex to the pons and lower brain stem in the monkey and chimpanzee. *J. Comp. Neurol.* 110:221–255.

———. 1958b. Corticobulbar connexions to the pons and lower brain stem in man: An anatomical study. *Brain* 81:364–389.

———. 1962. Corticospinal connections: Postnatal development in the rhesus monkey. *Science* 138:678–680.

Langworthy, O. L. 1928. The behavior of pouch young oppossums correlated with the myelinization of tracts in the nervous system. *J. Comp. Neurol.* 46:201–240.

Lawrence, D. G., and Hopkins, D. A. 1972. Developmental aspects of pyramidal motor control in the rhesus monkey. *Brain Research* 40:117–118.

Lipsitt, L. P. 1969. Learning capacities of the human infant. In *Brain and early behavior*, ed. R. J. Robinson, pp. 227–249. New York: Academic Press.

Luria, A. R. 1966. *Higher cortical functions in man.* New York: Basic Books.

Luria, A. R., and Homskaya, E. D. 1964. Disturbances in the regulative role of speech with frontal lobe lesions. In *The frontal granular cortex and behavior*, eds. J. M. Warren and K. Akert, pp. 353–371. New York: McGraw-Hill.

McConnell, J. V. 1962. Memory transfer through cannibalism in planarians. *J. Neuropsychiat.* 3(Suppl.1):S42–S48.

Milner, B. 1964. Some effects of frontal lobectomy in man. In *The frontal granular cortex and behavior*, eds. J. M. Warren and K. Akert, pp. 313–334. New York: McGraw-Hill.

Milner, B. 1967. Brain mechanisms suggested by studies of the temporal lobes. In *Brain mechanisms underlying speech and language*, ed. F. L. Darley, pp. 122–145. New York: Grune and Stratton.

Peiper, A. 1963. *Cerebral function in infancy and childhood.* New York: Consultant's Bureau.

Piaget, J. 1952. *The origins of intelligence in children.* New York: Norton and Co.

———. 1954. *The construction of reality in the child.* New York: Ballantine Books.

———. 1966. *The psychology of intelligence.* Totowa: Littlefield.

Pohl, W.; Butters, N.; and Goodglass, H. 1972. Spatial discrimination systems and cerebral lateralization. *Cortex* 8:305–312.

Pribram, K. 1971. *Languages of the brain.* Englewood Cliffs: Prentice Hall, Inc.

Robinson, R. J. 1966. Cerebral function in the newborn child. *Developmental Medicine and Child Neurology* 8:561–567.

———. 1969. Cerebral hemisphere function in the newborn. In *Brain and early behavior*, ed. R. J. Robinson. New York: Academic Press.

Romanes, G. J. 1947. Prenatal medullation of the sheep's nervous system. *J. of Anatomy* 81:64–81.

Sackett, G. P. 1966. Development of preference for differentially complex patterns by infant monkeys. *Psychon. Sci.* 6:441–442.

Sperry, R. W. 1963. Chemoaffinity in the orderly growth of nerve fiber patterns and connections. *Proceedings of the National Academy of Sciences, U.S.* 50:703–710.

Stamm, J. S. 1964. Retardation and facilitation in learning by stimulation of frontal cortex in monkeys. In *The frontal granular cortex and behavior*, eds. J. M. Warren and K. Akert, pp. 102–125. New York: McGraw-Hill.

Ulett, G.; Dow, R. S.; and Larsell, O. 1944. The inception of conductivity in the corpus callosum and
    the corticopontine-cerebellar pathway of young rabbits with reference to myelinization. *J. Comp.
    Neurol.* 89:1–10.
Windle, W. F. 1944. Genesis of somatic motor functions in mammalian embryos: A synthesizing
    article. *Physiological Zoology* 17:247–260.

4 | A Piagetian Model for Describing and Comparing Socialization in Monkey, Ape, and Human Infants

SUZANNE CHEVALIER-SKOLNIKOFF

■ INTRODUCTION

Primatologists are in general agreement that socialization is a process whereby an individual's future behavior is shaped by his social experiences and the culture or protoculture of his group. Socialization ensures that all adaptive behavior need not be rediscovered by each member of each generation (Poirier 1972, this volume). They also agree that socialization is the product of an interaction between the biological makeup of the individual and the social influences that impinge upon him. However, description of the socialization process and cross-species comparisons of socialization have proved difficult. One of the major reasons for this is the large number of variables, both biological and social, that contribute to the process (Hinde 1974). The other reason is the lack of methods for systematically describing the socialization process—the changes that occur through time during development—and for describing differences in the process between species.

Piaget's model of human intellectual development during the sensorimotor period (birth–2 years of age) provides a systematic and objective framework for comparative studies. The power of Piaget's model is its scope and its systematic approach. In scope it encompasses the simplest to the most complex behavior; it transects age levels; and it incorporates descriptive, functional and causal[1] approaches to behavioral analysis. The model also provides a systematic and relatively objective framework for behavioral studies for it defines a series of six qualitatively different sensorimotor stages that appear sequentially in human infant development. Each stage incorporates increasing voluntary control, increasing complexity (the addition of new motor patterns and more contextual variables), and increasing flexibility (nonstereotyped motor patterns and more varied contexts) in successive stages. Each stage in the series is characterized by a new level of functioning based on new abilities that becomes incorporated into the infant's behavior.

This is a preliminary description and comparison of the socialization process in primates using methods derived from Piaget's model (Piaget 1951, 1952, 1954). Behavior relevant to two of Piaget's Sensorimotor Series, the Imitation Series and the Sensorimotor Intelligence Series, are examined since they appear to be most relevant

159

to the socialization process. The paper is based upon the premise that potential for socialization is dependent upon intellectual capacity, and that this capacity changes during development, and may vary across species. I will examine the intellectual capacities of representative species from three phylogenetic levels within the primates and will attempt to analyze and compare across species aspects of the socialization process in terms of Piagetian abilities.

■ METHODS

SUBJECTS

This study is based on data on old world monkeys (stumptail macaques), apes (lowland gorillas and chimpanzees) and human infants.

The data on stumptail macaques (*Macaca arctoides*) are derived from the study on intellectual development by Parker (1973; see also Parker, this volume) and on my previous study on communicative development (Chevalier-Skolnikoff 1971, 1973, 1974a). Parker observed the intellectual development of one infant for 137 hours from birth to 5 months. The infant was living with his mother and most of the time also with an older infant. My study, consisting of 500 hours of observation of three infants—one male and two females—from birth to 6 months, living in a social group consisting of seven adult and subadult animals, focused on their communicative development.

The ape data are derived from a study currently in progress. Since the initiation of the study in January 1975, 215 hours of data have been collected. The ape sample consists of five infant and five adult lowland gorillas (*Gorilla gorilla gorilla*), and six infant and six adult chimpanzees (*Pan troglodytes*). The ape subjects belong to the

TABLE 1				
SPECIES	SUBJECT	SEX	AGES OBSERVED	INSTITUTION
Gorilla	*Mkumbwa	M	Birth–1 year	San Francisco Zoological Society
Gorilla	**Binti	F	7 months 8 months	San Diego Zoological Society
Gorilla	*Sunshine	M	13 months–2½ years	San Francisco Zoological Society
Gorilla	***Jim	M	18 months 20 months	San Diego Zoological Society
Gorilla	***Koko	F	4 years	San Francisco Zoological Society and Stanford University (Being taught sign language by Penny Patterson)

continued

TABLE 1 / CONTINUED

Gorilla	Dolly (Binti's mother)	F	Adult	San Diego Zoological Society
Gorilla	Missus (Sunshine's mother)	F	Adult	San Francisco Zoological Society
Gorilla	Jacqueline (Koko & Mkumbwa's mother)	F	Adult	San Francisco Zoological Society
Gorilla	Pogo	F	Adult	San Francisco Zoological Society
Gorilla	Bwana	M	Adult	San Francisco Zoological Society
Chimpanzee	*Betty	F	4 months–8 months 11 months	Stanford Outdoor Primate Facility
Chimpanzee	*Palita	F	12 months–16 months	Stanford Outdoor Primate Facility
Chimpanzee	*Delta	F	32 months–3 years	Stanford Outdoor Primate Facility
Chimpanzee	***Mowgli	M	About 2 years	Stanford Outdoor Primate Facility
Chimpanzee	***Babu	M	About 5 years	Stanford Outdoor Primate Facility
Chimpanzee	***Zippy	M	About 6 years	Stanford Outdoor Primate Facility
Chimpanzee	Bido (Betty's mother)	F	Adult	Stanford Outdoor Primate Facility
Chimpanzee	Poly (Palita's mother)	F	Adult	Stanford Outdoor Primate Facility
Chimpanzee	Gigi (Delta's mother)	F	Adult	Stanford Outdoor Primate Facility
Chimpanzee	Rock	M	Adult	Stanford Outdoor Primate Facility
Chimpanzee	Shadow	M	Adult	Stanford Outdoor Primate Facility
Chimpanzee	Bandit	M	Adult	Stanford Outdoor Primate Facility

*Reared with mother in social group.
**Mother-reared.
***Nursery- or home-reared.

San Francisco Zoological Society, the Stanford Outdoor Primate Facility, Stanford University, and the San Diego Zoological Society.

Two of the five infant gorillas are being reared by their mothers in a social group, one is being reared alone by her mother, and two have been nursery-reared. Four of the five adults are living in a social group, and one is living with her infant only. Three of the six infant chimpanzees are being reared by their mothers in a social group. The other three infants, home-reared during their early months, are now living in a social group composed of both infant and adult animals. The six adult chimpanzees are all living in social groups (see Table 1).

The data on human infants are derived from the literature, primarily from the work of Piaget (1951, 1952, 1954).

DATA COLLECTION PROCEDURES

Both cross-sectional and longitudinal methods were used in collecting the ape data. Twenty cross-sectional samples, plus three short (four-month) longitudinal samples have been collected. Two infant apes (Sunshine and Mkumbwa) were followed longitudinally for 1½ and 1 year respectively.

Several methods were employed to gather the ape data. Observational data were collected in continuous running narrative form. A focal animal sampling technique was employed. All behaviors of focal animals distinguished as relevant to the Sensorimotor Intelligence Series and Imitation Series were recorded. Animals living with conspecifics were observed as they behaved spontaneously in their captive environments. Objects were also introduced to group-living animals in order to give them better opportunity to demonstrate their intellectual capacities. For example, hanging swingable objects, drums, and gongs were introduced to test for *secondary circular reactions* (repeated attempts to reproduce environmental events initially discovered by chance); and balls, tubs of water with cups and floatable objects, and buckets with objects that could be put into them were introduced to test for *tertiary circular reactions* (trial-and-error experimentation with object-object, object-space, object-force and object-gravity relationships). Animals were observed when they were visible and active. The animal on whom the fewest hours of data had been collected was chosen for observation whenever there was a choice of visible, awake subjects. Focal periods lasted as long as the animal was visible and awake (or until the day's observation period was terminated), and then another focal subject was chosen.

Hand-reared animals were formally tested, presented with problems (such as how to obtain an out-of-reach object with a tool), or tested for imitative ability as well as being presented with objects that could be freely manipulated. The testing procedures were derived from those of Uzgiris and Hunt (1966), modified for use on apes. Behavior of gorillas was recorded on film; six and a half hours of film have been collected. Test results were recorded on check sheets. However, descriptions of the animals' responses were also recorded on audiotapes, since detailed observations of the animals' behavior may be as important as the final outcome of the tests. (For example, an infant chimpanzee, presumably in the fifth stage of the Sensorimotor

Intelligence Series [the stage of the *tertiary circular reaction*—the stage during which infants experiment with various object-object and object-space relationships], failed to thread a ring onto a string but tried to get the ring onto the string by repeatedly *pushing it against the middle* of the string. Although she failed to manifest the expected *tertiary circular reaction*, she clearly showed that she had a partial understanding of the conceptual relationships involved.)

DATA ANALYSIS

The data have only been partially analyzed. The more complete gorilla data have been more thoroughly analyzed than the chimpanzee data. All recorded gorilla behaviors that could be distinguished as relevant to the two Sensorimotor Intelligence Series under consideration have been abstracted from the narrative record. Based on Piaget's model of human sensorimotor development, behavior has been categorized according to the series to which it pertains: the Sensorimotor Intelligence Series or the Imitation Series.

Piaget has found that human infants pass through six developmental stages[2] in each series, between birth and 2 years of age. Each stage in each of the two series can be defined in terms of a specific set of behavioral parameters that are characteristic of that stage. These parameters involve such developmental changes as increase in voluntary control, increase in the number of motor patterns and contextual variables involved in a single act, and increased variability of motor patterns (see Table 2). However, one cannot ascertain prospectively whether this exact model, which is based on human infant development, will be suitable for studying nonhuman species. Therefore, Piaget's framework has been used as a model, but modified to accommodate the behavioral parameters of monkeys and apes when these differ from those of human infants. Following Parker's approach to the study of stumptail monkeys,[3] stage classification was made according to the behavioral parameters observed, and new stage categories, with new behavioral parameters, were developed if the observed behavior did not conform to the traditional Piagetian stages seen in human subjects. This approach made it possible to detect stages characterized by different behavioral parameters, or by some but not all of the behavioral parameters seen in the human stages.

Following this method, the abstracted behaviors were classified according to stages. Ages at which the first behaviors characteristic of each stage in each series appear were noted. Behaviors were then categorized according to sensory modality[4] and body part. Stage behaviors and ages at onset of stage were then compared between the different sensory modalities.

TABLE 2 / CHARACTERISTICS OF THE SENSORIMOTOR INTELLIGENCE SERIES* AS MANIFESTED BY HUMAN INFANTS

STAGE	AGE (MOS.)	DESCRIPTION	MAJOR DISTINGUISHING BEHAVIORAL PARAMETERS	EXAMPLE
1 Reflex	0–1	Stereotyped responses to generalized sensory stimuli	Involuntary	Roots and sucks
2 Primary circular reaction	1–4	Infant's action is centered about his own body (thus "primary") which he learns to repeat ("circular") in order to reinstate an event.	Repetitive coordinations of own body First acquired adaptations occur Recognition of various objects and contexts	Repeats hand-hand clasping Exhibits conditioned reflexes
3 Secondary circular reaction	4–8	Repeated ("circular") attempts to reproduce environmental ("secondary") events initially discovered by chance	Environment-oriented behaviors Establishment of object/action relationships Semi-intentional (initial act is not intentional, but subsequent repetitions are). Active attempts to effect changes in environment Simple orientations toward a single object or person	Swings object and attends to the swinging spectacle, or to the resulting sound; repeats
4 Coordination of secondary behaviors	8–12	Two or more independent behavioral acts become intercoordinated, one serving as instrument to another	Intentional Goal is established from the outset Establishment of relationships between two objects Coordination of several behaviors toward an object or person Objects explored as well as acted upon	Sets aside an obstacle in order to obtain an object behind it

5 Tertiary circular reaction (experimentation)	12–18	Child becomes curious about an object's possible *functions* and about object-object relationships ("tertiary"); he repeats his behavior ("circular") with variation as he explores the potentials of objects through trial-and-error experimentation	Familiar behaviors applied to new situations Infant begins to attribute cause of environmental change to others Behavior becomes variable and nonstereotyped, as the child invents new behavior patterns Repetitive trial and error experimentation begins Interest in novelty for its own sake Coordination of object-object, person-object, object-space and object-force relationships Considers others entirely autonomous	Experimentally discovers that one object, such as a stick, can be used to obtain another object
6 Invention of new means through mental combinations (insight)	18 +	The solution is arrived at mentally, not through experimentation	The child can mentally represent objects and events not present	Mentally figures out how one object can be used to obtain another object

*Abstracted from Piaget, with modifications from Sugarman Bell, in press, and adapted from Chevalier-Skolnikoff, in press[a].

■ RESULTS AND DISCUSSION

COGNITIVE ABILITIES: SENSORIMOTOR SERIES

Stumptail Macaques

An examination of the intellectual abilities of stumptail macaques, analyzed accord-ing to sensory modality and body part, shows that these monkeys complete all six stages[5] of the Sensorimotor Intelligence Series in the tactile/kinesthetic mode and in the visual/body mode. The six stages appear ontogenetically in the same sequence in stumptail macaques as in human infants between birth and about 6 months of age. For example, in the tactile/kinesthetic mode, stumptail monkeys show stage 1 reflex-ive behavior (stereotyped responses to nonspecific or generalized stimuli), as man-ifested by the Moro reflex, or the grasping and clinging reflexes; they manifest stage 2 *quasi-primary circular reactions* (repetitive "circular" self-directed "primary" be-havior), such as thumb- and foot-sucking (although this behavior is rare and only the sucking movements are repetitive, the hand-mouth and hand-foot coordinations are not);[6] they show stage 3 *secondary circular reactions* (repetitive "circular" attempts to reproduce environmental "secondary" events initially discovered by chance), as in mutual interanimal mouthing; they show stage 4 *coordinations of secondary circular behaviors*, as, for example, they mutually mouth and mutually grab each other during rough-and-tumble play; they show stage 5 *tertiary circular reactions* (repetitive "circu-lar" experimentation with the interrelationships, which are "tertiary," between ob-jects, persons or animals, space, gravity and force either in play or to attain a goal), as in trial-and-error approach-avoidance play, and they evidently manifest stage 6 *inven-tion of new means through mental combinations*, as they mentally figure out tactics for approach-avoidance play. Parker (1973, this volume) has observed that the higher stage "body reactions" are animal-oriented and not object-oriented as they characteris-tically are in human infants. However, stumptail monkeys evidently do not complete the six sensorimotor stages in other sensory modalities. In the visual/facial mode and in the vocal mode they manifest only the reflexive stage 1 behaviors that satisfy the criteria for the human Piagetian stages, and in the visual/gestural (manual behaviors received visually) and auditory modes,[7] they do not manifest any of the six stages. But stumptail macaques manifest stages that differ from those of humans. Their behaviors are nonrepetitive, but satisfy some of the requirements for stage 2 and stage 3 in the visual/facial and vocal modalities. Parker (1973, this volume) has called these be-haviors "linear." For example, their early lipsmacking in response to the appearance of a familiar monkey face incorporates stage 2 recognition of contexts (this may be a conditioned reflex) and they show behaviors incorporating stage 3 environmental orientation and semi-intentionality when they initiate environmentally oriented activ-ities, such as approaching peers with a play face. However, they do not show repetitive (circular) self-directed behaviors (stage 2) or repetitive (circular) attempts to reinstate changes in the environment (stage 3) in these modalities as human infants do. This

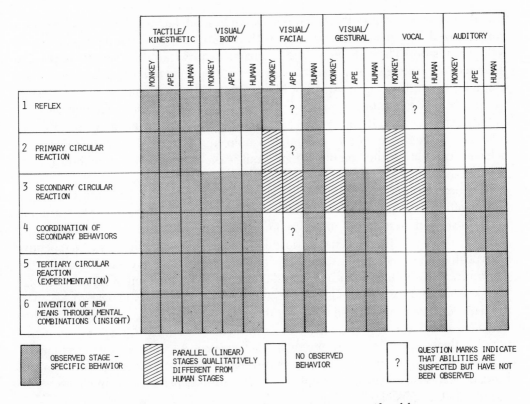

FIGURE 1. *The Sensorimotor Intelligence Series as manifested by three primate species. (Adapted from Chevalier-Skolnikoff in press a.)*

lack of repetitive environmentally directed behavior patterns in all but the body modes, and the lack of repetitive object-directed behavior in any mode, indicates that stumptail macaques evidently do not learn to voluntarily effect changes in objects or in the actions of objects or to voluntarily effect changes in the behavior of conspecifics through facial expression, gesture, or vocal or auditory behavior. Consequently, stumptail macaques evidently cannot voluntarily make facial expressions or vocalizations they do not feel or voluntarily make hand gestures or environmentally induced sounds in order to obtain a goal from a conspecific. In addition, not having learned to manually and voluntarily effect changes in objects or in the actions of objects in stage 3, stumptail macaques evidently cannot proceed to stages 4 and 5 and cannot learn to relate objects to one another as is necessary for tool-use (see Figure 1).

Great Apes

The great apes, like the macaques, also complete the entire Sensorimotor Intelligence Series in the tactile/kinesthetic and visual/body modalities. But unlike the macaques, and like human infants, they also complete all the higher stages in the visual/gestural modality. Consequently, they manifest *primary circular reactions*, characterized by repetitive coordinations of their own bodies, and manual *secondary circular reactions* characterized by repetitive interactions with objects such as swinging a branch and watching it sway. Through such stage 3 behaviors, apes learn how objects respond to their actions. Apes also manifest stage 4 *coordinations of secondary behaviors*, as, for example, when they pick up sticks or branches and put them in their mouths during running and chest-beating displays. They also show stage 5 *tertiary circular reactions* as they experiment with object-object relationships, as when they use a stick to poke otherwise inaccessible food out through a hole in a box. And they display stage 6 behavior (insight) as they mentally figure out without prior experimentation how to poke food out of a container.

The apes also complete the higher stages in the visual/facial modes (Chevalier-Skolnikoff, in press [a]). However, in the vocal mode the apes are similar to macaques and radically different from humans, completing only stage 3 in the linear series and probably stage 1 in the Piagetian series. Thus they fail to manifest stage 2 repetitive self-vocalizations (cooing), or stage 3 repetitive vocal attempts to effect changes in the environment (babbling, or vocal "games"), or stage 4 combination of sounds, or stage 5 experimentation with sounds. In the auditory modality, apes appear to be intermediate between monkeys and humans, showing stage 3 and stage 4 behaviors as they make noises by repeatedly banging objects in their environments (stage 3) and incorporating these repetitive noises into their displays (stage 4). Ontogenetic development is slower in the apes than in macaques, and after the first year it is slower than in human infants; apes evidently complete the sensorimotor period sometime between 20 months and 4 years of age.

Human Infants

Human infants complete the whole Sensorimotor Intelligence Series in all sensory modes by the end of the second year.

COGNITIVE ABILITIES: IMITATION SERIES

Stumptail Macaques

In the Imitation Series, stumptail macaques achieve only stage 3 in the visual/facial, the visual/gestural, and the vocal modes (see Table 3). They do not imitate *secondary circular reactions* in these modes but may do so in the visual/body mode. Consequently, they only show imitation of familiar motor patterns already in their repertoires when these behaviors are performed by others. A macaque can "imitate" another animal: if he sees another animal resting, he lays himself down to rest; if he sees another animal eating a particular kind of food, he does so himself; or if he sees another animal make a fear expression or alarm call, he makes a fear expression or an alarm call himself. This kind of imitation is called "stimulus enhancement" when the focus of the observer's attention is a particular object such as a food item, and called "social facilitation" in the case where the focus is a conspecific's actions. Kummer (1971) has emphasized the significance of this kind of behavior in the social organization of hamadryas baboons. However, while macaques are capable of socially facili-

TABLE 3 / CHARACTERISTICS OF THE IMITATION SERIES* AS MANIFESTED BY HUMAN INFANTS

STAGE	AGE (MOS.)	DESCRIPTION
1 Reflexive contagious imitation	0–1	Reflexive behavior (e.g., crying) is stimulated by the behavior in a model.
2 Sporadic "self- imitation"	1–4	Imitation by the model of the infant's own motor patterns (self-imitation) stimulates diffuse vocal activity in the infant. It is unclear whether the infant distinguishes the model's behavior from his own.
3 Purposeful "self- imitation" (social facilitation, or stimulus enhance- ment)	4–8	Self-imitation in which matching becomes more precise. For visual imitation, the infant's acting body part must be visible to him.
4 Imitation using unseen body parts	8–12	Imitation using unseen body parts. Infant attempts to imitate new behavioral acts, but often fails to precisely match the model.
5 Imitation of new behavior patterns ("true" imitation)	12–18	Imitation of new motor acts, precisely accommodating behavior to that of the model through repeated attempts at matching.
6 Deferred imitation	18 +	Child precisely imitates new motor acts, without preliminary attempts at matching, through symbolic representation. Child manifests deferred imitation.

*Abstracted from Piaget, and adapted from Chevalier-Skolnikoff, in press[a].

tated "imitation," they are evidently unable to imitate motor patterns or vocalizations not already in their repertoires. In the auditory modality, macaques appear to manifest no imitation at all, and evidently do not imitate nonvocal noises (see Figure 2).

Great Apes

The great apes complete all six stages of the Imitation Series in the visual/body, visual/facial and visual/gestural modalities as human infants do. Consequently, they are able to learn new body, facial, and manual motor patterns through repeated attempts at imitative matching (stages 4 and 5) and they can imitate new motor patterns on the first try without practice (stage 6). In addition, they manifest delayed imitation, imitating new motor patterns for the first time after the model is no longer present (stage 6). However, in the vocal modality, gorillas, like macaques, are capable only of stage 3 "socially facilitated" imitation of emotional vocalizations (Chevalier-Skolnikoff, in press[a]). In the auditory modality the apes, unlike the macaques, are

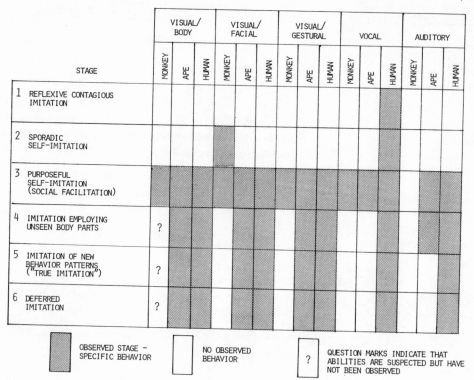

FIGURE 2. *The Imitation Series as manifested by three primate species. (Adapted from Chevalier-Skolnikoff in press a.)*

able to imitate nonvocal noises (stage 3) and to incorporate noises into their displays (stage 4).

Human Infants

Human infants complete the whole Imitation Series in all sensory modalities.

ONTOGENY OF COGNITIVE ABILITIES AND SOCIALIZATION

The intellectual development of macaques, great apes, and human beings defines certain potentials and certain limitations for their socialization. In terms of intellectual abilities the developing infant is a qualitatively different kind of animal at different periods during its development. Similarly, infants of different species may differ qualitatively. Consequently, the effects of the social environment upon the infant are also qualitatively different at different developmental periods, and for different species.

Stumptail Macaques

During stage 1, the reflex stage, (0–2 weeks), the sensorimotor abilities of the neonatal macaque provide him with the potential for making reflexive responses such as rooting, nursing, grasping, and clinging in response to tactile stimuli and for *irregularly* responding to *generalized* visual and auditory stimuli (such as moving body parts [legs] of other monkeys, swinging tires, shiny objects, and loud noises) (Chevalier-Skolnikoff 1974a). The nonspecific nature of the eliciting stimuli indicates that at this stage the infant does not recognize specific "social" stimuli and cannot yet "learn" the social significance of the stimuli around him. The communications that he sends are the reflexes which are received by others as tactile/kinesthetic communications (e.g., nursing and clinging), visual/body communications (e.g., jerking), visual/facial communications (e.g., pucker lips or lipsmacking expressions), and vocal communications (e.g., geckers and screams) (Chevalier-Skolnikoff 1974a).

Already during stage 1 the infant may initiate self-motivated reflexive behavior (e.g., rooting and sucking when he is hungry) as well as reacting to environmental stimuli (Chevalier-Skolnikoff 1974a). Therefore, even though his behavioral repertoire is limited to only a few, mostly tactile, reflexes, and even though he does not yet purposefully reach out to his environment, he still has an active role in determining his social interaction with his mother.

During stage 2 (from about 2 weeks), the macaque infant characteristically manifests *primary linear reactions*. *Primary circular reactions* are rare. During this stage, the macaque infant becomes capable of making associations between things in his environment (especially other animals) which impinge upon him (mainly through tactile/kinesthetic stimulation) and certain contextual results. He becomes capable of recognizing and responding to his mother's ventrum and nipple area specifically. During this stage he also becomes capable of learning simple signals from her such as the touch-cling signals by which his mother indicates with a touch that she is about to move (Chevalier-Skolnikoff 1974a, in press[a]). Most of these responses do not actually differ from the reflexes in form, and are, in fact, conditioned reflexes. He also begins to make emotional facial expressions (friendly lipsmacks and pucker-lips) and vocalizations (e.g., the trilled whistle, a contact vocalization) in friendly "social" contexts as when other lipsmacking animals approach him.

Stages 1 and 2 are both characterized by the infant's self-orientation and lack of attention to the world outside himself; he reacts to the outside world only as it impinges upon him.

The third stage, the stage of the *secondary linear* and *secondary circular reactions* (from weeks 2 to 4),[8] is characterized by the infant's reaching out to his environment. *Secondary linear reactions* in the tactile/kinesthetic and visual/gestural modes included reaching and touching or grabbing objects or other monkeys and responding to visual contact signals which are made by the mother or by other adults with approach and clinging (Chevalier-Skolnikoff 1974a). Visual/body behaviors include (playful) hopping and scampering toward or away from other animals (Parker 1973, this volume). Visual/facial behaviors include facial expressions of emotion, (friendly) pucker-lips, and lipsmacking, open-mouth-eyelids-down (play), and open-mouth (threat) expressions that often accompany approaches towards other monkeys (Chevalier-Skolnikoff 1971, 1974a, 1974b; Parker 1973, this volume). Vocal behaviors include emotional vocalizations such as chirl (threats) and (friendly) staccato grunts that also accompany approaches towards other animals (Chevalier-Skolnikoff 1971, 1974a; Parker 1973, this volume).

The first rudimentary sexual behaviors (evidently stage 3 linear behaviors) were observed during this period as infants, both male and female, approached other animals, especially copulating pairs of monkeys, and then climbed onto them (Chevalier-Skolnikoff 1974a).

During the third stage infant macaques first start to reach out and investigate their inanimate environments, and sample solid food items. They also begin to learn their troop's repertoire of food items during this stage as they "imitate" their mothers' and other troop members' feeding behavior, attending to what others eat (stimulus enhancement—see Figure 3), and grabbing bits of foods from others' mouths and picking and eating what others eat (social facilitation). Through these means, macaque infants learn to eat the foods their mothers and other troop members customarily eat. These early environmentally oriented behaviors appear to be linear, and they do not involve repetitive elicitation of environmental events.

Secondary circular reactions have been observed only in the tactile/kinesthetic, visual/body modalities and only towards other monkeys. Repetitive interanimal mouthing and play-biting are the most prevalent *secondary circular reactions* observed (Chevalier-Skolnikoff 1974a, 1974b). Branch-shaking is another *secondary circular behavior* that occurred (Chevalier-Skolnikoff 1974a). Branch-shaking was accomplished kinesthetically and was generally received as a visual/body communication by others although it was also occasionally received as an auditory communication. I suspect this may be a by-product of the visual display, since no other *secondary circular behaviors* were observed in the auditory modality. During this period the first rudimentary grooming gestures, consisting of repetitive parting movements of the two hands (applied to soft mushy substances as well as to fur), occur (Chevalier-Skolnikoff 1974a).

Stage 3 is a particularly significant stage in macaque adaptation for it is the highest stage attained in most sensory modalities in both the Sensorimotor Intelligence Series and the Imitation Series. It is the first "truly social" stage in macaque development since it is the first stage during which infants voluntarily reach out into their environments. It is stage 3 linear sensorimotor and stage 3 imitative abilities that define macaque facial and vocal communicative potentials. It is also stage 3 cognitive abilities that define macaque feeding techniques and macaque potential for the acquisition and transmission of "protoculture." It is through stage 3 behavior that Japanese macaques have been seen to acquire their troops' traditional food repertoires (Frisch 1968; Itani 1958; Itani and Nishimura 1973; Kawai 1965; Kummer 1971; Tsumori 1967; Yamada 1957).

The fourth stage, the stage of *secondary linear* and *secondary circular coordinations* (from about 2 months), is characterized by behavior patterns incorporating and combining two or more secondary behaviors characteristic of stage 3 in order to attain predetermined goals. *Secondary linear coordinations* include pushing one animal away while holding an object and transferring an object from hand to foot and then storing it there (Parker 1973, this volume).

Coordinations of secondary circular reactions include rough-and-tumble play which occurs primarily in the tactile/kinesthetic modality but which may also be reacted to visually as players make multiple tactile and kinesthetic adjustments to visual signals emanating from the other player (Chevalier-Skolnikoff 1974b, in press[a]). During the fourth stage, macaques begin to coordinate picking movements of the second digits with parting movements and lipsmacking into a more complex grooming behavior. During this period, sexual behavior incorporating *secondary circular reactions*[9] as well as coordinations was first observed. The male infant began to incorporate repetitive pelvic thrusts into his mounts, and his evident attempts at obtaining cooperation from his partners suggests that he was trying to voluntarily reinstate and control events in his (animate) environment. However, these mounts were still incomplete, often being incorrectly oriented, and not incorporating intromissions. The male infant also began to manifest homosexual behavior, coordinating mounts with anogenital stimulation of his partner; this first occurred with the aid of the partner, and was not accomplished entirely through the infant's own initiative.

The development of the potato-washing and wheat-sifting food preparation techniques reported in Japanese macaques (e.g., Itani and Nishimura 1973) suggests that macaques, or Japanese macaques specifically, may have some limited stage 4 sensorimotor abilities in interactions with objects not seen in my or Parker's studies on stumptail macaques. These techniques do appear to involve an understanding of stage 4 object-object and object-substance relationships. However, such behaviors appear to be rare; they appear to be learned through stage 3 or "self-imitation" because they incorporate familiar rather than new motor patterns (Chevalier-Skolnikoff 1975).

The fifth stage, the stage of *tertiary circular reactions* (from about 4 months), is characterized by trial-and-error experimentation and invention of new means through the application of variable behavior either in play or to achieve a desired end. In stumptail macaques, stage 5 behaviors also occur primarily in the tactile/kinesthetic sensory mode in social interactions, visual/body communications sometimes growing out of the tactile/kinesthetic ones. Trial-and-error approach-avoidance play and chasing and hiding "games" are among the most prominent *tertiary circular behaviors*. Trial-and-error sexual behavior also occurred during this period in the male infant as he experimented with other males in mutually stimulating homosexual interactions. These resulted in mutual manual and oral stimulation that was accomplished through a number of different motor patterns (see Chevalier-Skolnikoff 1974a, 1974c, in press[b]). Experimentation also probably occurs during heterosexual interactions and probably contributes to the perfection of mounting that ultimately results in the consistent orientation that permits intromission. Intromission combined with a long series of pelvic thrusts (but not ejaculations) was not observed until the end of stage 5 or the beginning of stage 6 when the male infant was 6 months old.

The sixth stage, the stage of *invention of new means through mental combinations* (rather than through experimentation), probably appears in stumptail macaques sometime after the fourth month. Stage 6, like stage 5, is also manifested only in the tactile/kinesthetic modality and only in social interactions as, for example, when the macaque mentally figures out tactics for approach-avoidance play.

Stages 4, 5, and 6 are evidently significant for stumptail macaque socialization only in the body modalities. But here they play a role in one of the most critical biological functions, sexual reproduction. The acquisition of copulatory skills—and they appear to be true skills—are evidently acquired through stage 4, 5, and 6 sensorimotor abilities. Effective fighting abilities may also be based on stages 4–6 abilities, and both copulatory and fighting skills are to some extent learned in the context of play.

Great Apes

Among the great apes the first stage of development, the reflex stage, is similar to that of the monkey except that it is more protracted, lasting about one month (see Table 4).

During stage 2 (1–3 months), the infant ape makes advances similar to the macaque. However the ape infant also shows other developments during this stage. For him this is truly the stage of the *primary circular reaction*. Being motorically immature until beyond the third month, the ape infant passes much of his time lying supine in his mother's arms or on her lap, occupying himself by performing repeated hand-mouth, hand-hand, and hand-foot behaviors (Figures 3a, 3b, 3c). It is possible that through *primary circular reactions* infant apes acquire a higher degree of hand-

TABLE 4 / APPROXIMATE AGES OF DEVELOPMENT OF THE
SENSORIMOTOR INTELLIGENCE SERIES AND THE IMITATION
SERIES IN THREE PRIMATE PHYLA

STAGE NO.	STUMPTAIL MACAQUE[1]	GORILLA[2]	HUMAN[3]
1	0–2 weeks	0–1 month	0–1 month
2	2 weeks–3 months	1–3 months	1–4 months
3	*about 3 weeks (1–4)– 3 months	3–less than 7 months	4–8 months
4	*2 months–	Less than 7– about 14 months (13½–14½)	8–12 months
5	*about 4½ (4–5) months	About 14–more than 20 months and less than 4 years	12–18 months
6	?	Less than 18 months– less than 4 years	18 months–

[1]Abstracted from Chevalier-Skolnikoff, 1974a; Parker, 1973, this volume.
[2]This study.
[3]Abstracted from Piaget, 1951, 1952, 1954.
*Stages in most sensory modalities are linear and differ from those observed in human infants by Piaget.

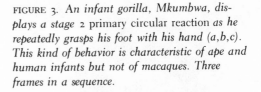

FIGURE 3. An *infant gorilla, Mkumbwa, displays a stage 2 primary circular reaction as he repeatedly grasps his foot with his hand (a,b,c). This kind of behavior is characteristic of ape and human infants but not of macaques. Three frames in a sequence.*

eye coordination than is ever achieved by macaques. This in turn may provide the basis for the subsequent higher cognitive and social achievements of the apes.

During stage 3 (from 3–<7 months), the stage of the *secondary linear* and *secondary circular reactions* and socially facilitated "self"-imitation, the infant ape also makes the cognitive and social advances seen in macaques, developing emotional facial expressions and vocalizations, and beginning to acquire the food repertoire of his mother through stage 3 *secondary linear reactions* (Figures 4a, 4b) and stage 3 socially facilitated imitation (Figures 5a, 5b). However, during the third stage, as during the second, ape development clearly diverges from that of the macaque because apes develop visual/gestural and auditory *secondary circular reactions*. Though these early *secondary circular reactions* do not appear to be social (they are essentially

FIGURE 4. *An infant gorilla, Mkumbwa, displays stage 3* secondary linear behaviors *as he reaches for a leaf (a) and later puts one in his mouth (b). This kind of behavior is characteristic of stage 3 monkey, ape, and human infants.* (Photographs by William Townsend.)

FIGURE 5. *An infant gorilla, Sunshine, shows stage 3* imitation of familiar motor patterns *(e.g., social facilitation). (a) He imitates his mother's body orientation, posture, and eating behavior. (b) He imitates his mother's body orientation, posture, and hand-to-mouth behavior. This kind of imitation is characteristic of monkey, ape, and human infants.* (Photographs by William Townsend.)

FIGURE 6. *Koko, at four years of age, manifests stage 3 imitation of a* secondary circular reaction. *(a) Penny Patterson bangs a gong and Koko watches. (b) Koko bangs the gong. This kind of imitation is characteristic of ape and human infants but not of macaque infants.* (Photographs by William Townsend.)

solitary activities), they clearly set the stage for many aspects of future social development. By the end of stage 3, *secondary circular reactions* are seen in play, and infants sometimes imitate the *secondary circular reactions* of their playmates (Figures 6a, 6b).

By the fourth stage (from <7 to about 14 months), ape development has strikingly diverged from that of the macaque. Various *secondary circular reactions* are combined, and by the end of stage 4, infants are incorporating these behaviors into their social interactions such as when they wave branches and beat noise-making objects in play and during displays (Figure 7). During stage 4, young apes begin to attempt to imitate new motor patterns, and they imitate behavior using body parts they cannot see (Figure 8).

During the fifth stage (from about 15 months until sometime between 20 and 48 months), the stage of the *tertiary circular reaction,* ape infants begin to manifest experimental nonstereotyped behavior in the visual/facial and visual/gestural modes unlike anything reported for macaques. Much of this behavior initially occurs in nonsocial contexts as, for example, when an infant experiments by itself with the effects of gravity by repeatedly placing a stick on an inclined surface and letting it roll down, or by repeatedly tossing sticks down a slope, or as he experiments with object-object and object-space relationships by repeatedly trying to poke a stick into a hole. However, *tertiary circular reactions* also become incorporated into social play and aggression as, for example, when animals use sticks as tools to poke and hit each

FIGURE 7. *An infant gorilla, Sunshine, manifests stage 4* coordinations of secondary behaviors *as he holds a string in his mouth, beats his chest, and makes a display run.* (Photograph by William Townsend.)

FIGURE 8. *Koko shows stage 4 imitation using body parts she cannot see as she imitates Penny Patterson touching her ear.* (Photograph by William Townsend.)

other. *Tertiary circular reactions* are also employed to solve problems (Figures 9a, 9b, 9c). Among the captive gorillas at the San Francisco Zoo, the use of tools for obtaining out-of-reach food is a regular occurrence and appears to be a social tradition passed from one generation to the next through the socialization process. It was probably first "invented" by one of the animals through stage 5 trial-and-error experimentation, or through stage 6 "insight" (i.e., mental comxinations) and then passed on to other group members through stage 5 or 6 imitation of the new motor pattern.

Tool use and other kinds of *tertiary circular reactions* and behaviors incorporating the methods of the *tertiary circular reaction* are prominent in the behavior of wild chimpanzees. Observers at the Gombe Stream Reserve have found that sticks, twigs, and stems are used for obtaining and eating termites and ants and as investigatory probes; sticks, miscellaneous vegetation, and stones are used for throwing during displays and in aggression (although the chimpanzee's aim is poor); leaves are employed for wiping; and leaves may be chewed, made into a wad, and then used as sponges for drinking (Goodall 1963, 1964; van Lawick-Goodall 1968, 1970, 1973; McGrew 1974, this volume).

Although only two incidents of tool use by wild gorillas have been reported (Pitman [1931], and Philipps [1950] observed gorillas using sticks to obtain food that was out of reach), wild gorillas evidently do manifest other sorts of *tertiary circular reactions* quite frequently. Fossey (personal communication) has seen wild gorillas throw branches in aggression (they throw them underhand and have good aim). She also has seen them collect branches for nesting by climbing into a tree, breaking a

FIGURE 9. *Koko manifests a* tertiary circular reaction *as she uses trial-and-error experimentation to obtain a goal: to get honey out of a narrow-mouthed container. (a) She tries a shovel; it is too wide and cannot enter the mouth of the jar. (b) She tries a narrow stick. (c) Success—using the narrow stick as a tool, she eats the honey.* (Photographs by William Townsend.)

series of boughs, and dropping them so that they fall in a pile below the tree where the nest is to be made. Fossey has also frequently observed *tertiary circular reactions* in play; infants often experiment with object-gravity relationships by tossing objects— such as branches or stones—off the edges of ravines and watching them fall; they also play with clumps of moss, tossing them into the air and catching them.

In addition, *tertiary circular reactions* have been observed in mother-infant interactions in the wild as mother chimpanzees (van Lawick-Goodall 1967) and gorillas (Fossey, personal communication) use insight to figure out how to distract infants from undesirable activities (such as playing with their very young siblings); the mothers often initiate tickling games under such circumstances.

The social process of protocultural transmission whereby chimpanzees learn to use tools has been well documented in the wild. Goodall (1973) has described the acquisition of termiting techniques, and McGrew (this volume) has described the acquisition of dipping for ants at the Gombe Stream Reserve. Infants were first observed orienting to their mother's activities (stage 3 stimulus enhancement), then they were seen imitating their mother's choices of termites or ants for food (stage 3 social facilitation) followed by imitation of the mother's object-grasping behaviors; i.e., grasping sticks their mother used as tools (stage 3 social facilitation). Infants subsequently unsuccessfully attempted to imitate their mother's tool-using and tool-making behaviors (stage 4 attempted imitation of new motor patterns through matching). Finally they were able to successfully imitate the new motor patterns through trial-and-error matching (stage 5 imitation). Five- and six-year-old chimpanzees were observed effectively performing these behaviors on their own, without directly imitating their mothers (stage 6, deferred imitation). The ages at which wild chimpanzee infants have been observed performing particular Piagetian stage behaviors in these studies closely correspond to the developmental stages observed in the captive animals in this study.

Further evidence to support the significance of the role socialization plays in the transmission of ape protoculture is the variability of protocultural traditions between wild chimpanzee populations. In some areas chimpanzees fish for termites, modifying tools for this purpose; in other areas they dip for arboreal ants, using several kinds of modified tools; and some populations dip for subterranean ants also using modified tools. Some populations have more than one such tradition, while others have not been observed to use tools (Beatty 1951; Goodall 1964; Itani and Suzuki 1967; Jones and Pi 1969; van Lawick-Goodall 1968, 1970, 1973; McGrew 1974; Nishida 1973; Rahm 1971; Pi 1974; Savage and Wyman 1843–44; Struhsaker and Hunkeler 1971; Suzuki 1966).

These data indicate that the major difference in the socialization potential of the great apes from that of macaque infants is development in the visual/gestural modality; development upon which ape tool-use and tool-making abilities are based.

Human Infants

During the first stage (0–1 month), the reflex stage, the potentials of the human infant appear to be similar to those of infant apes.

During stage 2 (1–4 months), the stage of the *primary circular reaction*, human infants make advances similar to those of the apes. However, by stage 2 they are manifesting a new potential: vocal *primary circular reactions* as they coo repetitively to themselves (Chevalier-Skolnikoff, in press[a]). Though these behaviors are initially nonsocial they, like the manual hand-hand and hand-foot grasping *primary circular reactions* in the tactile modality (also seen in apes), appear to set the stage for subsequent social behaviors in the vocal sensory modality.

Another striking difference between ape and human socialization during stage 2 is the human incorporation of eye contact and the smile into the socialization process. While the apes, like the macaques, appear to become attached to their mothers through the tactile/kinesthetic mode (Bowlby 1969; Harlow 1961; Harlow and Harlow 1969) and possibly the olfactory mode, by the beginning of the second month the smile is a major aspect of the human infant's attachment process (Bowlby 1969; Spitz and Wolff 1964; Wolff 1963).

During stages 3–6 human infants make advances similar to those of apes in most sensory modalities, most notably in the visual/gestural mode, in which human infants perform manual visual/gestural *secondary* and *tertiary circular reactions* which eventually, in stages 5 and 6, result in effective tool-use and tool-making. However, during stage 3 (4–8 months), human infants, unlike the apes, build on their stage 2 vocal advances and begin to use vocalizations voluntarily to elicit and prolong social interactions as they perform vocal *secondary circular reactions* in their back-and-forth babbling "games" with adults (Chevalier-Skolnikoff, in press[a]). During stage 3, smiling and eye-to-eye contact are also prominent in *secondary circular* interactions, as when babies play peek-a-boo games with their mothers and other adults.

During stage 4 (8–12 months), as well as in stages 5 and 6, the most notable advances of human infants are again vocal ones. Infants begin to combine their babbling sounds during social interactions (*coordinations of secondary circular reactions*), and they begin to attempt to imitate new sounds and words, manifesting echolalia (Chevalier-Skolnikoff, in press[a]).

During stage 5 (12–18 months), trial-and-error vocal matching becomes effective, and human infants speak their first words. They also utilize words as means to attain predetermined ends (i.e., as tools; therefore words function as *tertiary circular reactions*). During this stage voluntary facial expressions and nonvocal auditory behavior (e.g., object-banging) are also used as tools to attain desired ends, as, for example, when a baby bangs on his highchair tray in order to indicate that he wants to be put down (Chevalier-Skolnikoff, in press[a]).

During stage 6 (18–24 months) human infants begin to regularly send and receive symbolic communication as they begin to combine words into new two-word phrases (Slobin 1971) and to use these phrases in order to attain desired goals. After 18–24 months the socialization of human infants occurs primarily through symbolic communications in the vocal channel, that is, through vocal language.

Human socialization differs most notably from that of apes in the development of vocal skills which eventually lead to vocal language (Chevalier-Skolnikoff, in press[a]). However, human socialization during the sensorimotor period probably differs from that of apes in other less salient and less obvious respects as well. In the vocal mode human infants learn to sing while apes do not. In the auditory mode human infants may learn to play rhythms and to play tunes on musical instruments and apes do not and apparently cannot.[10] Although apes have the ability to perform stage 5 and 6 visual/facial behaviors, variability in ape facial expression is not obvious in social interactions, and socialization does not play an obvious role in the development of facial communication as it does in humans. Humans apparently make more nonemotional socially and culturally determined facial expressions than do apes (Chevalier-Skolnikoff 1973; Ekman 1972, 1973). Probably human abilities for tactile/kinesthetic and body behaviors (e.g., dancing) also surpass those of apes, although the differences have not yet been precisely described.

■ CONCLUSIONS

A preliminary analysis of primate potential for socialization in terms of cognitive abilities suggests that the socialization process is qualitatively different at different age levels, and that it is also qualitatively different in different species at different phylogenetic levels.

Neonatal monkeys are capable of only reflexive behaviors both in initiating and in responding to social situations. Their future behavior is probably only minimally affected by their social environments, and their potential for socialization is therefore minimal.

During stage 2 as the infant macaque is maturing, he becomes capable of the "social learning" of conditioned reflexive responses. However, his world is still self-centered and, at this stage also, his potential for socialization is extremely limited.

As the macaque infant completes stage 3 he probably nears the limit of his potential for socialization. He reaches out towards, explores, and attends to his social environment, and he initiates and responds to emotional communications. He also explores his inanimate environment, including potential food resources. During this period, through self-initiated exploration and social facilitation, or stage 3 imitation, he learns to eat solid foods, and acquires the food repertoire, or the macaque "protocultural" acquisitions, of his social group.

In the macaque, the major social advances that appear to transpire after stage 3 develop in the body modalities as infants learn sexual (and possibly aggressive) skills during tactile/kinesthetic interactions with peers as well as with adults through stages 4–6 cognitive mechanisms.

Stumptail macaques show no evidence of higher stage intellectual abilities in the visual/facial, vocal, or visual/gestural modes. They cannot invent or imitate new nonemotional facial expressions or vocalizations, nor can they voluntarily effect changes in the vocal behaviors of others as human infants do in their vocal cooing and babbling "games"; and they cannot learn new object-object relationships or new manual motor patterns (which are essential for tool-use, tool-making, and the cultural transmission of these skills) through trial-and-error experimentation or advanced imitation.

The social potentials of the stage 1 neonate and stage 2 infant ape are, like those of the infant macaque, relatively limited. However, the nonsocial *primary circular reactions* of the stage 2 ape and the quasi-social visual/gestural *secondary circular reactions* of stage 3 evidently lay the foundation for the more advanced social developments that occur during stages 4–6.

During stage 3 the infant ape evidently acquires social skills comparable to those of macaques, learning to interact socially on an emotional level, and acquiring a propensity to eat the foods eaten by other group members. In addition, he begins incorporating visual/gestural and auditory *secondary circular reactions* into his play patterns and into his object manipulations, which monkeys do not do.

During stages 4, 5, and 6, ape socialization far surpasses, and is qualitatively different from, that of macaques. Visual/gestural and auditory *secondary circular coordinations* are incorporated into play and into aggressive displays in stage 4. Stage 5 experimentation characteristic of the *tertiary circular reaction* is also incorporated into play and aggression in stage 5 as apes use sticks to poke and strike with and throw stones and other objects. Also during stages 4–6, apes learn the ape protocultural traditions of advanced food-getting techniques involving tool-use and tool-making through stages 4–6 imitation.

Human socialization during the sensorimotor period (birth–2 years) differs from that of the apes most saliently in the development of vocal skills and incorporation of these skills into the socialization process. Stage 2 nonsocial vocal *primary circular reactions* (cooing) are the foundations for the social vocal *secondary circular reactions* (babbling) of the third stage that are characterized by back-and-forth vocalizing and vocal "games" between mother and infant. During stage 4, as their babbling becomes more advanced, human infants begin to combine sounds, attempt to imitate new sounds, and attempt to regulate social interactions vocally. During stage 5 their imitation becomes more precise and they speak their first words, and during stage 6 they begin combining words, and language soon becomes the principal mode of further socialization.

As one examines the socialization of monkey, ape, and human infants in terms of their intellectual potentials, the striking findings are that macaque socialization is based on only limited cognitive abilities (stage 3) in most sensory modalities but on advanced stage (4–6) sensorimotor abilities in the tactile/kinesthetic modality. These advanced tactile/kinesthetic abilities provide the basis for the acquisition of macaque copulatory skills. Ape socialization differs from that of macaques in the elaboration of the auditory and visual/gestural channels. These advances provide apes with the ability to make auditory displays and to make and use tools. Human socialization during the sensorimotor period is unique in its development of the vocal channel, permitting human infants to learn and become socialized through vocal language.

■ ACKNOWLEDGMENTS

I thank the San Francisco Zoological Society, Saul Kitchner, William Mottram, Robert McMorris; the San Diego Zoological Society, Clyde Hill and James Dolan; Stanford University, David Hamburg, Patrick McGinnis, and Penny Patterson for providing access to the animals as subjects. I also thank Penny Patterson, John Alcaraz, Richard Cuzoni, and Sue Shrader for the time and effort they so graciously gave to this project. I thank Sue Parker for collecting a portion of the ape data on which this study is based. I thank William Townsend for taking most of the photographs. I thank Harriett Lukes for typing the manuscript.

This research was supported by two grants from the Academic Senate Committee on Research, University of California, San Francisco, and by a grant from the Endowment Fund of the George Williams Hooper Foundation, University of California, San Francisco.

■ NOTES

1. The model especially addresses environmental causes of behavior. However, sensorimotor development is undoubtedly related to the maturation of the nervous system, and Gibson (1970, this volume) and Chevalier-Skolnikoff (1974a) have examined some of the presumed relationships between behavioral and neurological development.

2. The issue of stages has been addressed in Chevalier-Skolnikoff (in press[a]). Until frequency analysis has been applied to the data, in order to determine whether stages actually exist, the issue of whether development is a continuum or a series of discrete stages cannot be solved. However, time

periods characterized by the manifestation of the behavioral parameters seen in the human stages described by Piaget can be determined when these exist in other species, so I will, for the present, follow Piaget's stage classification.

3. See Parker (this volume) for a discussion of the behavioral parameters (termed "metabehavioral parameters") considered in her classification of stumptail macaque behavior.

4. Behavior was analyzed according to sensory modality and body part in this study because in another study on the development of intelligence and communication, it was found that some primates attain different levels of intellectual capacity in different sensory modalities (Chevalier-Skolnikoff, in press[a]).

5. By completing all six stages, I mean that they show behaviors incorporating the behavioral parameters defining all six stages as human infants do. However, many of the specific motor patterns differ from those manifested by human infants.

6. The repetitive hand-mouth, hand-hand, and hand-foot coordinations that are so prominent in ape and human infants are absent in the macaque.

7. See page 172 for one possible exception.

8. Parker's infant manifested stage 3 at 4 weeks; the three infants I observed showed stage 3 behaviors (reaching towards other animals and distant objects and approaching other animals with lipsmacking expressions) at 4, 2, and 2 weeks respectively. The infant who had no peers manifested the stage at 4 weeks and those with peers at 2 weeks.

9. This may have occurred earlier even though it was not observed.

10. I have tested home-reared apes for this ability. They would beat a drum or a xylophone but would not precisely imitate a rhythm. However, further testing for these skills might, with longer testing sessions, yield different results.

■ REFERENCES

Bowlby, J. 1969. *Attachment and loss, vol. I—Attachment.* New York: Basic Books.
Beatty, H. 1951. A note on the behavior of the chimpanzee. *Journal of Mammalogy* 32:118.
Chevalier-Skolnikoff, S. 1971. The ontogeny of communication in *Macaca speciosa.* Ph.D. dissertation, University of California, Berkeley.
_____. 1973. Facial expression of emotion in nonhuman primates. In *Darwin and facial expression: A century of research in review,* ed. P. Ekman, pp. 11–89. New York: Academic Press.
_____. 1974a. *The ontogeny of communication in the stumptail macaques* (Macaca arctoides). *Contributions to primatology,* vol. 2. Basel: S. Karger.
_____. 1974b. The primate play face: a possible key to the determinants and evolution of play. *Rice University Studies* 60:9–29.
_____. 1974c. Male-female, female-female, and male-male sexual behavior in the stumptail monkey, with special attention to the female orgasm. *Archives of Sexual Behavior* 3:95–116.
_____. 1975. A proposed model for the analysis of primate "protoculture" in terms of intellectual abilities. Paper read at the meetings of the American Anthropological Association, San Francisco.
_____. in press[a]. The ontogeny of primate intelligence and its implications for communicative potential: A preliminary report. New York: Annals of the New York Academy of Sciences. 280:173–211.
_____. in press[b]. Homosexual behavior in a laboratory group of stumptail macaques (*Macaca arctoides*); its forms, contexts, and possible social functions. *Archives of Sexual Behavior.* 5(6):511–527.
Ekman, P. 1972. Universals and cultural differences in facial expressions of emotion. In *Nebraska symposium on motivation, 1972,* ed. J. Cole, pp. 1–62. Lincoln: University of Nebraska Press.
_____. 1973. Cross-cultural studies of facial expression. In *Darwin and facial expression: A century of research in review,* ed. P. Ekman, pp. 169–222. New York: Academic Press.
Frisch, J. E. 1968. Individual behavior and intertroop variability in Japanese macaques. In *Primates: Studies in adaptation and variability,* ed. P. C. Jay, pp. 243–252. New York: Holt, Rinehart & Winston.

Gibson, K. R. 1970. Sequence of myelinization in the brain of *Macaca mulatta*. Ph.D. dissertation, University of California, Berkeley.

Goodall, J. 1963. Feeding behaviour of wild chimpanzees. A preliminary report. *Symposia of the Zoological Society of London* 10:39–48.

――――. 1964. Tool-using and aimed throwing in a community of free-living chimpanzees. *Nature* 201:1264–1266.

Harlow, H. F. 1961. The development of affectional patterns in infant monkeys. In *Determinants of infant behaviour*, vol. 1, ed. B. M. Foss, pp. 75–88. New York: John Wiley.

Harlow, H. F., and Harlow, M. 1966. Learning to love. *American Scientist* 54:244–272.

Hinde, R. A. 1974. *Biological bases of human social behavior*. New York: McGraw-Hill.

Itani, J. 1958. On the acquisition and propagation of a new food habit in the troop of Japanese monkeys at Takasakiyama. *Primates* 1:84–98.

Itani, J., and Nishimura, A. 1973. The study of infrahuman culture in Japan. *Symposium IVth international congress of primatology*, 1: *Precultural primate behavior*, pp. 26–50. Basel: S. Karger.

Itani, J., and Suzuki, A. 1967. The social unit of chimpanzees. *Primates* 8:355–381.

Jones, C., and Pi, J. S. 1969. Sticks used by chimpanzees in Rio Muni, West Africa. *Nature* 223: 100–101.

Kawai, M. 1965. Newly acquired pre-cultural behavior of a natural troop of Japanese monkeys on Koshima Island. *Primates* 6:1–30.

Kummer, H. 1971. *Primate societies: Group techniques of ecological adaptation*. Chicago: Aldine-Atherton.

Lawick-Goodall, J. van. 1967. Mother-offspring relationships in chimpanzees. In *Primate ethology*, ed. D. Morris. London: Weidenfeld & Nicholson.

――――. 1968. The behaviour of free-living chimpanzees in the Gombe Stream Reserve. In *Animal behaviour monographs* 1:165–311.

――――. 1970. Tool-using in primates and other vertebrates. In *Advances in the study of behavior*, vol. 3, eds. D. S. Lehrman, R. A. Hinde, and E. E. Shaw, pp. 195–249. New York: Academic Press.

――――. 1973. The behavior of chimpanzees in their natural habitat. *American Journal of Psychiatry* 130:1–12.

McGrew, W. C. 1974. Tool use by wild chimpanzees in feeding upon driver ants. *Journal of Human Evolution* 3:501–508.

Nishida, T. 1973. The ant-gathering behaviour by the use of tools among wild chimpanzees of the Mahali Mountains. *Journal of Human Evolution* 2:357–370.

Parker, S. 1973. Piaget's sensorimotor series in an infant macaque: The organization of nonstereotyped behavior. Ph.D. dissertation, University of California, Berkeley.

Philipps, T. 1950. Letter concerning man's relation to the apes. *Man* 272:168.

Pi, S. J. 1974. An elementary industry of the chimpanzees in the Okorobikó Mountains, Rio Muni (Republic of Equatorial Guinea), West Africa. *Primates* 15:351–364.

Piaget, J. 1951. *Play, dreams and imitation in childhood*. Translated by C. Gattigno and F. M. Hodgson. New York: W. W. Norton & Company, Inc.

――――. 1952. *The origins of intelligence in children*. Translated by M. Cook. New York: International Universities Press, Inc.

――――. 1954. *The construction of reality in the child*. Translated by M. Cook. New York: Ballantine Books.

Pitman, C. R. S. 1931. A game warden among his charges. Cited in *The mountain gorilla*, G. B. Schaller, 1963, Chicago: Chicago University Press.

Poirier, F. 1972. Introduction to *Primate socialization*, ed. F. Poirier, pp. 2–29. New York: Random House.

Rahm, U. 1971. L'imploi d'outils par les chimpanzees de l'ouest de la Côte-D'Ivorie. *Terre et la Vie* 25:506–509.

Savage, T. S., and Wyman, J. 1843–1844. Observations on the external characters and habits of the *Troglodytes niger* and on its organization. *Boston Journal of Natural History* 4:362–386.

Slobin, D. I. 1971. *Psycholinguistics*. Glenview: Scott, Foresman and Company.

Spitz, R. A., and Wolf, K. A. 1964. The smiling response: A contribution to the ontogenesis of social relations. *Genetic Psychology Monographs* 34:57–125.

Struhsaker, T. T., and Hunekeler, P. 1971. Evidence of tool-using by chimpanzees in the Ivory Coast. *Folia Primatologica* 15:161–316.

Suzuki, A. 1966. On the insect-eating habits among wild chimpanzees living in the savanna woodland of western Tanzania. *Primates* 7:481–487.

Tsumori, A. 1967. Newly acquired behavior and social interactions of Japanese monkeys. In *Social communication among primates*, ed. S. A. Altmann. Chicago: University of Chicago Press.

Uzgiris, I. C., and Hunt, J. Mc V. 1966. An instrument for assessing infant cognitive development. Unpublished manuscript.

Wolff, P. H. 1969. The natural history of crying and other vocalizations in early infancy. In *Determinants of infant behaviour* IV, ed. B. M. Foss, pp. 81–109. New York: John Wiley.

Yamada, M. (Kawabe). 1957. A case of acculturation in the subhuman society of Japanese monkeys. *Primates* 1:30–46.

The Social Ecology of Defects in Primates

GERSHON BERKSON

■ INTRODUCTION

People with significant chronic defects have been part of human society at least since Neanderthal times (Goldstein 1969; Stewart 1958) and constitute at least 5% of the population of modern Western societies. There is variability both within and between societies in the character of response to abnormality, ranging from euthanasia (which is rare) through neglect (which is common) to complex institutions of education and rehabilitation aiming to keep disabled individuals in the mainstream of the culture (Edgerton 1970). No doubt this variability is a consequence partly of different economic, kinship, and value systems interacting with the behavior of individuals who have different kinds and degrees of disability. Considering the many sources of variability, one might not expect to find any general features of human society which describe human responses to disability. However, there are certain universal aspects of human social organization that suggest common patterns of response to the handicapped.

A major characteristic shared by human societies is that they tend to be permanent; groups live together for long periods and each individual is familiar with others in the group. Elaborate patterns of unlearned and learned behaviors keep the group together and protect individuals within it. Kinship, role, and status are aspects of social organization affecting how any individual, normal or handicapped, participates in the group. Humans also have low fertility rates and a long infant growth period. Each individual is therefore important for the group's reproductive potential, and protection of the infant during the entire growth period is well-developed.

These universal characteristics are also seen in the higher primates (Reynolds 1966). If one assumes that response to individuals with defects is a reflection of normal social organization, it becomes possible to ask whether there are certain predictable features of response to the handicapped in higher primates, in which cultural factors are minimized. This paper first considers the relationship between aspects of higher primate social organization and response to the handicapped. Then, the results of some studies of macaque monkey responses to visually impaired members of their group are presented. Finally, the results are discussed in relation to an ecological concept of handicap.

SOCIAL SELECTION AND THE DEFINITION OF HANDICAP

Within a simplistic concept of natural selection, one would not expect animals with defects to survive very long. Not only are they likely to be more vulnerable to predation and less able to compete successfully for limited food resources, but it has been suggested that the group itself might tend to eliminate them (Darwin 1871). Intraspecies competition for resources has been shown by Errington (1967) to increase mortality of muskrats who are ill, and Wynne-Edwards (1962 chapter 22) has reviewed other instances of intraspecies social selection. On the other hand, Darwin also cited instances of animals caring for individuals with defects, and Schultz (1956) and Berkson (1974) have reported on the existence of defects in natural animal populations. Social selection against defects is therefore not necessarily a feature of all animal societies at all times.

Before proceeding, it is important to differentiate between the terms "defect" and "handicap." Although an individual may be abnormal in some way, the abnormality will not be a handicap to adaptation unless the individual's particular environment requires significant use of the abnormal part. Thus, a blind raccoon (Sunquist et al. 1969) can show normal movement in its home range, and a one-armed gorilla (Schaller 1963) can be dominant in his group. The concept of handicap therefore is an ecological statement describing both the individual and its environment. A consequence of this is that defects and their effects are one potential way of describing the adaptive requirements of any environment (Berkson 1974). It also means that different defects are not expected to be equivalent in producing reactions from members of the social group. Defects in sense modes or response systems which are most important in normal social organization probably have more dramatic effects than those in less frequently used systems. More severe and extensive defects should show more obvious effects.

A further corollary, which is most important to the discussion following, is that social responses to the same defect can depend on the environment in which the group lives. This is not only because the behavior of the defective individual will tend to be more or less normal in various environments, but also because the group's response to deviance may vary with the environment. Social response to a defect, therefore, can be a consequence of severity of the defect; the degree to which the defect is related to normal social communication; and the extent to which the environment both differentiates the behavior of the defective individual from that of other group members and modifies tolerance of the group to deviance.

PERMANENT AFFILIATION, MUTUAL PROTECTION, SUBGROUPING, STATUS

A second group of considerations is related to the notion that higher primates live in permanent, socially differentiated groups functioning for mutual protection. Other things equal, it is likely that the increased viability that a normal individual obtains from being a member of a *mutually supportive social group* will also benefit the handicapped individual (see Fedigan and Fedigan, this volume). If the defect is incompatible with life (e.g., so that the individual cannot feed) or completely eliminates social responsiveness, the group will respond to the defective individual for a short time (Kling et al. 1970). But without a mutual response on its part, maintaining the social relationship, the group's responsiveness will decline.

Animals in permanent groups provide *general protection* by chasing off predators and strange conspecifics, uttering alarm calls, and locating food that the whole group eats. The larger the group, the greater the number of animals available to provide such general support. Also, in a large group, there is a greater number and variety of animals available to socialize with. Therefore, it is likely that a handicapped animal will be more likely to survive and be integrated into a large, as compared with a small, group. A possible limit on this principle might be a high density population. However, animals with defects survive in crowded monkey populations (Furuya 1966; Berkson 1973). It is unlikely that density by itself limits viability of the handicapped individual. Only if density interferes with normal social integration of the group (perhaps via pathologically high aggression levels) would an effect on the handicapped animal be seen. However, instances of such a breakdown occurring in primates are rare (Hall 1963; Turnbull 1972) and are not usually associated only with density.

To the extent that the society is socially differentiated, responses to the handicapped individual by other group members may be determined by the rules of differentiation of that particular species and group. *Kinship, specific age-sex groupings* and *dominance status* are the main dimensions along which enough is known to make predictions. Many animal societies provide protection in proportion to their kinship relationship (Hamilton 1964; Trivers 1971). To the extent that kinship is an organizing principle in primate societies (Sade 1965), one would expect closely related individuals to provide care and be otherwise socially related to the handicapped to a greater degree than less closely related individuals.

To a lesser extent, relationships would also be expected to the degree that other individuals interact under special circumstances. In most primate groups, young animals play together. Although a mother ordinarily has the closest relationship with her offspring, adults other than the mother often care temporarily for young infants. Young male macaques generally stay together in less permanent subgroupings. Adult female macaques tend to be part of their natal group while males may leave it.

The generality of the dominance principle in animal social organization has repeatedly been questioned (Gartlan 1968; Rowell 1972). However, there is little doubt that in situations where resources are limited, certain group members predictably have greater priority to them. In competitive situations, an individual with a defect, in general, would be expected to be handicapped and, through repeated failure, might learn that it is subordinate and behave accordingly. However, the picture is undoubtedly more complex. In competitive circumstances where a defect is not handicapping, no reduction in dominance is expected, and this might also be true much of the day when members of the group are not competing. Furthermore, through the kinship principle, if an animal who is dominant protects it, the handicapped animal might not be subordinate at all.

The second group of principles, therefore, are that, to the extent that groups are stable and based on permanent social affiliations, individuals with defects will survive if the defect does not handicap vegetative functions or sociability. Enduring kinship relationships are a major avenue of group integration, and the handicapped individual may also benefit from more specific interactions with certain other animals in the group. A defective individual will generally tend to have subordinate status. However, this effect may be attenuated to the extent that the defect is no handicap in competitive situations; to the extent that it is protected by dominant animals; and to the degree to which its group lives in an environment which does not evoke competition.

LOW FERTILITY AND PROLONGED DEPENDENCE

Monkey, ape and human births rarely exceed one infant, and births are spaced a year or more apart, depending on the species. Since primate groups are ordinarily small, death of an individual prior to reproductive age constitutes a significant waste in the resources available for supporting group size. In addition, most infant primates are immature at birth and grow slowly for a significant period of their life span. During growth, they depend on the group, and protection of them is vigorous.

In this context, care compensating for both temporary and chronic handicaps is expected, especially in infancy, and should have the same character as responsiveness to normal infants. That is, to the extent that a handicapped animal behaves in an awkward manner, the mother, and to some extent other group members, will respond as if to the awkwardness of an infant (see Fedigan and Fedigan, this volume).

That there are limitations to this compensatory responsiveness, both on the part of group members and by the infant, has been made clear by Rosenblum and Youngstein (1974; cf. also Rosenblatt 1963). A dead or sedated infant continues to evoke care but usually only for a limited period. Furthermore, normal reduction in care (which is a product both of maternal hormonal changes and age alterations in how the infant looks and behaves) predicts declining compensatory care with increased age of the infant.

Thus, a handicapped individual should remain part of a primate social group indefinitely, to the extent that it is socially responsive. The group gives care compensating for the handicap early in development, but this tendency decreases with age. As an adult, the handicapped animal will be present in the group, but isolated to the degree that its defect handicaps normal social interaction.

■ SOCIAL DEVELOPMENT OF HANDICAPPED MONKEYS

These concepts have been the main concern of our research during the last few years. The studies have looked at social responses to visually handicapped monkeys in wild (Berkson 1970), free-ranging (Berkson 1973) and laboratory (Berkson and Karrer 1968; Berkson 1974; Berkson 1975; and Berkson and Becker 1975) groups of macaques. Thus far, our studies have shown that blind infants survive and remain with their groups in all habitats at least for some time and live indefinitely when predators are absent and food plentiful. Their groups provide qualitatively normal care that compensates for their handicap during their first 2 years, and preliminary information indicates that they remain part of the group into adulthood. The social status of the disabled animals after two years of age depends on their degree of vision and the requirements of the habitat in which they live. The remainder of this paper provides the basis for these conclusions and a discussion of their implications.

Various types of defects have been used in analyzing social interactions of defective primates. Major natural field observations have included van Lawick-Goodall's (1968) description of a polio epidemic in a chimpanzee group. Amygdalectomy (Kling et al. 1970); phocomelia produced by thalidomide (Lindburg 1969); spontaneous generalized defects (Fedigan and Fedigan, this volume) and deafness (Talmage-Riggs et al. 1972) have also been studied. In the present series, partial and total blindness have been used because primate sensory adaptation is primarily via a visual mode, and it was therefore expected that animals with visual defects would be maximally handicapped in their relationship with their social and physical environment. The experiments to be described were carried out in wild, free-ranging and laboratory populations in order to gain an impression of the ways in which habitat variation affects survival and social responsiveness to handicapped animals. In all of the studies the defect was imposed on infants soon after birth, and they and their groups were studied longitudinally.

WILD POPULATION

A first study of partially blind animals (Berkson and Karrer 1968) had shown that their social development in the first 6 months was essentially normal when they were housed with their mothers in small cages. This led us to ask whether partial blindness would be a handicapping condition in the complex environment of a natural habitat. We observed two partially blind monkeys and two controls in a wild group of crab-eating monkeys (*Macaca fascicularis*) living on an arid island in the Gulf of Siam (Berkson 1970). The island supported two natural groups of 35 monkeys, although free water was absent for part of the year. On the island, there also lived a species of large monitor lizard which was a potential predator of young monkeys. The major issue was whether the blind animals could survive at all and, if so, how long they could live, and whether anyone in their group would protect them in such a severely selecting environment.

The results showed that the infants lived for 7 months but were handicapped, i.e., their movements were awkward and they were not able to stay with the group as efficiently as the sighted controls. If an observer approached the group, the group would ordinarily escape deep into the forest. However, if one of the blind infants had been left behind, the group stayed in the vicinity threatening at the observer. If he withdrew a short distance and hid himself, the mother quietly returned and retrieved her blind infant, and then the whole group left the area.

This suggested that a complex environment handicaps a partially blind infant, and that behaviors of the mother and the rest of the group which keep a young normal infant with the group also compensate for a handicap in an older infant. Although the blind infants survived until seven months of age the reason for their disappearance was not clear. It was plausible that a predator might have taken them, that they had simply been left behind as they grew older, that the group had killed them, or that they had died of thirst during the arid period during which they had disappeared.

FREE-RANGING POPULATION

Because the reason for the disappearance of the blind infants was not clear, and particularly to separate social selection on the one hand from predation and nutrition on the other, the study was repeated with a group of rhesus monkeys (*Macaca mulatta*) living under crowded conditions on an 80-acre arid island (La Cueva) in La Parguera, Puerto Rico. In this free-ranging colony, artificial food and water are provided *ad libitum* and there are no predators. In this habitat the two male and one female partially blind animals (T-9; V-4; VO) survived until 3 years of age when the project was terminated. Just at 3 years of age, the female disappeared from the group and the males were still alive. The three controls (T-0; V-7; V-8) were still alive and one of them (a male) was living outside the group.

During the first year (Berkson 1973; 1974) the blind infants stayed somewhat closer to their mothers than controls and were protected by her in feeder stations. There was no difference, however, between experimentals and controls in proximity to their mothers during the breeding season (when they were 6 months old) since the mothers of the blind animals often left them with the group. However, the blind infants were never left alone, and specific members of the group tended to stay near them and carry them if the group moved. Although intragroup aggression was high during the breeding season, this did not adversely affect the blind infants. At 1 year of age there was no difference between the normal and blind yearlings in proximity to their mothers.

During their second year the mother of one blind male died and the mother of a control was removed from the group for another study. The son of this control female was the animal that began living outside the group in its third year; the motherless blind animal continued with the group until termination of the experiment but tended to be socially isolated. During the second year, 2 mothers of blind infants had new normal babies. The presence of the new babies did not apparently affect responses to the blind infants, and the mothers carried both the blind and the new babies when necessary.

At 2 years of age controls who had mothers were seen with their mothers on occasion but were generally quite independent. The two blind animals with mothers were with them most of the time and were groomed by them frequently. The motherless blind male appeared socially isolated although he occasionally engaged in short play and grooming bouts with peers. The two blind males had lesions and scars on their faces which were subsequently seen to result from their bumping into branches in the mangrove forest where the groups spend most of their time. The blind female had no such lesions. No instance of aggression toward the blind animals was seen; however, they often cringed from an approaching animal.

Thus, in the second year, two blind animals still had close relationships with their mothers, and the one without its mother was socially isolated but remained with the group. At the end of the third year all three of the blind subjects were alive. The two males still had facial scars; the female did not; all were apparently in good health. Nevertheless, the female disappeared from the group at that time. It was not possible to determine the reason why this was so; however, it may have been related to the fact that the 80-member group split at about this time and approximately 20 animals swam to a nearby island (Guayacon). Perhaps she became separated from the group during the transition.

During the third year the situation with the two blind males was quite clear. The group spent most of their time around one of the four feed-water stations on the island. The two blind animals were seen there somewhat more than other animals. They did not interact with each other. The one blind and two control animals who still had mothers were seen with them occasionally. Ordinarily, all blind and sighted subjects sat, groomed, and played with a number of other animals. There was no dramatic difference between the two groups in the number of contacts with other animals.

The two blind animals were clearly subordinates. They ordinarily cringed from an approaching animal and on only one occasion did a blind subject displace another animal—a yearling. At the enclosed feed stations, the blind animal (who had lost its mother) entered the feeder only if the area was quiet and there were no more than one or two animals in it. The other blind animal was somewhat bolder, but still usually wary. The group contained about 10 peripheral males ranging from 2–5 years of age who often traveled separately from the main group and fed together after the main group had had its fill at the feeder. It is perhaps remarkable that the blind males were part of the main group rather than this subordinate peripheral male subgroup.

SUMMARY

It is possible to draw some general conclusions from the studies of partially blind animals in wild and free-ranging macaque groups. The first is that a visual acuity defect is handicapping in both environments. The group, and especially the mother, provides compensatory care in the early period. This care is adequate to protect infants in a natural environment in the first half year but not thereafter. In the absence of predators, and where food and water are available, the blind animals survive until young adulthood at least. During the first two years, the mother and other animals sporadically provide protection which assures that the infant can obtain food in competitive situations. After three years, the special relationship with the mother has waned almost completely. Nevertheless, the handicapped juveniles travel and maintain a degree of social relationship with the group. However, they are apparently not independent enough either to be full-fledged members of the main group or to join with peripheral males.

LABORATORY STUDIES

Although the studies in natural and free-ranging populations allowed some tentative general conclusions, it was necessary to perform more intensive studies in a laboratory environment to achieve a more precise picture. Detailed observations and clarifying experiments were possible. Perhaps most important, the laboratory provided the opportunity to determine the character of the social relationship of blind animals in a dense population when the physical environment placed minimal demands on vision.

Four standard groups of crab-eating monkeys (*Macaca fascicularis*) were studied in 4.8 × 4.0 × 1.7-meter cyclone fence cages. Each group initially consisted of an adult male, four pregnant females, and two juvenile males. During the succeeding years, the groups were allowed to grow to a maximum size of 15 animals which is somewhat smaller than the natural size for this species (Kurland 1973). Within 18 days after birth, ($\overline{X}$ = 9 days) the offspring were made totally blind (n = 5), partially blind as in the previous studies (n = 6) or underwent a sham operation (n = 11). One of the control animals had a generalized growth deficit, and one partially blind animal died at 2 years 3 months of age in an accident probably unrelated to its visual defect. The data of the infant with the growth defect are excluded from the analyses. The composition of the groups as of December 10, 1974, is shown in Table 1.

The experimental design was therefore a longitudinal study of the blind and sighted infants. Data presented here are primarily for their first 2 years (75 hours of observation per group per year). Twenty-four momentary observations of each of the 50 monkeys were made in each 2-week period of the 2 years employing a 38-category behavior checklist. For each observation, *proximity* and *behavior* were scored. The proximity measures included an identification of the closest and second closest animal to the one being observed and the distance of the closest animal (in contact; within two feet but not in contact; and further than two feet). The behaviors recorded are listed in Table 2. If the behavior was social, the individual with whom it occurred was identified. The interrater reliability of the observations averaged 98% agreement for proximity and 95% during observations in which at least one observer saw a behavior occur.

At this writing, the youngest animal in the experiment is 2 ½ years of age and the oldest of the animals born into the groups is a partially blind male who is 5 years, 5 months old. There was no mortality among the 22 monkeys of the control and two experimental groups in the first 2 years, and since then, only the one accidental death alluded to above. This indicates clearly that high population density alone (i.e., one animal/2.33 cubic meters) does not increase mortality of animals with defects.

Perhaps this is partly due to the fact that the familiar environment did not grossly handicap even the totally blind animals. They were familiar with all parts of the cage and easily located the food and water which were freely available. Figures 1–3 portray the development of behavior of the infants and their relationships to other animals over the first 2 years of the infants' lives. To derive these curves the data of each infant, for each category, were summed over four 2-week periods. Many categories occurred infrequently and some were therefore combined. Repeated measures analyses of variance for the three groups were performed, and the 1% level of confidence was used for all tests.

Figure 1 plots the distribution of the infants' major activities during the first two years. When a significant effect for blindness or its interaction with age was evident, the graph shows the data for the three groups separately. When there was no effect of

TABLE 1 / COMPOSITION AND CHARACTERISTICS OF THE GROUPS ON DECEMBER 10, 1974

No.	Group	Sex	Birth	Mother	No.	Group	Sex	Birth	Mother
		GROUP I					GROUP II		
01	Adult	M			01	Adult	M		
02	Adult	F	died 4/28/74		02	Adult	F		
03	Adult	F			03	Adult	F		
04	Adult	F	died 4/25/72		04	Adult	F		
05	Adult	F			05	Adult	F		
06	Juv	M			06	Juv	M		
07	Juv	M			07	Juv	M		
08	Partial	M	1/3/70	04	08	Control	F	9/12/71	02
09	Control	M	1/29/70	02	09	Partial	F	10/31/71	03
10	Partial	M	4/17/70	03	10	Total	M	8/6/72	04
11	Control	M	7/28/70	05	11	Partial	M	12/2/71 (died 3/8/74)	05
12	Control	M	8/19/71	03	12	Total	M	9/18/72	02
13	Control	M	5/10/72	05	13	Control	F	10/17/72	05
14	Adult	F	Replaced 04	7/20/72	14	Control	F	4/6/73	03
15	Total	M	12/10/72	02	15	Control	F	4/13/74	02
16	Control	F	9/3/74	05					

No.	Group	Sex	Birth	Mother	No.	Group	Sex	Birth	Mother
		GROUP III					GROUP IV		
01	Adult	M			01	Adult	M		
02	Adult	F			02	Adult	F		
03	Adult	F			03	Adult	F		
04	Adult	F			04	Adult	F		
05	Adult	F	died 2/12/74		05	Adult	F		
06	Juv	M			06	Juv	M		
07	Juv	M			07	Juv	M		
08	Control	F	1/12/72	04	08	Total	M	6/10/72	04
09	Partial	F	1/24/72	05	09	Partial	F	8/2/72	03
10	Control	M	5/19/72	03	10	Control	M	11/25/72	05
11	Total	M	2/17/72	02	11	Control	F	11/30/72	02
12	Control	F	10/31/72	05	12	Control	F	8/27/73	04
13	Control	M	6/23/73	02	13	Control	F	5/1/74	03
14	Control	F	7/20/73	04	14	Control	M	6/20/74	02
15	Control	F	8/25/74	03					

blindness, the data for the three groups are combined. In each case presented the age effect was significant.

The developmental data are generally consistent with those found for other macaques (summarized by Hinde 1971). A gradually declining relationship with the mother over the first two years is accompanied by increments in interaction with the

TABLE 2 / LIST OF BEHAVIOR CATEGORIES AND DEFINITIONS

CATEGORY	DEFINITION
contact	body contact
proximal	within 2 feet
beyond	beyond 2 feet
mount-copulate	mount with feet on partner's legs
being mount-copulated	being mounted with partner's feet on S's legs
present	present for copulation without aggression
groom	groom
being groomed	being groomed
play-wrestle	play-fight, rough-and-tumble play
social exploration	lift tail and look at perineal region
embrace	holds arm around other animal's shoulder
eat	hold or put food in mouth
drink	lick or suck water
locomote	walk or climb more than 1 foot without social interaction
explore	grasp or nose inanimate object
play-object	play with inanimate object
sleep	motionless and eyes closed or head between legs
displace	displace another animal's position
give-way	give-way position
stare	stare at other animal
mount	mount with feet on floor
present	present to aggression
lip-smack	lip-smack
fear-grimace	grimace
yawn	
shake	shake environment
gape	gape-threat
lunge	rapid movement toward another animal in context of aggression
grab	manual contact associated with other evidence of threat
bite	
vocalized threat	
punitive deterrence	punishes infant
cling	baby holds on to mother while in ventro-ventral position
dorsal	mother carries infant on back
retrieve	mother moves to grasp infant
restraint	mother holds infant to prevent movement away
being restrained	infant being restrained

physical and social environment. Reduction of clinging to the mother and locomotion were retarded in the totally blind animals. However, this retardation was not dramatic and was apparently limited to the first year. No gross abnormalities occurred in the development of grooming; being groomed; play-wrestling; eating and drinking; nonspecific social behaviors (social exploration and embrace); manipulate environ-

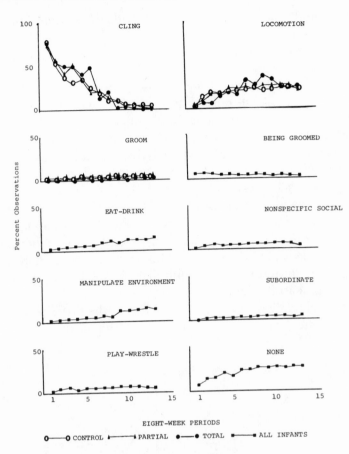

FIGURE 1. *Proportions of time spent by infants in various activities at different ages.*

ment (nonvisual exploration and play-object); or subordinate behaviors (present to aggression, lipsmack, fear grimace, give away).

The totally blind animals therefore stayed with their mothers somewhat more during the first year and moved around somewhat less. However, their developing interaction with the environment was quantitatively normal. This picture is reinforced by the proximity data. General and specific proximity measures are presented in Figure 2. General proximity is a weighted combination of the estimates of the first and second closest animal to the infant in question on every observation of the infant. This score was derived by determining the number of observations in which each animal of the group was counted as closest and second closest to the infant in question. A weighted score was derived for each animal's relationship with each other animal according to the formula provided in Figure 2. Then, the mean of this general proximity score was obtained for each age-sex class in the group. The infant's major relationship with the mother is again revealed from this score, with the totally blind animals closer to their mothers in the first year than controls. The infants' relationships with other adults were much lower than with the mother but also gradually declined with age. The relationships of the infants with the juvenile males and other infants increased. No general blindness effect on the relationship with animals other than the mother was noted.

With one exception the same result was seen from a consideration of specific proximity. This was an analysis only of those scores in which another animal in the group was counted as the closest animal to the infant in question. The frequency of occasions in which the closest animal was seen in contact with (Prox 1), less than two feet but not in contact with (Prox 2), and more than two feet (Prox 3) from the infant was determined and the specific proximity score derived as indicated in Figure 2. Again the mean for each age-sex class in each age interval was obtained. The differential declining relationship of the blind and sighted groups with the mother was apparent, and no blindness effect was shown in the relationship to other animals. There was a general increase over age in the relationship with other infants and the juvenile males. Employing this measure, however, it appeared that there was a somewhat higher relationship with the adult male during the first year than with the adult females other than the mother.

The major social relationship of the infants during the first 2 years, therefore, was with the mother. The data indicate that the relationship of the blind infants to their

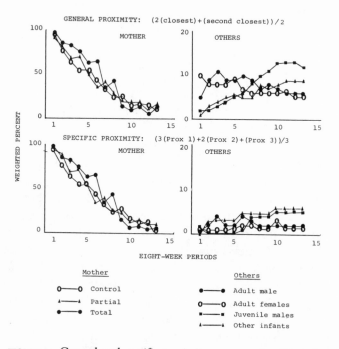

FIGURE 2. *General and specific proximity measures describing the degree of relationship of blind and control infants with mother and other animals classified according to age and sex.*

mothers was somewhat more enduring than that of controls. They clung and were closer to her for a longer period of time than were sighted infants. Figure 3 shows that, perhaps as a consequence of this greater proximity, the mother groomed the blind infants more than the controls. The studies in natural and free-ranging habitats have emphasized that the mother provides compensatory care for the blind infant by retrieving it in circumstances where it is unnecessary to do so for controls. However, the observations in this laboratory study show no increased incidence of maternal behaviors (i.e., retrieving and restraining) other than grooming toward the blind compared with the sighted infants. (See Figure 3.) This may be due to the fact that the environment was relatively simple. The cage not only prevented the blind animal from becoming separated from the group, therefore reducing the necessity of maternal retrieval, but also maximized the possibility that the blind infant would find its mother when it wished to (viz., Berkson and Becker 1975).

It should be emphasized that the discrepancy between social relationships of normal and blind animals was limited primarily to differences in the relationship with their mothers. Differences between the experimental groups in their relationship with other age-sex classes were negligible.

The discrepancy in the picture of the relationship of the infants with the adult females in the two proximity measures should be explained. It is probably due to the fact that with the general proximity measure the females were scored as close to the infant because they tended to sit close to the mother at the times she was carrying the infants, while this intervening relationship was not true of the specific proximity measure. On the whole, the relationship of females to infants other than their own was low. They were ordinarily tolerant of infants near them and occasionally groomed them. On rare occasions a female carried another infant with her own. One female in the experiment attempted to steal the babies of other females.

Adult males developed enduring social relationships with some of the infants. Three of the four males retrieved infants in the ventro-ventral position when caretakers entered the cage. They played with infants and in two instances developed relationships with individual infants lasting a number of months which involved play,

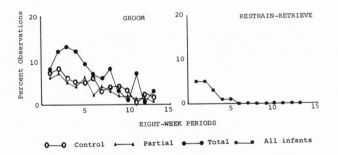

FIGURE 3. *Maternal behaviors directed by mother toward her infant.*

grooming, carrying in the ventro-ventral posture, and retrieving them. There was no apparent tendency for these relationships with the adult males to be different for blind and sighted animals; they did tend to occur in the middle of the first year.

There was also no difference between the blind and sighted animals in the emerging development of play and other social relationships with peers and the older juvenile males.

Thus, no obvious deficit in the general level of social relationships of the blind were seen between the first and second years from these general observational data. This picture is reinforced by more intensive studies (Berkson and Becker 1975) showing that in this familiar environment totally blind animals initiate proportionally as many social contacts as do sighted controls. On the other hand, although the blind animals are capable of grossly normal facial expressions, they show fewer threatening and fear expressions during aggressive social interactions and a generally lower level of facial expressions per social interaction. This means that although the integration of the blind animals in the group is generally normal in a simple environment, there are probably subtle abnormalities in their social relationships.

SUMMARY

The laboratory studies have shown that blind animals survive indefinitely in an environment which requires minimal use of vision. Despite a concentrated population, aggression directed toward them is no higher than toward controls. Their relationship with mothers is somewhat more intense in the first year, but their relationships with the familiar physical and social environment develop fairly normally, although they tend to be somewhat more passive and subordinate in their social interactions.

■ CONCLUSIONS

Taken together with other studies of defective animals in primate groups, these studies attest to the wide range of individual variability that higher primates tolerate within their groups. Visually handicapped monkeys are susceptible to predation or may die when they cannot obtain adequate food. However, they evidently are not killed by their groups at any age. There is also no evidence that the somewhat increased aggression levels or abnormal structure of crowded captive groups increases their mortality. Even totally blind animals benefit from the kinds of social relationships which assure survival of normal individuals.

The evidence also indicates that the degree to which a defect is handicapping, and the ways in which the group responds, depend on both the severity of the defect and the characteristics of the environment. In a natural group, partially blind monkeys were given compensatory care and then disappeared from their group at about 7 months of age. In an environment with no predators, individuals with the same defect survived until 3 years of age but became increasingly socially isolated in their groups. In a laboratory environment, partially blind animals were essentially normal, but totally blind individuals had somewhat abnormal social relationships.

Protection from predation is provided by the group as a whole, and blind individuals benefit from the sporadic relationships all young animals have with animals other than the mother. However, it is the mother who compensates most obviously for the defect by keeping the infant with the group, keeping it out of trouble, and maintaining its access to food. As the mother-infant bond wanes, the blind infant increasingly interacts with the physical and social environment. When the environment does not require much use of vision, its social development is relatively normal. However, in more demanding environments, to the extent that it must spend time foraging and following the group, the blind infant probably has fewer opportunities for interaction with other group members. Since it also does not initiate social interactions readily and since its responses tend to be fearful, it becomes increasingly isolated and subordinate within the group.

The major concepts supported by this research include survival and social tolerance of individuals with defects; ecological relativity of the degree of handicap; compensatory care of handicapped infants, especially by the mother but to a certain degree by the whole group; and the emergence of social isolation and a subordinate status. These are all consistent with the main characteristics common to human and subhuman primate social organization which were outlined in the introduction. They are also features of the social psychology of human disability. This makes plausible the notion that some broad aspects of human response to disability are a consequence not only of cultural variables but also reflect the evolution of primate social organization.

■ ACKNOWLEDGMENTS

The studies reported here were supported by funds from the Illinois Department of Mental Health, University of Illinois Research Board, Delta Regional Primate Research Center and the Caribbean Primate Research Center. The observations in the laboratory study were made by Linda M. Massen and Lilian Tosic. The enucleations were performed by W. C. Delannoy, Jr.

■ REFERENCES

Berkson, G. 1970. Defective infants in a feral monkey group. *Folia Primatologica* 12:284–89.
_____. 1973. Social responses to abnormal infant monkeys. *American Journal of Physical Anthropology*. 38:583–86.
_____. 1974. Social responses of animals to infants with defects. In *The effect of the infant on its caregiver*, eds. M. Lewis and L. A. Rosenblum, pp. 233–50. New York: John Wiley.
_____. 1975. Social response to blind infant monkeys. In *Aberrant development in infancy*, ed. N. R. Ellis. Potomac: Lawrence Erlbaum.
Berkson, G., and Becker, J. 1975. Facial expression and social responsiveness of blind monkeys. *Journal of Abnormal Psychology*. 84:519–23.
Berkson, G., and Karrer, R. 1968. Travel vision in infant monkeys: Maturation rate and abnormal stereotyped movements. *Developmental Psychobiology*. 1:170–74.
Darwin, C. 1871. *The descent of man*. New York: Modern Library.
Edgerton, R. B. 1970. Mental retardation in non-Western societies: Toward a cross-cultural perspective on incompetence. In *Socio-cultural aspects of mental retardation*, ed. H. C. Haywood, pp. 523–59. New York: Appleton-Century-Crofts.

Errington, P. L. 1967. *Of predation and life.* Ames: Iowa State University Press.

Furuya, Y. 1966. On the malformation occurred in the Gagyusan troop of Wild Japanese monkeys. *Primates.* 7:488–92.

Gartlan, J. S. 1968. Structure and function in primate society. *Folia Primatologica.* 8:89–120.

Goldstein, M. S. 1969. Human paleopathology and some diseases in living primitive societies: A review of the recent literature. *American Journal of Physical Anthropology.* 31:285–94.

Hall, K. R. L. 1963. Variations in the ecology of the Chacma baboon *Papio ursinus. Symposium, Zoological Society, London.* 10:1–28.

Hamilton, W. 1964. The genetical evolution of social behavior, II. *Journal of Theoretical Biology.* 7:17–52.

Hinde, R. A. 1971. Development of social behavior. In *Behavior of nonhuman primates,* eds. A. M. Schrier and F. Stollnitz, pp. 1–68. New York: Academic Press.

Kling, A.; Lancaster, J.; and Benitone, J. 1970. Amygdalectomy in the free-ranging vervet (*Cercopithecus aethiops*). *Journal of Psychiatric Research.* 7:191–99.

Kurland, J. A. 1973. A natural history of kra macaques (*Macaca fascicularis,* Raffles, 1821) at the Kutain Reserve, Kalimantin Timur, Indonesia. *Primates.* 14:245–62.

Lawick-Goodall, J. van. 1968. The behavior of free-living chimpanzees in the Gombe Stream Reserve. *Animal Behavior Monographs.* 1:165–311.

Lindburg, D. 1969. Behavior of infant rhesus monkeys with thalidomide-induced malformations. *Psychonomic Science.* 15:55–6.

Reynolds, V. 1966. Open groups in hominid evolution. *Man.* 1:441–52.

Rosenblatt, J. S. 1963. The basis of synchrony in the behavioral interaction between the mother and her offspring in the laboratory rat. In *Determinants of infant behaviour, III,* ed. B. M. Foss. London: CIBA Foundation.

Rosenblum, L. A., and Youngstein, K. P. 1974. Developmental changes in compensatory dyadic response in mother and infant monkeys. In *The effects of the infant on its caregiver,* eds. M. Lewis and L. A. Rosenblum, pp. 141–62. New York: John Wiley.

Rowell, T. 1972. *Social behaviour of monkeys.* Hammondsworth: Penguin Books.

Sade, D. S. 1965. Some aspects of parent-offspring and sibling relations in a group of rhesus monkeys with a discussion of grooming. *American Journal of Physical Anthropology.* 23:1–18.

Schaller, G. B. 1963. *The mountain gorilla.* Chicago: University of Chicago Press.

Schultz, A. H. 1956. The occurrence and frequency of pathological and teratological conditions and twinning among non-human primates. *Primatologica.* 1:965–1014.

Stewart, T. D. 1958. Restoration and study of the Shanidar I Neanderthal skeleton in Baghdad, Iraq. *Yearbook of the American Philosophical Society,* pp. 274–78.

Sunquist, M. E.; Montgomery, G. G.; and Storm, G. L. 1969. Movements of a blind raccoon. *J. Mammalogy.* 50:145–47.

Talmage-Riggs, G.; Winter, P.; Ploog, D.; and Mayer, W. 1972. Effect of deafening on the vocal behavior of the squirrel monkey (*Saimiri sciureus*). *Folia Primatologica.* 17:404–20.

Trivers, R. L. 1971. The evolution of reciprocal altruism. *Quarterly Review of Biology.* 46:35–57.

Turnbull, C. M. 1972. *The mountain people.* New York: Simon and Schuster.

Wynne-Edwards, V. C. 1962. *Animal dispersion in relation to social behavior.* Edinburgh: Oliver and Boyd.

6

The Social Development of a Handicapped Infant in a Free-Living Troop of Japanese Monkeys

LINDA MARIE FEDIGAN
LAURENCE FEDIGAN

■ INTRODUCTION

The Arashiyama West troop of 150 Japanese monkeys (*Macaca fuscata*) was transported to a ranch near Laredo, Texas, in February 1972, after having been studied in Arashiyama, Kyoto, Japan, for nineteen years (Hazama 1964; Norikoshi 1971; Koyama 1967, 1970). In Texas the Arashiyama West monkeys range over a 108-acre native brushland area surrounded by an electric fence. Human intervention is minimal and, although given supplemental provisions, the monkeys forage on local vegetation for most of their food. Since these monkeys have been studied by Japanese scientists since 1954, individual identification, genealogies, and troop historical data are available. Originally all troop members were given individual names by Japanese researchers; however, beginning in 1958 offspring were given the name of the mother plus the year of birth. Thus Wania 66 was born to Wania in 1966, and Wania 6672 was born to Wania 66 in 1972. The Arashiyama West Primate Research Ranch provides the unique research opportunity of studying free-living monkeys, in an intact troop, about which a great deal is known.

The present paper describes the results of a natural experiment in the social development of an infant monkey who had evidenced extensive brain damage, but who survived for thirteen months and interacted in a large, free-living troop. Of what interactions with his social and physical environments was this infant capable? How do troop members deal with a young, abnormal individual? How does a monkey mother cope with a disabled baby? How does the social development of a handicapped infant monkey compare with that of a normal infant monkey? These questions will be explored in the pages that follow.

■ FIRST OBSERVATIONS AND PATHOLOGY

On May 17, 1972, a male infant, Wania 6672, was born into the Arashiyama West Troop, being first observed at 18:05 hours when Wania 66 ran out of thick brush clasping the wet infant to her chest. The infant, her first, was covered with fluid and

FIGURE 1. *Wania 6672, handicapped* Macaca fuscata *infant.*

was still attached to a moist umbilical cord and placenta. Wania 66 sat down near her own mother and sister in a clearing and began to lick the infant and gnaw on the cord and placenta. For half an hour she held the infant upside down in her lap, his head pushed down under her legs, despite jerking movements on his part. During the first few days of his life it was noted that Wania 6672 could not cling to his mother's ventrum with his feet and hind limbs and that his mother continually held him to her ventrum while walking. Because some other infants of this troop have been observed to have initial difficulties in clinging, no special attention was paid to this problem other than noting the extra care of the mother who almost never let the infant drag on the ground. During the next two weeks Wania 66 walked continually on three limbs while supporting the infant with the fourth. We were absent during the summer and when we returned to study the troop on September 1, 1972, it was immediately apparent that Wania 6672 suffered from gross locomotor deficiencies.

Wania 6672 frequently bumped into objects in his environment, such as bushes, cactus, and other monkeys. His movements were spastic, ataxic, and accompanied by continual head dipping. He grasped food clumsily between his thumb and the rest of his hand, showing little manual dexterity, and often sought food items with his mouth rather than his hands. When the infant walked slowly and deliberately, the characteristic head-dipping movement and a lack of coordination with occasional falls were observed. However, when the least bit excited or hurried, as he often was when trying to catch up with his mother, he spun, lurched, and fell continually, his hind limbs being especially stiff and out of control. He developed a gait in which both legs were swung forward simultaneously outside his arms, giving him the appearance of "hopping" like a rabbit. Only when he had been lost for a time and had become extremely agitated did Wania 6672 attempt to climb. He climbed determinedly but with great difficulty, as he was unable to grasp with his toes and would hang by his hands as his legs slipped and flailed seeking a support for his whole foot. By four months of age the infant was able to cling ventrally without support from his mother, but his feet would slowly lose their grip on her fur and have to be repositioned frequently. He never rode dorsally, which most monkeys of this troop can do easily from approximately six weeks of age onwards.

The infant frequently sat in a hunched posture reminiscent of depressed (orphaned or sick) young monkeys, or else he leaned on something or someone, preferably his mother. He would topple over if his support moved and have difficulty getting up again. His eyes appeared crossed and while he seemed able to perceive the outline and motion of large objects, he appeared unable to see detail and to perceive depth. He would react fearfully if we approached him directly or quickly, although he was not afraid of other monkeys, and he would orient toward us and look *up* if we swayed our bodies or waved our arms from a distance of up to ten feet. He would also orient toward passing monkeys, but if his mother passed silently at more than ten feet he would not recognize her. Wania 6672 consistently sniffed at monkeys who approached closely, possibly using olfactory cues to identify or discriminate between individual monkeys. Preparatory to, and during movement he would repeatedly move his head in a dipping or arching half-circle as though trying to compensate for poor depth perception by a "scanning" gesture. He consistently held food directly and closely in front of his eyes before putting it in his mouth—a gesture very uncharacteristic of Japanese monkeys; and he had trouble locating earth-colored food items (e.g., monkey chow) on the bare ground. When the troop was provisioned with such food, Wania 6672 would rub his hands over the ground very much like a blind individual trying to locate an object. Yet he could reach up and, with fair accuracy, grasp a branch which hung down closely in front of him.

Dr. W. G. Tatton, a neurophysiologist at the University of Calgary Health Science Centre with an interest in primate motor deficiencies, reviewed segments of the several hours of video-tape film we have of Wania 6672 engaged in various locomotor and social activities. He suggested that the infant showed many characteristics seen in human children with spastic cerebral palsy. Cerebral palsy may include various degrees of muscular weakness, spasticity, ocular disorders, and mental deficiency, which may hinder normal cognitive and social development unless compensatory treatment is given. The above syndrome was recognized in Wania 6672, and although it is not possible to specifically diagnose his condition as resulting from spastic cerebral palsy, the latter does provide a useful model for understanding his behavior.

Dr. Tatton surmised that Wania 6672's lesions involved widely separated areas of the central nervous system as evidenced by his poor visual function, bilateral convergent strabismus (crossed-eyes), motor deficits, and ataxia. The infant's strabismus seemed so severe as to preclude binocular vision; hence, his characteristic head-dipping behavior may have developed as a compensatory mechanism for depth perception, as in many birds, for example. However, his visual function appeared so poor that a lesion in the visual system also was postulated. Motor deficits showed in extreme spasticity, especially in the hind limbs which were weak and poorly coordinated. In the little climbing he did, Wania 6672 depended upon his upper limbs for most support and he had developed a "hopping" gait, described above, to compensate for his lack of coordination when moving at any speed. Although his upper limbs were stronger and better controlled than his hind limbs, they too showed signs of deficits. Wania 6672 lacked independent finger movement which made it difficult to grasp small objects or to engage in grooming.

Mental deficiency in human children with spastic cerebral palsy frequently is characterized by increased aggressiveness and hyperactivity. Dr. Tatton felt that Wania 6672 exhibited both tendencies in segments of video-tape in which the infant showed extreme agitation and aggression towards animals grooming his mother. He continually tried to insert himself between the groomer and his mother, clumsily climbing and pulling at the animal, and in one sequence, repeatedly threatening and biting the groomer. As Wania 6672 became more excited and agitated his movements became more spastic, leading him to frequent falling and rolling on the ground.

■ QUANTITATIVE DATA

Both quantitative and qualitative data are used in this paper to provide as clear a picture as possible of the behavior of this abnormal infant. Descriptive data are used to amplify and extend the quantitative data which are presented in comparative form in order to emphasize the difference between the behavior of Wania 6672 and that of a normal infant. Thus, Ran 72, a male infant of similar rank and age and, like Wania 6672, with no close siblings, was chosen. Fifty ten-minute test sessions of data were tallied for each infant covering the period of September 1972 to June 1973, that is, from age four months to thirteen months. Values for Ran 72 thus show behavior of one other infant in the troop and do not represent typical or mean infant scores. However, since Ran 72 appeared to be a healthy and socially well-adjusted infant, the comparison seems as useful as it is striking.

■ ACTIVITY PATTERNS

Table 1 shows the percentage of time spent in major activities by the two infants. Activity categories are taken from an ethogram developed for this troop (L. M. Fedigan 1974) although some related categories have been combined (e.g., play here includes social and solitary play, object manipulation, and exploration).

Wania 6672 spent over 30% of his time sitting or lost calling, often doing the latter while sitting or staggering apparently aimlessly around. Sometimes he would wander

TABLE 1 / PERCENTAGE OF TIME SPENT IN
MAJOR ACTIVITIES BY WANIA 6672 AND RAN 72

ACTIVITY	WANIA 6672 PERCENT	RAN 72 PERCENT
SIT OR LIE	16.3	5.7
LOST CALL	15.3	1.5
NURSE/SEEK COMFORT	12.7	7.7
GROOMED	14.5	1.2
FOLLOW MOTHER	7.6	1.0
CARRIED VENTRALLY	3.7	0.0
CARRIED DORSALLY	0.0	0.4
TRAVEL	6.0	8.7
FORAGE	19.6	27.2
GROOM	0.0	5.3
PLAY	1.8	40.5

off, lost calling, when his mother was sitting quite close. Ran 72 in contrast, spent 40% of his time playing, manipulating objects, and exploring his environment. Wania 6672 probably could not engage in vigorous locomotor activities like wrestling or climbing and seemed to avoid close physical contact with monkeys other than his mother. He spent 38.5% of his time following his mother, seeking to nurse, nursing, being carried, and being groomed (almost always by his mother). Thus, more than one third of his time was spent with, or seeking to be with, his mother, whereas Ran 72 spent only 9.1% of his time in mother-oriented activities. Wania 6672 was carried ventrally by his mother (3.7%) in comparison to Ran 72 who was very independent, riding dorsally only rarely (0.4%) and traveling mostly on foot. For Ran 72, the play or peer group seemed to be the main focus of activity.

■ MOTHER-INFANT INTERACTIONS

Infants of the Arashiyama West troop have been known to survive without mothers from the age of six months (T. Mano, personal communication; L. M. Fedigan, unpublished data); however, the survival of Wania 6672 seems to have been completely due to his mother's close attention and support. Wania 66 consistently aided her handicapped infant more than other mothers aid their normal infants. Some primiparous mothers in this troop are incompetent—stepping on their infants, leaving newborns in the hot sun for long periods, and being unable to place and secure the infants properly on the chest for traveling and nursing. Other primiparous mothers immediately handle and care for their infants appropriately. In the Arishiyama West troop a mother's skill in caring for her infant seems to depend on several factors, not only previous experience as an "aunt" but also the individual temperament of the mother and response of the infant. Wania 66 not only gave her first-born infant competent care, she gave him *extra* care. For at least two weeks after birth she continually held him to her ventrum while walking, whereas other mothers let their weak, newborn, infants drag on the ground or even left them behind for short periods. Wania 66 carried her infant ventrally for thirteen months, while most other infants have converted to dorsal riding by four to six months, and by eight to ten months are carried mainly during emergencies.

Wania 66 frequently waited for, or returned to, her infant on those occasions when he attempted to walk on his own. The troop often moved out quickly to forage and travelled rapidly during the foraging trip. At such times Wania 66 often was seen to initially hurry off with the body of the troop and then stop to look back at her infant, calling and staggering down the path after her. She would then sit and wait until he had caught up to her, or she would go back to him and pull him on her ventrum and then hurry to catch up with the troop. Wania 66 retrieved her infant more than any of the other 23 mothers of 1972 infants observed, even when he was in no obvious danger, but simply giving lost calls as he searched for his mother. When the infant was threatened by high-ranking monkeys, his mother would run up, fear-grimacing, clasp the infant to her chest, and run.

Nevertheless, Wania 66 did occasionally leave her handicapped infant behind, and after he was about seven months we occasionally found the infant giving lost calls while far from the main body of the troop. On three of these instances we located the mother and surmised that she could not hear the infant's cries—unless her hearing was far better than ours. However, one to several hours later we found her reunited with her offspring. Once she left the troop and returned to the location where she had last been in contact with her infant, found him, and carried him back to the group. The infant had been silent until she was very close and it appeared that she had remembered where she had last seen him.

Wania 6672, for his part, concentrated all of his energy on keeping up with his mother. Faced with the choice of eating some food item she had dropped or following her, he would invariably choose the latter. The struggle of this handicapped infant to maintain close contact with his mother led us to observe that Japanese monkey mothers do not appear to respond vocally to the lost calls of their infants, behavior which could aid a stray infant in locating its mother immediately. All of the monkeys, however, do give sporadic "contact calls" which may function as locating aids, although the social stimuli for such vocalizations are difficult to determine.

All Japanese monkey mothers and offspring are reported to recognize one another's vocalizations (Itani 1963), and field observations indicate it is probable that at least by adolescence, a monkey recognizes the vocalizations of all, or most, of the other monkeys in the troop. On numerous occasions we were with a monkey mother who looked up when she heard an infant lost call in the distance, continued her activity for a moment, and then headed toward the area of the call, whereupon her infant hurried up to her. Many times we observed a lost infant locate its mother after hearing her contact call. Wania 6672 could recognize his mother's voice and relied strongly on her vocalizations to maintain contact. On more than twenty occasions we observed Wania 6672 locate and catch up to his mother over a long distance through dense brush, stopping to listen periodically for her contact calls. Yet when she sat silently ten feet away, he often was unable to find her.

■ INTERACTIONS WITH OTHER TROOP MEMBERS

Table 2 compares the type and frequency of behaviors performed in the 500 minutes of test data, for each of the two infants, Ran 72 and Wania 6672.

Table 2 shows that Ran 72 performed 27 *different* behaviors compared to only 15 for Wania 6672, and over the same time span performed more than three times as many behaviors (309 compared to 81). As might be predicted from the activity data, most of Ran 72's behaviors occurred in locomotor and play activities. Wania 6672 showed almost a complete absence of play-related behaviors, indeed, on the few occasions in which play behaviors were recorded for him, it was questioned whether he was in fact intending play or aggression. He never was seen to exhibit a play face but was capable of threatening and did, on occasion, attempt to bite and displace animals grooming his mother. He appeared very possessive of his mother and sometimes when she was being groomed he would try to push between the two animals. Often the groomer would take the intrusion to be playful and try to wrestle with the infant. However Wania 6672's unusual or unpredictable behavior seemed to "disconcert" the other animal who usually would withdraw or ignore Wania 6672's advances and go back to grooming. It is impossible to know whether Wania 6672 was capable of, or had developed, a complete range of signals, since his restricted activity pattern reduced the need to use many of them. Most of the behaviors recorded were directed toward his mother.

Table 3 compares the type and frequency of behaviors *received* by Wania 6672 and Ran 72 during the 500 minutes of test data for each infant.

The difference in the number of kinds of behaviors received (14 for Ran 72, 12 for Wania 6672) is small; however, the distribution of behaviors again is quite different. Ran 72 receives most behaviors in the context of social interactions, generally play. Behaviors directed at Wania 6672 are mainly in the context of infant retrieval and comforting. Wania 6672 was retrieved 20 times compared to twice for Ran 72, and was comforted 12 times, a behavior that was never seen directed at Ran 72. Although she generally was more patient and attentive than other mothers, Wania 66 did pinch and reject Wania 6672 far more often (22) than Ran rejected her son, Ran 72 (6). However, the ratio is similar to that of nursing-related behaviors *initiated* by the infants (see Table 2), so that attempts at nursing seemed to be rejected by both mothers in about the same proportions.

Sometimes when Wania 6672 sat and gave infant lost calls, as he so often did, he was retrieved by another troop member. Nearby monkeys, almost always young females, would approach the distressed infant and attempt to carry him ventrally. All infants in the troop are occasionally retrieved by someone other than their mother, but the frequency with which Wania 6672 was retrieved by nonrelatives (20 in the 500 minutes) was remarkable. The other interesting aspect of these interactions is that

TABLE 2 / FREQUENCY OF BEHAVIORS
PERFORMED BY WANIA 6672 AND RAN 72

BEHAVIORS	WANIA 6672 NUMBER	RAN 72 NUMBER
INITIATE PLAY	0	20
WRESTLE	5	37
ROUGH-AND-TUMBLE	0	40
BRIEF GRAPPLE	0	11
DISPLAY	0	19
MOUNT	0	39
BLUFF CHARGE	0	11
CHASE	0	8
JOIN IN	0	1
CLIMB	2	38
OBJECT MANIPULATION	1	11
RUN/HURRY	2	15
RUN WITH TROOP	0	6
HARASS	0	6
BIPEDAL/LOOK	0	12
LIP SMACK	0	2
FRIENDLY WARBLE*	1	4
MUZZLE	4	5
GENITAL INSPECT	0	3
SOLICIT GROOM	3	5
THREATEN	0	7
FEAR-GRIMACE	2	2
SCREAM	5	1
CONTACT AGGRESSION	0	1
CONTACT CALLS	4	0
APPROACH MOTHER	20	5
SEEK NIPPLE	15	5
INFANT GECKERS	11	5
BODY JERKS	2	0
WEANING TANTRUM	4	0

*Vocalizations of a tremulous quality—given when two monkeys encounter one another and some tension exists in the situation.

frequently he was retrieved by peers, and the retrieving often was accompanied by comforting behavior. These young infants, also usually females, would put their arms around Wania 6672, thrust their chests forward as though to be suckled, huddle with him, pat him, groom him sporadically, and sometimes attempt to initiate play. On a few occasions Wania 6672 was comforted by young male peers. By comparison, Ran 72 was only retrieved twice in the 500 minutes—both times by his mother—and never was seen to be comforted. Infants in the Arashiyama West troop spend most of their

TABLE 3 / FREQUENCY OF BEHAVIORS
RECEIVED BY WANIA 6672 AND RAN 72

BEHAVIORS	WANIA 6672 NUMBER	RAN 72 NUMBER
PLAY INITIATION	3	14
MOUNTED	2	10
CHASED	0	10
JOINED	0	3
LIP QUIVER	1	1
FRIENDLY WARBLE	0	1
MUZZLED	2	0
SEX PRESENT	0	1
GROOMED	1	0
THREATENED	4	11
SUPPORTED	0	2
CONTACT AGGRESSION	4	2
APPROACHED	0	13
RETRIEVED (MOTHER)	12	2
RETRIEVED (OTHER)	8	0
COMFORTED	12	0
REJECTED	18	3
PINCHED	4	3

time with their mothers or in large rough-and-tumble play groups, and it was always extraordinary to see such a young monkey leave its companions to hurry over and cradle an infant hardly smaller than itself.

One apparent reason for the recurrent attention Wania 6672 received was the great amount of time he spent giving infant lost calls (15.5% compared to 1.5% for Ran 72). At approximately three to five months when infants begin to achieve some measure of independence from their mothers and spend more and more time in peer play groups, infant lost calls are a quite common occurrence. However, as the infants become adjusted to not having a mother (and a nipple) on demand, the lost calls become more and more infrequent. By the time the infants are about seven months old these calls are most often made in the evening when the troop settles down to sleep, and they serve to reunite the mothers and their offspring. Wania 6672, on the other hand, spent a large percentage of his day giving lost calls throughout his thirteen months of life, reflecting the fact that he never did achieve independence from his mother.

The social organization of *Macaca fuscata* is based on female matrilineages (Kawamura 1958; Yamada 1963; Koyama 1967) and, as a result, infant Japanese monkeys spend much time, and forge close bonds with, their matrilineal kin. The Wania matrilineage is not as tightly knit as some uterine groups in the Arashiyama West troop, but Wania 6672 did spend a considerable amount of time with members of the Wania family. On a few of the occasions when Wania 6672 could not find his mother, he was seen to locate his grandmother or one of his aunts and attempt to follow and sit with them. Two of his peers, Wania 72 and Wania 6572, were close relatives and also among his most frequent companions and "retrievers."

Wania 6672's only other enduring social bond was with a young unrelated female, Rotte 6671, the daughter of Wania 66's closest companion. The friendship between the mothers may have influenced the close relationship that developed between the offspring. Wania 6672 showed some of his most positive, affinitive behaviors toward this young female. He would present to her for grooming, and make occasional attempts to groom with crude stroking gestures of his hand. Grooming was rarely attempted by Wania 6672 (see Table 1), although he received far more grooming, principally from his mother, than Ran 72, and reacted to grooming with the same postures as other monkeys. He never developed a grooming technique beyond the simple stroking or pulling with one hand, gestures which are similar to those described in the early stages of development in infant stumptails (*Macaca arctoides*) (Chevalier-Skolnikoff 1974). However, since Wania 6672 had no independent finger movement it was not possible to decide whether his lack of grooming and poor technique were due to arrested social development or to his physical deficits. Most importantly, Wania 6672 sometimes showed an interest in Rotte 6671's activities and seemed stimulated to try to engage in them, even those that were beyond his physical abilities. He would follow Rotte 6671 about and grab hold of the branch she was holding, peer in the ground squirrel hole she was investigating, or touch the stick with which she was playing. It was only in his interactions with Rotte 6671 and another young female of his own age (Rheus 626772) that Wania 6672 showed any of the widening focus of attention and increased number of social relationships which are necessary for the socialization of the infant into the troop.

Rheus 626772 appears to have triggered a temporary "blossoming" in Wania 6672's social life during March and April of 1973, when he was ten to eleven months old. During this time Rheus 626772 regularly sought out Wania 6672 and engaged him in gentle play in which the two infants would touch and hold each other—Wania 6672 frequently "exploring" the other infant's face and limbs with his hands. This gentle holding and touching was in great contrast to the rough-and-tumble play so characteristic of the other infants. Sometimes more infants would join these two and, contrary to his habit of avoiding groups of playing infants, Wania 6672 would stay and even attempt to play. This period of increased social activity and peer interest lasted only a month and seems to have been provoked by the special attention of Rheus 626772. When she no longer sought him out, Wania 6672 returned to his pattern of mother-centered activity.

One kind of unintentional aggressive interaction that Wania sometimes initiated resulted from his tendency to bump or stagger into other monkeys. However, most animals ignored him or simply pushed him firmly out of the way. Even leader males, who can be especially intolerant of any intrusion into their personal space, showed little reaction to Wania 6672, who was seen to stumble into three out of the four leader males on one or more occasions. Only one of these males actually "punished" the infant—shaking him by the scruff of the neck and rubbing him against the ground, which is the characteristic treatment given to "errant" infants. At the other extreme, Wania 6672 once lost his balance while going down an incline and tumbled head on into the groin of a leader male. The male merely looked down while the infant sniffed at the large male's scrotum and scrambled about in his lap trying to right himself. On another occasion Wania 6672 tripped repeatedly over the alpha male's foot while staggering about in search of his mother. The alpha male (Dai) was lying down being groomed and could not see who was tripping over his feet, but raised his eyebrows (lid threat) each time he was touched. Finally, he sat up with a threat face, looked at who had fallen onto him, and lay down immediately. Seconds later, Wania 6672 tripped on him again and Dai made no response.

The possibility that Wania 6672 received different, even special, treatment from the troop members because of his habitually abnormal behavior is further supported by the following example. The handicapped infant always consumed a great deal of food during provisioning and his frequent presence on, or next to, the alpha male's food pile was puzzling. Careful observation showed that frequently at the commencement of provisioning Wania 66, with Wania 6672 riding ventrally, approached the alpha male, attempted to obtain food from his pile, and when he threatened her, hurried off leaving the infant with him. The infant, oblivious to most of the social interactions around him, would begin to cram his mouth and cheek pouches with food. Occasionally Dai would raise his eyebrows at the infant or push him away, but the infant's response was to hunch over (sometimes fear-grimace) and continue to eat. Dai seldom continued to threaten the infant if it maintained a low profile. This, then, resulted in a spatial and social pattern of the alpha male and the handicapped infant eating from a large pile of food while other monkeys ate from smaller neighboring piles or tried to snatch food from the edges of the alpha male's source.

It is not uncommon for mothers to leave their infants during the excitement of provisioning, but no other mother we observed consistently left her infant in the vicinity of the alpha male. The Wania matrilineage does have a special close relationship with Dai, and perhaps this had some bearing on Wania 66's choice of a place to leave her infant.

When infant monkeys of the Arashiyama West troop are threatened by high-ranking monkeys they may scream in fear, and usually run away or cease the offending activity. If threatened by a low-ranking monkey, infants usually scream defensively in an attempt to receive support from their mother or sympathetic bystanders. Wania 6672 did not always perceive that he was being threatened, but when he did, he would react in some confusion. He would fear grimace or scream and begin crashing aimlessly about, or he would continue the offending behavior, apparently not recognizing why he was being threatened. When confronted with this strange response, most monkeys stopped threatening immediately and many peered at the abnormal infant as if confused by the unusual response. Thus the troop in general was very tolerant of Wania 6672's lack of appropriate social responses and showed a measure of acceptance of his deviant behavior. While Wania 6672's unpredictable social behavior seemed to create some confusion and tension, it did not make him the focus of increased aggression that would have threatened his survival.

■ FOOD-GETTING

All infants in the troop rely heavily on their mothers to learn appropriate food items. Although the Arashiyama West troop is new to the Texas environment, the monkeys already show very definite troop-wide preferences for the native vegetation, only eating the leaves of certain shrubs, the bark of certain trees, the seeds of certain grasses, and so forth. During foraging trips a young infant is carried by its mother ventrally or dorsally and when the mother stops to eat, the infant sits in her lap or just beside her. From as early an age as two to three weeks the infant may show great interest in what its mother eats, continually nuzzling her mouth while she chews, grabbing at food in her hand, or picking up dropped pieces to smell and chew. Later when the infant forages away from its mother, it can often be seen to carefully observe another foraging monkey, nuzzle the monkey, and then eat the same food item. Infant monkeys of this troop appear to have a repertoire of appropriate food items adequate to nourish themselves by as early as six months of age.

Wania 6672 relied heavily on his mother and other troop members to locate and obtain appropriate food. Whenever possible he foraged with his mother, eating right in her lap or under her hands. He preferred to eat the food she dropped or to eat from the very same branch or tuft of grass from which she was eating. Whereas many mothers in the troop became intolerant of the sustained close presence of their older infants while they ate, Wania 6672 continued to allow her infant to eat right from her food source throughout his life, seldom punishing him or pushing him away. When Wania 66 foraged on items found above ground level, such as hackberries (*Celtis pallida*) or mesquite beans (*Prosopsis julifora*), Wania 6672 sat directly below her and ate what she dropped onto him. If he was unable to keep up with his mother during foraging, he would approach a small group of monkeys actively consuming some food item and, after a period of observation, begin to eat what they were eating. Other monkeys were usually very tolerant of his presence although he might receive occasional pinches or pushes for attempting to eat too close to their chosen branch or cactus pad. When he was left totally alone, Wania 6672 sometimes showed himself capable of choosing an appropriate type of grass seed, ground plant, or shrub leaf to eat. However, he seldom ate when alone, but rather, spent his time calling and moving about in search of his mother.

Many items of the Arashiyama West troop diet require some preparation or dexterous motor ability, for example, seeds are removed from pods, prickly pear fruit (*Opuntia englemannii*) is rolled and batted on the ground to remove spines, and tough dead outside leaves are rapidly peeled off grass stems to obtain the sweet hearts. Wania 6672 was incapable of these types of preparation and usually relied on eating leftovers of these food items. He often obtained food with his mouth, whereas others used their hands, and was seen fairly frequently to munch dead grass stems, perhaps because he could not separate the dead from the live. In general, however, Wania 6672 had adequate powers to locate and obtain enough native food to sustain himself. At one time or another he appears to have eaten almost every item in the Arashiyama West diet with the possible exception of insects and bird eggs, which he could not obtain. However, Wania 6672 foraged a good deal less than Ran 72, and spent more time seeking and being nursed (Table 1).

Since Wania 6672 nursed more than Ran 72 and, indeed, nursed regularly up until his disappearance at thirteen months of age, the question of whether his mother, Wania 66, lactated longer than other mothers was raised. All Arashiyama West infants born in 1972 nursed long after the six months lactation time previously indicated for *Macaca fuscata* (Napier 1967). In addition, several 1971 infants continued to nurse until October 1973. At that time the capture of the troop for testing and tatooing allowed some lactation data to be gathered. The handicapped infant had been dead for four months; nevertheless, the data collected does shed light on the lactation question. Unfortunately not all mothers were tested; however, of those tested: (a) all (7) mothers of infants born in 1973 were lactating (three to six months duration); (b) all mothers of surviving infants born in 1972 tested (five out of twenty-one) were lactating (sixteen to seventeen months duration); (c) four of the eight mothers of 1971 infants who did not have additional infants in 1972 were tested, and two were found to be lactating (approximately 29 months of lactation). The two mothers of 1971 infants who were lactating had been allowing their offspring to nurse regularly until October 1973, while the two mothers of 1971 infants who were *not* lactating had *not* been allowing their infants to nurse.

It would appear that, in the Arashiyama West troop at least, Japanese monkey mothers continue to lactate as long as offspring nurse regularly. Why some Arashiyama West mothers allowed their offspring to nurse up to 29 months is unknown; however, it should be noted that the troop was transported from Japan to Texas in February 1972, when these infants were only 8 to 10 months old. The disturbance of capture and transportation, then adaptation to a completely new environment, may have affected nursing. Thus, since both the handicapped infant and Ran 72 continued to nurse regularly throughout the data collection period, we can perhaps assume that Wania 66 and Ran also continued to lactate.

■ INTERSPECIFIC ENCOUNTERS

The Arashiyama West troop shares its enclosure, water, and food resources with a herd of peccaries (*Pecari tacajou*) varying in number from 12 to 25 animals. Monkey-peccary interactions occur quite frequently in the winter when the peccaries are more diurnal, and the monkeys give way during these encounters (L. Fedigan 1973). On December 28, 1972, the troop was having a mid-day rest at one of the waterholes when the peccary herd approached, apparently intent on going to the water. A predator alarm call was given and all the monkeys ran and climbed into distant trees. Wania 6672, however, had become separated from his mother and, obviously agitated but unaware of which direction the monkey troop had taken, began to stagger toward the peccaries giving lost calls. Then two monkeys reappeared at the edge of the waterhole—one of the leader males and Wania 66. Leader males in this troop have a tendancy not to run in emergencies and frequently head toward the source of the scare, but Wania 66 had never shown herself to be a particularly brave female. On this occasion she cautiously approached the open area where her infant was calling and stumbling among the peccaries. Repeatedly, she lowered her head to the ground and peered through the line of brush which separated her from her infant, then turned to look back in the direction the troop had taken.

The peccaries we observed did not have aggressive temperaments and were never seen to charge at humans or monkeys. They seemed undisturbed by, and largely disinterested in, the monkeys, although they would run from humans. Peccaries are very short-sighted and easily startled, and when disturbed, tend to rush in many directions, creating a rather frightening effect on all primates. The Japanese monkeys react to them as to predators (L. Fedigan 1973). Thus although Wania 6672 was in no danger of being attacked by the wild pigs, his mother showed signs of fear. She moved silently closer to him nonetheless, and finally, apparently through good fortune, he stumbled in her direction. Then she ran the last few feet that separated them, clasped the infant to her chest, and fled towards the troop. A few peccaries snorted in surprise but soon resumed eating and wallowing in the mud. The leader male who had been watching this event slowly walked back in the direction of the troop.

▪ SOCIABILITY

"Proximity"—the number of stationary animals within a distance of ten feet from the subject at any given time—can be used as a rough measure of sociability. The number of nontraveling monkeys within ten feet of Ran 72 and Wania 6672 was tallied at the beginning of each of the 100 test sessions; the results are shown in Table 4.

Again, Ran 72 showed a greater peer orientation, being recorded in proximity to 21 out of the total of 25 infants born in 1972, as opposed to only 8 for Wania 6672. Further, over the 50 test sessions, a total of 91 peers were recorded within ten feet of Ran 72 as opposed to only eleven for Wania 6672. Indeed, in only 50 tallying sessions, 9 different infants of his own age were found in proximity to Ran 72 *at least five times*. In the case of Wania 6672 only 3 infants were noted in proximity more than once, and none more than twice.

Wania 6672's mother was found within ten feet more than twice as often as was Ran 72's mother (34 against 18), and more different subadult and adult animals (24 compared to 15) were recorded near Wania 6672 than near Ran 72. The total number of subadult and adult animals recorded in proximity to Wania 6672 during the 50 test sessions also was higher (76 compared to 32). At the beginning of all 50 test sessions there was at least one animal within ten feet of Ran 72, whereas at the beginning of 8 of the 50 test sessions for Wania 6672, there were no animals within ten feet. In addition, there were 14 sessions in which only his mother, Wania 66, was within ten feet.

TABLE 4 / PROXIMITIES: NUMBER OF STATIONARY MONKEYS WITHIN TEN FEET OF SUBJECT AT BEGINNING OF EACH TEST SESSION

	WANIA 6672 NUMBER	RAN 72 NUMBER
Number of Test Sessions	50	50
Number of different infants (peers)	8	21
Number of possible different infants	25	25
Total number of proximal infants	11	91
Number of different juveniles (1–4)	21	30
Number of possible different juveniles	63	63
Total number of proximal juveniles	30	48
Number of different subadults and adults	24	15
Number of possible different subadults and adults	58	58
Number of times mother in proximity	34	18
Total number of proximal subadults and adults	76	32
Total number of different monkeys	53	66
Total number of proximal monkeys	117	171

Ran 72's greater sociability is even more remarkable when the number of monkeys interacted with during the testing sessions is compared. During the 500 minutes of testing Wania 6672 interacted with a total of only 19 different monkeys, whereas during a comparable 500-minute sample, Ran 72 interacted with 51 different monkeys. Again, Ran 72's strong peer orientation is evident when it is noted that 42 of the 51 monkeys interacted with were infants or juveniles, compared to 10 out of 19 for Wania 6672. In adult monkeys, grooming may be used as a measure of sociability; however, the little grooming that infants attempt often ends in wrestling, so grooming would not appear to be a good indicator of infant sociability.

■ **DISAPPEARANCE**

Wania 6672 disappeared on June 15, 1973, apparently having become lost and unable to find his mother and the troop. His disappearance may have been precipitated by a change in the troop provisioning schedule which caused the troop to forage further and longer than had been their habit. Wania 6672's mother usually carried him out on these trips, but then would put him down and forage on without him for long periods. By this time Wania 6672 was thirteen months old and should have been quite independent. On two occasions just before his final disappearance, Wania 6672 was missed at provisioning time and was located and returned to the troop by the researchers who found him sitting and lost calling in the dense brush. The day of his disappearance he was seen being carried off by his mother as the troop foraged. She climbed a tree with him, but then pushed him away from her so that he fell to the ground a distance of approximately ten feet. It was the first time such a fall had been observed. He did not appear hurt, but sat in his characteristic huddled, or depressed posture. Later in the day when he was missed, the 108-acre area was repeatedly searched, but no trace of the infant was ever found.

■ CONCLUSION

Returning to the model of human cerebral palsy, such widespread central nervous system damage can be caused by a genetic deficit or mutation, perinatal asphyxia, birth injury, or contact with a toxic agent in utero. Dr. Tatton suggested that Wania 6672's widespread lesions were not typical of those caused by birth trauma or anoxia, and were more consistent with those due to a genetic deficit or a toxic agent. The Arashiyama West troop in Texas has suffered fairly extensively from a poisonous shrub, *Karwinskia humboldtiana.*[1] The toxin is concentrated most highly in the berry, ingestion of which causes progressive neural damage (demyelinization) and resulting weakness and paralysis beginning in the hind limbs and moving forward in the body. We discussed with Dr. Tatton the possibility that Wania 66 might have consumed a sufficient number of these berries while pregnant to prove toxic to the fetus while not visibly affecting the mother. A description of the progression of the physical symptoms of *Karwinskia humboldtiana* poisoning, plus review of video-tape of adult and subadult victims in motion, led Dr. Tatton to conclude that *Karwinskia humboldtiana* causes peripheral nerve damage, while Wania 6672's deficits involved the central nervous system. However, this does not rule out the plant as a cause of Wania 6672's deficiencies, as its effects on the fetal nervous system are not known. Consequently the cause of Wania 6672's handicap still is open to conjecture, even though genetic deficit seems most probable.

Wania 6672 showed a restricted range of behaviors and did not develop any measure of independence from his mother. He did not socialize into his peer group, remaining generally fearful of the often rough play groups. Other infants showed a development away from a self, self and mother, toward a peer, and then a role group stage (L. M. Fedigan 1974). However, Wania 6672 did develop a sufficient repertoire to survive, albeit with the constant help of his mother, for thirteen months. Further, although he remained mother-centered throughout his life and never initiated social interactions, he was able to respond to friendly attention from Rotte 6672 and Rheus 626772. What prompted the latter to seek out Wania 6672 and interact with him in such an unusual manner for four or five weeks is a mystery, as is the cause of her terminating the special relationship.

Nevertheless, in these interactions Wania 6672 did show some potential for social development, perhaps indicating that with special compensatory attention and help, he could have achieved more complete socialization. Again, the spastic cerebral palsy model is suggested, for with help, human infants with cerebral palsy can and do achieve high levels of cognitive and social development. The concept of sensory stimulation as an intrinsic reinforcer in early exploration and play (Baldwin and

Baldwin, this volume) also suggests that Wania 6672's ataxia, interfering as it would have with his mechanisms of sensory arousal, may have prevented him from developing the exploratory and play skills so important in early development. All Wania 6672's exploration seemed to be in response to extrinsic (e.g., food, mother-seeking) stimuli, and he never showed the "approach-withdrawal" behavior of infant monkeys cited by the Baldwins to illustrate cycles in sensory arousal.

Piaget (1954) suggests that in human infants, social and cognitive development occur with, and are largely dependent upon, physical development. He postulates a dual process of adaptation to, and assimilation of, the environment as the basis of this development. Exploration and play are seen to be the principle means of assimilation of the environment and, hence, an inability to engage in these activities can lead to incomplete social and cognitive development. Much compensatory education for the retarded or handicapped focuses on the development and refinement of fine muscle control, coordination, and manipulative skills. The suggestion is that normal social learning is retarded or arrested when an animal is unable to engage in the exploration and play activities accompanying normal physical development. Thus, Wania 6672's inability to engage in vigorous play or exploration may have contributed to the poor perceptions of rank, affiliation, and support relationships, manifest in his lack of appropriate responses to threats. In many ways his apparent unawareness when entering the social space of a central male or dominant female, and his confused reaction to threats, was reminiscent of the actions of a very young infant still in its black natal coat in that period of grace when all monkeys retreat from it. Wania 6672 never did seem to learn the complex web of relationships of rank, family, role, and his own place in the troop organization. Thus, he remained socially "naive" until his disappearance.

Temporary or even permanent injuries or deformities which do not seriously hinder sensory and motor development seem to have little or no effect on social development (Berkson 1970, this volume). In the Arashiyama West troop a young male with a very severe cleft palate showed no evidence of impaired social development. The occurrence of deformities (fused fingers and toes) in other troops of Japanese monkeys is reported by Itani (1963) and Furuya (1966). Furuya states that the cause of the deformities is unknown, and that the survival rate for such malformed monkeys, "cannot be said to be much lower than that of normal individuals," but does not mention any effects on socialization or relations within the troops.

The troop *did* appear to tolerate Wania 6672's unusual behavior, even seeming to be aware that he was "different," hence, the large amount of retrieving and comforting that he received, far exceeding that directed at even an orphaned infant. That Wania 6672 survived even thirteen months is due to the extraordinary care that his mother gave to him, although why she retrieved him and carried him so long after other infants were totally independent is as inexplicable as why, after thirteen months, she suddenly began to leave him. Wania 66 became pregnant again during the 1974 season but had a miscarriage in July 1974 (H. Gouzoules, personal communication). Thus, Wania 6672 remains the one offspring she has borne.

▪ ACKNOWLEDGMENTS

This research was supported in part by Grants-in-Aid of Research from the Explorers Club, The Society of the Sigma XI, and a National Science Foundation Graduate Fellowship (No. 26-1140-5850).

▪ NOTE

1. Other poisonous plants (e.g., *Solanum elaeagnifolium*) also occur within the monkey enclosure.

▪ REFERENCES

Berkson, G. 1970. Defective infants in a feral monkey group. *Folia Primatologica* 12:84–89.

Chevalier-Skolnikoff, S. 1974. The ontogeny of communication in the stumptail macaque (*Macaca arctoides*). In *Contributions to primatology*, vol. 2, ed. H. Kuhn et al. Basel: Karger.

Fedigan, L. 1973. Classification of predators by a transplanted troop of Japanese macaques (*Macaca fuscata*) in the mesquite-chapparel habitat of south Texas. Paper read at the annual meeting of the American Association of Physical Anthropologists, Dallas, Texas.

Fedigan, L. M. 1974. *Role behaviors in the Arashiyama West troop of Japanese monkeys* (*Macaca fuscata*). Ph.D. dissertation, The University of Texas, Austin.

Furuya, Y. 1966. On the malformation occurred in the Gagyusan troop of wild Japanese monkeys. *Primates* 7:488–492.

Hazama, N. 1964. Weighing wild Japanese monkeys in Arashiyama. *Primates* 5:81–104.

Itani, J. 1963a. Vocal communication of the wild Japanese monkey. *Primates* 4:11–66.

———. 1963b. The social construction of Natural troops of Japanese monkeys in Takasakiyama. *Primates* 4:1–42.

Kawamura, S. 1958. The matriarchal social order in the Minoo-B troop. A study on the rank system of Japanese monkeys. *Primates* 1:149–156.

Koyama, N. 1967. On dominance rank and kinship of a wild Japanese monkey troop in Arashiyama. *Primates* 8:189–216.

———. 1970. Changes in dominance rank and division of a wild Japanese monkey troop in Arashiyama. *Primates* 11:335–390.

Napier, J. R., and Napier, P. H. 1967. *A handbook of living primates.* London: Academic Press.

Norikoshi, K. 1971. Tests to determine the responsiveness of free-ranging Japanese monkeys in food-getting situations. *Primates* 12:113–124.

Piaget, J. 1954. *The construction of reality in the child.* New York: Basic Books.

Yamada, M. 1963. A study of blood relationships in the natural society of the Japanese macaque. *Primates* 4:43–65.

Part Two

SOCIAL INFLUENCES AND SOCIALIZATION

SOCIAL
INTERACTIONS
AND SOCIALIZATION

7 Perceptual Properties of Attachment in Surrogate-Reared and Mother-Reared Squirrel Monkeys

JOEL KAPLAN

■ INTRODUCTION

The primary object of emotional attachment for the infant primate is the mother. Its attachment to her is initiated by the comfort and security she provides through physical contact but expands with time to include other maternal attributes as well. It is the role of these other factors in the attachment process with which this chapter is mainly concerned.

We began our work by asking how simple perceptual characteristics of the mother were involved in the development of filial attachment. As a first step, we decided to look at the response of infant squirrel monkeys (*Saimiri sciureus*) raised with artificial, substitute mothers, or surrogates, instead of natural mothers. In this way, basic perceptual properties of the surrogates could be manipulated easily, and the results would not be confounded by how real mothers treated their young. After these initial studies, responses to perceptual characteristics of natural mothers were examined.

■ EXPERIMENTS WITH SURROGATES

Our experiments with surrogates were designed to provide some clues as to how infants identify their mothers during the early stages of life and the effects these perceptual qualities have on attachment. The importance of physical contact in establishing a strong filial bond and in providing emotional security to the infant had already been well documented (Harlow 1958; Harlow and Harlow 1966; Harlow and Suomi 1970; Harlow and Zimmermann 1959). Infant rhesus monkeys raised on surrogates covered with cloth that facilitated clinging, spent several hours each day on these substitute mothers, regardless of whether feeding occurred on the surrogates. The infants also showed less emotionality in fearful situations when the cloth-covered surrogates on which they were raised were present and available for contact. In addition to the importance of physical contact, the initial experiments with surrogate-reared macaques also suggested that the sight alone of the surrogate also was attractive and capable of providing a certain amount of comfort (see Rosenblum and Alpert, this volume). For example, infants raised in the presence of both cloth-covered and wire-framed surrogates for about 50 days, when given the opportunity to choose to view different objects, chose to look at the cloth surrogates and other infant monkeys to about the same extent, which was more often than they chose to view wire surrogates or "nothing" (Harlow 1958). Moreover, infants that had been separated from cloth mothers at about six months of age and tested periodically with their familiar surrogates were less disturbed in a strange environment if they could see the surrogates than if the surrogates were not present (Harlow 1958). Apparently, the mere presence of the surrogate associated with contact comfort in the home cage was sufficient to reduce distressful behaviors, although it was clear that a surrogate available for contact was more effective in relieving stress.

Except for the initial brief report of this effect, the importance of perceptual factors in the development of filial attachment has generally been ignored. The one exception appeared in a recent report by Mason et al. (1971), who systematically examined the effectiveness of familiar and unfamiliar surrogates and living environments in reducing distressful responses of one- and two-month-old rhesus monkeys. The relative importance of visual cues, as compared with combined visual and tactile cues, was examined by encasing the surrogates in clear plastic containers to prevent physical contact on some trials. In general, the results of this experiment for infants that were one month old showed that both the familiar and unfamiliar surrogates were similarly effective in reducing distress, and much more effective when they could be both seen and touched. By two months of age, infants received more security from the sight of familiar surrogates than from the sight of unfamiliar surrogates, although the security was still substantially less than that received when infants could both see and contact both familiar and unfamiliar surrogates. These results, therefore, confirm and extend the earlier findings on the role of visual factors. They show that the sight of a familiar object associated with contact comfort can provide some degree of emotional reassurance to the infant rhesus by the second month of life. Nevertheless, as was the case in the earlier studies, sight alone was much less effective than physical contact.

Because it is generally assumed that vision is the most important sense for all primates, we decided to begin our studies by manipulating the color of our infants' surrogates. The 12 infants discussed in chapter 17 were the ones used. They were divided into four groups of three and were raised for six months on surrogates covered in either green, black, or red fur material. One of the four groups received a different colored cover every three or four days (when the covers were changed for cleaning). Each of the other three groups, however, was raised with the same color throughout the rearing period.

One of our first observations had nothing to do with the infants' response to the color of the surrogates, but rather with their sensitivity to changes in the surrogate's smell. The first or second time the soiled covers were replaced with clean ones in the home cage, several of the infants refused to climb onto the surrogates. After sniffing the newly changed covers for a few seconds, these infants typically retreated to the back of the cage, each huddling into a ball and vocalizing continuously. After a period of time, in some cases as long as one hour, the infants again approached, sniffed and, finally, climbed on the surrogates. The hesitation behavior lasted for only one or two weeks, however, and by the time the hesitant infants were a month old and had experienced several cover changes they no longer responded adversely to the replacement of their dirty covers with clean ones.

Beginning when the infants were four weeks of age and continuing at four-week intervals until they were six months of age, we tested them for preference for their own odor and rearing color in the four-alley apparatus described in chapter 17. In one procedure, we gave them a choice between two surrogates that differed in odor or color and were situated next to each other in one of the alleys. In the odor tests, surrogates that had been lived on for a few days were paired with clean ones of the same color. For the color tests, the color on which the infant was raised was paired with a different colored surrogate, both of which were clean. Two five-minute trials were conducted for each test on each occasion, and the amount of time infants spent on the different surrogates was used as an indication of their preference. The position of the surrogates was reversed on the second trial to control for the fact that some infants consistently approached and climbed on either the left or right surrogate.

The behavior of the infants in these two-choice tests dramatically showed that the infants recognized and preferred surrogates containing their own odor. For example, in many instances, before selecting its own odorous surrogate in the odor tests, an infant would carefully smell both. If it *did* climb on the clean one initially this was generally followed by its crossing over to its own surrogate within the first minute of the trial. Some infants also responded strikingly to odor in the color tests with the two clean surrogates, refusing to climb on either of the clean surrogates, but staying close to them throughout the period, smelling each in turn. Many of the infants did, however, climb on a clean surrogate, but they showed random choice of color from trial to trial, particularly during the first three months. After three months of age, infants tended to become more stable in their color preferences, although the choices were never as striking as those for odor, which remained unchanged over the entire six-month period (Figure 1). It should be pointed out, however, that a few infants did display color preferences when they were as young as eight weeks. Especially interesting was the observation that the infants appeared to be calmed most by the odorous surrogates. Although distress-type calls typically subsided whenever an infant climbed on any of the fur-covered surrogates, the calls stopped abruptly and completely only when the covers contained the infant's odor.

A few of the infants were also tested for their ability to discriminate their own odorous surrogate from that of another animal. Each of two infants tested at four weeks chose its own surrogate on each of two trials and spent its entire time on it. One of these subjects continued to show a strong preference for its own odor at eight and twelve weeks of age, but the other's response diminished (80% preference at eight weeks and 60% at twelve weeks). One of the two additional animals tested at twelve weeks for such specificity also preferred its own odor and spent over 90% of its time on that surrogate. The other infant, however, showed no preference.

In addition to the two-choice tests, other visual tests were conducted in which the different colored surrogates were placed behind windows in separate stimulus cages of the preference apparatus (see chapter 17). On these occasions, the infants typically ran

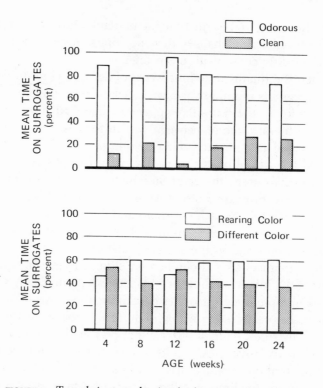

FIGURE 1. *Two-choice tests showing (top) responses to surrogates in the rearing color that either contained an infant's odor or were clean, and (bottom) responses to clean surrogates whose colors were the same and different from the rearing surrogate.*

from alley to alley without showing any particular color preference. The inability to contact the surrogates seemed to produce even less of a preference for the rearing color than occurred in the two-choice tests, where the surrogates were available for contact. The few infants under three months of age that showed a preference for their rearing color in the two-choice tests did not show any preference here.

Although these results appear to be clear, their interpretation is somewhat confounded because the odorous surrogates with which the infants were tested were those from the home cage that were also of the rearing color. It is possible, therefore, that the infants were responding to the combined effect of odor and color, not just to odor alone. Nevertheless, the results suggest that color did not influence odor preference among the animals reared on all of the colors throughout the study. These infants showed as strong a preference for their own odorous surrogate, regardless of its color, as did those that were reared on only one color for the entire six-month period. However, as a further check on possible interaction between color and odor, we plan to separate these variables in future experiments by raising infants on colored surrogates that contain artificial odors. Preference tests could then be conducted with the familiar odors on either familiar or unfamiliar colors.

Of course, these initial findings only suggest that the sense of smell may have meaning for the developing infant squirrel monkey. They do not answer the question of whether an infant can actually recognize its own mother by smell. In fact, it might be argued that the data from our tests with surrogates cannot represent natural conditions because infants were responding to only their own odors deposited on surrogates. It should be pointed out, however, that under natural conditions the recognition of a mother probably represents the ability of an infant to detect its own as well as its mother's odor. For example, young infants that defecate while on their mothers' backs normally move their hind-quarters away from her body so that feces are not deposited on her. However, they are not always successful, and the mothers' backs are likely to become soiled (as were the surrogates). Urine, also, is often deposited on the mother's back when infants urinate or engage in urine-washing (see chapter 17 for description of urine-washing behavior). Finally, general body odors from the infant's ventral surface are combined with the mother's own odors whenever the infant is on her back.

■ EXPERIMENTS WITH NATURAL MOTHERS

We have tested the responses of infants to their natural mothers' odors in much the same manner as responses to surrogates' odors. Moreover, other tests that provided additional sensory exposure of mothers were conducted to determine how much the different sensory modalities contributed to the responses of infants to their mothers. Again, all tests were conducted in the preference apparatus described in chapter 17.

The recognition of mothers by odors was examined by presenting each infant with its own and another, anesthetized mother. The mothers' heads were covered with gauze hoods to prevent the infants from seeing their faces. This technique was used so that the mothers' bodies could be sniffed at close range and could also be available for the infants to contact. We did not think that infants would be able to visually recognize their mothers merely by seeing only their bodies because of the similarity in coloration of most adult females. We were concerned also that, unless the infants were able to make contact with the bodies, they would remain too excited to pay attention to the different odors. Using anesthetized stimulus animals prevented the possibility of infants responding to behavioral cues. One problem that generally exists when measuring one animal's preference for another is determining whether the behavior of the stimulus animal is a salient factor. Is one animal attracted to another because it perceives certain physical qualities in the other, or because it is attracted by certain behavior that the other animal displays, or both? This is particularly important in the case of mothers and infants, where infant preferences may be based to some degree on responses of mothers upon seeing their own infants. We have noticed, for example, that when an infant not yet completely weaned is picked up and held in view of several mothers recently separated from their own infants, the infant's own mother is generally the only one that becomes excited and attempts to retrieve it. In the present series of experiments, preference tests with unanesthetized mothers as stimuli were also conducted in order to further clarify this issue.

In our initial experiments, the anesthetized, "hooded" females were placed next to each other inside one of the alleys of the choice apparatus. Identical tests without hoods were also conducted to determine whether viewing the mothers' faces enhanced recognition. Five infants between the ages of thirteen and fifteen weeks, that had lived alone with their mothers, were tested. Four of the five showed strong and equivalent preferences for their own mothers in the "hooded" and "unhooded" tests, suggesting that odors associated with the two mothers could be distinguished by the infants without the need for visual identification.

It is, of course, possible that the infants were able to recognize their mothers on the basis of tactile and visual cues even when the mothers were wearing hoods. Judging from the infants' behavior, however, this did not appear to be the case. As in our earlier surrogate study, the infants tested here did a great deal of body-sniffing, regardless of whether the females had been covered with "hoods." They did not appear to respond particularly to either visual or tactile stimuli associated with the females. There was another problem, however, that arose from the way in which these tests were conducted. Since infants had complete access to the mothers' bodies, attempts to nurse often occurred during physical contact, making it possible that those who tried to nurse were able to recognize the taste of the mother's nipple or milk.

This possibility was eliminated in a more recent study which also compared the preferences of infants subjected to different amounts of sensory exposure. Thirteen infants (eight males, five females), ranging in age from four to twelve weeks, were the subjects in this study. They lived alone with their mothers but interacted with each other and the other mothers for four two-hour periods each week. Three types of two-choice tests, each administered three times at approximately two-week intervals, were conducted in our preference apparatus with "own mother" and "different mother" as stimulus choices. In the tests, the two adult females were placed in separate holding cages in different, randomly-selected alleys. In each of two ten-minute trials, the females' locations were reversed. The amount of time the infants spent in each of the alleys was used as an index of preference.

In the odor tests, the mothers were anesthetized and strapped to the holding cages with their backs toward the inside of the preference chamber; their heads were covered with gauze hoods. In other tests, the stimulus females were not anesthetized, but the degree of access to them for infants was varied. In one of these, a clear plastic partition was inserted in front of the holding cages, preventing the infants from physically contacting the females. This condition also reduced the extent to which infants could smell the females. In the other condition the partition was eliminated, allowing maximal exposure to sensory information and behavioral cues.

Relative preferences for mothers over other females are shown in Figure 2. It is clear that the highest preferences of mothers occurred in tests in which mothers were awake and in cages *without* partitions. This means that combined sensory and/or behavioral cues had the greatest impact on the infants' choices. The next highest scores occurred when females were awake and in cages *with* partitions, suggesting that the inability to physically contact and/or smell the females at close range reduced the infants' ability to discriminate between the two choices. Lowest scores occurred when the mothers were anesthetized. This indicates that the elimination of behavioral

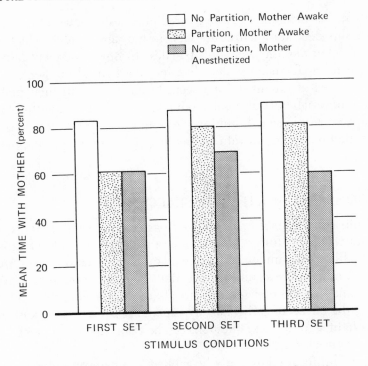

FIGURE 2. *Response of infants to their mothers in preference ap-*
paratus when allowed to choose between their own mother and a
different mother under different degrees of sensory exposure. Each
set of tests was separated by approximately two weeks, and infants
ranged in age from four to twelve weeks when first set was con-
ducted.

and/or visual cues further reduced differential responding. Even in this last condition, however, the response to mothers was still greater than would be expected by chance.

The reactions of different infants were much more variable in the odor tests than in the other tests, but the responses of individual infants were consistent from trial to trial. Some infants held on tightly to their anesthetized mothers through the cage bars and buried their noses in their fur, while others merely sat near the mothers or moved back and forth from one alley to the other. Sniffing the mothers' backs was the most prevalent behavior as infants approached. In the no-partition tests, infants generally stayed close to the mothers' cages, often reaching through the cage bars to touch them. Mothers varied considerably in reactions to their infants in both partition and no-partition tests. Some became very excited and, in the no-partition tests, clutched at their infants through the cage bars. Others appeared less concerned and did not interact with infants during the trials. It was difficult to quantify the effect of the mothers' behavior, but our general impression was that maternal responsiveness was not a major determinant of how the infants reacted. General behavioral cues associated with an awake state, as compared with an anesthetized state, seemed to be more important. As mentioned earlier, our initial tests showed that infants chose hooded and unhooded, anesthetized mothers to the same extent, which was greater than choice of other anesthetized adult females. In our subsequent tests, however, there was a greater preference for alert than for anesthetized mothers.

The age and sex of infants did not seem to correlate with their preferences. A greater amount of variability was observed in infants under two months of age the first time the tests were administered, but this effect appeared to be less related to perception and recognition than to individual differences in emotionality and motor coordination.

■ DIRECTIONS FOR FURTHER RESEARCH

Using the results of our experiments with surrogate and natural mothers, we plan to concentrate much of our future research on elucidating the role of early olfactory experiences in filial attachment and subsequent social relationships. Are olfactory cues from the mother more significant at certain stages of development? What is the relative importance of different sensory modalities to infants at different ages? How permanent are the effects of early olfactory experiences? Can such experiences affect behavior later in life? All of these questions can be examined using procedures similar to those we have used to date.

For example, artificial odors could be applied to differently colored, or otherwise visually distinct, surrogates. The relative importance of olfactory and visual cues to infants at different stages of development could then be examined by testing infants with different combinations of these stimuli, e.g., a familiar odor with a strange color. The olfactory and visual qualities of natural mothers could also be altered during testing, as we did to a limited extent when we masked the mothers' faces.

We would also like to develop procedures for rendering infants temporarily or permanently anosmic, that is, unable to smell. Use of temporary anosmia, possibly produced by the intranasal administration of an anesthetic solution, enables the experimenter to use an infant as its own control, i.e., under anosmic and nonanosmic conditions. Infants raised with permanent anosmia would be useful in determining whether the development of social bonds with mother and peers is in any way affected by the inability to smell, and how well infants adjust to their surroundings without this sense.

In our experiments with natural mothers, we were unable to determine the extent to which visual cues, separate from those associated with the mothers' behavior, influenced recognition and preference. One way of separating these factors would be by testing infants who could see their anesthetized mothers completely. In addition, tests with awake mothers behind a one-way mirror (Rosenblum and Alpert, chapter 15) would be useful for comparing the significance of visual cues associated with awake and anesthetized states. Such tests would also be helpful in evaluating the importance of maternal behavior specifically directed toward an infant, as opposed to that merely associated with being awake.

■ CONCLUSIONS

The results of our experiments show that infants can recognize and prefer their mothers when olfactory cues appear to be the major source of information. Moreover, the addition of behavioral cues and information from other sensory modalities tends to enhance this recognition. Several reports have recently appeared in the literature on olfaction in the young of various phylogenetically "lower," nonprimate species (Carter and Marr 1970; Gregory and Pfaff 1971; Leon and Moltz 1972; Mainardi et al. 1965; Marr and Gardner 1965; Mykytowycz 1970; Rosenblatt 1972; Shapiro and Salas 1970). Infant animals, including kittens, nestling rabbits, guinea-pigs, rats, and mice, recognize and are strongly influenced at later ages by odors associated with their mothers, littermates, and home area. Our data on both surrogate-raised and mother-raised infants represent the first clear demonstration of the use of this modality by a young primate.

It remains to be determined, however, whether primates other than the squirrel monkey would react in a similar fashion. The infant squirrel monkey in its early life is extremely suited to being influenced by olfactory stimuli, more so than many other species. It rides on its mother's back almost continuously during the first month (see chapter 17), a position that prevents it from viewing the mother's face. While on her back its nose is often pressed close to the mother's body, providing an excellent opportunity to acquire olfactory information associated with the mother. However, other species, whose ontogeny and/or evolutionary background differ greatly from that of the squirrel monkey, may be less sensitive to olfactory stimuli and more dependent on other sensory modalities.

■ ACKNOWLEDGMENTS

This research was supported by USPHS Grant No. HD04905. I wish to thank Curt Busse and Emily Lee for their assistance with testing infants and Ann Ball for her helpful comments on the manuscript.

■ REFERENCES

Carter, C. S., and Marr, J. N. 1970. Olfactory imprinting and age variables in the guinea-pig, *Cavia porcellus. Anim. Behav.* 18:238–244.

Gregory, E. H., and Pfaff, D. W. 1971. Development of olfactory-guided behavior in infant rats. *Physiol. Behav.* 6:573–576.

Harlow, H. F. 1958. The nature of love. *Amer. Psychol.* 13:673–685.

Harlow, H. F., and Harlow, M. K. 1966. Learning to love. *Amer. Sci.* 54:244–272.

Harlow, H. F., and Suomi, S. J. 1970. Nature of love—simplified. *Amer. Psychol.* 25:161–168.

Harlow, H. F., and Zimmermann, R. R. 1959. Affectional response in the infant monkey. *Science* 130:421–432.

Leon, M., and Moltz, H. 1972. The development of the pheromonal bond in the albino rat. *Physiol. Behav.* 8:683–686.

Mainardi, D.; Marsan, M.; and Pasquali, A. 1965. Causation of sexual preferences in the house mouse. The behavior of mice reared by parents whose odour was artificially altered. *Atti Soc. It. Sci. Nat.* 54:325–338.

Marr, J. N., and Gardner, L. E. 1965. Early olfactory experience and later social behavior in the rat: Preference, sexual responsiveness and care of young. *J. Genet. Psychol.* 107:167–174.

Mason, W. A.; Hill, S. D.; and Thomsen, C. E. 1971. Perceptual factors in the development of filial attachment. In *Proceedings of the third international congress of primatology*, vol. 3, ed. H. Kummer, pp. 125–133. Basel: Karger.

Mykytowycz, R. 1970. The role of skin glands in mammalian communication. In *Communication by chemical signals*, eds. J. W. Johnson, D. G. Moulton, and A. Turk, pp. 327–360. New York: Appleton-Century-Crofts.

Rosenblatt, J. S. 1972. Learning in newborn kittens. *Sci. Amer.* 227:18–25.

Shapiro, S., and Salas, M. 1970. Behavioral responses of infant rats to maternal odor. *Physiol. Behav.* 5:815–817.

8 A Preliminary Report on Weaning among Chimpanzees of the Gombe National Park, Tanzania

CATHLEEN B. CLARK

- ## INTRODUCTION

This is a preliminary, descriptive report on weaning among six pairs of chimpanzee (*Pan troglodytes schweinfurthi*) mothers and their infants at the Gombe National Park in Tanzania, East Africa. Although the term "weaning" has been used in the literature to describe the entire process from birth of the growing independence of the infant, I restrict the term to mean weaning from the nipple. I will also discuss the effect of weaning on behaviors such as travel, play, grooming, and nesting.

The chimpanzee mother-infant relationship is one of great duration and intensity. Monkeys in general are weaned by 1–2 years of age and then, moving away from the mother, enter into relationships with others in the troop. The young chimpanzee, however, remains with the mother for 9–12 years, often spending most of its time with her as its only companion. The monkey infant has a younger sibling usually by the time it is two; the Gombe chimpanzee infant has no younger sibling until at least 4, more often 5, years of age, and suckles to within a few months of the next birth. This is an extremely long duration of parental investment on the part of the chimpanzee mother.

From birth the chimpanzee infant becomes more independent, and the process is a gradual and very gentle one. The mother may begin to reject her infant from the nipple by the infant's second year of life; however, at first the rejections are infrequent. They intensify with the resumption of the mother's estrous cycle which usually occurs by the time the infant is 4 years old. However, the 2–4 year period of gradual weaning from the nipple does not completely soften the blow to the infant when it finds the milk dried up. Weaning from the nipple is the first serious break in the bond with the mother, and the infant's behavior is affected when it is denied suckling and is supplanted by a younger sibling at its mother's breast. As weaning progresses, the infant may display many elements of depression; it may often sit huddled for long periods of time; its interactions with peers may decrease, as may its appetite, and its response to environmental stimuli; it may also insist on greater physical contact with its mother. Some infants may come out of this period of depression within a few months, whereas others may remain depressed for a year or longer.

■ STUDY AREA AND METHODS

The study area consists of a narrow stretch of rugged, mountainous country running some ten miles along the eastern shores of Lake Tanganyika and inland three miles or less to the tops of the peaks of the rift escarpment. The rift is intersected by many steep-sided valleys which support permanent streams. Riverine gallery forest is found in the valleys; between the valleys the slopes are often more open, supporting deciduous woodland. Many of the higher ridges and peaks are covered only with grass. The area supports a population of between 100 and 150 chimpanzees. This population is divided into 3–4 communities of individuals who recognize and may interact with each other (van Lawick-Goodall 1973).

A longitudinal study of one of these communities was initiated by Goodall in 1960, and an artificial feeding area was set up in 1963. Consequently, from 1963 on, researchers had the option of making observations "in camp," that is, in the feeding area, or "out of camp," where chimpanzees could be followed as they foraged. One of the major foci of Goodall's study has been the mother-infant relationship (van Lawick-Goodall, 1967, 1968). The 765 hours of data upon which this study is based were collected in the field by several research assistants, including myself from 1968 to 1969.

Due to the nature of the field situation, it was not always possible to obtain data consistently on every mother-infant pair due to a lack of human observers and/or the disappearance of the chimpanzees for periods of time. I have analyzed only part of the available data for each mother-infant pair and have used only data collected outside the artificial feeding area with the exception of 50 hours of narrative data collected on Flo and Flint during Flo's period of sexual receptivity in 1968. There are no statistical tests on the data due to the small sample size (N = 6) and the differing rates of maturation among the 6 infants. Therefore, this report must be viewed as a preliminary, descriptive one, bearing in mind the fact that the method of data collection in the field was not specifically designed for this particular study.

■ THE REPRODUCTIVE CYCLE

It is important to look at the reproductive cycle because of its influence on weaning (see Table 1). The mother's association pattern and rejection of the infant changes with the resumption of her estrous cycle, and the infant will only cease to suckle after she is again pregnant. This is not uncommon among nonhuman primates. Most monkeys do not wean their offspring completely, but nurse them through the following pregnancy for decreasing amounts of time (Rowell 1972).

The Gombe chimpanzee female typically begins to exhibit adolescent sexual swellings of her perineum at about 9–10 years of age. She experiences the first full sexual swellings and menarche at approximately 11–12 years of age at which time she

TABLE 1 / AGE OF INFANT AT CRITICAL POINTS IN WEANING

						AGE OF INFANT IN MONTHS WHEN:			
Infant	Sex	Mother	Infant's Date of Birth	Older Siblings	Tot. Hrs. Obs.	Mother resumes estrous cycle **	Mother becomes pregnant	suckling last observed	younger sibling born
Flint	m	Flo	1 March 1964	Faben, Figan, Fifi	300	46	46	50	54
Goblin	m	Melissa	7 Sept. 1964	0	55	15	56	57	63
Pom	f	Passion	12 July 1965	0	143	52	68	72	76
Mustard	m	Nope	28 Oct.–22 Nov. 1965	0	65	83	84	86	92
Atlas	m	Athena	19 Sept.–2 Oct. 1967	0	77	26*	62	62	70
Moeza	f	Miff	20–22 Jan. 1969	0	125	44	49	50	58
					765 tot.	44.3 aver.	60.8 aver.	62.8 aver.	68.8 aver.

*Athena had a sexual swelling once when Atlas was 26 months old, became pregnant but lost the infant. She did not cycle again until Atlas was 40 months old.
**No menstrual data available.

enters into sexual relations with adult males. There then follows a period of infertility often referred to as adolescent sterility—not uncommon among primates, including humans (Döring 1969)—which lasts approximately 1–2 years. The female then becomes pregnant when she is approximately 13–14 years of age and, after an average eight months of gestation, she gives birth. After a lactation interval of 3–4 years without the estrous cycle, the mother resumes the cycle while still lactating. (Gombe baboons, *Papio anubis*, resume the cycle while still lactating after a 10-month interval [Ransom and Rowell 1972]). Although engaging in sexual activity, she generally will not become pregnant again until her infant is 4–5 years of age. When she does become pregnant, her infant ceases to suckle within a few months; she gives birth, and the cycle begins again.

A female may produce, with a birth interval of five years, approximately five to six offspring during her life (not including those infants that die very young). The birth interval for monkeys is generally 1–2 years (e.g., Ransom and Rowell 1972; Hall and DeVore 1965; Jay 1965; Tanaka et al. 1970). Schaller estimates the interval to be 3½–4½ years among the mountain gorilla, *Gorilla gorilla beringei* (Schaller 1963). Among a human group, the San (Bushmen) people of the Kalahari, the birth interval averages 4–5 years (DeVore and Konner 1970). The oldest Gombe mother, Flo, produced five offspring; three survived to maturity. Two old females, Olly and Flo, discontinued sexual swellings during the last few years of their lives; it appears that menopause may occur late in the female's life.

Of special interest is the individual variation among mothers. For example, Melissa resumed her estrous cycle when Goblin was only 15 months old and cycled irregularly for 44 months before menstruation was observed and she bore another infant. Nope, however, did not resume her estrous cycle until her son was 83 months old; she then became pregnant within 2 months (no menstruation data are available). One female (Athena) became pregnant when her son was only 2 years old; however, no living offspring ensued. As yet, no Gombe mother has been observed with a new offspring before her infant is at least 4 years of age.

If a female becomes pregnant and loses the fetus, has a stillborn infant, or if her infant dies, she usually resumes cycling, menstruates within two months, and becomes pregnant soon after. Zuckerman (1931) has reported among captive chimpanzees, if the mother does not nurse, the menstrual cycle begins about two weeks after parturition. Among Japanese macaques, *Macaca fuscata*, births followed by the loss of young were more frequently followed by subsequent births in the following year than were those having no infant loss (Tanaka et al 1970). In Rwanda, the birth interval in nonlactating human mothers who had a stillbirth or whose baby died within a week was 15 months, whereas lactating mothers averaged 27 months (Saxton and Serwadda 1969).

Another interesting phenomenon among Gombe chimpanzee females (although not all females necessarily exhibit this) is the continuation of sexual swellings into pregnancy, accompanied with sexual activity. Generally, swellings occur during the first half of the pregnancy and they may be irregular. In the Budongo Forest a chimpanzee female was observed in estrous when approximately 5 months pregnant (Sugiyama 1969). Sexual swellings (with maximum swelling) during pregnancy have been noted among captive chimpanzees (Graham 1970; Tutin and McGrew 1973). This also occurs in Old World monkeys; for example, among the rhesus, *Macaca mulatta*, of Cayo Santiago, distinct estrous cycles occurred after conception, some with copulation (Conaway and Koford 1965), and up to four periods of maximum perineal swelling were observed in pregnant baboons, *Papio ursinus*, in captivity (Gilbert and Gillman 1951).

■ ASSOCIATION AND TRAVEL PATTERNS

Chimpanzees associate in ever-changing groups within a community (referred to as "regional population" by Japanese primatologists [Sugiyama 1972]) of individuals that recognize each other. The most stable group is that of the mother-offspring. The son typically remains with the mother until approximately 9–10 years of age whereas the female usually stays with the mother until 11–12 years of age when, upon becoming sexually attractive (with full swellings) to adult males, she wanders off on her own for long periods of time. However, even after social maturity (approximately 13–14 years of age for both males and females), the offspring will still frequently travel with the mother, and the bond between mother and offspring remains strong through life.

The six infants in this study were only separated from their mothers when lost, which was seldom and usually for never more than a few hours. Times such as these brought about whimpering and loud screaming from the infants who would even climb tall trees to search for their mothers. The chimpanzee mother-infant pair are often alone, unlike monkey infants such as young baboons and macaques who, after about the first six weeks, come to spend most of their waking time in play with other infants and juveniles, including siblings (Rowell 1972).

Individual differences existed among the mothers as to group preference (see Figure 1). Flo and Passion were loners, and were usually observed with offspring only. Flo's two mature sons, Figan (approximately 15–16 years old) and Faben (approximately 20–21 years old) occasionally traveled with her, and Fifi (approximately 10–11 years old) was often with Flo and Flint, especially when her infant sister Flame was alive. Passion's only companion was usually her daughter, Pom. Flo interacted with older adult males, whereas Passion avoided them. In a group Passion would stay by herself on the fringes, whereas Flo would join the older adult males and her own offspring in long grooming sessions. Melissa, Athena, Nope, and Miff were sociable females, often traveling with and interacting with others. Melissa seemed to prefer the company of adult males whereas Athena and Miff interacted with, and traveled with, both males and females. Nope avoided adult males and interacted for the most part with other mothers and young females. Figure 1 does not include data on Nope. Of the six mothers, she was the only one who was unhabituated to human followers, especially avoiding them when alone. Thus, most of the data collected on Nope and Mustard were taken while Nope was with others.

FIGURE I. **ASSOCIATION**

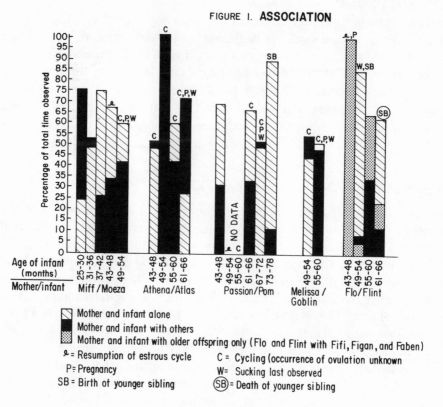

Mother and infant alone
Mother and infant with others
Mother and infant with older offspring only (Flo and Flint with Fifi, Figan, and Faben)
♀ = Resumption of estrous cycle   C = Cycling (occurrence of ovulation unknown
P = Pregnancy   W = Sucking last observed
SB = Birth of younger sibling   ⑤Ⓑ = Death of younger sibling

With the resumption of the estrous cycle, the mothers' association patterns changed, and they all associated more often with adult males, even Nope and Passion. Then, with the advent of pregnancy, all the females spent more time alone with their offspring and avoided groups, especially those including adult males. Even the usually very sociable mothers such as Athena kept on the fringes of groups while pregnant.

In general, the mothers and infants did not cover as great a range as did the males and nulliparous females. Three mothers, Flo, Passion, and Melissa, spent most of the time in the community under observation. Miff, Athena, and Nope, however, disap-

TABLE 2 / AVERAGE NUMBER OF MINUTES/BOUT OF SUCKLING

MONTHS	WK	PF	FD	GM	MZ	AP	AVERAGE	RANGE (BY MONTH)
1–6	2.0	2.3	2.9	—	—	—	2.3	1.6–3.0
7–12	2.1	2.9	2.1	—	—	—	2.4	1.6–3.7
13–18	—	3.5	2.6	2.6	—	—	3.0	2.2–3.9
19–24	—	2.7	2.6	2.7	—	—	2.7	1.4–3.3
25–30	—	—	2.4	2.0	2.3	—	2.3	1.8–2.7
31–36	—	—	—	2.1	2.1	—	2.1	1.5–2.6

(based on minimum of 5 bouts observed per month)

TABLE 3 / AVERAGE NUMBER OF TIMES TO NIPPLE/HOUR OBSERVED

MONTHS	WK	PF	FD	GM	MZ	AP	AVERAGE	RANGE (BY MONTH)
1–6	2.8	2.9	2.4	—	—	—	2.7	1.6–4.9
7–12	1.2	2.0	1.1	—	—	—	1.5	0.8–2.1
13–18	—	1.6	1.0	0.6	—	—	1.2	0.5–2.0
19–24	—	1.3	1.0	0.8	—	—	1.0	0.5–1.6
25–30	—	—	1.0	1.0	1.0	—	1.0	0.8–1.2
31–36	—	—	—	1.1	1.5	—	1.3	0.6–2.4

(based on a minimum of 100 minutes or more of observation/month)

peared for months at a time and probably joined other communities. Of the 6, Flo, Passion, Melissa, and Athena were observed with sexual swellings among the Gombe community males 8 months before giving birth to the younger siblings of the infants in this study.

■ WEANING

SUCKLING

In order to ascertain how suckling rates changed through time, it was necessary to establish a baseline of data for the first 3 years of life, as there were insufficient quantitative data on the six infants in this study. The data on 5 younger infants at Gombe were examined (see Tables 2, 3, and 4). Three of these infants, Prof, Gremlin, and Aphrodite, are siblings of three of the subjects, Pom, Goblin and Atlas. The number of minutes in an average bout of suckling during the first 3 years was 2.5. The number of times per hour that the infant was observed to make nipple contact was highest during the first 6 months at an average of 2.7, decreasing to 1.5 during the second 6 months of life. This was also the time when the infant was beginning to break contact with the mother (and she with it) to explore the environment and was doing some supplemental feeding. There was then a decrease to 1.0 times per hour to the nipple where the rate remained throughout the second and third years of life. The percentage of total time observed spent on the nipple was highest during the first 6

TABLE 4 / PERCENTAGE OF TOTAL MINUTES OBSERVED DURING WHICH SUCKLING OCCURRED

MONTHS	WK	PF	FD	GM	MZ	AP	AVERAGE	RANGE (BY MONTH)
1–6	9	9	11	—	—	10	10	5–14
7–12	5	10	4	—	—	—	6	2–12
13–18	—	8	4	3	—	—	5	2–10
19–24	—	5	4	3	—	—	4	3–7
25–30	—	—	4	3	3	—	3	2–5
31–36	—	—	—	3	5	—	4	2–8

(based on a minimum of 100 minutes or more/month of observation)

TABLE 5 / NUMBER OF MINUTES IN AN AVERAGE BOUT OF GROOMING WITH MOTHER AND IN AN AVERAGE BOUT OF SUCKLING (# BOUTS IN PARENTHESES)

age (mos.)	FLINT groom	FLINT suck	GOBLIN groom	GOBLIN suck	POM groom	POM suck	MUSTARD groom	MUSTARD suck	ATLAS groom	ATLAS suck	MOEZA groom	MOEZA suck
28											2.5 (10)	2.3 (9)
29											2.3 (2)	2.0 (3)
30												
31											0.7 (3)	1.8 (2)
32											1.5 (10)	1.8 (5)
33											2.1 (20)	2.5 (11)
34											2.2 (6)	2.2 (6)
35												
36												
37												
38											6.4 (8)	2.3 (9)
39											3.6 (11)	3.0 (8)
40											5.5 (15)	2.8 (12)
41												
42									1.3 (7)	1.0 (7)	2.0 (6)	2.3 (6)
43					1.8 (21)	2.0 (7)			8.3 (12)	2.5 (6)	9.5 (10)	5.1 (9)
44									5.3 (8)	1.7 (3)	4.7 (29)	4.4 (13)
45					1.0 (11)	2.8 (6)			7.3 (6)	2.0 (2)	5.1 (15)	3.9 (9)
46					0.8 (7)	2.0 (2)			2.5 (5)	0.8 (3)	3.0 (28)	3.0 (8)
47	4.3 (15)			8.2 (15)	5.0 (2)	2.7 (2)			7.3 (4)	2.0 (3)	2.2 (13)	4.0 (2)
48					2.5 (8)	1.5 (7)			2.0 (2)	3.0 (6)		
49												
50									6.7 (3)	1.6 (9)	1.3 (3)	3.5 (4)
51											10.5 (8)	0.0 (0)
52												
53	2.6 (13)	0.0 (0)										
54	8.7 (77)	0.0 (0)	5.7 (3)	5.0 (1)					3.5 (8)	2.5 (8)	2.2 (17)	0.0 (0)
55	0.5 (3)	0.0 (0)	6.3 (8)	3.4 (8)								
56	0.9 (6)	0.0 (0)	5.0 (5)	4.0 (3)					4.7 (3)	2.5 (2)		

	1	2	3	4	5	6	7	8	9	10
57	2.0 (1)	0.0 (0)	8.0 (16)	4.0 (3)					7.5 (4)	5.5 (2)
58	3.8 (3)	0.0 (0)	9.0 (7)	0.0 (0)						
59	4.4 (13)	0.0 (0)	9.0 (4)	0.0 (0)					5.2 (5)	3.3 (3)
60	0.2 (2)	0.0 (0)								
61					1.0 (1)	1.7 (3)				
62	10.0 (2)	0.0 (0)			1.0 (2)	5.0 (1)			3.6 (5)	6.5 (2)
63	7.0 (19)	0.0 (0)			3.3 (3)	2.3 (4)			4.4 (10)	0.0 (0)
64	10.0 (5)	0.0 (0)			5.3 (7)	2.8 (6)			10.0 (5)	0.0 (0)
65					0.0 (0)	4.0 (1)			12.0 (4)	0.0 (0)
66	3.4 (13)	0.0 (0)			1.6 (4)	2.5 (5)				
67					4.3 (10)	3.4 (6)	2.0 (5)	2.3 (4)		
68					9.0 (6)	7.4 (5)	1.5 (13)	2.3 (11)		
69					15.0 (1)	0.0 (0)				
70					3.0 (6)	13.0 (2)	2.0 (2)	3.0 (1)		
71							2.3 (3)	2.0 (2)		
72							2.0 (1)	2.0 (1)		
73							3.9 (10)	3.5 (2)		
74					13.6 (7)	0.0 (0)	5.5 (2)	4.0 (1)		
75					6.8 (35)	0.0 (0)	18.7 (3)	2.3 (3)		
76					6.8 (8)	0.0 (0)	1.7 (3)	0.0 (0)		
77					5.2 (12)	0.0 (0)	17.3 (4)	6.0 (1)		
78										
79										
80										
81										
82										
83										
84							3.3 (15)	5.8 (10)		
85							10.0 (2)	6.0 (1)		
86							20.6 (5)	6.8 (5)		
87							8.8 (11)	0.0 (0)		
88							12.5 (2)	0.0 (0)		
89							1.0 (1)	0.0 (0)		
90										

TABLE 6 / FREQUENCY (#/HOUR OF OBSERVATION) OF GROOMING* & SUCKLING* BOUTS

FLINT

	Age (mos.)	# Hrs Obs.	#Times/Hr Suckle	#Times/Hr Groom with Mother
	47	14	1.1	1.1
SB. Born	53	13	0.0	1.1
	54	121	0.0	0.5
	55	4	0.0	0.7
	56	10	0.0	0.6
	57	4	0.0	0.2
	58	10	0.0	0.3
	59	14	0.0	0.9
SB. Dies	60	10	0.0	0.2
	61	3	0.0	0.0
	62	11	0.0	0.2
	63	23	0.0	0.8
	64	7	0.0	0.7

GOBLIN

Age (mos.)	# Hrs Obs.	#Times/Hr Suckle	#Times/Hr Groom with Mother
52	3	0.0	0.7
53	2	0.0	0.0
55	10	0.8	0.8
56	7	0.5	0.7
57	16	0.2	1.0
58	11	0.0	0.7
59	2	0.0	0.0
60	3	0.0	1.2

POM

	Age (mos.)	# Hrs Obs.	#Times/Hr Suckle	#Times/Hr Groom with Mother
	43	9	0.8	2.3
	45	7	0.9	1.6
	46	3	0.8	2.7
	47	4	0.6	0.5
	48	8	0.9	1.0
	64	2	0.5	1.0
	65	3	1.3	0.9
	66	3	1.8	1.9
	67	4	0.2	0.0
	68	4	1.3	1.0
	69	6	1.1	1.8
	70	6	0.9	0.9
	71	4	0.0	0.3
	72	5	0.4	1.3
	73	2	0.0	1.3
	74	10	0.0	0.7
SB Born	75	27	0.0	1.3
	76	11	0.0	0.8
	77	13	0.0	0.9
	78	5	0.0	0.8

*Grooming bouts may include suckling, and suckling bouts may include grooming.
Italicized Numbers: Infant last observed on nipple
Note: # hrs. observed has been rounded off to the closest whole number.

MUSTARD				ATLAS				MOEZA			
Age (mos.)	# Hrs Obs.	#Times/Hr Suckle	#Times/Hr Groom with Mother	Age (mos.)	# Hrs Obs.	#Times/Hr Suckle	#Times/Hr Groom with Mother	Age (mos.)	# Hrs Obs.	#Times/Hr Suckle	#Times/Hr Groom with Mother
42	2	0.9	1.3	42	6	1.2	1.2	25	2	1.2	1.8
60	2	0.9	1.7	43	7	0.9	1.7	26	3	0.8	0.4
62	3	1.5	1.9	44	5	0.6	1.5	27	4	0.9	0.7
63	6	1.7	2.4	45	4	0.5	1.6	28	9	1.0	1.1
67	4	0.3	0.8	46	5	0.6	1.0	31	2	0.9	1.4
69	2	1.1	2.2	47	2	1.4	2.0	32	8	0.6	1.3
71	8	0.5	1.5	48	3	2.0	0.7	33	5	2.4	3.7
74	3	1.1	1.8	50	5	1.9	0.7	34	3	2.0	2.0
76	4	0.0	0.9	54	8	0.8	1.1	38	6	1.5	1.3
77	4	0.3	1.0	58	3	0.8	1.5	39	7	1.1	1.5
84	7	1.4	2.5	60	4	0.9	1.4	40	11	1.1	1.5
86	4	1.3	2.1	62	4	0.5	1.2	42	3	1.8	1.8
87	7	0.0	1.7	63	5	0.0	2.1	43	7	1.2	1.3
88	2	0.0	1.1	64	5	0.0	1.0	44	9	1.4	3.1
89	3	0.0	0.3	66	12	0.0	0.3	45	10	0.9	1.5
								46	12	0.7	2.3
								47	8	0.2	1.6
								50	3	1.2	1.2
								51	3	0.0	2.8
								54	9	0.0	1.9

months at approximately 10%, decreasing to approximately 6% within the second six months, and falling to 3.5% by the third year.

During the 4–5 years while the infant is still suckling, it gradually learns the various food plants of its community and how they are collected and eaten. By the time the infant is in its fourth year, it is most probably no longer nutritionally dependent on the mother's milk, but is still psychologically dependent. This psychological dependence can be seen most dramatically when the infant is upset and runs to the mother for a quick reassuring suckle after such an event as rough play.

As weaning progressed, the length of the suckling bouts of the six infants increased (see Table 5). Moeza's average bout length during her final months rose from 2.6 minutes to 4.1, the longest lasting 5.1 minutes; Atlas' rose from an average of 2.2 minutes to 4.3, the longest lasting 10.0; Pom's rose from 2.1 minutes to 5.1, the longest lasting 20.0; Mustard's bouts rose to 6.1 minutes during his final months of suckling, four of which were over 10.0. During Flint's final months on the nipple, he was averaging 8.2 minutes per bout; Goblin suckled the least of the six infants, yet his bout length increased to an average of 3.7 minutes in the final months of suckling, the longest lasting 6.4.

The infants averaged 3–4% of the total time observed on the nipple until the final year of suckling when all but one of them spent greater time suckling. In the last month of suckling to be observed, Atlas' nipple contact time had risen to 5%, Moeza's had risen to 9%, Pom's to 10%, Flint's to 14%, and Mustard's to 15%. Goblin was the only infant whose nipple contact time decreased and was observed to be 1% during his final month.

Whereas the bout length and the percentage of total time spent suckling increased towards the end of suckling, the number of times per hour that the infant made nipple contact generally decreased (see Table 6). In the final month observed on the nipple, Moeza, Mustard, and Flint were still making nipple contact approximately once per hour observed. Atlas, Pom, and Goblin, however, fell from approximately once per hour to 0.5 and less per hour observed.

LACTATION AND THE RESUMPTION OF THE ESTROUS CYCLE

Rejections by the mother generally became more frequent and intensive (and in some cases were observed for the first time) at the resumption of the estrous cycle. This also was observed among some of the Gombe and Nairobi baboons who resumed cycling when their infants were 35–40 weeks of age; although among Uganda baboons, mothers began to cycle before any significant amount of rejection was noticed (Ransom and Rowell 1972).

During weaning, a particular type of behavior was observed in all six infants and in other infants previously observed by Goodall (van Lawick-Goodall 1968) with the resumption of the estrous cycle. It was characterized by the infant showing distress (hoo, whimper, squeak, scream, grin, tantrum) after having made nipple contact without any discernible rejection by the mother. In fact the mother often drew her distressed infant to her breast, offered her nipple, and yet the infant, upon suckling again—often with rapid nipple switching—backed off grinning (fear face), whimpering, and/or screaming. The mother often seemed as confused as the infant and tried to reassure her offspring with an embrace, grooming, playing with the infant, or even presenting her bottom as in Melissa's case (see "Special Behaviors" below). The most severe of these observed incidents occurred at the time of estrous swelling when the mother was evidently ovulating and probably became pregnant, and during the month following impregnation. Although in some cases when the infant went to the nipple after the period of sexual activity and probable impregnation had passed (and it was impossible to tell whether the infant was, in fact, getting any milk), suckling ceased within a few months. The following is an example of one of these incidents which occurred at the time of Athena's sexual swelling in October 1972 when she most likely became pregnant, giving birth to Aphrodite eight months later.

> Atlas sat in contact with Athena in the tree. He put his mouth to her nipple but didn't actually touch it and began whimpering. He continued this behavior repeatedly during the next eight minutes. He then was seen to put his mouth so it was just touching the nipple and he began to become more excited. His whimpers would turn to squeaks as he backed away from Athena showing a full closed grin (fear face), but then he would again put his mouth to the nipple just barely, while whimpering. He repeated this over and over again for nearly ten minutes, despite Athena's attempts at grooming him, extending her hand and pressing him to her other nipple. She finally seemed to calm him somewhat by grooming him. Atlas then left to play with Mustard. As the play bout gradually moved down towards Athena, Atlas broke it to put his mouth to Athena's nipple. Very suddenly he began screaming. Athena extended a hand, touched and embraced him. She then groomed him and continued grooming for several minutes after he'd stopped screaming. (Merrick 1972)

This was the last occasion on which Atlas was observed making nipple contact.

MATERNAL REJECTION

Chimpanzee mothers play an active role in the weaning of their infants as do other nonhuman (and human) primates. Weaning tactics (or rejection from the nipple) have been observed among the following (these are examples—the list is by no means complete): squirrel monkeys, *Saimiri sciureus* (Baldwin 1969); hanuman or common langurs, *Presbytis entellus* (Jay 1965; Sugiyama 1965); nilgiri langurs, *Presbytis johnii* (Poirier 1968); patas monkeys, *Erythrocebus patas* (Hall 1965); vervet monkeys, *Cercopithecus aethiops* (Struhsaker 1971); baboons, *Papio cynocephalus* (Hall and DeVore 1965; Ransom and Rowell 1972; Anthoney 1968); rhesus macaques, *Macaca mulatta* (Southwick et al. 1965; Lindberg 1971; Hinde 1964, 1967); bonnet macaques, *Macaca radiata* (Simonds 1965; Rahaman and Parthasarathy 1969; Rosenblum and Kaufman 1967, 1969); pigtail macaques, *Macaca nemestrina* (Rosenblum and Kaufman 1967, 1969); Gibraltar macaques, *Macaca sylvana* (Burton 1972); Japanese macaques, *Macaca fuscata* (Itani 1959); mountain gorilla, *Gorilla gorilla beringei* (Schaller 1963).

The maternal rejections among Gombe chimpanzee females begin two to three years before the cessation of suckling, and they intensify and become more frequent through time. Captivity almost always lowers the age of onset and heightens the intensity of rejections (see Nicolson, this volume). At Gombe it is not yet clear what sets off the rejections. They do begin among some mothers before the resumption of the estrous cycle, although other mothers have not been observed to reject their infants until after the estrous cycle resumes.

The mothers all, with the exception of Flo during her sexual swelling, were remarkably gentle during weaning. There was a great deal of individual variation among the six Gombe mothers as to tactics and frequency of rejection. (There were insufficient data to do a quantitative analysis of the frequency.) Flo was the oldest female and the only mother of the six with experience, having previously weaned Faben, Figan, and Fifi. She was observed employing the most tactics in prevention of suckling, and was the only mother observed using physical aggression (she was not observed to do so with Fifi, however). The following methods of rejection were observed among the mothers:

1. hold infant out from chest (Flo)
2. twist body away from infant (Flo)
3. pull infant off ventral position (Flo)
4. hold infant tightly ventral to restrict movement (Flo)
5. walk away from infant while it is suckling (Passion, Flo)
6. aggression—cough threat, hit, bite, grab, slap, kick, shove the infant (Flo)
7. push infant gently from the nipple (Flo, Miff)
8. arch back away from infant and/or jerk nipple out of infant's mouth (Flo, Melissa, Passion)
9. take infant ventral and travel (Flo, Nope, Miff)
10. block nipple access with arm, hand, knee or by pressing chest against the ground or a branch (all six)
11. avoid infant by leaving, moving away as infant approaches (all six)
12. play with and/or groom the infant when the infant tries to suckle (all six)

RESPONSE OF THE INFANT TO MATERNAL REJECTION

The infants were very disturbed whenever denied access to the nipple. When the rejections first began, a whimper from the infant was generally sufficient to induce the mother to allow suckling. However, as the infant grew older, it evidently became more difficult for it to get the mother to give in. The infant might slip under the mother's arm or pull gently at the blocking appendage. If this did not meet with success, the infant might work the whimper into a scream or a tantrum as did Flint and Moeza. This worked with Miff (primiparous) on the three occasions when Moeza had tantrums over nipple access. Flo (multiparous), however, would often react to Flint's tantrums by biting or slapping him; Flint would then resort to violence himself, and they would roll about biting and hitting each other. Flint would then return to Flo who would groom or play with him, evidently to soothe him. Flint was the only infant of the six with a mother experienced in weaning offspring. Thus, his tactics to obtain what he wanted often ended in frustration and tantrums due to her ability to fend off demanding offspring.

The two female infants were observed approaching their mothers and grooming them around the nipple while slowly moving in to the nipple. This has recently been observed in a younger male infant, Freud, at Gombe (Goodall, personal communication), although not in the four males in this study. Moeza was whimpering almost every time she approached Miff to suckle or even only to groom during her final months of suckling. The females appeared more fearful of their mothers and exhibited more caution in their dealings with them than did the males, who were more direct and demanding. When Flint or Mustard wished their mothers to groom them, they would tug at the mother's hand until she responded. Flint would shove Flo when he wanted her to travel and was even observed on one occasion to shove her over onto her back to suckle. The two females, Pom and Moeza, were never observed shoving or tugging at their mothers. However, it is not possible with such a small sample size to make any statements about sex differences at this point, as the differences may be due to individual variation. That rhesus mothers treat their infants differently by sex has been demonstrated under controlled conditions (e.g. Hansen 1966; Mitchell 1968). There are several infants approaching weaning age at Gombe now, and we should know more about sex differences soon.

MOTHER-INFANT AGGRESSION

There was very little aggression (physical punishment such as hit, bite, grab, slap, kick, push, cough threat) in any context observed among the Gombe chimpanzee mothers and infants. Of the six mothers, Melissa and Nope were never observed being aggressive towards their offspring. Passion and Athena were only seen to display aggression once, and Miff twice. Flo and Flint, however, were observed 300 hours (including 50 in the artificial feeding area) during which 92 incidents involving aggression were recorded. Flint was the only infant observed who was ever aggressive towards his mother. During Flo's sexual receptivity in January 1968, 24 aggressive incidents several minutes in length, involving hitting, biting, slapping, kicking, and shoving by both, occurred during 44 hours of observation (or 1 per 1.8 hours). The aggression usually occurred after Flint repeatedly tried to gain access to the nipple or when he exhibited distress while on the nipple by screaming, whimpering, leaning back and rocking, and nipple switching rapidly (seemingly due to lack of milk). Flo usually would first use the more gentle forms of prevention by blocking her nipples, arching her back, twisting her body away, or by leaving Flint. When he persisted with whimpers and screams, she then resorted to physical punishment which was often answered with Flint's aggression towards her. A third of the aggressive incidents during estrous ended with Flo grooming and/or playing with Flint; none ended with Flint on the nipple. During this time Flo would often display a full grin (fear face) when Flint approached her, trying to get in to the nipple. When her swelling subsided she again allowed her infant to suckle, but he soon ceased as she was pregnant. Of the six infants, Flint had the most difficult weaning and was making brief nipple contact for at least two years after he actually stopped suckling.

TANTRUMS

Tantrums were not a common occurrence and appeared regularly in one individual only. Pom, Atlas, and Goblin were never observed in a tantrum during the hours of this study, Mustard in only one, and Moeza in three. For Flint, however, 38 tantrums were recorded (and many were unrecorded), half of which were coupled with physical aggression by Flo. Flint's older sister Fifi and brother Figan were also known to have had tantrums when they were younger and could not get their way (Goodall, personal communication), but certainly not nearly as frequently as their younger sibling. Flint had so many tantrums during weaning that it was considered an ordinary daily event to observers, whereas, for the other five infants, a tantrum was considered an extraordinary event.

SPECIAL BEHAVIORS

There were certain behavior patterns observed for the first time during the final months of weaning that usually disappeared after suckling had ended. Moeza developed the behavior characterized by crouching and bobbing in front of Miff with a full closed grin (fear face). Miff would eventually respond by raising her arm allowing nipple access, and by embracing and grooming her infant. Pom was observed to whimper and bend away (a submissive gesture) from Passion during her final year on the nipple. When Flint was very distressed Flo would grab and hold him tightly ventral; she appeared to be trying to stop the distress with a tight hug. This behavior was recorded on 38 occasions, half of which were accompanied by mother aggression (usually a bite).

A very interesting behavior that was observed among five of the six pairs was that of the mother turning and presenting her perineum to her distressed infant, and the infant mounting her (in the case of the two female infants), or mounting with intromission (in the case of the males). When Goblin was extremely upset during a suckling situation, Melissa was observed to present her bottom to him on three occasions after trying to reassure him in other ways. Goblin would then penetrate and thrust several times with an erection (once thrusting over 30 times). Melissa, after Goblin had ceased thrusting, would gradually move away from him, breaking contact. Neither she nor any of the other mothers (with the exception of Flo during her 1968 sexual swelling) were with swelling.

Special behaviors at the time of weaning have been observed in other nonhuman primates. Among Nairobi baboons there was a special moan used only in conjunction with weaning and not observed at any other time in the infant's life (DeVore 1963).

INFANT DISTRESS AND MOTHER'S RESPONSE

As weaning progressed, the infants displayed many elements of depression; they would often sit huddled for long periods of time; their interactions with peers decreased, as did their appetites, and their response to environmental stimuli; and they also insisted on greater physical contact with their mothers.

The six infants varied greatly in the amount of distress they displayed, from Moeza and Flint at one extreme (high), to Goblin (low). Moeza and Flint, who seemed to be the most visibly upset during weaning, were the two youngest to be weaned. Both infants continued to make nipple contact with whimpering (just briefly touching the nipple with the lips) after suckling had ceased. Goblin and Pom were not recorded as mking nipple contact after suckling had ended; there are insufficient data available on Atlas and Mustard.

During the final months of suckling there was generally a rise in whimpering among the infants and they regressed to infantile behaviors such as ventral riding on their mothers and long periods of nipple contact. The infants were observed exhibiting distress in the following situations: feeding with mother, grooming with mother, traveling off mother, evening nesting, rough play with others, fear (of sudden noise,

TABLE 7 / SITUATIONS DURING WHICH INFANTS WERE OBSERVED TO EXHIBIT DISTRESS (HOO, WHIMPER, SQUEAK, SCREAM, TANTRUM, GRIN, ROCK)

CONTEXT	FLINT	GOBLIN	ATLAS	POM	MUSTARD	MOEZA
Travel off mother	106	31	29	98	38	216
Groom with mother	18	0	0	0	0	7
Beg for food from mother	15	5	2	11	3	6
Context unknown	12	7	17	12	9	17
Evening nesting	2	0	0	0	0	0
Social excitement and/or aggression	3	4	1	2	3	8
Rough play with others	1	1	3	2	11	3
Fear (sudden noise, etc.)	1	2	1	2	1	1
Sex: mother and others	0	0	3	0	0	0
Suckling	weaned	3	4	10	2	36
Aggression by mother towards infant	9	0	1	0	0	3
Feeding (infant trying for mother's termite hole)	0	0	0	0	0	10
TOTAL	167	53	61	137	67	305
# months observed with 100 or more minutes of good observation	9	8	14	20	15	17

snake, etc.), suckling, sex (mother and adult male), and social excitement and/or aggression (see Table 7). Travel was by far the most distressful situation, with five of the six doing over 50% of their whimpering while traveling off the mother. Often the infant kept up a continuous bout of whimpering while traveling behind the mother during the final months of suckling. Pom did over 70% of her whimpering during travel; this was probably linked to the fact that Passion left her behind when she was only nine-months-old and continued to do so throughout Pom's infancy (van Lawick-Goodall 1968). After Pom's brother was born, she did very little whimpering during travel and kept farther away from Passion, appearing listless and depressed for over a year.

The mothers were, for the most part, remarkably tolerant and gentle with the infants during the weaning period. They would spend long periods of time grooming their infants and allowed the infants to suckle for as long as 20 minutes at a time. During weaning, even Passion, who in Pom's earlier years was often inattentive to her infant's needs, allowed Pom to cling and make nipple contact for great lengths of time, waited for her during travel, and was more attentive than ever before observed.

■ OTHER BEHAVIORS AFFECTED BY WEANING

TRAVEL

Chimpanzee infants are carried in the ventral position until approximately 6–8 months when the mother generally initiates dorsal riding by lifting or scooping the infant onto her back. By the second year of life the mother begins to signal to the infant to travel on its own by such tactics as avoidance, tipping or shaking it off, dropping a hip until it slips off, or stopping and/or sitting until it moves off on its own. In general, by the time the infant is five years of age, it is doing most of its traveling off mother but may revert to an increase in riding on her (even the infantile behavior of ventral riding) during the final months of suckling. After the sibling is born, the older offspring is rarely seen to travel on the mother.

There were interesting individual differences in this behavior. Goblin, after turning 4 years of age, was never observed riding on Melissa, even though not yet weaned. He whimpered infrequently during travel, and it often appeared that Melissa had to keep her eye on him, rather than he on her. Pom began dorsal riding when she was only 8 weeks old which was apparently due to the fact that at this time she sustained a bad injury to one foot and was unable to use it for gripping hair (van Lawick-Goodall 1968). Passion began tipping Pom off, dropping a hip or shoulder, long before the other mothers did so. Pom thus (like Goblin), by the time she was four years of age, was doing very little traveling on Passion (and was never observed to ride in the ventral position). Nope, an extremely tolerant mother and a highly anxious one, was first observed signaling Mustard to travel on his own when he was four years old. However, Mustard rode on his mother frequently and for long periods of time until he was eight—when Nope became pregnant and he stopped suckling. He ceased riding her after the birth of his younger sibling. Flint, even after having been weaned from the nipple, frequently rode on Flo during travel. Even when Flo tried to shake him off, especially during her pregnancy, Flint would insist upon riding her, and often rode ventrally. He continued to ride dorsally on Flo after the birth of his younger sibling who rode ventrally. After his younger sibling's death, Flint increased his riding on Flo, again often in the ventral position. Flo no longer rejected her son as frequently, and appeared to accept him in place of the lost five-month-old daughter.

During the final months of suckling, all the infants with the exception of Goblin (who never rode on Melissa at all), increased their riding on their mothers, and Mustard, Flint, Atlas, and Moeza reverted to occasional ventral riding, Moeza and Mustard suckling in this position (Moeza at 4 years of age and Mustard at eight).

PLAY

The amount of social play is dependent upon the availability of partners as well as the desire to play. Flint, Goblin, and Pom engaged in very little play with peers while out of the artificial feeding area, partially due to lack of partners, whereas Atlas, Mustard and Moeza, who were often in groups of mothers and infants out of the artificial feeding area, played more frequently. However, for all six infants, the percentage of total time spent in play dropped during the final months of suckling.

There was very little mother-infant play observed among the six pairs. This behavior appears to be linked to the age of the infant, occurring for the most part during years 2–4, and is also influenced by the mother's personality. Flo, the oldest female in the community, was only observed to play with Flint on seven occasions in his fifth year. Five of these occurred just after Flame's birth when Flo evidently used play to distract her son from his new sister. She was much more playful when younger (van Lawick-Goodall 1968), and her lack of play with Flint was most certainly partially due to her advanced age. Melissa and Passion were very rarely observed playing with their offspring. Athena and Nope were slightly more playful, and Miff, the youngest of the mothers, was the most playful of the six. However, during the final months of

TABLE 8 / PERCENTAGE OF SUCKLING BOUTS INVOLVING GROOMING

(TOTAL # SUCKLE-GROOM BOUTS/TOTAL # SUCKLE ONLY BOUTS
PLUS TOTAL # SUCKLE-GROOM BOUTS)

Age in Months	Moeza	Atlas	Mustard	Pom	Goblin	Flint
25–36	48% $(\frac{22}{46})$					
37–48	75% $(\frac{57}{76})$	67% $(\frac{20}{30})$		67% $(\frac{16}{24})$		27% $(\frac{4}{15})$
49–60		73% $(\frac{16}{22})$	81% $(\frac{17}{21})$	79% $(\frac{15}{19})$	73% $(\frac{11}{15})$	
61–72		100% $(\frac{2}{2})$	92% $(\frac{22}{24})$	65% $(\frac{22}{34})$		
73–84			79% $(\frac{11}{14})$			
85–96			100% $(\frac{9}{9})$			

TABLE 9 / PERCENTAGE OF GROOMING BOUTS INVOLVING SUCKLING
(TOTAL # SUCKLE-GROOM BOUTS/TOTAL # GROOM ONLY BOUTS
PLUS TOTAL # SUCKLE-GROOM BOUTS)

Age in Months	Moeza	Atlas	Mustard	Pom	Goblin	Flint
25–36	62% ($\frac{36}{58}$)					
37–48	42% ($\frac{57}{136}$)	45% ($\frac{20}{44}$)		31% ($\frac{16}{51}$)		27% ($\frac{4}{15}$)
49–60	0% ($\frac{0}{26}$)	70% ($\frac{16}{23}$)			24% ($\frac{11}{45}$)	
61–72		8% ($\frac{2}{24}$)	52% ($\frac{22}{42}$)	56% ($\frac{22}{39}$)		
73–84			37% ($\frac{11}{30}$)	0% ($\frac{0}{69}$)		
85–96			25% ($\frac{6}{24}$)			

suckling, even this pair decreased their play. Five of the mothers currently have second offspring, and with analysis we will be able to ascertain how play is related to the age of the infant, the age of the mother, and the personality of the mother.

GROOMING

From the beginning of life the chimpanzee infant experiences grooming in association with suckling. With age, infants develop an interest in being groomed and eventually in grooming. Grooming enables the infant to maintain contact with the mother, and she with it, as suckling ends; this contact behavior endures throughout their lives. In the months before suckling ended for the six subjects, approximately three-fourths of the suckling bouts involved grooming and over one-half of the grooming bouts did not involve suckling (see Tables 8 and 9).

Infant-other grooming

The infants did virtually no grooming (nor any other type of social interaction with the exception of play) with others until after weaning occurred. Almost 100% of their grooming was with their mothers, with the exception of Flint, who occasionally groomed his older siblings.

Mother-infant grooming

The percentage of the total time spent in grooming for the six study pairs is reflected in Figure 2. In the first 3 years of life the infants did very little (less than 2%) grooming on the mother. The mothers spent approximately 5–6% of their time grooming their infants. With the fourth year, the infants began to take more of an interest in grooming and spent anywhere from 2–11% of the time grooming their mothers. The mothers generally increased the amount of infant grooming as well. Some mothers (Athena, Miff, and Nope) groomed their infants more than others (Passion, Flo and Melissa). The still suckling infants in this study groomed their mothers less than the mothers did them, with the exception of the two youngest, Moeza and Flint. The two female infants, after suckling ended, spent more time grooming their mothers than their mothers did grooming them. The amount of grooming appeared to be linked to the weaning stage and the individuals concerned. The infant who was ceasing to suckle was spending more time grooming. Thus, Moeza, who ceased to suckle at approximately 50 months, was spending 20% of her time grooming with her mother, Miff, while Mustard at the same age was spending only 4% of his time grooming with his mother. However, in Mustard's seventh year when the end of suckling was approaching, the mother-infant grooming rose to 22% and then to 24% in his eighth year when he stopped suckling. The grooming both by mother and infant rose as suckling ceased, and there was generally a peak just as suckling ended. Then grooming lessened with the advent of a sibling. In Flint's case, the mother-infant grooming was very low (2%) while his younger sibling was alive; with her death, grooming with Flo increased to 9% of the total time observed.

Grooming bouts in the early years of life averaged approximately 2.5 minutes each (the same as the suckling bouts). Then, as suckling ended, the grooming bouts increased greatly in length (see Table 5) as did the suckling bouts. The frequency of the grooming bouts generally remained the same as that of the bouts of suckling and/or grooming preceding the end of suckling (see Table 6). During the final year of suckling, grooming bouts for Flo and Flint averaged 4.3 minutes, Miff and Moeza 4.5, Passion and Pom 4.7, Athena and Atlas 6.6, Melissa and Goblin 7.2, and Nope and Mustard 11.7. During the months immediately following the last observation of suckling, the average number of minutes in a grooming bout remained high. The

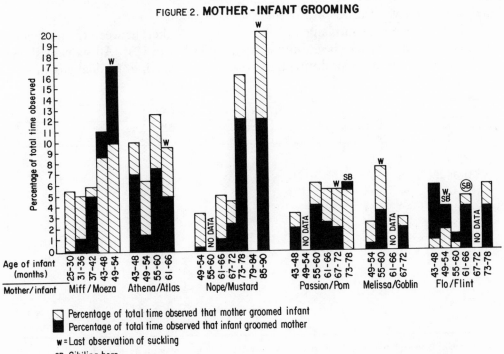

FIGURE 2. **MOTHER-INFANT GROOMING**

□ Percentage of total time observed that mother groomed infant
■ Percentage of total time observed that infant groomed mother
w = Last observation of suckling
SB Sibling born
Ⓢ️Ⓑ️ Sibling dies

TABLE 10 / NUMBER OF CONTACT BOUTS (GROOM AND/OR SUCKLE) PER HOUR OF OBSERVATION

AGE IN MONTHS	MOEZA	POM	MUSTARD	ATLAS	GOBLIN	FLINT
1–12						
13–24						
25–36	2.2					
37–48	2.1	1.9		1.7		1.8
49–60	2.7			1.5	0.9	0.6
61–72		1.3	1.8	1.2		0.7
73–84		1.0	1.7			
85–96			1.3			
within 4 months after the end of suckling (no younger sibling present)	2.4	1.0	1.1	1.1	0.6	0.7

number of bouts per hour observed ranged from 2.4 (Moeza) to 0.6 (Goblin) (see Table 10). Pom's and Flint's grooming bouts with their mothers decreased in length and in frequency with the birth of their younger siblings.

Mutual grooming (simultaneously grooming each other) between the mother and infant increased with age, beginning in the infant's fourth year. All six pairs showed an increase in this behavior during their final year of suckling. Mutual grooming, in

TABLE 11 / MOTHER-INFANT GROOMING

AGE IN MONTHS	MOEZA		ATLAS		MUSTARD		POM		GOBLIN		FLINT	
	%	# bts	%	# bts	%	# bts	%	# bts	%	# bts	%	# bts
25–30	91	11										
31–36	96	28										
37–42	97	33										
43–48	99	84	92	26			98	44			100	15
49–54	89	28	100	10							100	79
55–60			82	11					81	37	100	52
61–66			94	18	94	16	92	13			94	49
67–72					93	15	88	26				
73–78					86	7	84	55				
79–84												
85–90					77	13						

Of the total number of grooming bouts that occurred within a minute after an approach to within arm's reach by either the mother or the infant, the above percentages are infant approaches mother.

TABLE 12 / MOTHER-INFANT GROOMING: PERCENTAGE OF BOUTS THAT BEGAN WITH THE INFANT GROOMING THE MOTHER

	MOEZA		ATLAS		MUSTARD		POM		GOBLIN		FLINT	
AGE IN MONTHS	%	# bts	%	# bts	%	# bts	%	# bts	%	# bts	%	# bts
25–30	18	11										
31–36	32	28										
37–42	66	32										
43–48	76	83	33	26			66	44			80	15
49–54	86	28	0	9							80	79
55–60			50	8					27	37	83	52
61–66			41	17	25	16	23	13			65	49
67–72					29	14	32	25				
73–78					0	7	76	51				
79–84												
85–90					46	13						

particular, appeared to be very reinforcing to the mother-infant relationship during the time of stress just as lactation ceased and suckling ended. Often if the mother stopped during a mutual session, the infant would whimper, looking into her face, until she resumed. With Goblin and Melissa, mutual grooming was not recorded until two months before suckling ended, when 44% of the mother-infant grooming was mutual; this percentage fell to 17% during the year following weaning. Mutual grooming decreased with the advent of a younger sibling. In Pom's and Flint's cases, where there were data available for the months after the birth of a sibling, the mutual grooming decreased to nil.

Almost every bout observed for the six infants in this study began after an approach by the infant to within arm's reach of the mother (see Table 11). Three of the infants—Pom, Flint, and Moeza—consistently (during each month observed) did over 80% of the approaching with grooming following within a minute of the approach. Nope, Athena, and Melissa, mothers of three of the males, did more approaching than the two mothers of the females. These three mothers also began the grooming in more than 50% of the bouts observed each month (see Table 12). Flo did not, until she had recovered from her illness after the death of her infant daughter, at which time she, too, began grooming her son Flint, first, over 60% of the time observed, as well as approaching him more often with the intention of grooming. In general, the two female infants were doing more of the initial grooming than were the males (with the exception of Flint). There appear to be differences in grooming patterns between mothers and female infants and mothers and male infants, just as there are individual differences across the mother-infant dyads; however, a larger sample size is necessary before sex differences can be determined.

NESTING

Another behavior affected by the cessation of suckling and the advent of a younger sibling is evening nesting. Until the end of suckling, the infants spent the night in their mothers' nests. However, during the final year of suckling all six of the infants were building their own evening nests, although they were not recorded as actually sleeping in them until after the birth of a sibling. The data on the six subjects of this study did not indicate when the weaned infants finally and regularly began sleeping away from their mothers. Observers at Gombe are watching for this behavior among infants presently undergoing weaning. Flo was the only mother observed trying to eject her (weaned) son from her nest (which she was already sharing with her infant daughter), and many tantrums were heard in the evening hours at Gombe. After Flame's death, however, Flo no longer rejected Flint as often. Flint nested with Flo for over 4 years after weaning from the nipple; none of the other infants was observed to do this.

■ SUMMARY

The mother-infant relationship among chimpanzees is one of long duration and great intensity. The young chimpanzee remains with the mother for 9–12 years, often with her and a sibling as its only companions. The infant suckles for approximately 5 years until the mother becomes pregnant again, during which time very few interactions except play occur with others. The mother begins weaning the infant from the nipple by the infant's second year of life. At first, rejections are infrequent, but they intensify through time, especially with the resumption of the mother's estrous cycle, generally occurring when the infant is 3–4 years of age. Chimpanzee mothers and infants both play active roles in the weaning process which is typically nonaggressive. (For an interesting perspective on the weaning process, see Trivers 1974.) Infants may regress to infantile behaviors such as long suckling bouts, ventral riding on the mother, and increased whimpering. Individual differences exist among mothers and infants. The parity, experience, age, and personality of the mother are important factors, as are the infant's age, experience, and personality. There appear to be differences in behavior between male and female infants, and in the mothers towards male and female offspring. As the infant approaches the end of suckling, it goes through a period of depression during which time contact with the mother increases and interactions with peers decreases; the depression usually extends through the birth of the younger sibling and may last as long as a year or more.

■ ACKNOWLEDGMENTS

I wish to express my deepest gratitude to Dr. Jane Goodall for her financial and moral support which made this study possible and for her many helpful suggestions on the preparation of the manuscript. I am greatly indebted to Dr. David A. Hamburg for his interest and support and to Dr. Helena Kraemer for her assistance with the analysis of the data. I thank all those researchers who have contributed to the long-term study at the Gombe Stream Research Centre and whose data I have used in this study. I also wish to thank Dr. Robert A. Hinde for his advice on the writing of this paper. This work received partial support from USPHS grant no. RR00169. Last, but not least, I pay tribute to the chimpanzee mothers and infants who are the subjects of this study.

■ REFERENCES

Anthoney, T. R. 1968. The ontogeny of greeting, grooming and sexual motor patterns in captive baboons (superspecies *Papio cynocephalus*). *Behav.* 31:358–372.

Baldwin, J. D. 1969. The ontogeny of social behavior of squirrel monkeys (*Saimiri sciureus*) in a semi-natural environment. *Folia Primat.* 11:35–79.

Burton, F. D. 1972. The integration of biology and behavior in the socialization of *Macaca sylvana* of Gibralter. In *Primate socialization*, ed. F. E. Poirier, pp. 29–62. New York: Random House.

Conaway, C. H., and Koford, C. B. 1965. Estrous cycles and mating behavior in a free-ranging band of rhesus monkeys. *Jour. of Mammal.* 45:577–588.

DeVore, I. 1963. Mother-infant relations in free-ranging baboons. In *Maternal behavior in mammals*, ed. H. L. Rheingold, pp. 305–335. New York: John Wiley & Sons.

DeVore, I., and Konner, M. J. 1970. Infancy in hunter-gatherer life: An ethological perspective. In *Ethology and psychiatry*, ed. N. F. White, pp. 113–247. Toronto: University of Toronto Press.

Döring, G. K. 1969. The incidence of anovular cycles in women. *J. Reprod. Fert. Suppl.* 6:77–81.

Gilbert, C., and Gillman, J. 1951. Pregnancy in the baboon (*Papio ursinus*). *S. Afr. J. Med. Sci.* 16:115–124.

Graham, C. E. 1970. Reproductive physiology of the chimpanzee. In *The chimpanzee* vol. 3, ed. G. H. Bourne, pp. 183–220. Basel: S. Karger.

Hall, K. R. L. 1965. Behavior and ecology of the wild Patas monkey, *Erythrocebus patas*, in Uganda. *J. Zool. Lond.* 148:15–87.

Hall, K. R. L., and DeVore, I. 1965. Baboon social behavior. In *Primate behavior: Field studies of monkeys and apes*, ed. I. DeVore, pp. 53–110. New York: Holt, Rinehart & Winston.

Hansen, E. W. 1966. The development of maternal and infant behavior in the rhesus monkey. *Behav.* 27:107–149.

Hinde, R. A.; Rowell, T. E.; and Spencer-Booth, Y. 1964. Behaviour of socially living rhesus monkeys in their first six months. *Proc. Zool. Soc. Lond.* 143:609–649.

Hinde, R. A., and Spencer-Booth, Y. 1967. The behaviour of socially living rhesus monkeys in their first two and a half years. *Anim. Behav.* 15:169–196.

Itani, J. 1959. Paternal care in the wild Japanese monkey, *Macaca fuscata*. *Primates* 2:61–93.

Jay, P. 1965. The common langur of North India. In *Primate behavior: Field studies of monkeys and apes*, ed. I. DeVore, pp. 197–249. New York: Holt, Rinehart & Winston.

Kaufman, I. C., and Rosenblum, L. A. 1969. The waning of the mother-infant bond in two species of macaque. In *Determinants of infant behaviour* vol. 4, ed. B. M. Foss, pp. 41–59. London: Methuen & Co.

Lawick-Goodall, J. van. 1968. The behaviour of free-living chimpanzees of the Gombe Stream Reserve. *Anim. Behav. Monographs* 1:161–311.

———. 1967. Mother-offspring relationships in free-ranging chimpanzees. In *Primate ethology*, ed. D. Morris, pp. 287–346. Chicago: Aldine.

———. 1973. The behavior of chimpanzees in their natural habitat. *Amer. J. Psychiatry* 130:1–12.

Lindberg, D. G. 1971. The rhesus monkey in North India: An ecological and behavioral study. In *Primate behavior: Developments in field and laboratory research* vol. 2, ed. L. A. Rosenblum, pp. 1–106. New York: Academic Press.

Merrick, N. 1972. Notes from the Gombe Stream Research Centre field data record, Stanford University.

Mitchell, G. D. 1968. Attachment differences in male and female infant monkeys. *Child Development* 39:611–620.

Poirier, F. E. 1968. The nilgiri langur (*Presbytis johnii*) mother-infant dyad. *Primates* 9:45–68.

Rahaman, H., and Parthasarathy, M. D. 1969. Studies on the social behavior of bonnet monkeys. *Primates* 10:149–162.

Ransom, T. W., and Rowell, T. E. 1972. Early social development of feral baboons. In *Primate socialization*, ed. F. E. Poirer, pp. 105–144. New York: Random House.

Rosenblum, L. A., and Kaufman, I. C. 1967. Laboratory observations of early mother-infant relations in pigtail and bonnet macaques. In *Social communication among nonhuman primates*, ed. S. A. Altmann, pp. 32–42. Chicago: The University of Chicago Press.

Rowell, T. E. 1972. *Social behaviour of monkeys*. Harmondsworth: Penguin.

Saxton, G. A., Jr., and Serwadda, D. M. 1969. Human birth interval in East Africa. *J. Reprod. Fert. Suppl.* 6:83–88.

Schaller, G. B. 1963. *The mountain gorilla: Ecology and behavior*. Chicago: The University of Chicago Press.

Simonds, P. E. 1965. The bonnet macaque in South India. In *Primate behavior: Field studies of monkeys and apes*, ed. I. DeVore, pp. 175–196. New York: Holt, Rinehart & Winston.

Southwick, C. H.; Beg, M. A.; and Siddiqi, M. R. 1965. Rhesus monkeys in North India. In *Primate behavior: Field studies of monkeys and apes*, ed. I. DeVore, pp. 111–159. New York: Holt, Rinehart & Winston.

Struhsaker, T. T. 1971. Social behavior of mother and infant vervet monkeys (*Cercopithecus aethiops*). *Anim. Behav.* 19:233–250.

Sugiyama, Y. 1965. Behavioral development and social structure in two troops of hanuman langurs (*Presbytis entellus*). *Primates* 6:213–247.

———. 1969. Social behavior of chimpanzees in the Budongo Forest, Uganda. *Primates* 10:197–225.

———. 1972. Social characteristics and socialization of wild chimpanzees. In *Primate socialization*, ed. F. E. Poirier, pp. 145–163. New York: Random House.

Tanaka, T.; Tokuda, K.; and Kotera, S. 1970. Effects of infant loss on interbirth interval of Japanese monkeys. *Primates* 11:113–117.

Trivers, R. L. 1974. Parent-offspring conflict. *Amer. Zool.* 14:249–264.

Tutin, C. E. G., and McGrew, W. C. 1973. Chimpanzee copulatory behaviour. *Folia Primat.* 19:237–256.

Zuckerman, S. 1931. The menstrual cycle of the primates. Part 4. Observations on the lactation period. *Proc. Zool. Soc. Lond.* 1931:593–602.

9 | Socialization and Object Manipulation of Wild Chimpanzees

W. C. McGREW

▪ INTRODUCTION

The purpose of this paper is to focus upon the interface between a social process, socialization, and an individual activity, object manipulation. The subjects are members of a wild chimpanzee population of about 125 individuals inhabiting the Gombe National Park on Lake Tanganyika in northwestern Tanzania. The paper will cover four main areas; first, a specific type of object-oriented behavior, water contact avoidance, will be considered in detail to show how conclusions about proximal mechanisms of social transmission can be drawn in the field; second, preliminary norms for object manipulation by older chimpanzee infants will be presented; third, the performance of a specific tool-use activity, leaf sponging, will be related to properties of play behavior; finally, the ontogeny of two complex tool-use patterns, termite fishing and ant dipping, will be compared. Both involve the use of vegetational tools to feed upon social insects, but the techniques and their acquisition differ greatly. Throughout, the discussion will emphasize the implications of these findings as they relate to social traditions in nonhuman primates.

Socialization may be defined as the process whereby a developing individual interacts with and relates to other group members who contribute to its eventual adult social competence. In the case of the wild chimpanzee, the primary socializing agent is the mother; for at least the first five years of life the vast majority of the infant's social interactions are with her. This period of prolonged dependency insures that the infant is exposed to all of her activities at close range, including those which are community traditions. But the infant's acquisition of these skills is not by some simple sponge-like process of absorption, in spite of what is sometimes implied in the nonpsychological literature. Beck (1975), in his comprehensive review paper, has outlined several ways by which social traditions may be transmitted. Trial-and-error learning may occur from merely being placed repeatedly in certain contexts in which environmental contingencies may act on the individual. Observational (usually social) learning may act through social facilitation, stimulus enhancement or imitative copying when two or more individuals are simultaneously present. Truly insightful learning apparently may occur in situations in which the individual is temporarily free of preoccupying environmental pressures, e.g., the unweaned infant. The elaboration of the behaviors may be achieved by repetition and by both stimulus and response generalization. Therefore, socialization, as it relates to the acquisition of social traditions, is not limited to intentional "teaching" interactions (which are probably uncommon in nonhuman primates)[1] but is involved in all the behavioral "results" of social relationships.

There are several reasons for focusing on object manipulation. Apparently all primates (and to some extent, many mammals) frequently engage in this activity during the course of maturation. This propensity also manifests itself socially in prolonged bouts of social and self-grooming exhibited by many higher primate species. That it is of eventual basic importance to survival seems likely. Unlike more specialized herbivores, insectivores, carnivores, etc., the more omnivorous primates depend for nutrition upon the assimilation of many handable objects. Among widespread human cultures the ubiquity of object manipulation in the form of "toys" makes it likely that this activity is of basic phylogenetic significance. This relates to a second point, that of motivation. Exploratory behavior as an expression of curiosity in immature animals is commonly directed toward objects, especially novel ones. As Hutt (1966) has pointed out, *exploration* becomes *play* when "What does this object do?" changes to "What can I do with this object?" Object play in chimpanzees, a nonhuman species apparently possessing the rudiments of self concept (Gallup 1970), is particularly complex (Loizos 1967). Finally, object manipulation may develop into tool use. The manufacture and use of implements has traditionally been considered as a somewhat advanced intellectual ability, until recently restricted to hominids.[2] Goodall (1964) was the first to show that wild chimpanzees possessed a varied and complex tool-use repertoire, and numerous later field workers have augmented these findings (Nishida 1973; McGrew 1974). In all known cases, both in the wild and in captivity, spontaneously developing chimpanzee tool-use behaviors began as object manipulation in infancy (e.g., van Lawick-Goodall 1968; Menzel 1972).

■ NONHUMAN PRIMATE "CULTURE"?

The debate over whether or not culture is uniquely human continues (Lancaster 1975; Weiss 1973). This problem was first seriously raised by primatologists studying free-ranging Japanese monkeys (*Macaca fuscata*), and their findings remain the single most comprehensive body of literature (see review in Itani and Nishimura 1973). It is spurred on by continuing discoveries from field studies of nonhuman primate behavior and ecology (Kummer 1971), as well as advancing psychological research in the laboratory on the learning processes involved in complete cognitive processes (Menzel 1973b). As is also currently happening with another important "human" phenomenon, language, the discussion often follows traditional discipline lines, with concentration on arguments over qualitative versus quantitative differences. The results are bound to be fruitful, if for no other reason than that they force careful re-examination of long-held assumptions (and presumptions) and imprecise definitions. In the case of culture, the more recent development of multiple field studies of the same species in different habitats (e.g., the olive baboon, *Papio anubis*) raises the possibility of truly cross-cultural studies of nonhumans.

It is this author's view that the problem of nonhuman culture (or preculture, or protoculture, or infraculture), is largely semantic. No one would deny now that social traditions (in the sense of Kummer's [1971] "behavioral modification induced by the social environment") can and do exist in free-ranging nonhuman primate populations.[3] It can be further postulated that when social traditions are passed between generations the process of socialization is involved. The unanswered questions lie in the area of the proximal mechanisms of social traditions, i.e., how novel behaviors are acquired, disseminated, and transmitted (Beck 1974). Here several practical problems arise. First, it is difficult to examine experimentally such questions in nature without disruptive human intervention. Natural conditions rarely hold the appropriate variables constant. The Japanese research workers have succeeded in such investigations, but only with monkey populations well habituated to human presence and willing to interact "unnaturally" with humans (e.g., Itani and Nishimura 1973; Kawai 1967). A second problem arises with sampling. In addition to the controlled environment, continuous observation can be maintained with captive subjects, and the complete development and elaboration of new behaviors can be followed from inception. Key incidents, innovations, exchanges, etc., can be noted with certainty, e.g., ladder-using in a captive chimpanzee group (Menzel 1972, 1973; McGrew et al. 1975). In the wild no observer can count on this. All the tool-use traditions exhibited by wild chimpanzees in the Gombe population were apparently well established when studies began, and none have been added in the 15 years since. This relates to a third problem, which arises from the prolonged maturation and wide birth intervals of nonhuman primates in nature. If a field worker is unlikely to see the origin of a social tradition on a group or deme basis, then he or she must look at its acquisition and development in individual ontogeny. For higher primates, this means years of long-term longitudinal data. As most field workers are unable to spend unbroken periods of many years at one site, they are dependent upon on-going cooperative projects in which cumulative data are available.

■ METHODS

The wild chimpanzee (*Pan troglodytes schweinfurthii*) subjects of this paper have been studied since 1960, when Jane Goodall initiated research at Gombe. The development of the research project that emerged, the Gombe Stream Research Centre (GSRC), has been described elsewhere (van Lawick-Goodall 1968, 1973).

Field workers at Gombe follow the chimpanzees' daily activities at a distance of 5–10m, which is only possible because of the chimpanzees' tolerance. Only a small minority of these observations now occur in an artificial provisioning area (Camp); most take place in the surrounding forest. If at any point it tires of human company, the chimpanzee merely leaves, and no human follower can avoid being left behind if the subject is sufficiently motivated. Such observation periods ("follows") focus on a target individual chimpanzee and may last up to 13 hours, i.e., from nest to nest.

Observers work in pairs: one records details of specific behavior types, such as tool use, sex, agonism, etc., and the other records group composition and movements and the target individual's feeding. Virtually all data are taken in time-sampling check sheet form, either directly onto paper or indirectly by cassette tape recording and later transcription. All quantitative records are supplemented by descriptive notes and qualitative impressions.

Many of the findings presented are taken from the extensive back records contributed over the years by many Gombe observers. Most are accumulated descriptive notes, summaries, check sheets, and photographs maintained in the Gombe files. Some comprise standardized records forming a pool to which all GSRC researchers are obliged to contribute and from whence all may draw freely. Just one of these instruments, the Travel-and-Group-Chart, makes available thousands of hours of observation on feeding, travel, and social associations. In this sense, all Gombe findings are collective, and acknowledgment is taken for granted.

▪ WATER CONTACT BEHAVIOR

The process of imitation in the acquisition of new behaviors is difficult to demonstrate empirically in nonverbal organisms, especially immature ones, even under experimentally controlled laboratory conditions. The necessary presence of a model means that possible confounding social variables must be considered. The lower-order cognitive processes involved in social facilitation ("contagion") and social stimulus enhancement may be responsible for effects prematurely interpreted as imitation (Wechkin 1970). In this case, the model's presence is necessary, but not sufficient, for the human observer to classify the subject's behavior as imitative. Similarly, the demonstrator's behavior may be completely and misleadingly unnecessary, such that the subject is acquiring the new behavior by ordinary trial-and-error according to the law of effect. The fact that two individuals in close physical proximity perform similar actions in close temporal proximity does not prove any interaction between them. (Granted, if early responses are shown by an experienced and proficient subject, and later responses by an initially naive and inept subject, the interpretation is strengthened.) Finally, some combination of the two types of processes may be operating. But all of these do not demonstrate the existence of imitation, a higher-order cognitive process which requires the ability to acquire without direct instrumental reinforcement, behaviors which are not intrinsically stimulus- or socially-bound.

The categorization of novel behavioral acquisition as imitative is further complicated if the subject is immature. The occurrence of a new behavior may be due to maturational factors and be independent of specific individual social experience. To attempt to sort out the relative contributions of the genetic and experiential factors requires *Kaspar Hauser* experiments, but these are virtually impossible to do in nature. The chimpanzee "pant hoot" vocalization does not enter the behavioral repertoire until middle infancy. Its initial performances are rudimentary in form and usually coincident with adult performances, especially those of the mother. Does this mean that pant hooting is acquired by imitation? Studies of chimpanzees reared in species-isolation from birth refute this, as such individuals often pant hoot spontaneously.

This does not mean that strong circumstantial evidence for imitative processes cannot be produced for the acquisition of behaviors by wild chimpanzees. Consider the chimpanzee's avoidance of standing or running water on the ground, a phenomenon recently reviewed by Angus (1971). She stated that all field studies of wild chimpanzees report them to be extremely cautious of such water.[4] Fifteen years of study of the Gombe chimpanzees supports this conclusion of hydrophobia. All individuals avoid contact with the waters of Lake Tanganyika, the numerous small streams which flow down the valley bottoms, and even rain puddles along paths or in other bare areas. Chimpanzees have been seen a few times to drink from the lake shore but not to enter the water. They cross streams by leaping quadrupedally, using stepping stones or fallen log bridges, or by climbing over arboreally (see Goodall 1963: 294); they walk around puddles or step over them. These patterns are exhibited by independently moving individuals (from approximately four years old onwards). Such chimpanzees rarely immerse even their fingers in water, even when crouching beside a stream to drink. Wading or any other extreme water contact behavior has not been seen. Dependent youngsters are almost always carried over water by their mothers. If an infant is inadvertently left behind on one side of a stream when its mother crosses to the other, its response is highly predictable; it emits distress vocalizations which may begin as soft whimpers and escalate to loud screams. It does not cross the stream, even if it is narrow enough to be easily jumped or if an overhead arboreal avenue exists which it could easily climb; instead the mother recrosses the stream and retrieves the infant or reaches across to pull it to her belly. In short, the infant shows extreme fear of the water.

However, the chimpanzee infant shows no such extreme response when similarly separated from the mother by dry ground. A milder response may sometimes occur when mother and infant are separated in the trees by intervening empty space. Here the mother usually waits while the infant approaches and regains contact, often by employing precarious routes through the canopy high above the ground. Finally, there are exceptional occasions when even infant chimpanzees may show water play; these will be discussed below.

A simple explanation often put forward in older accounts is that chimpanzees have an "instinctive" fear of water. Angus (1971) says that this remains a moot question, as it has not been systematically investigated. Wind (1974) has linked ape hydrophobia to relatively low buoyancy and unfavorable anatomical features selected for in hominoid phylogeny. This may be true, but this explanation seems insufficient in view of accumulated observations of chimpanzees in captivity. Chimpanzees at Chester Zoo escaped repeatedly by wading to freedom (Mottershead 1963). Chimpanzees in a large outdoor enclosure at the Delta Regional Primate Research Center rolled and splashed in large rain puddles (personal observation). Pygmy chimpanzees in the San Diego Zoo repeatedly stood waist-deep, dunked their heads and drenched each other at their wading pool (Tutin, personal communication). A young chimpanzee at the University of Oklahoma Institute for Primate Studies immerses itself up to neck level in the moat surrounding her island home (Savage, personal communication). Two chimpanzees in the Holloman Chimpanzee Consortium waded into the water-filled restraining moat in response to threats from dominant animals (Angus 1971). Numerous other examples of water contact behavior by captive chimpanzees could be added, although this does not necessarily mean that they exhibit a generalized boldness toward water. A recurring factor seems to be that these individuals have been hand-raised by humans in captivity rather than mother-reared in the wild. Angus (1971) points out that no chimpanzee has been seen to swim (although this has not been systematically taught), and that chimpanzee drownings in captivity are an unfortunately all-too-common occurrence in moated enclosures. However, there seems to be a significant difference in water contact behavior of wild- and captive-reared chimpanzees, so that innate fear of water seems unlikely.

It also seems possible to exclude the explanation that Gombe water is somehow more dangerous than water in captivity. Under normal conditions, no Gombe stream exceeds 3m in width and 1m in depth (see Goodall 1963: 284); the vast majority of Gombe's streams are less than 1m wide and 10cm deep. The water is relatively fast flowing but clear, and the stream beds are thick with protruding rocks and debris and overhanging vegetation. After occasional heavy and prolonged rains, the streams may be temporarily swollen, murky, and choked with moving objects. They are then clearly dangerous even to humans, but this temporary state is easily discernible from normal conditions. There are a few isolated waterfalls and rapids, but none require more than a 25m detour to circumvent (which the chimpanzees do). The latter water courses are sensorially impressive, but they are constantly dangerous, permanently situated, and easily shunned. The lake water at Gombe is also normally clear and free of predators, and the lake shore comprises sand or pebble beach. The olive baboons there regularly enter the water and swim, even as youngsters. All types of Gombe surface water are *Bilharzia*-free, hard-bottomed, and devoid of entangling aquatic vegetation. In summary, it is difficult to see why Gombe waters should be perceived and treated as dangerous by Gombe chimpanzees.

Are, perhaps, the chimpanzees aware of dangers unknown to the human observer? Considerable negative evidence exists on this point. In tens of thousands of observation hours at Gombe, no chimpanzee infant drowning has been recorded, nor has near drowning nor even a near dangerous experience with water. Of course this could have happened when the chimpanzees were not being observed by humans, especially if one-trial learning were applicable here. (It is difficult to see how such a sampling bias would operate. If anything, one might expect any disturbance from human presence to exaggerate the rate of such accidents.) In thousands of observed stream crossings, nothing approximating a negatively reinforcing experience has been delivered to an infant chimpanzee. Moreover, no aversive experience has been observed to occur from which an infant chimpanzee might have vicariously received such reinforcement. Additionally, water has not been implicated in any of the score of known deaths of Gombe chimpanzees, while for example, falls from heights and violent death are known causes of mortality. These results concur with those obtained in other field studies of chimpanzees in east, central and west Africa.

Perhaps avoidance of water contact by wild chimpanzees is not occasioned by fear of danger but by discomfort through stimulus generalization from rainfall. Direct evidence exists on this point from Gombe and other field study sites. Wild chimpanzees almost always sit impassively through prolonged heavy rains which soak them to the skin (Goodall 1963); they only rarely seek shelter under thick vegetation although such shelter is available. In contrast, the human observers and probably sympatric wild baboons of Gombe make more frequent use of natural shelters of vegetation. Captive chimpanzees kept in indoor-outdoor housing take advantage of artificial shelters, thus indicating the species' ability to appreciate the principle involved. Young Gombe infant chimpanzees may take shelter in the mother's ventral embrace, but this drops in frequency long before they begin regularly to cross streams alone. Wild chimpanzees *do* show discomfort at being wetted by rain; they vigorously shake themselves and rub their bodies against vertical tree trunks. But hypothermia is not a problem in these tropical climates and the chimpanzees do not show fear of rainfall. In light of all this, it seems unlikely that the discomforting effects of water contact would be sufficient to account wholly for the extreme aversion to surface water.

Therefore, it is difficult to see how a chimpanzee growing up in Gombe could acquire the fear of surface water by a gradual trial-and-error process. According to Beck (1975), trial-and-error learning is the main process whereby nonhuman primates acquire tool-using behaviors, and the same line of reasoning would seem to apply to other novel behaviors oriented toward specific aspects of the physical environment. Its exclusion in the case of water contact avoidance leaves three types of observational learning remaining: (1) social facilitation, (2) stimulus enhancement and (3) imitative copying. The first applies to the contagious spread of the same behavior between individuals on a given occasion, e.g., Beck's (1975) example of baboon alarm calls. The process referred to is not one of acquisition, but of synchronized performance. It is not applicable to chimpanzee water contact avoidance, as the infant's behavior (i.e., refusal to approach, fearful expressions and vocalizations) differs from the mother's (i.e., stone-stepping, arboreally climbing over, impassivity).

Stimulus enhancement is almost certainly involved in the chimpanzee's acquisition of water avoidance. Even before the infant breaks physical contact with its mother, it is aware of, and interacts with, the inanimate surroundings. It seems likely that a chimpanzee infant could not avoid forming associations between that fraction of the substrate which is aqueous and the mother's special orientation to it. But as Beck (1975) points out, stimulus enhancement alone is insufficient. It only focuses attention on certain aspects of the environment, but the acquisition of the appropriate response requires more. This usually means the linking of socially originating stimulus enhancement with individual trial-and-error learning (see below in connection with tool use in obtaining social insects for food), but for the reasons given above, it seems unlikely to be operating in the transmission of water contact avoidance behavior.

Do experienced chimpanzees perhaps intentionally teach naive individuals to avoid the water? Leaving aside the problems of divining intentionality in nonverbal organisms, it seems unlikely that tuition is responsible. No chimpanzees have been seen to be socially punished for making contact with water, as such contacts are virtually absent, even in the young but mobile infants. Nor do older chimpanzees pull or herd younger ones away from water. Perhaps they would, if necessary, but wild chimpanzees do not enter the water even if apparently free to do so. No chimpanzee has been seen to be rewarded by another for performing stone-stepping, fallen log bridge-using, overhanging vegetation-climbing or simple frog-style leaping. Chimpanzee mothers *do* approach and reinitiate contact with their infants when the latter refuse to cross streams. This physical contact may be pleasurably reinforcing to any disinclination to move toward water, but it does not explain the appearance of later fearful circumventing behaviors oriented to water. Finally, water-avoiding chimpanzees have not been seen to mold the appropriate behaviors in naive individuals. Mothers do not lead infants across fallen log bridges nor urge them to climb across appropriate overhanging branches. Molding is known to be an efficient method of imparting novel behavior patterns to chimpanzees (Fouts 1972), but its absence is conspicuous in thousands of observed Gombe stream crossings.

Insightful problem solving by novel behavior vis-à-vis surface water (e.g., bridge building) has not been seen at Gombe or at any other chimpanzee field site. Perhaps the closest to this was an incident in which an adult female used her 9-year-old daughter's back as a "stepping stone"; as the daughter crouched beside a stream to drink, the mother stepped onto her back and then across to the opposite bank (Busse, personal communication). This would seem to be within the intellectual capacity of a proficient tool-using species, but there seems to be little environmental pressure to do so. There is limited suggestive evidence that captive chimpanzees exposed to the human use of boats may understand the rudiments of the principles of flotation. But until wild chimpanzees exhibit some such innovative behavior, it is useless to speculate about its genesis.

What remains by elimination, as a process of acquisition, is imitative copying.[5] That chimpanzees possess such abilities has been shown in studies of chimpanzees reared in human homes (e.g., Hayes and Hayes 1952). The other potential explanations for the consistent performance of water contact avoidance behavior—"instinct," trial-and-error, social facilitation, stimulus enhancement, tuition—cannot adequately account for the phenomenon. This explanation cannot be considered as definitive, if for no other reason than that it is based on argument by exclusion and negative evidence. But the cumulative circumstantial evidence for imitative copying would seem to be stronger here than for any other social tradition of the chimpanzee so far noted in nature.

That wild chimpanzee infants have ample exposure to models for stream crossing is empirically demonstrable. The author observed six chimpanzee mothers with eight offspring of 0–5 years of age for a total of 505 hours. During this time 145 stream crossings were seen, a mean of one crossing every 3.5 hours. Individual means showed little variation, ranging from 3.03 to 3.99 hours. The average chimpanzee waking day is about 12 hours long. An average of 3 crossings daily over 5 years gives a total of over 5,000 probable opportunities for a chimpanzee infant to observe its mother's behavior in stream crossing. (By implication this renders unlikely the explanation that chimpanzee hydrophobia is a specific instance of fear of the unfamiliar.)

Why should chimpanzees have begun avoiding water in the first place? Under certain circumstances most mammals will enter the water, even those mistakenly presumed not to do so, e.g., large felines. For chimpanzees, a number of factors may be involved. According to Angus (1971) they have a very high percentage (91%) of body weight as lean mass, making it difficult for a chimpanzee to float. Presumably, selection for this extra musculature (which perhaps enables a dual terrestrial-arboreal habitat exploitation) was achieved at the expense of amphibiousness. Such an hypothesis could be tested by correlating lean mass percentage with water use in nature across a number of primate species. Nonhuman primates readily show one-trial learning, so that one bad experience with drowning or aquatic predation might be sufficient to instill fear of the water in the others; Maxim and Buettner-Janusch (1963) found this to be the case with a single shooting in a wild baboon troop. Finally, chimpanzee society seems to be a very conservative one, in contrast to the usual generalizations that one hears about the flexibility and adaptability of nonhuman primates. For example, Gombe chimpanzees will not eat proffered mangoes, a prized chimpanzee food elsewhere in Africa, although these have been growing in the Park for decades. A social tradition once acquired might long outlive its usefulness. This is totally speculative, but perhaps the results of ongoing studies of captive-reared chimpanzees being rehabilitated to natural habitats (e.g., Hladik 1973) will eventually provide clues to its plausibility.

■ NORMS OF OBJECT MANIPULATION

It is commonly stated in the primatological literature that primates are very manipulative creatures in comparison to the rest of the mammals. It is pointed out that they have fleshily-padded, clawless digits and varying degrees of digital opposability, especially that involving the thumb. This results in a variety of potential grips, the most important apparently being the "power" and "precision" grips as distinguished by Napier (1956). From the psychological viewpoint, laboratory workers such as Harlow et al. (1950) have postulated the existence of a manipulative drive, at least in rhesus macaques (*Macaca mulatta*). This is based on the experimental demonstration that young monkeys will learn to perform a sequence of manipulative actions without any primary drive reinforcement. This adaptive blend of form (the morphology of the primate hand) and function (the varied manipulation of useful objects) is intuitively satisfying and apparently clear. It commonly features in speculation about hominid evolution because of its obvious relevances to questions about the origins of bipedalism, cortical structures, fine motor skills, etc. (Napier 1962).

Yet the fact remains that the vast majority of studies on nonhuman primate object manipulation have been done in the unnatural conditions of captivity. Most of these have used man-made objects presented in contrived situations to subjects living in conditions of sensorimotor deprivation. The few studies of object manipulation by wild nonhuman primates (e.g., Menzel 1966) have focused on responses to strange, artificial objects. Baseline norms of normal object manipulation by wild primates appear to be lacking.

The willingness to accept behavioral norms derived only from captive subjects is less and less tenable, as shown by the cumulative results of field studies. Two common patterns of behavior involving objects which are exhibited by captive anthropoids, coprophagy and depilation, are rarely shown by their free-living counterparts; both these excesses are often attributed to the boredom of captivity. It is possible that other types of object manipulation by captive primates may be just as atypical. Similarly, the dangers of inferring function from form are manifest. Physical anthropologists who until recently classified the great apes as brachiators based on the morphology of the upper limb and shoulder girdle have been forced to revise their conclusions. Observations of chimpanzee locomotion in a variety of natural environments have shown that brachiation is only exceptionally used. Similar cautions might apply to armchair inferences about primate object manipulation based on hand morphology.

With this in mind, the author made observations of three male and three female older chimpanzee infants (age 1 ½ to 5 years). (See Table 1.) In practical terms, this meant devoting two columns of a field check sheet (*KIDS*) to some simple measures of the treatment of objects. The rest of the check sheet dealt with the primary focus of study, infant socialization. In one column were coded the inanimate objects with which the subject made physical contact in certain defined ways. (Animate objects such as other chimpanzees were dealt with elsewhere.) In the second column were recorded the codes for four types of treatment: manipulation, transport, mouthing, and ingestion. Only the first will be considered here, and it was defined as prehensile (i.e., palmar) contact by at least part of one or both hands with an object. Thus, the questions posed were: How often do older chimpanzee infants handle objects? What types of objects do they manipulate? Are there individual or other differences in object manipulation?

TABLE 1 / INDIVIDUAL INFORMATION ON CHIMPANZEE INFANT SUBJECTS OF *KIDS* STUDY

SUBJECT	SEX	SIBLING RANK	AGE (MOS.)	MONTHS OF OBSERVATION	HOURS OF OBSERVATION	OUT CAMP (PERCENT)	OBSERV-ABILITY (PERCENT)
MZ	♀	1	71–72	2	33	98	91
AL	♂	1	62–69	7	70	91	87
SS	♀	1	33–40	7	66	80	93
PT	♂	1	27–30	3	39	95	87
GM	♀	2	25–29	4	60	86	90
FD	♂	1	18–25	7	100	83	89
	♀♀ = 3			♀♀ = 13			
	♂♂ = 3		x̄ = 39–44	♂♂ = 17	368	x̄ = 87%	x̄ = 90%

The results are based on 368 observation hours collected in 1973–75. Observation sessions averaged just over 2.5 hours in length, and data were taken in 60-second segments using the Hansen frequency method (Altmann and Wagner 1970). Procedural details will be given in a future publication, but it should be noted that no significant differences existed between subjects in amount of time spent outside the feeding area (*Out Camp*). Similarly, the six subjects showed little variation in "observability," either among themselves or over time (see Table 1).

On the average, the six chimpanzee subjects exhibited object manipulation during 75% of the observation minutes. (See Table 2.) This basically indicates how often their hands were nonsocially "busy" and reflects the performance over most of the waking day, as records were balanced over the period from 0730 to 1800 hours. Little individual variation and no age or sex differences emerged. The vast majority of objects in the physical environment which were manipulated comprised one category: 67% of observation minutes included handling of living and flexible, but attached (*in situ*) vegetation. These included the trunks, branches, twigs, leaves, flowers, fruits, bark, etc., growing as parts of trees, shrubs, herbs, grasses, vines, etc. (e.g., see Figure 1). When and if these items became detached and movable, they were classified separately.

An apparent difference emerges if one compares the ratio of manipulated non-attached objects to attached vegetative ones according to sex: male $\bar{x} = .41$ vs female $\bar{x} = .55$. The former include solid foods (fruits, palm nuts, bananas), detached vegetation (stick, leaf, grass), other natural objects (stone, soil), and occasional filched or found artificial objects (cloth, paper). (See Figure 2.) There is also a suggestion of an age difference and greater individual differences. (See Table 3.) The oldest male subject (AL) was weaned and ate much fruit that he processed alone. If one excludes his high monthly rates during the peak milkpod (*Diplorhyncus*) feeding season of May and June, his totals for the other months fall within the other males' range. With a few exceptions, the females appear to show a greater tendency to manipulate all

TABLE 2 / SOME NORMS FOR OBJECT MANIPULATION
BEHAVIOR OF OLDER INFANT CHIMPANZEES

BEHAVIORAL MEASURE	MZ	AL	SS	PT	GM	FD	GROUP MEAN
Overall object manipulation rate (% of min)	79	79	69	75	72	76	75
Manipulation of vegetation rate (% of min)	69	72	62	68	62	71	67
Ratio of manipulation of other objects to vegetation	.67	.53	.51	.37	.58	.33	.48

FIGURE 1. *Manipulation of attached vegetation: a female infant (GM) swings suspended from an oil nut palm frond.* (Photo by C. Tutin)

movable detached objects rather than growing vegetation. Great month-to-month variation reflects the seasonal changes in the type of fruits available.

Such an apparent sex difference must be taken with caution, as the number of subjects (N = 6) is small and the results could be due to individual differences.[6] However, they fit a coherent pattern when considered with other findings. Adult female chimpanzees at Gombe show more frequent tool-use in obtaining insect foods by termite fishing and ant dipping (see below) than do adult males (McGrew in press). Juvenile female chimpanzees and orangutans in captivity interact more frequently with inanimate objects than do males (Nadler and Braggio 1974). Numerous other accounts suggest female superiority in nonhuman primate manipulatory skills, e.g., the catching of thrown objects by Japanese monkeys (Kawai 1967). It is well known in human children that females are superior to males in manual dexterity from early childhood onwards (Garai and Scheinfeld 1968).

Before leaving the subject of object manipulation norms in chimpanzees, it is worth returning to a point made earlier. Given that wild chimpanzee infants spend so

FIGURE 2. *Manipulation of a detached object: a one-year-old male infant (WL) handles and mouths a fragment of oil nut palm frond.* (Photo by C. Tutin)

much time manipulating objects, it is not surprising that individuals reared in restrictive captive situations perform poorly on problem-solving tasks involving objects (Menzel et al. 1970). Since all restriction-reared subjects suffered massive general privation by wild standards, it is impossible to sort out specific cause-effect links. It seems likely that the sensorimotor privation of the bare environment would be sufficient to bring about the effects, regardless of missing socialization factors. Yet, even prestigious research institutions studying higher primate cognitive processes routinely house their subjects in bare cells devoid of movable objects (personal observation). It would seem that any conclusions drawn about primate intellectual capacities and limitations based on such conditions should be treated with caution.

TABLE 3 / RATIO OF THE RATE OF MANIPULATION OF NON-ATTACHED OBJECTS TO THE RATE OF MANIPULATION OF ATTACHED VEGETATION FOR OLDER CHIMPANZEE INFANTS

	SUBJECT	DEC.	JAN.	FEB.	MAR.	APR.	MAY	JUN.	JUL.	INDIVIDUAL MEANS
	AL	—	.38	.38	.39	.27	.84	.73	.44	.53
Males	PT	—	.34	.55	.17	—	—	—	—	.37
	FD	—	.33	.31	.27	.29	.42	.26	.54	.33
	MZ	.79	.56	—	—	—	—	—	—	.67
Females	SS	—	.32	.42	.29	.68	.65	.57	.58	.51
	GM	—	—	—	—	.55	.60	.57	.59	.58

Besides the general finding that object manipulation by wild chimpanzee infants is much more common than previously supposed, these results have other more specific relevance. For example, McNeill (1974) has recently proposed that chimpanzee linguistic structure differs significantly from that of humans. He argues that chimpanzee communication reflects primary concerns with social relationships and only secondary concerns with object relationships, the reverse of the priorities in humans. To support this, he states that wild chimpanzees show "what seems to be a remarkable lack of interest in physical objects" and that they lack "the intense human curiosity in objects." He asserts that chimpanzee infants "almost never play with physical objects the way human infants so avidly do, even from the earliest months of life" (Ibid.: 91). The findings presented here diametrically contradict these conclusions, and it would be useful to have some comparable norms presented for young human primates. Chimpanzees may have a different linguistic structure from humans (indeed it would be surprising if they did not, given their different niche adaptations), but it seems highly unlikely that a difference in object interest is the basis.

▪ LEAF SPONGING BEHAVIOR

This section combines two topics covered earlier: water contact and object manipulation. They combine in the use of tools to deal with water found in tree holes, i.e., leaf sponging. The data comprise descriptions of ten immature and six adult chimpanzees at Gombe exhibiting 23 bouts of leaf sponging (3 of which were unsuccessful) during 1973–74. Descriptions of the behavior pattern have previously appeared (Goodall 1964; van Lawick-Goodall 1968: 207–208), so it will be only summarized here. In brief, a chimpanzee picks green leaves by hand and then crumples them in its mouth. This loose jumble of leaves is inserted into a tree hole, then withdrawn and replaced in the mouth. The water-laden mass of leaves is then sucked for a few seconds, apparently by compression between tongue and palate. The sequence may be repeated 20–30 times in bouts which do not normally exceed 10 minutes in length. This activity appears to be a truly cultural one. Ten years of observation of another chimpanzee population 170km away has yielded no reports of this behavior there (Nishida, personal communication).

Chimpanzees are the only nonhuman species known to use tools in the wild to facilitate water intake. However, drinking behavior in other nonhuman primates is not limited to simply lapping or sucking up ground water. Hofer and Altner (1972: 90–91) reviewed various alternative methods of drinking exhibited by several species, e.g., immersion of the hand or tail which is then licked or sucked. Both wild hamadryas (*Papio hamadryas*) and olive baboons may dig (by hand) seepage holes for drinking water in riverbeds or lake shores during the dry season (Kummer 1971: 84). Marais (1969: 58–59) described how chacma baboons (*P. ursinus*) use an ingenious digging technique to cool drinking water from hot springs. In captivity many species of nonhuman primates acquire the habit of using water (as a tool?) to moisten dry commercial monkey chow pellets (Lorenz and McGrew, in preparation). Therefore,

Gombe chimpanzee leaf sponging behavior should not be considered in isolation, but rather seen as one of several in a spectrum of adaptations for coping with a basic physiological need.

Why should wild chimpanzees exhibit such an involved behavior pattern for obtaining water when more accessible sources appear to abound? Upon closer examination, it appears that the alternative sources are less convenient than might be imagined. As noted above, apparent fearful conservatism prevents the chimpanzees from using the lake water at Gombe, although this is a major water source for sympatric olive baboon troops. This aversion to surface water may generalize, so that nonthreatening alternative sources are preferred. The dendritic stream bed systems in each Gombe valley are extensive, but most have no flowing water except for brief periods after heavy rains. Even the main channels in the valley bottoms may shrink to intermittent pools in a few valleys during the dry season. Runoff from the steep hillsides and ridges is rapid, so that puddles are mostly limited to valley bottoms. Although the straight line distance from any point in the park to surface water is small (less than 1km), the metabolic cost during the dry season of moving over the rugged terrain to the nearest surface water source is probably not insignificant.

In these circumstances it would seem adaptive to utilize a convenient alternative water source if the cost of utilization is low. Tree hollows are formed when a tree limb has broken off at the bole and a cavity has rotted inwards. These holes fill with trapped rain water which apparently lasts (if undisturbed) for many weeks before evaporating. Nine of the 11 leaf sponging sites observed in use during 1973 were located at a height of less than 2m, so that chimpanzees using them worked easily on the ground beside the tree. If the hollows were brimming with water, or if the openings were sufficiently wide, the chimpanzee sipped directly with its lips or dipped its hand. If the water level was lower and the opening narrow (usually less than about 5cm), leaf sponging occurred. Chimpanzees did not seem particular about the type of leaf used, and no sponging individual had to move more than 2m to obtain plentiful supplies of leaves. Unlike other Gombe chimpanzee tool-use activities which involve interactions with prey species, leaf sponging is completely self-paced, so that chimpanzees presumably work at optimal rates until satiated or until the water is used up.

In summary, leaf sponging seems to be an easy and useful activity requiring little energy investment, although the resource is very limited and its use infrequent. There was a slight preponderance in 1973 of leaf sponging occurring during the dry season rather than the wet season, but the number of cases was too small to draw statistically reliable conclusions. Nevertheless, this conclusion concurs with the finding that leaf sponging sites were not randomly distributed over the habitat. The vast majority were on ridge tops or upper hillsides far from surface water. This also makes it unlikely that tree hole water is preferred for some reason other than slaking thirst.

In her discussion of the ontogeny of chimpanzee tool use, van Lawick-Goodall (1968) stressed the importance of observational learning and imitative copying. She constructed artificial hollows in tree trunks in Camp to facilitate observation of leaf sponging behavior, and her observations emphasize the variations which occur on the polished adult techniques, especially in efforts by infants. The observations collected by the author of leaf sponging outside Camp confirm and extend these. For example, a 5-month-old infant female showed prolonged interest in her mother's leaf sponging. While clinging ventrally, the infant repeatedly peered at close range at her mother's face and into the tree cavity. During the bout, a leaf dropped from the mother's sponge onto her leg; the infant picked it up and mouthed it. A 3-year-old male repeatedly inserted a leaf sponge normally, then removed and sucked on one leaf at a time. A 3-year-old female picked and inserted a single unmodified leaf into a water-filled hollow; she then withdrew the wet leaf and placed it in her mouth. Finally she chewed it (i.e., made the "sponge") but then directly abandoned it. A 9-year-old adolescent male made a new tool each time he sponged, then discarded each one after only a single use.

These variations might be construed as the rudimentary efforts of a "learner" whose skill at a complex motor task is being operantly shaped by an inefficient trial-and-error process. However, most object manipulation by immature chimpanzees appears to be largely playful.[7] The beginnings of what becomes in this case a very functional tool use sequence are no exception. Loizos (1967) listed five ways in which a sequence of motor patterns may be altered in a playful context: reordering, exaggeration, repetition, fragmentation, and incompletion. To these might be added: substitution, omission, syncopation, and innovation. Almost all of these characteristics are exemplified in the following episode, presented in the form of the original descriptive field notes. A young adult female, *FF*, and her 24-month-old son, *FD*, were feeding upon the fruit of the *Diplorhyncus* tree atop a low ridge at midday; the mother continued feeding more than 15m away from the leaf sponging site throughout the bout. All behaviors described below were performed by *FD*.

12:02 FD finds tree hole full of water. (No need to sponge.) Dips hand and licks off water. Splashes water with hand. Has play face and slightly upturned upper lip expression.

12:03 Splashes and sucks fingers. Convenient potentially spongeable leaves within 30cm of hole.

12:04 Stamps tree trunk. Sucks hand.

12:05 Leaves hole, then returns to stamp and splash. Licks water from hand.

12:06 Stamps and splashes. Has penile erection.

12:07 (Leaves hole.) Returns to hole, splashes.

12:08 Plays with water, with acrobatics and stamping. Inserts hand into hole.

12:09 Sucks and licks water from hand.

12:10 Moves back and forth between hole and branches. Play looks distracted. Splashes and sucks.

12:11 Returns and leaves hole. Mouths and carries piece of vine but does not use it in water.

12:12 Returns to splash, still with play face. Sucks.

12:13 Licks from hand. Attempts insertion into hole of two unmodified leaves (from vine). Removes these from hole after 2–3 seconds, then departs. Returns to hole carrying leaves in mouth.

12:14 Mouths and carries two green leaves.

12:15 Approaches *FF*. Still mouths and carries leaves. (Leaves hole.)

12:19 *FF* suckles *FD*.

12:22 Returns to hole and dips hand. Sucks from hand. Splashes.

12:23 Inserts hand and sucks finger. Play face.

12:24 Inserts hand again. Leaves hole. Returns with leaf.

12:25 Tries to insert leaf with mouth. Doesn't seem to know quite what to do. Drops leaf. Returns to dipping hand.

12:26 Sucks from hand. Mouths and carries twig to hole but drops it immediately.

12:27 Splashes and inserts hand.

12:28 Stamps. Inserts leaf into hole with lips. Has yet to make sponge by crushing leaves.

12:29 Tries to get lips into hole. Bites off leaf.

12:30 Leaves tree hole.

12:34 *FF* and *FD* leave area.

The examples of playful characteristics in the behavioral sequence are manifest. Dipping is exaggerated to splashing, and this is repeated throughout the bout. The infant interrupts the sequence with irrelevant motor patterns, e.g., stamping, or bursts of irrelevant activity, e.g., acrobatic play in the branches. Although most of the elements are singly present, the complete sequence of leaf sponging is never performed. An inappropriate object, a twig, is substituted for the efficient objects, leaves. The vital element for increasing absorbency, crumpling the leaves, is omitted. Parts of the sequence are abbreviated, e.g., the handling of the vegetational objects, while others are prolonged, e.g., sucking. Finally, a leaf is inserted with the lips rather than the hand, a novel variation not seen in the polished adult form of leaf sponging. It is impossible to say whether the infant is playfully sponging or spongily playing.

It has been frequently suggested that one of the functions of play is to provide opportunities for practicing motor patterns or combinations of motor patterns which will contribute later to the survival of the individual. This would seem to apply to much of the behavior exhibited by *FD* in the incipient leaf sponging session described above; it less directly applies to the few instances of other water play which have been seen in immature chimpanzees at Gombe. These exceptions to the usual avoidance of water occur on the uncommon occasions when chimpanzee mothers stay more than momentarily beside the surface water. For example, this may happen during the dry season when female chimpanzees attempt to fish for termites from mounds beside stream beds. This involves entering mostly dry gullies (*korongos*) which may make pools accessible to the infants. At these times infants usually remain in physical contact or within arm's reach and may engage in water play. This sometimes consists of merely dabbling the fingers and then licking or sucking them. More interestingly, the infant may transfer some or all of the leaf sponging sequence to this different situation. Presumably this involves stimulus generalization from water as found enclosed in tree holes, to open water found moving or standing on the ground. One

2-year-old infant (GM) was particularly inclined toward this misplaced sponging; she was seen to perform it on at least six occasions and appeared to lack the usual degree of caution about water. Response generalization may also occur: a 7-year-old male chimpanzee made what appeared to be a termite fishing tool from a blade of green grass but then used it to dip into a pool of water. Van Lawick-Goodall (1968) described similar behavior by younger individuals using artificially presented drinking bowls in Camp. In one case, a 4-year-old female appeared unable to retrieve the leaf sponge from the bowl; she then switched to longer twigs, which she stripped, inserted, and licked. It seems unlikely that imitation is involved in these generalizations, as adults do not use leaf sponges with surface water but, instead, drink directly with the lips.

■ ONTOGENY OF ANT DIPPING BEHAVIOR

One of the most important early findings to emerge from Gombe was that wild chimpanzees regularly make and use tools to feed upon social insects. The best known of these is termite fishing, initially described and illustrated over 10 years ago (Goodall 1963). Later publications (Goodall 1964; van Lawick-Goodall 1968) elaborated on this behavior, which involves the use of probes of vegetation (grass, vine, bark, twigs, etc.) to extract the termites from their earthen mounds. The chimpanzee predator opens a small hole on the mound surface and inserts the tool into it; the termites inside the mound attack the intruding object, affixing themselves to it with their mandibles. The chimpanzee carefully withdraws the tool with the insect prey still attached and plucks them directly from the probe with the lips.

Goodall's early papers also referred to tool use in feeding upon driver (safari) ants, *Dorylus (Anomma) nigricans*. Detailed description and illustration of this behavior, ant dipping, have appeared recently (McGrew 1974). Upon finding the temporary subterranean nest of these nomadic ants, the chimpanzee makes a tool of green woody vegetation which conforms to definite specifications of length, diameter, shape, and texture. This is inserted into the excavated ant nest entrance, whereupon hundreds of ants stream up it in massed antipredator defense. The chimpanzee watches their progress and when the ants have almost reached its hand, the tool is quickly withdrawn. In a split second the opposite hand rapidly sweeps the length of the tool in a loose power grip (the "pull-through") catching the ants in a jumbled mass between thumb and forefinger. These are then popped into the open, waiting mouth in one bite and chewed frantically.

Numerous commentators have confused the two tool-use techniques or classed them as equivalent; however, many differences exist between termite fishing and ant dipping, some of the most striking being in ontogeny. Summarizing Goodall, the ontogenetic sequence for chimpanzee termite fishing is approximately the following. Before the age of two years, infants may show all elements of the technique in rudimentary form; they eat termites, poke or mop at them on the mound's surface, manipulate discarded tools, and even modify tool raw materials. Between 2 and 4 years of age, infants increasingly attempt and succeed in termite fishing. During this period all infants spend considerable time clambering about on termite mounds, poking into holes and snatching at their mother's tools, often appearing to make a general nuisance of themselves. Their fishing form and sequential integration gradually improves over time, as does their choice of tool materials. By as early as 4 years, some individuals may exhibit the polished adult form of the behavior, but persistence in long feeding bouts lags behind this. By between 5 and 6 years of age, apparently all Gombe chimpanzees are capable of completely proficient termite fishing.

In 48 instances between 1968–1974 it was possible to observe the behavior of immature chimpanzees present with their mothers when driver ants were being successfully dipped. These involved 16 individuals ranging in age at the time of observation from 4–91 months. They were the offspring of eleven mothers, eight of whom had more than one immature offspring during the study period. In analyzing the data, an individual longitudinal approach would be preferable, but the behavior occurs too infrequently. Instead the subjects are pooled and a cross-sectional approach adopted. These fall into three natural developmental age phases: 0–1.5 years (N = 13), 1.5–3.5 years (N = 12) and 3.5–7.5 years (N = 23).

In the youngest phase, the eight infants observed (AP, FD, GM, MM, PF, SS, VL, WL) spent virtually all the time clinging in the ventral position while the mother dipped for ants. No infant was seen to break physical contact with its mother during dipping. As the infants were neither ill nor asleep and the mothers were not traveling, this was unusual behavior for periods averaging about 20 minutes in length. Normally, in other situations, at least during the second half of this phase, infants would

FIGURE 3. *A female infant (SS) watches from close range as her mother (NV) fishes for termites.* (Photo by C. Tutin)

FIGURE 4. *A family fishes for termites together. In the background is the mother (PS); in the foreground are her infant son (PF, left) and adolescent daughter (PM, right).*

be breaking physical contact and venturing out from their mothers for short distances. For example, an infant of this age whose mother is fishing for termites moves freely about and spends much time in brief spells trying all aspects of the fishing operation. They sometimes do succeed in catching and eating termites, even with their clumsy techniques.

The behavior of the older (1.5–3.5-year-old) infant chimpanzee while its mother dips for driver ants can be summed up as curious but cautious. In the most extreme case, the infant climbs into a nearby tree out of the ants' range immediately upon

FIGURE 5. *A seven-month-old male infant (MM) clings ventrally and watches his mother (MF) dip for ants.* (Photo by J. Moore)

FIGURE 6. *While MF is in the midst of a "pull-through,"* MM *reaches toward the ant-laden tool.* (Photo by J. Moore)

arrival at the nest site. It stays there throughout the feeding session, sometimes casually attending to the mother's activities but usually disregarding her to play alone. Sometimes in the beginning the infant remains on the ground or hangs overhead nearby and closely watches its mother but, typically, within a few minutes it grimaces and whimpers in response to the ants' attacks and retires. One male infant rushed

FIGURE 7. *When MF pauses in the ant dipping bout,* MM *manipulates and mouths the tool while clinging partially.* (Photo by J. Moore)

FIGURE 8. *While NV dips for ants below, SS remains out of range of the ants in the tree overhead.* (Photo by C. Tutin)

forward several times and flailed in the air at the ants but then fled without making contact. An infant does not try to dip but will attempt to perform, in isolation, certain elements of the operation. Usually these are the ground raking motion of opening the nest or manipulation of the mother's tool during one of her temporary pauses to self-groom. In one case, a 21-month-old infant effectively ended its mother's dipping session by almost throwing a tantrum when the mother carried it back to the nest site. She left soon after and the infant calmed immediately.

Of the seven older infants and juveniles (AL, FT, GB, MG, MU, MZ, PM) observed at ant dipping sites, four were seen to dip successfully and eat ants. The youngest individual seen to do so was a female infant age 46 months. But in general, juveniles only persisted briefly in ant dipping, showed clumsy and inappropriate techniques, and used inadequate tools, even (in some individuals) to the end of this phase. Juveniles attempted ant dipping significantly more often when they were alone with their mothers than when they were with their mothers in groups, but this result is confounded with sex differences. In the youngest known case, the infant persisted only for three minutes before retreating to the trees. Sometimes she merely dragged

FIGURE 9. *When NV joins SS in the tree between bouts of ant dipping, SS hangs by one hand from a tree limb and manipulates the tool with the other hand. The author tape records observations from a distance of 5–10m.* (Photo by C. Tutin)

the ant-covered tool from the nest and removed the ants singly by hand rather than use the normal pull-through. A 5-year-old male tried to use a short (30cm), crooked tool only half stripped of leaves. After withdrawing the stick laden with ants, he held it as though not sure quite what to do with it, and the ants swarmed up his arm. Eventually he executed pull-throughs but his action remained sloppy by adult standards. Older juveniles persisted for normal lengths of dipping time, but they remained unskilled, and some disastrously introduced inappropriate elements from termite fishing, e.g., attempting to take the moving ants from the tool directly with the lips and thus being bitten about the face. Other aberrancies were attempts to dip ants from the surface of tree trunks and attempts to mop up ants from the ground with the back of the hand. Van Lawick-Goodall (1973) described a five-year-old who performed the dipping sequence correctly but only in the presence of a few ants. In contrast, by halfway through this phase (i.e., at about 5½ years) all Gombe chimpanzees are proficient in all aspects of termite fishing techniques (unpublished data). In summary, the development of ant dipping lags behind that of termite fishing throughout chimpanzee infancy and childhood. Assuming that this sophisticated tool-use technique is a social tradition (and ant dipping has not been reported in any other wild chimpanzee population), then this disparity may reflect differences in acquisition of the skills. Why should this be so?

It seems likely that the differing ontogenies of termite fishing and ant dipping reflect alternative chimpanzee responses to the two prey species' anti-predatory behaviors. *To simplify, while the termites react to predation with limited, passive defense, the driver ants react with massed, active aggression.* Ant dipping is the only Gombe chimpanzee tool-use behavior which involves pain, and it is the only one in which proficiency is not achieved until adolescence. It may be that exposure to these ants induces a high arousal state which initially inhibits the learning of the technique. Unlike termite fishing, leaf sponging, body wiping, etc., the development of ant dipping involves no "play" at any stage. Young chimpanzees "fiddle" about on termite mounds, but nothing like this has been seen at driver ant nests. This implies that the acquisition of the sophisticated ant dipping technique relies much more upon vicarious observational learning than does termite fishing. Presumably, casual trial-and-error is minimized, and perhaps the operant shaping of the behavioral sequence is strengthened by a combination of negative reinforcement (being bitten by ants) and positive reinforcement (successfully obtaining food). So far, it has not been possible to examine alternative explanations, e.g., individual insight, as all chimpanzees of known age at Gombe have been exposed during infancy to their mothers' ant dipping activities. Chimpanzees at Gombe have not been seen to practice either the tool-making or tool-using motor patterns of ant dipping outside the immediate context of dipping. Nor do they attempt to feed upon all driver ants encountered. Whatever the learning processes involved, it seems that chimpanzees are capable of acquiring a complex tool-use technique with only limited exposure and practice.

Other possible explanations exist for the delay in development of skilled ant dipping. In comparison to termite fishing, it is much less frequently performed and for much shorter bouts. The infant has fewer opportunities to observe the example of the adult behavioral form. But such opportunities are no less frequent than for leaf sponging, which follows approximately the same developmental sequence as termite fishing. Ant dipping is a more complex behavioral sequence than is termite fishing, so that one would expect a longer period of skill acquisition, all else being equal. But this seems unlikely to explain the differences since, for example, infants do not even attempt ant dipping until late in their fourth year of life. Perhaps the maturation of some kinds of cognitive abilities is involved, of the type exemplified in many of the Piagetian stages described for human children. Just as children cannot correctly seriate blocks of wood before a certain age, no matter what their training, so perhaps chimpanzee infants are incapable of certain tool-use tasks until older. The operation of being able to assimilate a skill with a maximum of safe observation and a minimum of aversive first-hand practice would seem to be a more advanced one than playful trial-and-error. Perhaps non-cognitive aspects of maturation are involved, e.g., physical maturity, which permits a necessary minimal level of pain tolerance; emotional maturity, which allows an individual to persevere in spite of temporary unpleasantness; and sensory maturity, which permits enjoyment of the unusual acquired taste for formic acid. Whichever (if any) of these factors contribute to the differing acquisition of these tool-use social traditions, it is certain that socialization is involved, as the young offspring's exposure to the activity depends upon the mother. Some mothers (e.g., AL and AP's mother, AT) seem more inclined to dip for ants than others (e.g., GB and GM's mother, ML), but the numbers of cases are too few to draw definite conclusions.

■ CONCLUSION

Most of the discussion of topics covered in this paper has been included at appropriate points above and will not be repeated here. The main items to be stressed are that an inherently social process, socialization, affects the performance of an individual activity, object manipulation. It seems highly likely that the transmission of social traditions includes, in at least some cases, the higher-order cognitive processes of imitation as well as other simpler aspects of observation learning. Even if this is not always true in specific cases, it must obtain in the general sense. A necessary outcome of the mother-offspring relationship with its prolonged period of dependency is that an infant receives selective exposure to environmental stimuli. Viewed from the other aspect, it has been demonstrated that infant chimpanzees in nature are highly motivated to manipulate objects in their physical surroundings. Such inclinations provide the raw material for social traditions of tool-use and other more directly survival-oriented activities, e.g., the processing of solid food items before ingestion. These are not novel interpretations, but it is hoped that this paper has provided additional evidence for their existence in the life of wild chimpanzees.

Definitive conclusions about chimpanzee social traditions in the wild cannot be drawn on the basis of one small population. Nor can anecdotal observations, from brief field studies, however interesting and suggestive, provide the sort of substantial evidence now needed. What is required are long-term field studies which permit continuous detailed observations of the activities of known individuals over long periods. Only then can meaningful comparisons be made between separate great ape populations. Driver ants are distributed widely throughout Africa, yet the ant dipping technique for exploiting them in feeding has not yet been recorded other than at Gombe. Conversely, a different chimpanzee tool-use technique, apparently combining elements of termite fishing and ant dipping as well as novel elements, has been recently reported: Nishida (1973) found chimpanzees in the Mahali Mountains (170km south of Gombe) feeding upon *Camponotus*, an ant genus which is also widespread and almost certainly occurs at Gombe. These chimpanzees insert short lengths of bark or twigs into the arboreal nests of the ants, then ingest the prey directly from the tool. In 15 years of observation, Gombe chimpanzees have not yet been seen to feed upon these ants.

Particularly needed are field studies of wild chimpanzee populations which are undergoing significant changes in environmental stresses. Similarly, much may be learned from follow-up studies of captive-reared chimpanzees being repatriated into natural habitats (e.g., Hladik 1973). It seems most likely that the origins of new social traditions will be observable in such situations. For example, at least some members of a small relic population of chimpanzees near Gombe use Lake Tanganyika as a source of drinking water. (Neither Gombe nor Mahali Mountains chimpanzees have ever been seen to do this.) Expanding clearance of nearby forests by agriculturalists has apparently cut the chimpanzees off from alternative permanent water sources, forcing them to use the lake. It is hoped that future studies of such phenomena will provide a truly cross-cultural knowledge of the wild chimpanzee. Such studies should aid in the conservation of this fascinating species by reminding us, yet again, how tragic it would be if the chimpanzee in nature were to be allowed to disappear through our own short-sightedness.

■ ACKNOWLEDGMENTS

The author is grateful to: The Republic of Tanzania and the Tanzania National Parks for permission to reside and study in the Gombe National Park; J. Goodall for aid and facilities at the Gombe Stream Research Centre; Department of Psychiatry of Stanford University, W. T. Grant Foundation, Carnegie Trust for the Universities of Scotland, Royal Zoological Society of Scotland for financial support; Badische Anilin- & Soda-Fabrik AG for cassette tapes; J. Moore and C. Tutin for special observations and photographs; R. Campbell, A. Chamove, C. Henty, S. Savage for critical comments on the manuscript; E. Cross for numerous retypings of the manuscript; all the members of the GSRC for their encouragement and unfailing support.

■ NOTES

1. See, for example, Hall 1968.
2. With more recent findings this criterion appears less and less useful, e.g., Jones and Kamil (1973) have reported tool-making in blue jays (*Cyanocitta*).
3. This is not to imply that social traditions are limited to primates; for a well-documented example in passerines see Hinde and Fisher (1951).
4. However, Izawa and Itani (1966: 127–128) presented strong circumstantial evidence of chimpanzees wading across a 4m-wide stream in the Kasakati Basin of Tanzania. They did not see the behavior but found chimpanzee footprints leading to one stream bank and away from the opposite side. No stepping stones, bridges, or overhead vegetation were present (Itani, personal communication). More recently Nishida (personal communication) twice has seen chimpanzees wade a 50m wide river in western Tanzania.
5. This should somehow be distinguished from mere mimicry. Given the cliché of the contrast of the acquisition of spoken English by parrots and human infants, it would seem to be more akin to the latter.
6. The term "sex difference" is used in its broadest sense to indicate differences in the behavior of morphologically male and female individuals. It is possible that these differences may be "gender differences," but until the operation of differential socialization processes are demonstrated, the more conservative general term will be retained here.
7. The author is unaware of any wholly adequate functional or motivational definition of "play."

But as numerous observers have pointed out, it has a quality which is reliably recognized and some of its properties can be operationally defined (Loizos 1967).

■ REFERENCES

Altmann, S. A., and Wagner, S. S. 1970. Estimating rates of behavior from Hansen frequencies. *Primates* 11:181–183.

Angus, S. 1971. Water-contact behavior of chimpanzees. *Folia Primat.* 14:51–58.

Beck, B. B. 1974. Baboons, chimpanzees, and tools. *J. Hum. Evol.*, 3:509–516.

———. 1975. Primate tool behavior. In *Antecedents of man and after, III: Primate socio-ecology and psychology*, ed. R. H. Tuttle. The Hague: Mouton.

Crook, J. 1970. The socio-ecology of primates. In *Social behavior in birds and mammals*, ed. J. Crook, pp. 103–169. New York: Academic Press.

Fouts, R. S. 1972. Use of guidance in teaching sign language to a chimpanzee (*Pan troglodytes*). *J. Comp. Physiol. Psychol.* 80:515–522.

Gallup, G. G. 1970. Chimpanzees: Self-recognition. *Science* 167:86–87.

Garai, J. E., and Scheinfeld, A. 1968. Sex differences in mental and behavioral traits. *Genet. Psychol. Mono.* 77:169–299.

Goodall, J. 1963. My life among wild chimpanzees. *Nat. Geog.* 124:272–308.

———. 1964. Tool-using and aimed throwing in a community of free-living chimpanzees. *Nature* 201:1264–1266.

Hall, K. R. L. 1968. Social learning in monkeys. In *Primates: Studies in adaptation and variability*, ed. P. Jay, pp. 383–398. New York: Holt, Rinehart and Winston.

Hall, K. R. L., and Goswell, M. 1964. Aspects of social learning in captive patas monkeys. *Primates* 5:59–70.

Harlow, H. F.; Harlow, M. K.; and Meyer, D. R. 1950. Learning motivated by a manipulation drive. *J. Exp. Psychol.* 40:228–234.

Hayes, K. J., and Hayes, C. 1952. Imitation in a home-raised chimpanzee. *J. Comp. Physiol. Psychol.* 45:450–459.

Hinde, R. A., and Fisher, J. 1952. Further observations on the opening of milk bottles by birds. *Brit. Birds* 44:393–396.

Hladik, C. M. 1973. Alimentation et activité d'un groupe de chimpanzes reintroduits en forêt gabonaise. *Terre et la Vie* 27:343–413.

Hofer, H., and Altner, G. 1972. *Die sonderstellung des menschen.* Stuttgart: Gustav Fischer Verlag.

Hutt, C. 1966. Exploration and play in children. *Symp. Zool. Soc. London* 18:61–81.

Isaac, G. L. 1968. Traces of pleistocene hunters: An East African example. In *Man the hunter*, eds. R. B. Lee and I. DeVore, pp. 253–261. Chicago: Aldine-Atherton.

Itani, J., and Nishimura, A. 1973. The study of infra-human culture in Japan: A review. In *Precultural primate behavior*, ed. E. W. Menzel, pp. 26–50. Basel: S. Karger.

Izawa, K., and Itani, J. 1966. Chimpanzees in Kasakati Basin, Tanganyika. (I) Ecological study in the rainy season 1963–1964. *Kyoto Univ. Afr. Stud.* 1:73–156.

Jones, T. B., and Kamil, A. C. 1973. Tool-making and tool-using in the northern blue jay. *Science* 180:1076–1078.

Kummer, H. 1971. *Primate societies.* Chicago: Aldine-Atherton.

Kawai, M. 1967. Catching behavior observed in the Koshima troop—a case of newly acquired behavior. *Primates* 8:181–186.

Lancaster, J. B. 1975. *Primate behavior and the emergence of human culture.* New York: Holt, Rinehart and Winston.

Lawick-Goodall, J. van 1968. The behavior of free-living chimpanzees in the Gombe Stream Reserve. *Anim. Behav. Mono.* 1:161–311.

————. 1973. Cultural elements in a chimpanzee community. In *Precultural primate behavior*, ed. E. W. Menzel, pp. 144–184. Basel: S. Karger.

Loizos, C. 1967. Play behavior in higher primates: A review. In *Primate ethology*, ed. D. Morris, pp. 176–218. London: Weidenfeld and Nicolson.

Marais, E. 1969. *The soul of the ape*. Harmondsworth: Penguin.

Maxim, P. E., and Buettner-Janusch, J. 1963. A field study of the Kenya baboon. *Amer. J. Phys. Anthrop.* 21:165–180.

McGrew, W. C. 1974. Tool use by wild chimpanzees in feeding upon driver ants. *J. Hum. Evol.* 3:501–508.

————. In press. Evolutionary implications of sex differences in chimpanzee predation and tool use. In *Perspectives on human evolution 4*, eds. D. A. Hamburg and J. Goodall. New York: Holt, Rinehart and Winston.

McGrew, W. C.; Tutin, C. E. G.; and Midgett, P. S. 1975. Tool use in a group of captive chimpanzees. I. Escape. *Z. Tierpsychol.* 37:145–162.

McNeill, D. 1974. Sentence structure in chimpanzee communication. In *The Growth of Competence*, eds. K. Connolly and J. Bruner, pp. 75–94. London: Academic Press.

Menzel, E. W. 1966. Responsiveness to objects in free-ranging Japanese monkeys. *Behaviour* 26:130–150.

————. 1972. Spontaneous invention of ladders in a group of young chimpanzees. *Folia Primat.* 17:87–106.

————. 1973a. Further observations on the use of ladders in a group of young chimpanzees. *Folia Primat.* 19:450–457.

————, ed. 1973b. *Precultural primate behavior*. Basel: S. Karger.

Menzel E. W.; Davenport, R. K.; and Rogers, C. M. 1970. The development of tool using in wildborn and restriction-reared chimpanzees. *Folia Primat.* 12:273–283.

Mottershead, G. S. 1963. Experiences with chimpanzees at liberty on islands. *Zool. Garten* 28:31–33.

Nadler, R. D., and Braggio, J. T. 1974. Sex and species differences in captive-reared juvenile chimpanzees and orangutans. *J. Hum. Evol.* 3:541–550.

Napier, J. R. 1956. The prehensile movements of the human hand. *J. Bone Jt. Surg.* 38B:902–913.

————. 1962. The evolution of the hand. *Scient. Amer.* 207:56–62.

Nishida, T. 1973. The ant-gathering behavior by the use of tools among wild chimpanzees of the Mahali Mountains. *J. Hum. Evol.* 2:357–370.

Rumbaugh, D. 1970. Learning skills of anthropoids. In *Primate behavior: Developments in field and laboratory research*, vol. I, ed. L. Rosenblum, pp. 2–70. New York: Academic Press.

Wechkin, S. 1970. Social relationships and social facilitation of object manipulation in *Macaca mulatta. J. Comp. Physiol. Psychol.* 73:456–460.

Weiss, G. 1973. A scientific concept of culture. *Amer. Anthropologist* 75:1376–1413.

Wind, J. 1974. Human drowning: Phylogenetic origin. Unpublished manuscript.

Wrangham, R. W. 1974. Artificial feeding of chimpanzees and baboons in their natural habitat. *Anim. Behav.* 22:83–93.

10 Orang-utan Maturation: Growing Up in a Female World

DAVID AGEE HORR

■ **INTRODUCTION**

The Borneo orang-utan presents an unusual social organization for an anthropoid primate. Unlike most monkeys and apes, orangs do not live in cohesive groups or troops. Instead, Borneo orangs range more or less alone through the jungle, the only real exception being mothers and their young who form independent little social units. As a result, most parental care comes from the mother, even though the learning process is lengthy for orangs and vital to survival. This, too, is virtually unique for a higher primate.

In order to better understand this social system it is important to look closely at the way in which the orang-utan adapts to its environment and also the process whereby a young orang-utan becomes an adult. This is a circular situation; the adult ways of behaving have been selected for over time to produce an adapted system which permits survival and perpetuation of the species, while at the same time the process of behavioral maturation is tailored to mold the developing behavior of the young in such a way that they achieve the adult behaviors which promote survival. The two must be understood in terms of each other as well as in terms of the wider environmental context in which they function.

■ **A THEORETICAL APPROACH FOR INTERPRETING PRIMATE SOCIAL DEVELOPMENT**

Although biological maturation is always a basic factor during behavioral development, in higher organisms there is an ever-increasing emphasis on individual experience and learning, so that the proper functioning of the adult is highly dependent upon the input of other members of its species. This is particularly true of the nonhuman primates, and the abundant literature on social deprivation in macaques underscores the massive deleterious effects on normal functioning resulting from denial of social experience during development. Even seemingly minor deprivations can reduce the later social effectiveness of the individual, while withholding maternal inputs or the influence of peers at critical ages can result in reproductive nonfunctioning or even social nonfunctioning to the point of death. Hence, social interactions in primates are a powerful molder of normal behavior, and an understanding of adult behavior is incomplete without close inspection of the events and social mechanisms which surround and impinge upon the developing young.

As one ascends the primate scale, it is clear that specific environmental and social knowledge becomes increasingly important to adaptation. One important consequence of this is that the duration of postnatal maturation lengthens, and another is that the input from conspecifics may assume overwhelming importance to survival. As an example, the building of sleeping platforms in trees is probably crucial to the survival of young orang-utans who otherwise might be subject to rather heavy predation, yet young orangs raised by humans may show little propensity to construct such nests when placed in a jungle environment.

Social experience in the higher primates, then, is critical to the developmental process, the relative success of which directly affects the survival and proper functioning of the individual. As such, it is a crucial element in the reproductive strategy of the species so that there are strong constraints on adults which ensure that their postnatal parental investment is an important factor in their own reproductive fitness.

Viewed in this way, social experience operates on two main aspects of behavior: the developing individual learns to adjust its behavior to that of conspecifics in such a way as to maximize its survival and reproductive fitness, and, especially in the case of the primates, this experience aids in mastering the intricacies of the environment to further the same ends. The learning component has become so large in primates that one can in some cases speak of "enculturation," that is, the adoption of behavior patterns which are specific to a particular environment and group. These patterns may differ in detail from groups in other areas (Jay 1970), the local variations serving to tune the individual even more finely to its surroundings.

Traditionally, social development in primates is described as resulting from a process of socialization; a process whereby the development of the individual is enhanced or affected by social experiences of all kinds. Socialization theory as it now stands, however, represents only a partial understanding of the developmental process. Although the adult behavior of the individual may be better understood when viewed in terms of the historical experience each individual undergoes, the reasons for a specific type of socialization and the form the socialization process takes are incompletely understood and an understanding of the motivating forces for the instruments of socialization is largely or totally absent.

It should be mentioned that, in terms of the actual development of nonhuman primates, it is artificial to separate socialization from other kinds of learning. Although ideally one might designate some behaviors as existing only or primarily to facilitate social behavior and others serving only to adapt to the nonsocial environment, in the actual learning process adaptive behavior is simply learned and any behavior element contributes to social or nonsocial adaptation in varying degrees. What *is* important is that the behavior contributes to the adjustment of the organism to its environment—social or otherwise. Therefore, it is useful to consider behavioral maturation in terms of the inputs to the learning process because these reflect the basic adaptive structure of the species or group. An understanding of the sources of social learning should aid in understanding the form and structure of the end product of such learning, the adaptive behavior patterns of the species.

For the developing nonhuman primate there are three basic sources of experience, each having a somewhat different basis for acting as socialization or learning inputs. The first comprises the parents, generally the mother. Parental inputs, which may differ radically depending on whether they come from the mother or the father, relate directly to the reproductive success of the parent. Hence, an entire set of behaviors, whether social, individual, or a combination of both, are selected for in the parent because the resulting survival of the infant contributes directly to the number of genes the parent transmits to the next generation.

A second set of inputs may be characterized as resulting from inclusive fitness (Hamilton 1964). These are behaviors contributing to infant adaptation and "socialization" which come from closely related individuals, not the parents. Here would be included behaviors from siblings, "aunts" and "uncles." The fact that such individuals do in fact interact more frequently and intensively with young primates has become a commonplace observation in quantified primate behavior studies. These closely related individuals benefit reproductively from the correct adaptation of the developing young, both to environment and to the social group, because they share a certain number of genes with them.

The third set of inputs comes from all other conspecifics, generally those in the same bounded social group. These are the wider "socialization" inputs coming from peer group interaction and other situations in which the developing animal adjusts its behavior to the entire range of individuals with whom it will spend its life.

The maturation process of any primate takes place in these three social learning contexts. However, different species have different emphases depending upon the nature of the social adaptation promoting survival. If one is to understand fully the process whereby the developing individual becomes a full-fledged, functioning member of its society, it is important that the maturation process be viewed not only in terms of the behavioral interactions during development and the putative consequences for adult life, but also in terms of the relative emphasis on the types of social learning inputs. These latter lie much closer to the heart of the adaptive social system than do the cause-and-effect relationships of early experience vs. adult behavior.

Before proceeding to a brief discussion of inclusive fitness and parental investment theory, it will be useful to present an overview of orang-utan social organization and habitat, since these represent the conditions towards which orang-utan behavioral maturation must prepare each developing orang.

■ THE STUDY

The field data on which this paper is primarily based were gathered between September 1967 and November 1969 in Sabah, Malaysian Borneo. After several months

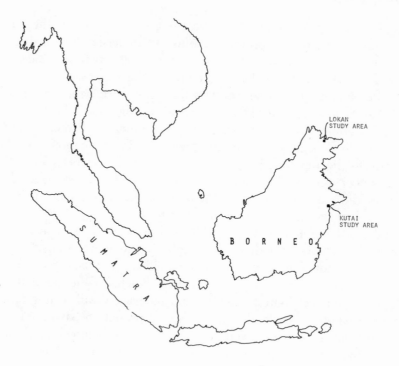

FIGURE 1. *Map of study areas*

of surveying a large tract of jungle to determine the basic nature of orang-utan distribution, some 1200 hours of direct observation were made on 27 orang-utans in an area of some 8 square miles along both banks of the Lokan River. This area is approximately 60 miles inland from the east coast of Borneo and is part of the Kinabatangan River drainage.

Between July and mid-September 1971, an additional 250 hours of observation were made on 7 previously habituated individuals in a 3 km² study site on the Sengatta River in the Kutai Nature Reserve in Kalimantan, Indonesian Borneo. The Kutai Study area is some 15 miles from the east coast of Borneo and approximately 350 miles directly south of the Lokan (see Figure 1).

Though the two study areas are not identical, they share many common features. Both are located along rivers and are composed of primary lowland dipterocarp rainforest on low ridges or hills. Elevations above sea level range from 50 feet to 450 feet on the Lokan and from 50 to 1000 feet in the Kutai. The Kutai site is relatively uniform in vegetation, while the Sabah site includes large areas of riverine seasonal swamp as well as some true permanent swamp and thus represents most of the normal orang-utan habitat types. Orangs also inhabit coastal Nipa-Mangrove swamp, but no observations were made in this habitat.

Despite differences in detail, behavioral data from the Kutai largely conform to findings made previously at the Lokan site. This increases confidence that these basic findings will be consistent for Bornean orangs living in lowland primary rainforest.

■ SOCIAL ORGANIZATION

The Borneo orang-utan (*Pongo pygmaeus*) has an unusually diffuse social organization for a higher primate (Horr 1968, 1972, 1975; MacKinnon 1971; Rodman 1973). Unlike most other higher primates, young orangs do not mature in the context of a group or troop of several individuals of all ages and sexes in which long term relationships are developed on the basis of daily, face-to-face contact. Not only is there no cohesive, geographically bounded "troop" such as one might expect for baboons or macaques, but there are apparently no instances in which a large number of orang-utans come together for coordinated movement from place to place or for feeding. In this respect the orang-utan has a far more fragmented group structure than has been reported for the chimpanzee (i.e., Sugiyama 1968; van Lawick-Goodall 1968). Since the structure and organization of individuals within the group affect the kinds of experiences which contribute to the socialization of the developing individual, it is important to outline the way in which orang-utans organize themselves in time and space.

The basic unit, in fact the only face-to-face social unit to endure any period of time, is the adult female and her still-dependent offspring (Figure 2). This unit minimally includes a female and one offspring, but often has two such offspring and in some instances might have a third in reasonably close association. Although twinning has been reported from zoos, selection in the wild has most probably

FIGURE 2. *Female and infant*

stringently favored single births, so the typical female-offspring unit is composed of individuals of different ages and in different stages of development. Reproductive data suggest that births are spaced at least 2½ and probably 3–3½ years apart. The mother-offspring unit sleeps, moves, and feeds together, though, as described later, maturing offspring ultimately separate from their maternal unit.

On the Lokan and in the Kutai, mother-offspring units ranged habitually over a relatively small area of jungle, perhaps ¼ square mile in size. These home ranges were apparently conservative in time and space. For example, over an 18-month time span, including a period of "drought-like" conditions which was quite unusual and in no way an annual event on the Lokan, the home range boundaries of females did not seem to alter. Although the home ranges were small, and a female could cross her entire range in one day, movement was generally more restricted with daily shifts in position of a few to a few hundred yards being the common pattern.

Home ranges of mother-offspring units are not exclusive, and varying degrees of overlap with home ranges of other mother-offspring units occur. However, these are usually limited to one or two other such units so that these overlaps do not result in large aggregations of mothers and their developing young.

Adult males range independently as solitary animals (Figure 3); they do not remain in close association with adult females for any period of time other than during the brief courtship or consortship, nor do males range together as all-male groups. Again, in this respect, the orang-utan's social organization is far more dispersed than that of any other anthropoid primate, including the chimpanzee. The only contacts observed or reported between adult males have been of an aggressive nature, and even these are extremely rare. Cooperative hunting among males, as reported for chimps (Teleki 1973; Wrangham, personal communication), would apparently be impossible for orangs given their current adaptive style.

Adult males range over much larger areas than do the adult females. These ranges are normally 2 or more square miles in area (Horr and Ester 1976) and overlap the ranges of several adult females. As adult males move through their ranges they may give loud trumpeting vocalizations which serve to announce their presence. These calls carry long distances in the jungle and may be heard for a mile or more. As locator calls, these vocalizations probably serve three purposes. They identify the caller to other orangs who most likely can recognize him by his "voice" since humans can learn to do so. They give females the opportunity of localizing the male by his call and choosing to avoid or contact him by moving away from or towards the call depending upon the female's state of sexual receptivity. Finally, they allow males to avoid or challenge each other depending upon their relative "dominance" position. In general, the function of the call vis-à-vis other males—from the viewpoint of the caller—is most likely to warn them away from him.

Although males probably challenge each other whenever they meet and may give ostentatious calls to warn other males away, there is no evidence to indicate that they are territorial in any classic sense. Male home ranges may overlap to an even greater extent than do female home ranges and, in any case, many of the home ranges are so large as to preclude any effective defense of "territory."

FIGURE 3. *Adult male*

   The final category of individuals in orang "society" are the developing juveniles. Depending upon whether the juvenile is male or female, at a certain point in their development they begin to move away from their maternal unit and establish their adult ranging patterns. As will be discussed later, juvenile females (Figure 4) move away somewhat later than their male counterparts, and it is likely that they frequently set up coterminus and probably overlapping home ranges with their mothers. Juvenile males (Figure 5), on the other hand, may move away earlier and farther from their maternal units. Juvenile males have been found moving alone in the jungle at ages when juvenile females tend to be with or in the very near vicinity of their mothers. These young males may form brief "social" associations with other juvenile males, but no evidence for lengthy juvenile male peer groups has been found. No evidence for associations of independent juvenile females has been reported.

   Summarized briefly, orang-utan social organization is based on mother-offspring units moving in conservative, partly overlapping home ranges, with adult males ranging over much wider areas to cover more than one female range. Juveniles transit from moving with their maternal unit to assuming adult ranging patterns, with an intermediate, semisolitary phase in between.

   A model interpreting orang-utan social structure and behavior in terms of reproductive strategy has been made elsewhere (Horr 1972, 1975). Briefly stated, orang females and their offspring range over small areas which are efficient in terms of feeding strategy. As will be described below, critical food items are thinly distributed through the forest, and large concentrations of orangs would rapidly exhaust the food

FIGURE 4. *Juvenile female*

in any given area. Female-offspring ranges represent the minimum area of support. Since males are unimportant in defense of the young against predators, and a system of ranging males ensures an adequate conception rate, association of an adult male with a female might be selected against because the male would only further deplete her food supply and force her to move over wider areas, which could make her and her offspring more vulnerable. Since females are sexually active only an average of once every 2½–3½ years, males further maximize their reproductive success by ranging over wider areas of jungle, which enables them to contact as many receptive females as possible.

By ranging in a highly dispersed fashion, orangs avoid overloading the food supply, yet achieve maximum reproductive efficiency through a system of calling males who make themselves available to females with sufficient frequency to insure an optimum reproductive rate.

It is, then, within this highly fragmented social situation that young orangs mature, and it is obvious that the socialization environment of the developing orang-utan is markedly different from that of most higher nonhuman primates who normally have available to them individuals of all age-sex categories. Not only must young orang-utans learn their entire range of social actions and responses in a limited social environment, but they have the further task of assimilating a lot of knowledge about the highly complicated habitat to which they are adapted. The complexity of this task is accentuated by the fact that orangs exploit a great deal of the jungle in which they live. Before moving to a discussion of orang development, it is important to review briefly some aspects of their habitat, in particular, as it relates to the learning tasks they face.

FIGURE 5. *Juvenile male*

■ HABITAT

Normal orang-utan habitat is mature primary rainforest, a highly complex environment both in its species composition and in the structure and distribution of its flora and fauna. Tropical rainforest in its mature state is characterized by a large number of plant and animal species and by a highly scattered distribution of these species. Large stands of particular plants or "herds" of animals are seldom found,[1] and a given area—even a relatively small one—may contain a surprisingly large number of species. In addition, individual members of any given plant species in an area may seem to be randomly distributed. Although microecological factors may in fact closely determine the distribution of a species in any given area, this "structure" is not obvious or easily discernible, and the location of specific food trees or other edible plants must largely be learned.

Some general patterns do, of course, exist, in the sense that riverine or seasonal swampy areas may contain a somewhat richer flora than ridges or hills, but within these microhabitats, distribution still must be learned.

■ FEEDING PATTERNS

A further complication of tropical rainforest flora is the relative lack of large-scale environmental variations which synchronize fruiting patterns. In northern latitudes, the alternation of seasons is part of an annual cycle which brings mild to marked variations in light periodicity, temperature and humidity. These overriding environmental conditions tend to limit optimum times of fruiting, so that fruiting seasons within and among species tend to be synchronized. This, combined with the tendency toward clumping of individuals of the same species, tends in northern latitudes to produce large-scale crops at given times in given places, alternating with periods of little or no fruit availability. In some extreme latitudes the availability of foliage and herbaceous material is also time-determined, so that at some seasons the availability of any kind of green plant food is severely curtailed.

In the equatorial forest, variations in temperature and light periodicity are minimized, and the main gross environmental variations involve seasonal differences in rainfall. This does affect fruiting in that there is a tendency for more plants to fruit during periods of increased rainfall. However, Janzen (1975) has pointed out that fruiting strategies of tropical plants are quite complex, noting that some species produce fruit very rapidly following flowering, others delay maturation up to six months or a year, and some flower before one rainy season but do not produce fruits until the next rains. In addition, there is not always strong synchronization among individuals of the same species in the same area. For example, three small "groves" of durian (*Durio grandiflorous*) on the Lokan fruited quite independently of each other; individuals within the same clump fruited independently, some of them "missing" years, while one tree apparently fruited twice in the same year, probably as a result of the delayed fruiting strategy mentioned by Janzen (ibid.).

While orang-utans are demonstrably omnivorous (they eat insects in the wild, and in captivity accept with some enthusiasm both eggs and ground meat), the bulk of their diet in the wild is composed of fruits, foliage, and other herbaceous matter. Despite the large mass of leaves and other relatively nonnutritious material consumed by orangs it is clear that the critical component of their diet is composed of fruits.[2] These are eaten whenever available, and the structure of the forest is such that during any given month something may be in fruit, though sometimes several weeks may pass with little or no fruit being available. The dispersion of fruits through the jungle in space and in time does, however, make them a difficult resource to exploit, and it is clear that orangs must learn a great deal about fruit distribution and maturation patterns in order to exploit their environment maximally. Not only must an orang learn which fruits to eat, but it must know the location of edible species within its area and further have some knowledge of the fruiting conditions of each tree. For example, on the Lokan, orangs were observed to pass through certain edible trees in a nearly random fashion until the tree showed the initial stages of fruit production. Visits to the tree then became more frequent until such a time that the orangs actually began to feed. Interestingly enough, although orangs ate ripe fruit, often they began to eat fruit just before it had ripened fully, though really unripened fruit was not normally eaten. This means that the orangs get at the fruit just prior to the time when they would face the most competition for it, i.e., when its full ripeness attracted other potential consumers. Considering that orangs ate more than 100 fruit types on the Lokan, this implies a certain expertise in fruit development on the part of the orangs. In some instances this expertise may not be trivial since some species may protect their immature fruit by including toxic secondary compounds during early stages, these compounds being replaced by edible ones or otherwise removed from the fruit as part of the ripening process (Janzen 1975).

■ LOCOMOTION

Although adult males spend large amounts of time on the ground, much orang-utan locomotion is arboreal. Their large size presents a number of problems in this respect which are only partially solved by their quadrumanous nature (Figure 6). Orangs spend much of their time in the continuous middle story of the jungle. Here, tree crowns tend to be contiguous so that movement from place to place is facilitated for an arboreal animal.

When large branches are unavailable, orangs spread their weight around among smaller twigs using their hands and feet (Figure 7), and to move from one small tree to another they often pull the intended tree in towards them, then climb onto it and "ride" its flexion across to the crown of the next small tree thus exploiting the springiness of the small tree trunk in a very sophisticated fashion. Orangs learn optimal routes through the trees and tend to move along the same route, in detail, whenever they reenter any particular place. Judging by the number of broken bones found in collected specimens (Schultz 1941), the learning of safe routes is important to orang survival.

Although overall forest structure probably dictates to a large degree which pathway is most adequate, there is the further factor of tree and wood characteristics which

FIGURE 6. *Orang-utan hands and feet*

FIGURE 7. *Juvenile hanging by all fours*

determines the relative strength, brittleness, springiness, etc. of a given tree or branch. It is probable that orangs learn to judge just such factors in predicting the optimum use of trees and vines in locomotion.

■ NESTING

The same conditions are even more obvious in the case of sleeping platforms. Orangs, like gorillas and chimps, build sleeping platforms, or "nests" at night and, in some instances, during the day. These structures are reasonably elaborate, made from branches broken inwards to form a frame and then lined with smaller branches and twigs often topped off with a mat of leaves and covered by a "roof" during rains. Since these nests support the weight of a large animal throughout the night and may be subject to considerable swaying during storms, some care must be used in selecting nest sites and trees. It is not surprising, then, to find that orangs prefer certain tree species for nest construction, a particular favorite on the Lokan being the belian (*Eusideroxylon* sp.). Belian is well suited for nesting for two reasons. The structure of belian limbs is such that several small branches often diverge at the same point, providing good structural foundation for a nest (this is true of some other preferred nest species as well). In addition, the quality of belian wood is ideally suited to nest building. In good nest construction, the branches forming the basic structure are bent inwards and broken only *half* off, so that some percentage of the wood fibers still attach the branch to the tree to impart strength to the structure. However, by breaking some of the wood fibers the tendency of the branch to spring back is reduced or removed. This is important since the base branches are not normally woven together in any sophisticated way and any residual springiness might result in unstable nests which could come apart in erratic fashion, perhaps at inopportune moments. Since not all wood has this combination of strength and brittleness, orang understanding of tree characteristics and their ability to locate appropriate nest trees may not be trivial in terms of successful adaptation to their requirement.

■ PREDATION

Nonhuman predation on orangs in Borneo is probably not heavy. The only potential chronic predator is the clouded leopard (*Felis nebulosa*), and this animal is large enough to be a threat only to independent or unwary juveniles. The size of the orang female, plus her ability to move out onto small branches and to nest far out on limbs, make her quite adequate in the protection of the young. I have postulated elsewhere that this makes the continued presence of the adult male unnecessary and has allowed him to pursue his more solitary life (Horr 1972). The situation in Sumatra is apparently different—several species might prey on females and young and there is evidence that males may remain close to females with young (MacKinnon 1971).[3]

Young juveniles, however, *are* susceptible to clouded leopard predation, and hence the protective role of adult females is important for young orangs. Not only do mothers protect their young from predators, but they also provide an opportunity for young orangs to learn the habits and ranges of the predators living in their mother's home range. This knowledge is probably not trivial in juvenile survival, especially as the juveniles move out of their mother's range. The implications of this for differential survival of male and female juveniles is discussed in the summary of this chapter.

■ PARENTAL CARE

Orang-utan habitat predisposes this species to increased parental care. A universal theory of parental investment is currently being developed by evolutionary biologists (Gadgil and Bossert 1970; Trivers 1974; Wilson 1975). In general this theory seeks to understand the evolution of parental care as an adaptive mechanism which is brought about by the interplay between a set of basic environmental conditions and the demographic characteristics of populations. Some combinations will strongly select for increased parental care as part of the reproductive strategy of the species.

Without going into great detail about the theory, stable, predictable environments with a large amount of complicated structure tend to select for larger adult size, longer life, and periodic reproduction with small numbers of young (= K selection, MacArthur and Wilson 1972). If, in addition, the food sources are scarce, specialized, or otherwise difficult to exploit, selection favors prolonged immaturity to provide greater opportunity for learning. Predation pressure predisposes towards increased parental protection, and all these factors combine to favor increased amounts of parental care as a survival mechanism.

By contrast, environments which are short-lived phenomena, or which have widely varying seasonal conditions (such as in the Northern latitudes), may favor production of as many offspring as possible at the same time. With little parental care possible, survival depends upon flooding the environment with offspring, only a small portion of which may survive to perpetuate the species (=r selection, MacArthur and Wilson 1972).

The factors described for K selection tend to be self-reinforcing. Longer life, permitted by "guaranteed" resources over time select for longer development and larger size, both useful in exploiting a stable environment more fully. This, however, brings pressure for reduced numbers of offspring and greater spacing of births to allow parental care to aid in learning the complicated environment. Also, since offspring now live to an age where they come into direct competition as adults with their parents, there is a point at which production of young reduces the reproductive fitness of the parent, and this reinforces a lowered birth rate. For any given species these factors balance out to produce its characteristic birth rate, brood size, and parental investment pattern and, by extension, they influence the social organization and social behavior of the species.

The orang-utan, as a large animal in a highly K-selective environment, has perforce expanded parental care and parental investment to an extreme state and it has modified it in special ways specific to the unique adaptive characteristics of the species.

▪ PARENTAL INVESTMENT

Crucial to the reproductive strategy of a species is the amount and form of parental care and the way in which the burdens of parental care are apportioned between the sexes. This is usually described in terms of the relative investment made by each sex in the survival of their offspring; the amount of investment is measured in terms of the degree to which it contributes to the reproductive success of the parent.

Parental investment has been defined as the energy devoted to one offspring which cannot therefore be devoted to another (Trivers 1972). In sexually reproducing verte-brates, parental investment amounts at least to contribution of sperm by the male and ova plus gestation time by the female. This generally results in a predisposition towards greater female parental investment in individual offspring, since males can presumably be investing elsewhere while the female is devoting her energies to the gestation process already underway. In his analysis of this phenomenon, Trivers (1972) has shown that continued parental care by one sex or the other is to some extent determined by the amount of investment already made by each sex, since this prior investment represents a loss in reproductive fitness if it is abandoned before the offspring is independently viable.

As a consequence, it has been shown that a long gestation period leads to extended immaturity and prolonged postnatal care. This, in turn, emphasizes lowered fertility in terms of a lower reproductive rate, e.g., extended intervals between births and reduced number of offspring per birth.

The orang-utan fits a model of heavy parental investment proceeding from the above conditions. However, while prolonged immaturity and emphasis on parental investment is not surprising for an anthropoid primate, the orang-utan has carried this to an extreme degree. When combined with the male ranging pattern this has skewed the investment pattern very heavily toward the female. With this in mind, we can now proceed to a consideration of the actual structure of parental investment in orangs and, concomitantly, to the socialization process which accompanies it.

■ PARENTAL INVESTMENT, INCLUSIVE FITNESS, AND SOCIALIZATION IN ORANG-UTANS

Viewed in terms of relative parental investment, the contribution to the development of the infant orang by male and female parents is quite different. Most daily contact with developing orangs is provided by its mother-offspring unit, usually including other siblings, with only passing influence exercised by adult males or other individuals. In terms of parental investment this represents a certain economy, since females are certain that they are investing in their own offspring whereas a male would have far less certainty that a given orang was in fact its own. With the exception of gibbons and certain prosimians, this is generally true of the nonhuman primates; however, in other species, the male is often induced to give a certain amount of parental care to several developing offspring and, in theory at least, thereby contributes some parental care to those who are actually his own offspring. In the case of orangs, the male investment pattern permits him virtually to disregard questions of paternity and to maximize his reproductive success in other ways. This has massive implications for the socialization process and, by extension, for the inner structure of orang-utan society.

■ FEMALE INVESTMENT PATTERN AND THE UNIT OF INCLUSIVE FITNESS

Although females assume the burden of socialization of young orangs, this is achieved in the context of the female-offspring unit. Socialization here is more complex, however, than simple inputs from the female, and it consists, in fact, of the input from herself and other individuals who share a high percentage of her genes. The different inputs may be described in terms of the individuals involved but, taken together, these individuals form a unit of inclusive fitness, that is, these individuals also benefit genetically to some extent by promoting the survival of the infant in question.

This unit, which I call the uterine kin group (opinion is divided on the appropriateness of this term for nonhuman primates), is theoretically made up of all existing individuals, male and female, who are related through some female antecedent, probably encompassing no more than three generations in any operational sense (Wilson 1975). At the simplest, this would consist of all the offspring of a given female (or the siblings of any *ego*). However, in the wider sense this "unit" would include siblings of the mother and all their offspring as well. In the case of orangs, in terms of the experience for any given offspring, this unit would be composed of individuals

TABLE 1 / ORANG-UTAN AGE STAGES AND THEIR
DEFINING CHARACTERISTICS

STAGE

Gestation   (Duration approximately 275 days)

Pregnancy is indicated by a swelling in the female
perineal region and, in later stages, by ventral
distension.

Infant 1   (Birth to 12 or 14 months)

In both sexes this stage is marked by near-total
dependence upon the mother for feeding and
transport. (At 12 months deciduous dentition is
complete. Brandes 1939.)

Infant 2   (12/14 months to ca. 2½ or 3 years)

In both sexes this stage is marked by increasing
independence from the mother and culminates
by weaning from the breast and ability to build
nests.

	Female	Male
Juvenile	(2½/3 years to ca. 6 years)	(2½/3 years to ca. 4 or 5 years)

In both sexes this period encompasses the time up
to complete independence from the mother and
establishing an independent range. (At 3½ years
permanent dention begins to erupt. Throat pouch
in males now complete. Brandes, 1939.)

	Female	Male
Sub-adult	(6 years to 7 or 8 years)	(4/5 years to 10 or 11 years)

In both sexes this is the period between establishing an independent range and first effective
breeding.

	Female	Male
Adult	At 7/8 years the first conceptions probably occur, though some females may not con- ceive until later.	At age 9-10 the canines develop. At age 10 or 11 the cheek callosities start to develop their adult form.
		Male now has equip- ment for effective breeding competition.

ranging nearby, largely consisting of females due to the nature of orang ranging patterns. In the mother-offspring unit there is a 50% probability that any younger or older sibling ranging with the mother is a female; all other individuals resident nearby with whom a young orang would come into reasonably frequent contact would probably be either its own older female siblings who have established contiguous ranges, or female siblings of the mother ("aunts") who might also have contiguous ranges, plus their associated offspring (again at least 50% of whom would be female).

The wider socialization context of unrelated conspecifics is quite limited for young orangs. With this social learning context in mind, we can now proceed to a basic discussion of the actual process of behavioral maturation in terms of the experiences which impinge on young orangs at different stages of development. Hence, the bulk of social contacts for the young orang is with its mother and own siblings; most of the remainder are probably with other members of the same inclusive fitness group, and a remaining small number of contacts are with "unrelated" conspecifics. In any case, the probability that most contacts are with females is quite high.

Actual age stages are difficult to draw precisely for free-ranging orangs because, to date, no one has reliably followed the life histories of an adequate sample of wild orang-utans of known age. While zoo and other captive data is useful (e.g., Brandes 1939), maturation—both physical and behavioral—may be strongly affected. The following chart (Table 1) gives a working approximation of stages and their defining characteristics for the purposes of this paper. A more detailed description is being developed and will be published elsewhere.

■ **INFANT DEVELOPMENT**

MOTHER-INFANT INTERACTIONS

The birth of an orang-utan in the wild has not been reliably reported, so maternal behavior towards a newborn wild orang is not known. Joan, a semi-wild orang at the Sepilok Center, was observed to blow into her newborn infant's nostrils following birth and prior to inducing it to suckle (deSilva 1971), and one of Rodman's guides reported seeing a female orang with what may have been a newborn infant down near a stream where she appeared to be putting water on the infant, but this report was not confirmed (Rodman, personal communication).

It is not known how soon the infant can grasp with sufficient skill to allow the mother free movement. In zoos, clinging has been noted in the first few days (Brandes 1939). All observed infants have been able to support themselves on their mother's body with little or no help and the exigencies of orang locomotion dictate that the infant must support itself at least during some maneuvers. It is not impossible that this might be delayed a few days after birth, at least, since I observed an adult male orang for five continuous 24-hour periods during which time he did not leave his nest except briefly to eliminate. He was able to do this because he had made his nest in a tree having edible leaves, and he simply snacked on his nest throughout his stay. A second possibility is that the female might spend a few days on the ground, but this might increase the vulnerability of the infant in unacceptable ways.

For the first few months of life, the mother is the sole means of transport for the infant orang. An infant male orang estimated to be 3–6 months of age was quite adept at clinging to his mother's hair as she moved about in the trees. At this age he was not observed to leave his mother's body, but he was quite active in crawling over her as she fed or rested. On occasion, infants of this age dangled from their mothers by one hand and one foot or by two hands, the other two appendages hanging free. The mother allows this dangling to continue for a few seconds; she then scoops the infant back and may cradle it to her chest. If she begins to move she scoops the infant to her and it grabs on with all fours. A characteristic position during locomotion is for the infant to cling to the mother's side, below and slightly to the front of her arm pit. Here the infant seems quite at ease even though its mother may be moving quite violently or erratically. In this position it may look around apparently quite unconcerned by its mother's actions. Interestingly, there was some evidence that infants preferred to cling to the mother's left side during locomotion. Clinging infants did not seem to hamper their mother's movements in any general way, though females were sometimes observed to readjust their infant's position prior to locomotion. Infants generally cling by locking their fingers around their mother's hair; occasionally they clutch to her skin but the preponderance of support is gained from her hair alone (Figure 8).

When the mother of a very young infant is resting the infant may crawl around on her body or the mother may cradle it in her lap. At times the infant may grasp twigs or leaves but at any sign that it may be leaving her body, the mother restrains it and prevents it from overbalancing off her. If the infant is resting on the mother and she

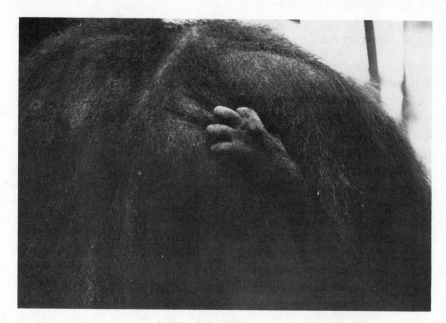

FIGURE 8. *Hand clinging to mother's hair*

begins to move, it clutches at her. Whether the young infant actually leaves the mother's body at night in the nest is unknown, but it is perhaps unlikely until it is several months old. In a zoo study an infant of 3½ months was capable of hanging from a diagonal branch, but not moving on it. At 6 months of age it was very active and climbed everywhere (Brandes 1939).

During the first 6 months the infant is slowly introduced to solid food. While the mother is moving and feeding, infants are seen to mouth and, most likely, suck at the nipple. As the mother feeds, the infant may manipulate the food item, but whether it puts it in its mouth is quite random and it does not seem to eat. Infants of this age have been observed to place edible leaves on the mother's breast, but otherwise take no interest in them. Infants do manipulate their mothers' faces and mouths while the mothers are eating and at this age females have been observed to put masticated food and juices on their hands and place the hand under the infant's mouth. Brandes (1939) reports a captive mother giving water to her infant in this fashion during the first month and premasticated food shortly thereafter.

Between an estimated 8 and 12 months of age the infant is more oriented to its environment. Although it still suckles (Figure 9), it plucks and mouths food objects, occasionally biting into them, though it is not clear to what extent an infant actually eats. These feeding behaviors may alternate. There is far more manipulation of the

FIGURE 9. *Suckling infant*

environment, and when the mother is resting she may allow the infant to leave her body and hang on nearby branches or twigs.

Over a period of about 15 months the infant becomes progressively independent of its mother. An infant estimated to be 3–4 months of age stayed constantly on its mother and was restrained from leaving her. Five months later it was grabbing onto twigs and vines and seeming to pull itself up from her body. By 11–12 months the infant was hanging near its mother while she was in a day nest or feeding. At approximately 15 months the same infant was moving from its mother to climb briefly onto an older female sibling while resting, and at an estimated 18 months it was moving around in the tree near its mother while she fed.

During the first year of life, other kinds of observational learning are developing. For example, an infant male estimated to be about 1 year of age was seen to mimic its mother during threat behavior. The adult female—an unhabituated one—had been threatening the observer in a desultory fashion, including some branch waving and dropping. Her young infant was seated beside her on a large limb. It watched the proceedings with great interest for some time, then grabbed onto a tiny twig with a single leaf which was sprouting up in front of it from the limb, and after waving this back and forth several times managed to break it off and let it drop. As is so often the case with adults, it then leaned over and watched the slow drifting fall of the leaf to the forest floor below. Here it had mimicked almost to perfection the adult threat response, lacking only in intensity and of course in any locomotion or throat pouch inflation.

At about this time infants also begin to mimic nest building motions by the mother in preparation for later independence from her. Only glimpses were caught of this behavior at this young age.

Grooming and play by the mother are largely confined to the infants. During daytime rest, mothers occasionally groomed their infants for short periods though these grooming sequences have none of the long intensive nature as seen for baboons and some other higher primates. Mothers also played idly with their infants for brief periods. As infants matured, they initiated much of the playful behavior. At 1 year or more, when infants rested or moved about away from the mother's body, they frequently swung towards her or reached out towards her, then withdrew in playful activity. This would sometimes stimulate the mother to play briefly. Infants also initiated play sequences with older juvenile siblings, though here the juveniles sometimes initiated the play behavior. Actual contact in play is usually brief and much play consists of swinging towards and away, including some grabbing and mouthing of parts of the other individual. Again, the contrast with monkeys such as baboons and macaques is striking, not only in terms of the quality of play, but also the degree of parallel play as opposed to interactive play, and the marked low frequency of this latter behavior.

Weaning from the breast is apparently over by the end of infancy. Although orangs in captivity have been reported to nurse until 6 years of age and the mother was known to produce milk throughout (Brandes 1939), this process must be complete by 2–2½ years of age in the wild. Not only would nursing of an older offspring be difficult to sequence with nursing of a newborn or young infant, but orang-utan receptivity may be closely tied to lactation levels, in which case conception would probably not be possible as long as a previous offspring was still nursing full time.

From 4 or 5 months of age, infants are heard to whine after the mother has retired to the nest with them, and sometimes this is heard during daytime rest. This whining could, of course, occur in a number of contexts, but probably some of it is indicative of maternal withholding of the breast. The many observations of females putting masticated food near the mouths of very young infants may indicate that weaning from the breast is a long process which starts fairly early. In view of the many food plants and food types which orangs must learn, this long transition process may be quite adaptive. Chronic deprivation of the breast could assist in the gradual acceptance of many different food types, so that at the time of total weaning the young orang would have considerable experience with its varied menu. It is interesting to speculate that early breast deprivation might reinforce the relatively affect-less behavior of adults.

JUVENILE INTERACTIONS WITH THE INFANT

From an early age the infant is subjected to the attentions of any older siblings still with its mother. During the day juveniles often approach or reach out to touch or manipulate young infants, and they are normally rebuffed by the adult females. Juveniles become quite adept at reaching for or touching infants while their mothers' backs are turned or they are involved in feeding. The surreptitious nature of these contacts is often quite comical to watch. Attempts by young infants to go over to an older sibling are discouraged by their mothers. If the older siblings become insistent about taking an infant to them, or if they are too persistent about wanting to manipulate a clinging infant, the mother may threaten them off with grunts and smack threats. Juveniles were never seen to carry young infants or to have them on their bodies during the day. Of course, during night nesting it is entirely possible that infants play with and crawl on their older siblings. However, what transpires at night is not known. Some of the play contacts with siblings include elements of grooming by the juvenile, but this is very limited in nature when compared with other higher primates. To summarize, young infants are constantly subjected to the attention of older siblings (usually only one), but actual contact during the day is severely limited, both by environmental conditions and by their mothers.

Older infants, ones now spending time off the mother's body and moving around in nearby branches, may climb onto older siblings. This does not seem to develop into transport of older infants by juveniles, and infant orientation is still towards the mother. This lack of opportunity for caretaking is highly unusual for an advanced primate species.

By the time the developing infant approaches its juvenile stage (2½–3 years) older siblings will have moved away from the mother-offspring unit so little or no long-term contact or play is possible.

Older infants are still transported by their mothers, but much of the time is spent moving around in the tree near them. Actual clinging tends to occur only when the mother is moving and, ultimately, the main role of the mother in offspring transport is to hold tree crowns close together to form a bridge for the infant to scramble across. In one instance a mother and older male infant were both agitated at the presence of the observer and both were moving around rather actively in the crown of a tree. The mother finally moved rapidly towards the edge of the crown and, as she began to cross over into a neighboring tree, her infant charged towards her and clung tightly to her just as she swung out and across. Once across, he abandoned her body to move along with her across the new tree crown and into the next tree.

INFANT INTERACTIONS WITH OTHER ORANG-UTANS

Direct inputs from other conspecifics are probably largely nonexistent. Infants are not handled by other females or juveniles, and adult or subadult male contact with infants would be avoided. An exception to this would be those instances in which a male forcefully attempts a copulation with a female who has a climbing infant (MacKinnon 1971). This, however, scarcely constitutes a positive or "male parental care" situation.

▪ JUVENILE DEVELOPMENT

MOTHER-JUVENILE INTERACTIONS

The juvenile stage represents a real watershed in orang development. Transport by the mother, especially if she has a new infant, is now nonexistent, although she might still "bridge" between two trees for an older offspring. Grooming and play have dwindled, nursing seems to be a thing of the past, and independent night nesting is within the ability of the juvenile. The juvenile, then, is capable of independent existence and probably possesses all adult skills (except mating) to some degree at least. However, the juvenile remains in close proximity to its mother for some years, and this time is a period of increasing refinement of its adult skills.

The juvenile phase is an interesting combination of continued dependence upon the mother in ways crucial to adult development and of increasing distancing from the mother; it is best described in terms of these behavioral shifts.

During an 18-month period from May 1968 to October 1969 we were able to watch the critical processes of the distancing of a juvenile female from her mother. At the beginning of this process the mother had a male infant, estimated from its size to be 3–4 months of age. The juvenile was estimated to be approximately 3½ years of age.

At the outset the relationship between the juvenile and her mother was quite close. The juvenile maintained close contact with her mother at all times observed and generally did not stray more than one tree away from her. At this stage the juvenile was extremely sensitive to her mother's movements. If the adult female began to move in such a way indicating that she was going to change location, the juvenile responded almost immediately, and when the female did move off in some direction, the juvenile followed her in a very accurate fashion. By this I mean that her behavior mirrored that of her mother. She tended to follow exactly the same route through the trees, using the same twigs, vines, and other steadying elements, and she also tended to follow at a "fixed" distance behind the mother. The distance depended upon the nature and complexity of the pathway they were using. For example, in more or less open branches, the juvenile might follow at a distance of 20–30 feet, while in dense areas or where the going was harder, the juvenile might stay right behind her mother. This mimicking of movements was at times quite comical, and the juvenile often behaved as though she were a pull-toy being dragged along on a cord—when the mother stopped, she stopped, etc., and she tended to follow exactly the same route as the mother. This was not always the case. However, the nature of the pathway was not a limiting factor; in many instances alternate routes were clearly available and within the locomotor competence of the juvenile, yet she tended to mimic her mother's actions, even under these circumstances. Mainly, then, the burden is on the offspring imitating its mother rather than on any active "socialization" by the adult female.

Although the adult female seemed to move without regard to the juvenile, upon occasion she would look back at her offspring, and in one instance when the juvenile was unable to make a crossing and whined, the female returned to help her across. In another instance the juvenile slipped and fell about 15 feet in a tree, resulting in some distress vocalizations. The female went to her and cradled her briefly before proceeding. This behavior was seen in other pairs; in particular, a female with a juvenile male was seen to bridge across two trees to permit her son to cross. Path specificity was observed elsewhere in the Lokan and in the Kutai in other mother-juvenile pairs.[4]

During this period of juvenile development the adult female was observed to offer food to her juvenile and share feeding bouts. In one instance, the female broke off a portion of a termite nest. While feeding, she held out the nest piece to the juvenile who fed from it. The female retracted the nest piece, fed from it, and then offered it again to the juvenile who fed. Here the female clearly controlled the interaction. This feeding association continued for the next 10 months with the juvenile often feeding nearby, sometimes face to face with her mother—including eating on opposite sides of the same epiphytic fern leaf in some instances.

During the 18-month observation period, the juvenile female became more exploratory and independent. In the morning she began to leave the nest before her mother—sometimes 30 minutes to an hour before—and during this time she fed or moved around in the canopy near her mother's nest. During the day, she moved further distances from her mother during feeding. Although during the earlier stage she continued to follow quite accurately and closely her mother's movement through the canopy, she began to move with increasing independence and sometimes she preceded her mother during travel.

At times, however, the juvenile clearly became "lost," in the sense that she became apprehensive when she was away from her mother and could not relocate her immediately. In one instance the juvenile was moving through some low trees paralleling the path of her mother who was 30 or so yards away from her. The mother descended to move along the ground out of sight of the juvenile. After some minutes the absence of locomotion sound from the mother became obvious to the juvenile who became increasingly agitated and nervous. She finally became somewhat frantic in her searching motions and began to emit a distress vocalization which was never heard again in that or any other circumstance. This continued for a minute or two until tree sounds were heard off to the right and the juvenile moved rapidly in the direction of her mother who was now coming back up from the ground into a low tree.

As the juvenile matured, rather violent rejection interactions took place between her and her mother. These were observed in two specific types of situations—during daytime feeding and at the time of nest building. The behavior in both situations was largely identical. It consisted of a violent exchange of vocalizations accompanied by much physical activity from the juvenile. The juvenile approaches the mother, giving a light wailing vocalization, and is rebuffed by her with arm gestures and/or deep grunting or roaring sounds. The juvenile responds by screaming—a loud, high-pitched vocalization which rises in intensity to sound like a human being in the extremes of hysteria. While screaming, the juvenile swings towards or approaches the mother who again roars, sometimes with slightly inflated throat pouch, and makes brief intention movements of hitting or shoving the juvenile. The vocal and physical action takes on a pendulum-like quality with the juvenile swinging in and screaming, then recoiling as though buffeted back by the female's vocalizations. The resultant exchange grows in intensity until it literally reverberates through the jungle. The exchange may continue for several minutes before it subsides.

The daytime occurrences of this tantrum behavior usually centered around feeding, or the infant, or both. The juvenile would approach the mother while she was feeding and attempt to feed and/or touch the clinging infant. This combination of circumstances provoked "warding off" behavior by the adult female, leading to the tantrum sequence.

The other setting for tantrum behavior was at nest-building time. The pattern of behavior here was for the juvenile to approach the mother's sleeping nest and try to enter it. Sometimes this was successful; at others the female prevented the juvenile from entering.

By the juvenile stage, orangs are quite capable of independent nesting. One activity juveniles engage in as play behavior is nest building (Figure 10), although these nests are often only partially made and then abandoned as the juvenile's attention is directed elsewhere. On many occasions, however, young juveniles sleep in the mother's nest even though she may have another, smaller offspring. As juveniles mature and grow larger, females become less tolerant of their nighttime presence in the nest. Under these circumstances, juveniles adopt two strategies. One is to approach the mother's nest and attempt to enter. If rebuffed they may go off a short distance and build their own nest, then later return to the mother's nest and try to gain entry. Another strategy is simply to make a nest at sundown, remain in it some time, then try to enter the mother's nest.

Ultimately, the adult female may resist totally the entrance of her older offspring into her nest, leading to tantrums as described earlier. In the early stages of such tantrum interactions, juveniles may continue with their displacement strategy. One juvenile female was observed to approach her mother's nest on five successive occa-

FIGURE 10. *Orang juvenile in nest*

sions in the space of about an hour, each time being rebuffed and each time retreating
to make a new nest of her own. After remaining a few minutes in the new nest she
repeated the process with each successive nest being somewhat closer to that of her
mother. She was left finally, an hour after sunset, apparently bedded down for the
night in her own nest; however, the next morning found her emerging from her
mother's nest where she had probably spent the bulk of the night. Later in this
juvenile's development truly spectacular nest tantrums occurred, and shortly afterwards she seemingly gave up all attempts to sleep with her mother and nested independently. At this point in her development she began to spend entire days away from
her mother and was clearly becoming an independent individual with her own
habitual range. Another juvenile female, observed in the Kutai, who was estimated to
be perhaps a year older than the one described above, spent most of her time far
removed from her mother, ranging and nesting independently and associating only
infrequently with her mother and younger male sibling.

Older juveniles range at increasing distances from their mothers. Juvenile females
may remain in fairly close proximity until they are young adults. In the Kutai an older
and a younger adult female, both with young infants, had closely overlapping home
ranges, and it is not unlikely that the younger may have been the offspring of the other
(Rodman, personal communication). In any case, ranging patterns apparently become sexually dimorphic in the late juvenile phase, with older juvenile males already
embarking on their larger home ranges.

Whether the earlier distancing of juvenile males is accompanied by more violent
or earlier tantrum behavior is not known; however, in one instance a juvenile male
was located in the jungle engaging in both sides of a tantrum. This male was observed

in mid-morning swinging violently back and forth in the crown of a small tree. He was emitting the shrill whining and screaming heard from other juveniles, but he was also inflating his throat pouch to a small degree and giving the roaring grunt normally used by the mother when chasing off a juvenile offspring. Careful searching showed that he was quite alone in the jungle.

INFANT SOCIALIZATION OF JUVENILES

As noted previously, juvenile females persistently try to interact with younger siblings. Although this is prevented by the mother during the early stages of infant development, juveniles have much greater access to older infants. Nonetheless, no instances were observed in which a juvenile female actually carried her younger sibling around or even spent time in close association during feeding bouts. In other words, "aunting" behavior was not observed, and it does not seem that juvenile females are able to "practice" mothering behavior directly prior to giving birth to their own offspring. Of course, juvenile females do have several years of observational learning available.

Free-ranging play behavior between siblings seems to be limited by age separation; by the time an infant is beginning to be fully independent from its mother, older siblings are already ranging at some distance.

INTERACTION BETWEEN JUVENILES

Opportunities for juvenile interaction and play are limited in the wild. The association of mother-offspring units does offer some occasions in which juveniles may come together, but this is random for any given juvenile, and the age structure of offspring

FIGURE 11. *Two young orangs playing together*

FIGURE 12. *Juveniles in parallel play*

militates against it being a uniform happening. The meetings of juveniles observed in the wild showed that they exhibited interest in each other but did not engage in long play sequences. One juvenile male, ranging by itself, was encountered by an adult female with an older juvenile female. Both females evinced great interest in the juvenile male, who seemed to be agitated and somewhat upset by this attention. After several minutes of inspection and touching by the two females, he moved away from them and disappeared.

Juveniles of several ages were observed interacting at the Sepilok Rehabilitation Center near Sandakan (Figure 11). These animals had been raised from an early age by human beings and had been taken to the Center in an attempt to reintroduce them to the jungle. They engaged in a fair amount of play with each other; however, even here there was a high percentage of parallel play when compared with contact play behavior (Figure 12).

JUVENILE INTERACTIONS WITH OTHER ORANG-UTANS

Opportunities for contact with other orangs are somewhat greater for juveniles due to their growing independence from their mothers. In instances when mother-offspring units met in the same tree, the offspring evinced interest in each other but not much physical interaction ensued.

MALE INVESTMENT

The role of the adult male as a socializing agent is problematical at best. As described earlier, adult males only infrequently associate with adult females, and the frequency of real interaction with adult females is at its lowest when she is accompanied by a young infant.

On the Lokan the reaction of an adult female with a young infant to adult male contact calls was to ignore or move away from the caller. Consortships are not reported for any orang females with young infants and, in addition, adult males do not have access to young infants to groom or to otherwise interact.

Paternal behavior towards infants, then, is largely or totally nonexistent in Bornean orangs. Not only is there no close interaction of adult males with infant orangs, but there is in fact very little contact with or observation of adult males and adult male behavior by infant orangs of either sex.

With regard to juveniles, the observed data is roughly the same with, however, some exceptions. Orang females with juvenile offspring are more often in contact with adult males unless the female has an infant with her as well. If the juvenile is of sufficient age and independence, the female may consort with the male. Here, however, the contact is largely negative for the juvenile in that males may ignore or attack juveniles to get with the female, and if a consortship forms the juvenile will be excluded.

Juvenile and subadult males probably have only negative experiences with adult males. They are not observed together, and they have been observed to avoid the approach or presence of adult males. Some of the scars seen on adult males may come from damage suffered at the teeth of other adult males while they were juveniles.

In toto, the direct adult male contribution to socialization is near zero. In terms of reproductive strategy it is clear that the adult male does not serve any protective or nurturing role towards offspring. Beyond this, males are not important in the complex environmental learning tasks which face young orangs, and it is not clear whether they serve any function in socializing to the "group" or in the learning of any social behaviors. They may, however, serve as remote role models, for example, by demonstrating male calls, male aggressive behavior exhibited towards other orangs, and sexual behavior. In this case, the male is a "socializing agent" in a peculiar sense of the term, serving as a "role model" for the kind of independent, basically asocial and sexually-serving behavior which developing males will assume as they mature.

▪ **SUMMARY AND CONCLUSIONS**

Bornean orangs have a highly fragmented social system in which the population as a group is only tenuously in social contact. Environmental and phylogentic factors have led to a need for high parental investment in orang-utans, but the nature of the social group has drastically reduced male investment in the young and has resulted in a highly skewed parental investment pattern in which females compensate for the absent males. To better understand orang social organization it may be useful to review this parental care pattern in terms of some of its implications for orang-utan social behavior and adaptation.

The intensity of female investment is shown in a number of ways, starting with the total length of time each offspring stays with its mother. However, the intensity of energy channeled to each offspring is revealing. Not only do females sacrifice a lot of energy in transporting their young infants, but there is also a very large energy investment made in terms of food sharing. Food sharing starts during infancy in the form of feeding premasticated food from the lips, but of more significance is the fact that mothers continue to share important foods—e.g., termites, which probably provide proteins unavailable elsewhere in the diet—with their *juvenile* offspring, and also they allow their juveniles to feed side by side with them on fruit. In other words, although "weaning" fights over environmental foods do occur—in particular near the point when the juvenile is starting the final separation from its unit of natality—mothers actually mitigate their food competition behavior in favor of their juvenile offspring. Although this energy sacrifice on the part of a female for her own offspring probably exists to some extent among many other higher primates, it is perhaps significant in the orang that it persists into the juvenile phase and that it is carried to the point of actual direct sharing of scarce foods. In gorillas, for example, such food sharing does not occur (Fossey, personal communication). In chimps food sharing with older offspring is quite common, and it is interesting that among the higher primates, chimps alone also show many other social behavioral elements reminiscent of orangs.

The social microcosm in which young orangs mature is composed largely of related individuals. The basic group, the uterine kin line proceeding from the mother, is the primary unit of association for developing orangs but even most wider contacts are on an average probably with individuals related to the mother.

The uterine kin group in orangs thus probably forms a selective unit. For adult females the uterine kin group is a unit of inclusive fitness in the sense that the survival of young orangs is enhanced by the association with other kin—either younger or older siblings and possibly with the female siblings of their mother. In view of the lengthened learning period required for orangs, and the heavy emphasis of this burden on the adult female, this association is probably selected for. In this sense young orangs *do* mature and develop behaviorally in a small bounded group composed largely of closely related individuals, and this small unit is perhaps critical to orang survival in the same way the larger troop is to other primate species. The contribution of the adult male to this unit of direct and inclusive fitness is problematical and, at best, highly attentuated.

It might be argued that such a system provides a very biased role-model process for developing orangs. Both males and females mature over long periods of time amounting to several years, but the bulk of any socializing experience and other learning is provided by adult females. It is true that some behavior models are presented by older siblings; however, these are not adult animals so, at best, they function as "next stage" models for infants, and there is no guarantee that they will be of the same sex.

Contact between adult female units may be relatively frequent so that developing young may observe other adults, but again these are female role models. It is true that adult males periodically associate with female-offspring units. These contacts, however, are brief in terms of the total life history of developing orangs, and the only "socialization" provided is observation of certain types of male behaviors, e.g., copulation, courtship, and, in some instances, aggression.

In animals whose behavior is so molded by learning it is interesting that virtually all social and survival models can be provided by adult females. While it may be argued that adult males are around with sufficient frequency to provide role models for developing males, the behavior of males and females varies so widely in some respects that it is surprising that juvenile males can adapt so well to the far-ranging behavior appropriate to adult male life.

This may imply that, even in a primate as phylogenetically advanced as the orang, the genetic control (or hormonal mediation) of male vs. female behavior is far greater than one might suspect. This is not to imply that male behavior is dictated in *detail* by unlearned biological processes, but that overall parameters of behavior—such as activity levels, aggression levels, etc.—effect greater controlling influence than one might expect. When behavior styles develop in the absence of obvious "training" or learning situations it draws attention to the extent to which such behaviors are basic.

Given the low level of male parental investment, the high degree of female investment in the young is of obvious importance. It is more difficult to understand why a system of such low male parental investment has not been selected against in a species requiring so much learning and in which the adult behavior of males and females shows some significant differences. The answer may lie in the reproductive structure of the species.

In looking at the orang-utan system of parental investment, it would seem that maturing females are adequately "trained" for all aspects of survival in their mother-offspring unit. As a juvenile female moves out of her mother's home range she slowly establishes her own range which is small and probably contiguous to her mother's. This gives the maturing juvenile female an opportunity not only to learn the particular structure and botanical configuration of her new home range but also to learn the ranges and patterns of any resident predators.

The developing male, however, is schooled in only part of his adult survival behavior. As he moves out into his much larger range he must assimilate and cope with a much wider range of habitat variation because it is unlikely that his mother's range encompasses all the variation to be found in an area four to eight times its size. This also increases the chance of falls due to unfamiliarity with the new, larger terrain. In addition the wider ranging individual is more subject to predation partly because he covers the ranges of more predators and also because he has less opportunity to learn their location and ranging patterns. This move into uncharted waters should make the older juvenile male subject to a higher potential mortality, and this disadvantage is exacerbated to some extent because he apparently does not have the luxury of adult male assistance in learning appropriate skills. In fact, his new independent existence may be made more precarious by aggressive overtures from adult or subadult males he may encounter in the jungle.

Although this state of affairs might seem nonadaptive for orangs, a somewhat higher mortality among male orangs than females may not be disadvantageous since the system of ranging males ensures optimal conception rates even without a 50:50 sex ratio (Horr and Ester, 1976). In fact, other aspects of orang survival might favor a somewhat reduced ratio of adult males to adult females especially if the surviving males have been stringently selected for crucial characteristics. Such selection would thus remove "superfluous" males from food competition, perhaps enhancing the survival of young orangs. Whatever these "group" benefits, the potential increase to individual male reproductive fitness is a sufficient cause for wider male range.

Whatever the differential nature of the male vs. female maturation experience, the relative lack of play and other interaction with the young of both sexes is quite striking. That is not to say that orang mothers do not play with their children or that siblings do not interact with each other; however, the level of interaction when compared with other species is not high. This pattern is, however, quite consistent with the development of adult behavior in orangs.

The slow progression towards semisolitary life is not only clearly accompanied by an increase of negative interaction with the mother-offspring unit but also includes a slow shift of positive behaviors. Although grooming of the infant is fairly frequent this behavior dwindles as orangs mature, so that juveniles seldom groom or are groomed by their mothers, and grooming is not a persistent aspect of juvenile interactions with each other. Hence, the increasing independence of juveniles is marked not only by negative behaviors—such as tantrums—but also by a lessening of overt positive behaviors. Instead of an increasing tendency towards social activity as seen in juvenile peer group interactions in other primate species, orang development is instead marked by a continuously decreasing tendency to associate with other animals. This constantly reducing amount of social interaction along parameters such as grooming, play, and social contact, clearly prepares for the semisolitary adult nature of orangs.

With all its unique aspects the single most salient feature about orang-utan social organization is its extreme emphasis on the uterine kin group. This small social unit is the main locus for social interaction, maturation, and social learning. It represents the final distillation of the social group in higher primates—the sine qua non of primate social survival. It also serves to emphasize the central nature of the adult female in primate society, who serves as the primary focal point for kin-related behavior as well as the primary agent of behavioral development. That the female is so centrally important to primate social structure is often masked in other species where many other individuals are in contact with each other on a daily basis. However, in the case of the orang-utan the others have been dissected away by circumstance, yet the adult female remains adequate to the task of rearing her offspring to survival.

■ **ACKNOWLEDGMENTS**

The field work on which the foregoing is based was supported by NIMH grant MH–13156. Additional support was obtained from the William F. Milton Fund, Harvard University.

Particular credit must go to: Dr. Richard K. Davenport, Jr., who first surveyed the Lokan area in 1964 and who gave the author invaluable advice and assistance; Professor Irven DeVore, who sponsored the initial research; Mr. J. E. D. Fox, who provided invaluable botanical insights; Mr. G. S. De Silva, who enabled us to work in Sabah; and Mr. Walman Sinaga, who enabled us to work in Kalimantan.

The staff of the Sabah Forest Department and the Lamag District Office were crucial in enabling us to work in the Lokan area, and the field stations of Pertamina and C.G.G. served the same crucial role in the Kutai. Peter and Carol Rodman, who joined the study to investigate the ecological relationships of the sympathic primate species, kindly assisted me in the Kutai Nature Reserve where they had relocated the study.

I want to thank especially Mr. Chuck Ebert, Anthropology Department, Brandeis University, and Dr. Sarah Blaffer Hrdy, Anthropology Department, Harvard, for their careful reading and thoughtful comments.

■ NOTES

1. An exception is formed by some monkey species which occur in troops of 40 (*Macaca fascicularis*) to 80 (*Nasalis larvatus*).

2. The large number of fruit species eaten by orangs and the distances orangs travel compared with other frugivores probably make them a major seed disperser agent for many tree species.

3. MacKinnon has reported leopard predation on juvenile orangs in Sumatra, and clouded leopards have been known to kill large proboscis monkeys in Borneo (Davis 1962).

4. This is also the case among Nilgiri langurs and has been interpreted as the process whereby an arboreal animal learns the best travel routes (Poirier 1973).

■ REFERENCES

Brandes, G. 1939. *Buschi; vom Orang-Säuglung zum Backenwülster*. Verlagsbuchhandlung Quelle & Meyer in Leipzig. vii + 135pp.

Davis, D. D. 1962. Mammals of the lowland rainforest of North Borneo. *Bull. Nat. Hist. Mus. Singapore*, No. 31.

de Silva, G. S. 1971. Notes on the orang-utan rehabilitation project in Sabah. *Malay Nat. Journ.* 24:50–77.

Gadgil, M., and Bossert, W. H. 1970. Life history consequences of natural selection. *American Naturalist* 104:1–24.

Hamilton, W. D. 1964. The genetical evolution of social behavior. *Journ. Theor. Biol.* 7:1–16.

Horr, D. A. 1968. Progress to date. Application for extension of research grant MH–13156, NIMH.

————. 1972. The Borneo orang-utan. *Borneo Res. Bull.* 4:46–50.

————. 1975. The Borneo orang-utan: Population structure and dynamics in relationship to ecology and reproductive strategy. *Primate Behavior* 4:307–323.

Horr, D. A., and Ester, M. 1976. Orang-utan social structure: A computer simulation, in *The measures of man: Methodologies in biological anthropology*, eds. E. Giles and J. Friedlaender, pp. 3–56. Cambridge: Peabody Museum Press.

Janzen, D. H. 1975. *Ecology of plants in the tropics*. Studies in Biology no. 58, Edward Arnold (Publ), London, vi + 66pp.

Jay, P. 1968. *Primates*. New York: Holt, Rinehart and Winston. xii + 529pp.

Lawick-Goodall, J. van. 1967. Mother-offspring relationships in free-ranging chimpanzees, in *Primate ethology*, ed. D. Harris, pp. 287–346. Chicago: Aldine.

MacArthur, R. H., and Wilson, E. O. 1967. *The theory of island biogeography*. Princeton: Princeton University Press. xi + 203pp.

MacKinnon, J. R. 1971. The behavior and ecology of wild orang-utans (*Pongo pygmaeus*). *Animal Behavior* 22:3–74.

Poirier, F. E. 1973. Primate socialization and learning. In *Culture and learning*, eds. S. Kimball and J. Burnett, pp. 3–14. Seattle: University of Washington Press.

Rodman, P. S. 1973. Population composition and adaptive organization among orang-utans of the Kutai

Reserve, in *Comparative ecology and behavior of primates*, eds. R. P. Michael and J. H. Crook, pp. 171–209. New York: Academic Press.

Schultz, A. 1941. Growth and development of the orang-utan. *Carnegie Inst. Wash. Contrib. Embryol.* 29:58–111.

Sugiyama, Y. 1968. Social organization of chimpanzees in the Budongo Forest, Uganda. *Primates* 9:225–258.

Teleki, G. 1973. *The predatory behavior of wild chimpanzees*. Lewisburg: Bucknell University Press. 232pp.

Trivers, R. L. 1972. Parental investment and sexual selection, in *Sexual selection and the descent of man, 1871–1971*, ed. B. Campbell, pp. 136–179. Chicago: Aldine Press.

———. 1974. Parent-offspring conflict. *American Zoologist* 14:249–264.

Wilson, E. O. 1975. *Sociobiology: The new synthesis*. Cambridge: The Belknap Press. ix + 697pp.

11 | Incest Avoidance among Human and Nonhuman Primates

WILLIAM J. DEMAREST

■ INTRODUCTION

The universal existence of incest taboo among human societies intrigues anthropologists interested in human evolution; its patterning of sexual behavior interests those who are psychologically oriented, and the sociological benefits it confers upon human society have made it an object of study for structural-functionalists.

Many of the above agree that the imposition of incest prohibitions lies at the point in time when nature becomes culture. In their minds the incest taboo is a watershed separating human and nonhuman evolution. The following statement by Claude Lévi-Strauss (1956: 278) reflects this viewpoint:

> Indeed, it will never be sufficiently emphasized that, if social organization had a beginning, this could only have consisted in the incest prohibition since, as we have just shown, the incest prohibition is, in fact, a kind of remodeling of the biological conditions of mating and procreation (which know no rule, as can be seen from observing animal life) compelling them to become perpetuated only in an artificial framework of taboos and obligations. It is there, and only there, that we find a passage from nature to culture, from animal to human life, and that we are in a position to understand the essence of their articulation.

A different view of incest, first proposed by Edward Westermarck in 1898, is that incest prohibitions are not strictly cultural artifacts but are based on an aversion to sexual intercourse which automatically arises between people reared together from childhood. This hypothesis leads to Westermarck's conclusion that outbreeding existed prior to its symbolization. His argument that a symbolization of aversion would then automatically arise is not easy to see, but his contention that there was an antecedent set of behaviors upon which the taboo was built allows the construction of a nondisjunctive evolutionary sequence.

This paper attempts to vindicate Westermarck's position by compiling evidence from both human and nonhuman primates. It will be shown that the incest taboo is not "a kind of remodeling of the biological conditions of mating and procreation," but resonates with and reinforces a biological tendency to outbreed—a tendency which has apparently arisen in response to prolonged maturation, group life, and socialization of the young.

■ INCEST AVOIDANCE IN HUMANS

It is difficult to find situations in human groups which can test the Westermarck hypothesis. In many societies marriage and intercourse between adult pairs raised as closely as siblings is precluded by exogamy rules and, where significant degrees of endogamy are practiced—as among Latin American barrios, Indian castes, or groups which in-marry for religious reasons—the childhood association of marrying couples is not so intense as in the sibling relationship. Two situations have been studied, however, which illustrate what happens when children are socialized together as siblings and then allowed to satisfy mutual sexual attraction.

The collective child-rearing practices of Israeli kibbutzim pose one such situation. In one kibbutz, Melford Spiro (1954) observed no cases of marriage within a peer group although there were no sanctions against such a practice. Parents would actually have preferred such marriages since these would tend to strengthen the integrity of the kibbutz. In another study, Yonina Talmon (1964) studied 125 couples in two kibbutzim, and states there was "not one instance in which both mates were reared from birth in the same peer group" (ibid.: 492). She states (ibid.: 504) further that young people themselves believe that overfamiliarity breeds disinterest.

> They refer to the curiosity, excitement and anticipation that unfamiliar people evoke in them and to the exhilarating sense of discovery and triumph they get when they establish a relationship with one of them.

In a large-scale study, Joseph Shepher (1971) computer-analyzed census data from all Israeli kibbutzim. The analysis revealed that of 2,769 marriages which took place within the kibbutzim, only thirteen couples were from the same peer group. In eight of those, partners became members of the same peer group after the age of six, while in the other five cases the time spent together in the same peer group was never more than two years of the period from birth to six years of age.

It is interesting that a great deal of prepubertal sex play is allowed peer group members, even to the extent that young children sometimes simulate intercourse. With the beginning of puberty, however, these same children find each other quite distasteful and sexual interests are directed outside the peer group. It is not entirely clear, however, that sexual indifference is arising only as a result of cosocialization. Spiro (1954: 846) writes, "When they are asked for an explanation of this behavior, these individuals reply that they cannot marry those persons with whom they have been raised and whom they, consequently, view as siblings." The avoidance may thus result from an extension of the incest taboo to a situation which is considered sociologically similar to the sibling relationship. Such an explanation is doubtful, however, since peer group members do not call each other by brother and sister terms. Furthermore, the nuclear family continues to associate at least once a day, making clear the distinction between sibling and peer.

In a carefully conceived series of investigations Arthur Wolf (1970) has looked for indices of sexual dissatisfaction in an ethnographically unusual marriage pattern which has occurred frequently in China, the *sim-pua* marriage. This pattern, which Wolf calls minor marriage because it carries relatively low social prestige, is unique in that children are betrothed at a young age, the future daughter-in-law usually coming to live with her husband's family when she is less than one year old. In the more prestigious major marriage, the bridal pair does not normally meet until the night of their wedding: a wedding which contrasts sharply with the minor-marriage ceremony in being publicly and conspicuously celebrated. Chinese marriage practices thus provide a situation where marriage partners socialized together as siblings can be compared with marriage partners who are for the most part strangers.

An initial comparison of the two marriage forms was made by the collection of anecdotal material which begins to give one an idea of what partners to minor marriages think of their situation:

> One old man told me that he had to stand outside of the door of their room with a stick to keep the newlyweds from running away; another man's adopted daughter did run away to her natal family and refused to return until her father beat her; a third informant who had arranged minor marriages for both of his sons described their reactions this way: "I had to threaten them with my cane to make them go in there, and then I had to stand there with my cane to make them stay." (ibid.: 508)

A more detailed comparison of the two marriage forms was made possible by the existence of household registration records kept by the Japanese during their occupation of Taiwan. Wolf's hypothesis was that if familiarity did indeed breed sexual disinterest there would likely be differences between minor and major marriages in frequencies of divorce, separations, and illicit sexual liaisons, as well as differences in fertility rates.

In his preliminary study (1970) utilizing the Japanese registration records, Wolf's hypothesis about differential rates of fertility was apparently confirmed; major marriages produced 30% more children than minor marriages. However, in order to make certain he was not measuring the relative fecundity of adopted daughters who are notoriously ill-treated, Wolf compared adopted daughters who had for some reason married in the major fashion with adopted daughters who had minor-married as intended. Adoption was found not to be the determinant of low rates of fertility since those adoptees who major-married, likewise, had nearly 30% more children than those who minor-married.

Similar comparisons were made measuring divorce and adultery rates. Of 132 minor marriages, 46.2% ended in divorce or involved adultery by the wife as opposed to 10.5% in the 171 major marriages surveyed. Again, Wolf was concerned that he might be measuring the results of adoption, rather than the results of childhood propinquity, so the marital careers of adopted daughters who had major-married were surveyed. Only 9.5% of these marriages ended in divorce or involved adultery as opposed to the 46.2% noted in the group above.

In work (Wolf 1974) on household registers which has continued since 1970, there has been an incorporation of data on a third form of marriage in China. In this marriage pattern the husband goes to live with his wife's father to fulfill economic obligations caused by his inability to pay a bride price. These uxorilocal marriages are at least as low in prestige as minor marriages, and men who so marry are called "barbarian's cows" (ibid.: 12). Comparing differential fertility rates for minor marriage, major marriage, and uxorilocal marriages therefore allows one to assess the effects of low marriage prestige on fertility. Prior to the incorporation of the uxorilocal marriage pattern into the study, the criticism had been made that minor marriages produced fewer children because of the low social status of the marriage. However, it has been found that uxorilocal marriages produce even more children than major marriages in spite of their exceedingly low prestige. The suggestion that low fertility in minor marriage is a function of low status therefore seems unlikely.

The incorporation of data on uxorilocal marriages has also made improbable other explanations not based on Westermarck's hypothesis that propinquity leads to disinterest. Since its partners are more likely to seek sexual satisfaction outside of marriage, the suggestion was made that minor marriage may harbor a higher incidence of venereal disease than major marriages, thus causing a decline in fertility. However, one would expect a similar situation in the uxorilocal case since such marriages are often made by prostitutes. Another reinterpretation of the data is that since minor marriage partners are usually about the same age the woman is in a relatively more powerful position than her peer in a major marriage and thus more likely to reject the sexual demands of her husband. Again, however, the data on uxorilocal marriages makes this possibility unlikely since women in those marriages also enjoy a relatively powerful position due to the husband's dependence on the wife's property. When the wife's father dies, the wife is given title to the land or it is registered with her sons who trace their descent through the mother. The husband owns nothing.

In his most recent work (personal communication), Wolf has established a linear relationship between the husband's age when the wife is adopted and the percentage of marriages ending in divorce. Of the marriages in which the husband's age was four or less when the wife was adopted, 16.4% ended in divorce; 12.0% ended in divorce when the husband was between five and nine at the time of adoption; and 5.4% ended when he was ten or more at the time of adoption. A similar linear relationship was established between the husband's age when the wife was adopted and the percentage of minor marriages never completed. Clearly, childhood association does not enhance sexual attraction, but erodes it.

Wolf has offered an alternative to a Westermarckian explanation of his data. Since the *sim-pua* couple address each other just as siblings do, "The negative reaction to the alternative form of marriage may be the result of an inadvertent extension of the incest taboo rather than an example of the conditions that give rise to the taboo"

(1966: 893). As was stated earlier, this same interpretation was entertained by Melford Spiro (1954) in his description of Israeli kibbutzim.

At this point in the argument evidence from nonhuman primates gains particular significance since they have no language-based prohibition to prevent inbreeding. If a significant amount of incest avoidance can be found among nonhuman primates it must be explained by a natural, rather than a verbal, inhibition. Next it must be shown that as in humans, avoidance occurs in conjunction with the continuous association made necessary by prolonged socialization. If this is indeed found to be the case, the suspicion aroused by both the Chinese and Israeli studies—that an extension of the incest taboo is insufficient to explain why propinquity causes sexual disinterest—will be substantiated.

■ INCEST AVOIDANCE IN NONHUMAN PRIMATES

The most detailed studies on the inhibition of incest mating among primates come from the colony of rhesus macaques (*Macaca mulatta*) on Cayo Santiago Island, Puerto Rico. Identification of individual monkeys and construction of genealogies have made it possible to determine mother-son and brother-sister matings. (Since paternity is impossible to ascertain, the frequency of father-daughter matings cannot be measured.)[1] During the test period, observers on the island witnessed 363 copulations, only 4 of which were between mother and son (Sade 1968: 18–20). This is surprising in light of studies on grooming interactions among rhesus monkeys which show that close relatives are a monkey's preferred grooming companions. Loy and Loy (1974) have observed that siblings groom each other up to five times as much as would be predicted from a random selection of grooming partners, while Sade (1965) observed that 40% of one male's grooming bouts were with his mother while 52% of his nongrooming, passive contact pairings were with her. These indicators of affect are also highly visible among consort pairs, yet they do not lead to mating in mother-son dyads.

The only time Sade personally observed a mother-son mating, a male succeeded in establishing dominance over his mother after a series of fights spread over the first three months of the breeding season. At the end of the fourth month, copulation occurred between the two, dominance roles having been reversed one month before when the male finally succeeded in beating up his mother. Incestuous relations did not occur between mother-son pairs where the male was subordinate. On the basis of this observation, Sade speculates that the mother-son inhibition rests on dominance relationships but notes that males do not hesitate to mate with other females of higher rank, which seems to point to a specific mother-son inhibition. Sade (ibid.: 37) concludes his study as follows:

> We may now speculate that a pre-existing condition became invested with symbolic content during hominization; the origin of at least the mother-son incest taboo may have been the elaboration of a phylogenetically older system, a system which can still be observed operating at the monkey level of organization.

Other studies on Cayo Santiago have shown a significant amount of intergroup transfer by males (Carpenter 1942; Koford 1966), but it has not been clear if these transfers were a result of overcrowded conditions or if they bore any relation to sexual behavior. Lindburg's (1969) study of rhesus monkeys in northern India, however, has shown that such transfers are not confined to dense populations and do indeed coincide with the mating season. When groups first detected a newcomer they responded aggressively, but such encounters usually abated after three or four days. Transfers were both temporary (lasting between 30 and 100 days) and permanent. While only 4% of the total rhesus population migrated, it is significant that one-third of the adult males changed groups.

The earlier investigations by Koford (1966) yield an interesting piece of information correlating troop size with the number of males who leave. During a four-year period, 10% of the males of a troop with 263 members migrated elsewhere. This is a small percentage when compared to the 21% in a troop of 72 or the 42% in a troop of 46 and suggests that where the probability of inbreeding is increased by small group size, the probability of transfers which prevent inbreeding also increases. One would, of course, expect this patterning of transfers if close association led to sexual disinterest.

Drickamer and Vessey (1973) have recorded male transfers in two island populations of rhesus macaques at La Cueva and Guayacan, Puerto Rico. Their data also shows that migrations peak during the mating season and that population density does not affect group-changing. Of the males that changed groups, one did so before the age of three and 50% did so during the ages of three and four. All of the males reaching seven years of age left their natal troop; mother's rank did not influence the age at which monkeys transferred.

Japanese macaques (*Macaca fuscata*) have also been closely observed for nearly 25 years and their incest activity noted. Again, establishment of monkey colonies and construction of genealogies have made possible the collection of such data. Although the Japanese reports on incest inhibition are not as detailed as the reports from Cayo Santiago, both Imanishi (1961) and Takuda (1962) state that a mother-son mating has never been observed among many hundreds of monkeys. Junichiro Itani (1972) summarizes the evidence for incest avoidance among primates in a recent article. He (ibid.: 166–167) describes a mechanism which helps insure nonincestuous matings among Japanese macaques:

> The detailed research of Koyama and others during the fission of the Arashiyama troop showed that the troop divided into two halves each consisting of one hundred animals. The sixteen consanguineal groups of the original troop were divided between A-troop and B-troop with no break in the consanguineous relationship. Many of the young males followed their consanguineous relatives; that is, they joined the troop to which their mothers belonged. However, within one year an unusual change occurred: The young males who first belonged to A-troop migrated to B-troop, and those of B-troop joined A-troop. We are unable to construct a satisfactory sociological theory to explain this phenomenon, but we

can note that the probability of incest occurring was significantly reduced in the situation where the males left their maternal troop, as opposed to the situation where they did not.

Incest avoidance among savannah baboons is also striking. Craig Packer (personal communication) has been observing one species of savannah baboon, *Papio anubis*, in the Gombe Stream Reserve. These animals live in troops of 40 to 80 individuals with as many as 30 males per group. Packer has observed 100% male transfer between these groups—movement taking place when the baboons are young adults or older subadults. Males are not reproductively active in their natal group in spite of the fact that they sometimes rise high in the dominance hierarchy.

Where intergroup transfer appears among nonhuman primates, it obviously reduces the likelihood of incestuous matings. But a difficult question is whether incest avoidance is the cause or the consequence of these migrations. In some circumstances it appears that young adults are peripheralized and forced from their natal troops by their low position in the dominance hierarchy which not only subjects them to the antagonism of more dominant members of the troop but limits their access to females. But low dominance cannot account for all cases of intergroup transfer since many young males are high in the hierarchy by virtue of their being sons of dominant females (Sade 1967; Marsden 1968). Among baboons and macaques, therefore, where nearly all young males emigrate, there are many individuals whose transfer is not prompted by low dominance status but occurs for another reason. Drickamer and Vessey (1973: 366) observed that

> Males born to high ranking females, and therefore high ranking themselves, left the group at the same mean age as those born to low ranking females. Therefore, males probably left the group voluntarily rather than being forced out due to low rank.

The fact that male macaques and baboons are not reproductively active adults in their natal troop would account for these voluntary migrations. And since this inactivity precedes emigration it indicates that migration is not simply a mechanism which prevents incest but may, in fact, be motivated by incest avoidance.

Indian common langurs, *Presbytis entellus* (Hrdy 1974; Mohnot 1971; Sugiyama 1965, 1966), display two types of troops. One is composed of females, their offspring, and an adult male who acts as genitor and defender and is removed after three to five years of leadership. The second kind of troop, from whose ranks come many of the successors to the adult males of the bisexual troops, is composed only of males older than infants. Although it is not yet clear how universal social change is among langurs, the replacement of the adult male of the bisexual troop by challenge or default (he may die or otherwise be forced to leave the troop) is widespread. Where the adult male is present and healthy, an adult from an all-male troop may succeed in driving him off. In case of default through death or accident, a successor may either emerge from an all-male troop or already be the adult leader of another bisexual troop. In either case the consequences are the same: the male progeny of the deposed leader are driven from their natal troop and both male and female infants are killed.

The likelihood of brother-sister and mother-son incest is dramatically reduced since all males are expelled from their natal troops at least by the time they are five years old, and father-daughter incest is limited by the removal of the adult male leader every three to five years. Since sexual maturity is reached in four years there are times when inbreeding could occur barring another preventive mechanism, but that possibility is reduced by the infanticide practiced by new leaders. In short, this pattern of langur social change appears to effectively prevent incest.

But while this is significant, it does not look as if langur social change is determined only by incest avoidance. Incest avoidance appears to be universal among higher primates, and while social change is widespread among langurs, there are areas of low population density where the existence of bisexual troops with more than one adult male indicates that social change in these areas has a different pattern (Yoshiba 1968). Another variation is found among Nilgiri langurs who appear to practice leadership change but not infanticide (Poirier 1974: 134). This variation in langur social change indicates that the phenomenon is dependent on more than a disinclination to inbreed. Instead, high population density, coupled with langur social structure, may be a determinant (Yoshiba 1968; Hrdy 1974) along with male-male intolerance when females are near (Poirier 1974: 132–133). Given the advantages and ubiquity of incest avoidance in other nonhuman primates, viable langur populations which do not display social change will probably express the disinclination to inbreed through another mechanism such as intergroup transfers by young adults. Although it prevents incestuous matings, social change is complex and may be masking the more usual expression of incest avoidance—migrations from the natal troop.

Studies of apes also illustrate the inhibition against intrafamilial mating, although slower maturation and fewer animals have made the accumulation of data a lengthy process. With her study reaching its tenth year, Jane van Lawick-Goodall (1971) reported that there was an apparent inhibition operating on mother-son incest, as a mother-son copulation was never witnessed. There was a somewhat less effective inhibition against brother-sister intercourse as well:

> We were interested to discover that Fifi was extremely reluctant to be mated by her brothers. She even prevented little Flint from mounting her—though in the days before her first true swelling she had shown no objection whatsoever. Moreover, though Faben and Figan were observed to mate with their sister—after much screaming on her part—subsequent sexual interactions between the siblings occurred rarely. (ibid.: 182)

Anne Pusey (personal communication) has also studied chimps in the Gombe Stream Reserve. Female intergroup migration was seen on a number of occasions— some adolescent females taking up permanent residence in a new group, while at least two others made temporary transfers.

Migration of female chimpanzees has also been observed by the Japanese at study sites in the Mahali Mountains (Nishida and Kawanaka 1972). Two groups of chimpanzees were studied and females from each were seen taking up temporary residence in the other group. Thirty-nine episodes of migration were recorded; other episodes undoubtedly took place but were not detected. Length of residence in the new group seemed determined by the sexual condition of the female, estrous, or potentially estrous, females staying longer than anestrous females with infants. The majority of females who had grown up in one of the groups moved out of it when they reached sexual maturity, prompting the authors (ibid.: 162) to make the following statement: "The life history of a chimpanzee female can be considered to be that she is born in a unit-group, grows there and transfers to another for the first time when she attains sexual maturity." These female migrations seen by Pusey and Nishida would be effective in preventing inbreeding among chimps just as the male transfers described by Lindburg, Itani, Drickamer, Vessey, and Koford would prevent inbreeding among macaques.

Among the monogamous gibbons, inbreeding is prevented by parents driving their same-sex offspring from the home range (Carpenter 1940; Ellefson 1968). Chivers and Aldrich-Blake (1973) observed the expulsion of a subadult male during study of one species of gibbon, *Hylobates syndactylus*. Initially, the subadult was kept away from the family group by the adult male but still remained within their territorial range. As he became increasingly peripheralized, however, the adult's aggressive behavior increased, and the subadult's calling bouts began lasting twenty minutes longer than those of his parents. A young female was eventually seen in the area, and after she took up residence with the subadult in a new territory, the authors speculated that the subadult's prolonged calls had been an attempt to find a mate. It is interesting that this entire episode "coincided with the onset of sexual activity between the adult pair" (Chivers and Aldrich-Blake 1973: 635).

Richard Tenaza (personal communication) has observed an adolescent female gibbon moving from her parental range to a contiguous range inhabited by an unmated male. The female, who was already peripheralized when observations began, would join the male in the late afternoon and sleep with him in the same tree. In the morning she was seen returning to her parent's range.

Although less work has been done on gorillas, certain observations have been made which may point to inbreeding avoidance among that species as well. Schaller (1963: 104–108) describes several instances of male and female intergroup transfer, while Itani (1972: 169) thinks a mobile male population is indicated since many groups have only one silverback male. Studies by Dian Fossey (personal communication) complement these observations, for she has seen males practicing "bride capture" by stealing young females from groups other than their own.

■ THE SIGNIFICANCE OF INCEST AVOIDANCE IN NONHUMAN PRIMATES

Wolf's study of Chinese minor marriages demonstrates that childhood association does not enhance sexual attraction between adult pairs, but erodes it. Along with a

similar lack of attraction between couples raised in the peer groups of Israeli kibbutzim, this makes likely the interpretation that the incest taboo is not primarily responsible for motivating individuals to make sexual choices outside the family. Nevertheless, the ubiquitous presence of the incest taboo and the human facility in language
make it very difficult to design a research strategy which will control for unintended
generalizations of the taboo. Peer group members or *sim-pua* couples may exhibit
sexual disinterest because they consider their relationship to be sibling-like. But the
lack of language facility among nonhuman primates makes it possible to effectively
control for these effects of the taboo, making the near absence of incestuous matings
among monkeys and apes particularly significant.

The similarities of incest avoidance in human and nonhuman primates promises
to be one way to arrive at an understanding of this behavioral complex which has had
such a profound influence on human institutions. In both groups, disinterest in
inbreeding develops after several years of close association; both groups share the
components of social organization which make close association inevitable; both have
maturation periods lasting many years; and both share a great deal of the
neurophysiological substrata which govern reproductive behavior. The enormous
complexity which underlies the similarities of all these traits attests to the validity of
making substantial comparisons between human and nonhuman incest avoidance
while searching for an explanation of the complex. All of the primate species described in this paper exhibit inbreeding inhibitions. Female versus male migration, a
wide extension of avoidance among baboons, gibbon intolerance of nonmate conspecifics, and the diverse patternings of human incest behavior are internal variations
on the same theme—the sexual attraction an individual feels for members of its own
natal unit is largely absent by the time adulthood is reached.

■ THE SOCIALIZATION OF INCEST AVOIDANCE

The preceding sections presented data showing that there is little primary incest and
inbreeding among human or nonhuman primates even when there are no jural
prohibitions against inbreeding. However, there have been few attempts to describe
the mechanisms that underlie this erosion of intrafamilial sexual attraction.

While Freud disagreed with Westermarck's idea that childhood association an-
nihilates sexual desire, his view of the dynamics of family life drew him into agreeing
on the need for incest avoidance. Although "the first choice of object in mankind is
regularly an incestuous one" (Freud 1920: 294), he felt that this sexual attraction was
doomed to frustration because of the opposition of parents and rival siblings. Their
hostility towards the child's desires finally culminates in his repressing the impulse.
Sexual love is split from affectionate love, and while incestuous feelings continue to
affect adult personality, they are not usually consciously or behaviorally apparent in
well-adjusted individuals.

On the other hand, Robin Fox (1962) thinks that permissiveness towards children's sex play is the component of socialization that eventually leads to sexual
disinterest. Sex play causes arousal, but Fox contends that the experience is necessar-

ily frustrating because consummation is impossible. If socialization practices do not interfere with prepubescent sex play, constant repetition of the frustrating experience will lead to "negative conditioning" and disinterest in the childhood partner. Fox tests the hypothesis by examining cross-cultural data on both severity of incest prohibitions and proscriptions on prepubescent sex play. He reasons that adults in societies that permit children's sex play will have few incest wishes and those societies will therefore have mild punishments for incest or no punishments at all. Conversely, individuals in societies that prohibit childhood sex play will not be subject to "negative conditioning" and will harbor incest wishes as adults; these societies will proscribe severe punishments for incest in order to counteract such wishes. The results of the survey support the original hypothesis of "negative conditioning" since cultures which permit childhood sex play tend to have mild incest prohibitions (e.g., Israeli kibbutzim and the Tallensi), while cultures like the Chiricahua Apache and Trobriand Islanders that prohibit childhood sex play exercise capital punishment in cases of adult incest. Nonhuman primates would also appear to support Fox's theory since sex play among infants is seldom interfered with and adult interest in incestuous matings is low.

But though Fox's cross-cultural data are intriguing, his interpretation of the socialization mechanism underlying avoidance is not persuasive. Fox assumes that preadolescents experience a coital drive which is stimulated, but then left unsatisfied, during play. Few would argue with Fox's observation that children exhibit a sex drive, but that it necessarily remains unsatisfied is debatable. Fox, for instance, cites Meyer Fortes's (1949: 251) description of an incident of childhood sex play among the Tallensi:

> They were both in a state of high excitement, panting and giggling and muttering to each other, with obvious sexual pleasure. They seemed to be oblivious of their surroundings. This game went on for about twenty minutes after which they separated and lolled back as if exhausted. These two children were most attached to each other.

If Fortes's description of the sex play sequence is accurate, it contradicts Fox's belief that childhood sex play must end in frustration and tears. Furthermore, the Kinsey studies (Kinsey et al. 1953: 104) on human sexual behavior conclude that the sexual responses of preadolescent children and even infants terminate in sexual orgasm. If children are able to achieve a type of climactic response in sex play, then Fox's theory of "negative conditioning" is most likely incorrect.

Arthur Wolf (1966) has proposed that the inhibition of aggression is the component of socialization that eventually leads to sexual disinterest. He cites Frank Beach (1951: 408) who has observed that "male mammals often fail to copulate in an environmental setting previously associated with punishment." Wolf (1966) then speculates that since the socialization of aggression is a universal aspect of child-rearing, it could account for the widespread occurrence of sexual disinterest within the primary family where children are punished not only for striking their parents but for fighting with each other as well. "We need only to assume that such natural impulses as sex and aggression have a common subjective component that can serve as a basis for generalization" (ibid.: 893). Paul MacLean (1963) has done work on squirrel monkeys (*Saimiri sciareus*) which illustrates the neurological base upon which such a common component might rest. During electrical stimulation of the limbic system, a structure found in all mammalian brains, MacLean (ibid.: 26) found that:

> Within the space of a millimeter, one may pass from a point at which stimulation results in erection and an apparent state of placidity to one at which the electrical current elicits erection in conjunction with an angry or fearful type of vocalization and showing of fangs. As one lowers the electrode a little deeper one may obtain only fearful or angry-like manifestations during stimulation, but see erection appear as a rebound phenomenon after stimulation is terminated.

A possible relationship between the socialization of aggression and incest disinterest can also be seen in Sade's (1968) description of the one case of incest described earlier in this paper where copulation occurred after the male succeeded in establishing dominance over his mother. Mounting in other dominance interactions also seems to reflect a similarity underlying both aggression and sex, supporting the hypothesis that the suppression of one of the behaviors may induce an inhibition in the other.

■ A NEUROPHYSIOLOGICAL MODEL FOR THE SOCIALIZATION OF INCEST AVOIDANCE

With Wolf's inhibition of aggression hypothesis in mind, a series of ablation and stimulation experiments on the amygdaloid complex prove to be intriguing. Kluver and Bucy (1937) performed bilateral lesions of the temporal lobe and reported the following set of symptoms: compulsive oral behavior, loss of usual fear and aggression, hypersexuality, and visual agnosia. It was then discovered that the entire syndrome associated with temporal lobe lesions was, in fact, due to the bilateral destruction of the amygdala (Pribram and Bagshaw 1953). The occurrence of hypersexuality along with loss of aggression remind one of Wolf's hypothesis, but the alterations in behavior caused by amygdalectomy have proven complex. First, attempts at finding discrete locations within the amygdaloid complex for specific behaviors failed, and it was also noted that behavioral alterations due to amygdalectomy tend to fade after years or months—hypersexuality being the first to disappear (Gloor 1960). One abnormality does not disappear, however, and that is the abnormal response to novelty demonstrated by attention and reaction to every visual stimulus. It is this that Karl Pribram (1971) has concentrated on in arriving at a description of amygdaloid function.

The normal response to novelty can easily be seen in all mammals including humans. When there is an unexpected flash of light or a sudden horn blast the organism immediately reacts to the stimulus; and there are physiological measures of its orienting, such as changes in galvanic skin response, heart rate, and respiration. If the stimulus is repeated frequently, this startle reaction fades and the organism is said to habituate to the stimulus. Eugene Sokolov (1960) demonstrated that this orienting reaction occurs not only when there is a sudden increase in noise, but also when there is a drop in stimulus intensity.

> Until Sokolov's demonstration the assumption was always made that habituation simply raises the threshold of the nervous system to input. Sokolov's findings mean that the person who has habituated must be matching the current sound against a stored representation of prior tone beeps—why else would a diminution in intensity call forth again the full-blown orienting response? . . . We have all experienced this surprising reaction to sudden silence. (Pribram 1971: 49)

When monkeys have their amygdalas removed, orienting and habituation to novel stimuli do not follow this normal pattern. The changes in galvanic skin response, heart rate, and respiration which characterize the normal orienting reaction are absent, and there is a failure to habituate to repeated stimuli. The consequences of this are profound. Schwartzbaum and Pribram (1960) have shown that amygdalectomized monkeys do poorly in tasks that require them to remember what they learned from a similar previous task. These monkeys treat the new task as completely novel and perform poorly whereas normal monkeys are able to make use of what they learn in the original task. In intact animals,

> Electrical changes have been recorded from the amygdaloid complex of the limbic systems whenever the organism is exposed to a novel event or one that has meaning in terms of reward and punishment. These electrical changes subside once the organism is familiar with the event. . . . (Pribram 1961: 571)

Upon the occurrence of a novel event the amygdala is at first disequilibrated but is stabilized during habituation to the event's continued presence. This habituation causes a configuration of the event to be stabilized in neural hardware which allows its registration in awareness and memory (Pribram 1967). When the amygdala is damaged this registration cannot take place.

> Such a lack of registration is a commonplace in clinical epileptic seizures originating from abnormalities around the amygdala, abnormalities which also produce the famous *deja vu* (inappropriate feeling of familiarity) and *jamais vu* (inappropriate feeling of unfamiliarity) phenomena. (Pribram 1971: 205–206)

But if the amygdala somehow monitors the organism's familiarity or unfamiliarity with events, how can this affect sexual behavior? Why is it that familiarity causes sexual disinterest? There are receptor-like surfaces near the third and fourth ventricles of the brain that are known to be sensitive to changes in glucose concentration, osmotic pressure, temperature, etc. These have been shown to be the sites that monitor and control some of the organism's more primitive functions such as hunger, thirst, and temperature regulation. The receptor-like qualities of this area of the brain are analogous to the receptor surface provided by skin. In fact, these periventricular receptors develop in embryogenesis from the same layer of ectodermal cells as does the skin. This periventricular area may also house receptors that monitor and control chemical substances that influence behaviors such as sex and perhaps fear and aggression (Pribram 1971: 183). If this is the case, we are in a position to explain the lowering of intensity during socialization of both aggression and sexual attraction. The amygdala has strong neural connections with the periventricular area. It is therefore possible that the electrical changes caused by novelty are carried from the amygdala to the periventricular area and disequilibrate receptors in a selective fashion creating feelings of sexual attraction, fear, or aggression, depending on the amount of novelty in the situation. This interpretation would account for the behavioral effects caused by electrode stimulation of the amygdaloid complex. "The most common behavioral responses elicited by stimulation of the amygdala are reactions of attention, fear and rage . . . upon increasing the intensity of stimulation, attention will merge into fear and finally lead to rage" (Gloor 1960: 1406).

Heath (1954) describes a similar response to increasing stimulation of the amygdala in a human patient. Low levels of stimulation elicited attentiveness and, as stimulation was increased, the patient first exhibited flirtatious behavior and, finally, fear. Thus, lower affective behaviors such as attention, exploratory behavior, and sexual arousal are elicited in situations of substantial novelty, whereas higher affective states such as fear and aggression are elicited by extreme novelty. A sexual object is one that is novel enough to cause arousal, but not so novel as to cause fear, although the two feelings are often present in the sexual experience. Extreme novelty—and those situations are the most uncertain and potentially harmful for the organism— elicits feelings of fear or aggression, predisposing the organism to flee or fight.

Among the higher primates, where maturation is an extended affair lasting many years, social organization provides constant care and protection for infants until they reach adulthood. But infants themselves must learn to interact successfully with other members of the group. One code of behavior must be learned with respect to peers, another for elders, and even another for other animals that may populate the region. This socialization period is far from peaceful but is often punctuated with punishment as the maturing individual is made to understand and remember the behavior expected of him. At first, most experiences are novel and frightening, but over the years socialization gradually imparts a familiarity that acts as a guide in predicting the behavior of others in the social unit and one's own appropriate response to any event.

Early in socialization, when the novelty of one's social universe is nearly total, activity in the amygdala will be correspondingly high as new events are registered and remembered. Memory is active; learning is quick. Input to periventricular receptors involved with feeling states is also high and their frequent disequilibration will be accompanied by sexual arousal, fear, and even rage. Parents are often amazed at the strength and range of their children's emotional expressions. They lack what adults call "self-control," and they frequently amuse us with exhibitions of ecstasy, grief, or rage that are so profound but ephemeral that they seem but caricatures of our own powerful emotions. Childhood is a time of seething and conflicting emotions that often subject the developing organism to feelings of fear and hate for those primary socializers whom he loves and whose love he needs. But as sexual maturity is reached, and one's social environment and the behavior of close and familiar individuals becomes predictable, disequilibrations of the amygdala, and the receptor sites it feeds, become less intense. Novelty, and the array of emotions it arouses, are replaced by the less emotional but more peaceful and predictable state of familiarity. Now it is individuals outside of the immediate social unit who are most likely to become enemies and lovers. Yet people, institutions, or events sometimes conspire to make a person feel incapable of moving outside the family to establish meaningful and emotional relationships with other individuals. In this very sad and difficult situation the sexual urge is occasionally directed toward other members of the family, causing frightening problems in interpersonal adjustment as well as severe genetic problems if children should result from the relationship.

■ THE GENETIC ADVANTAGES OF INCEST AVOIDANCE

One result of outbreeding is an increase in heterozygosity over homozygosity. Geneticists now postulate that a heterozygous condition is superior for at least two reasons: it provides an organism with biochemical versatility, and it provides a population with greater evolutionary plasticity. Recent findings have shown completely recessive alleles to be quite rare. Rather, every gene contributes to the genetic product even though that contribution may not be visible. A classic example of this is the resistance to malaria infection found among persons who carry the gene for sickle-cell anemia in a heterozygous pairing. When the gene is present in a homozygous pairing, the effects of anemia are devastating, but the recessive has nevertheless been retained in the gene pool by the superior adaptability of the heterozygote in areas where malaria is common.

Heterozygosity provides variability to a population by providing a greater assortment of genes and, thereby, a larger number of genetic products. When both parents are homozygous for a trait controlled by two alleles, for instance, there is only one possible resultant; where one parent is heterozygous two results are possible; when both are heterozygous there can be three genetic products. For traits which are determined by more than two alleles, a great deal of variability can result. A locus with eight alleles, for instance, holds thirty-six genetic possibilities.

Such versatility makes genetic variability available at all times for an immediate evolutionary response to a change in the environment. The population as a whole thus possesses great evolutionary plasticity. Furthermore, being composed of several genotypes, a population that is polymorphic owing to heterosis can better utilize different components of the environment (different subniches). (Mayr 1970: 138)

The genetic argument which is usually made in accounting for the existence of the incest taboo cites the high incidence of abnormality found among children born of incestuous relationships. Inbreeding permits the pairing of recessive genes and thereby allows their expression in the phenotype. This is dangerous since the majority of deleterious genes are found to persist among recessives where they can be hidden from the pressures of natural selection. When such genes become paired, however, their expression in the phenotype is made possible and their deleterious effects are exposed.

Frank Livingstone (1968) has analyzed this argument and concludes that "inbreeding does not increase the death rate from homozygosity, but rather decreases the frequencies of deleterious genes" (ibid.: 46). In other words, inbreeding ought to be advantageous since it would rid the gene pool of deleterious recessives by exposing them to natural selection. It is therefore incorrect, says Livingstone, to argue that increased death rates due to homozygosity are the basis for the formation of the incest taboo. Livingstone notes, however, that where large amounts of deleterious recessives have accumulated in the gene pool, it might take a very long time for them to be eliminated by natural selection when they become exposed through the occurrence of inbreeding. Lévi-Strauss (1956: 276–277) has similarly criticized this genetic explanation for the incest taboo. He points out that the incest taboo makes possible the accumulation of deleterious recessives by preventing inbreeding, and that if there had been no taboo, the dangers of inbreeding would not exist. Livingstone's and Lévi-Strauss's criticisms develop a major flaw, however, if one looks at the general evolutionary trend. Among lower animals deleterious recessives will accumulate, since the social ordering which makes a significant amount of inbreeding possible is absent. Thus by the time animals are making the adaptation to social life, a large amount of deleterious recessives will have already built up in the gene pool because of the outbreeding of nonsocial ancestors. At this point, a high rate of inbreeding finally becomes a real possibility since social ordering forces animals to live with their close relatives. The following studies indicate that if inbreeding were to then begin to remove deleterious recessives in the manner described by Livingstone, the process might prove too costly for the species, especially if it takes a very long time. It would lower the frequency of recessive alleles that are deleterious when homozygously paired, but it would also lower the frequency of advantageous heterozygote pairings and a great deal of genetic variability would be lost.

C. O. Carter (1967), an English physician, examined 13 incest cases. His observations began before the pregnancies came to term and ended when all the living children were between four and six years old. Father-daughter incest had occurred in 6 cases while brother-sister incest accounted for the remaining 7. One child eventually died of cystic fibrosis of the pancreas, a common autosomal recessive condition arising from inbreeding; another died of progressive cerebral degeneration with blindness; a third died of Fallot's tetralogy; and a fourth child was severely subnormal, able to speak only a few words at age five. Four of the children were subnormal with I.Q.'s between 50 and 75, while the remaining 5 were normal. Of the 13, then, 4 (31%) died or suffered from a major abnormality.

Adams and Neel (1967) have conducted a similar study, selecting 18 cases before the pregnancies came to term. Four of the children died; 2 were retarded in addition to having seizure disorder and spastic cerebral palsy; 1 had a bilateral cleft lip; and 3 had I.Q.'s of about 70. The remaining 7 were normal at six months. Of these 18 children, then, 6 (33%) died or suffered from a major abnormality.

A study conducted in Czechoslovakia by Eva Seemanova (1971) is particularly significant because of its large sample size and its use of the same mothers as controls. Of 161 children born of incest, 88 were from father-daughter, 72 from brother-sister, and 1 from mother-son relationships. Two of the children were stillborn while 21 others died after birth. Sixty were placed in the abnormal category, while 78 were placed in the normal group (mild retardation and slight abnormalities were included). For this sample, then, 83 (51%) died or were significantly abnormal. Forty-six mothers were studied separately so that children fathered by unrelated males could be compared to their half-sibs born of incest. Of 92 children not born of incest, 5 died and 3 suffered abnormalities (9%). The same mothers gave birth to 50 children fathered by primary kin. Six of these children died and another 20 were significantly abnormal (52%).

These high rates of death and abnormality suggest that the capacity for incest avoidance may have been selected for at the time animals begin living in social groups. But if attention, arousal, emotion, and memory in mammals involves processes like those outlined by Pribram and others, the capacity for incest avoidance may not have been selected for at the time animals became social and were exposed to the dangers of inbreeding. Instead it seems more likely that the capacity for incest avoidance did not have to be selected for on the basis of death and abnormality rates but was always present in the structure of the mammalian mind, and only became apparent when animals began living in social groups.

■ CONCLUSION

Wolf's (1970) study of Chinese minor marriages demonstrates that childhood association does not enhance sexual attraction between adult pairs, but erodes it. Along with a similar lack of attraction between couples raised in the peer groups of Israeli kibbutzim, this makes likely the interpretation that the incest taboo is not primarily responsible for motivating individuals to make sexual choices outside the family. But since both the Israelis and the Chinese have a taboo against brother-sister incest, it is difficult to control for unintended generalizations of the taboo to those situations which are similar to the brother-sister relationship. However, one can control for the incest taboo by looking at nonhuman primates who also avoid incest though they have no jural prohibition that makes them do so.

Due to the genetic advantages of heterozygote matings and the high death and abnormality rates found in children born of incest, it seems reasonable to assume that incest avoidance was selected for at the time mammals developed social organization and were exposed to the dangers of inbreeding. However, it appears that incest avoidance is caused by a mode of operation common to all mammalian brains and was therefore available for use before the need for it actually arose. Only when long-maturing mammals were organized into social groups did the capacity become behaviorally apparent. Because of the need to express this capacity, humans are known to suffer severe problems of psychological and interpersonal adjustment when people, institutions, or events conspire to prevent them from establishing meaningful relationships outside the family and thereby force a sexual choice to be made among primary relatives.

▪ ACKNOWLEDGMENTS

I am grateful to Dr. Suzanne Chevalier-Skolnikoff, Dr. Karl Pribram, and Dr. Arthur Wolf for their criticism and helpful suggestions.

▪ NOTE

1. An interesting case of incest-like avoidance is cited by Mitchell and Brandt (1972: 187):

The third-ranking male, Boris, in the Oregon troop of *Macaca fuscata* showed parental behavior toward several juveniles. He cared for a female, Gamma, for at least eighteen months, until she was four years old. He groomed, defended, and huddled with her; and his behavior was similar to the protective behavior shown by a mother to her two- or three-year-old offspring. When Gamma became sexually receptive, she and Boris each mated with other monkeys but, curiously, not with each other. Gamma delivered an infant the following birth season, and both before and after her delivery she remained close to Boris who contacted, groomed, and defended her.

▪ REFERENCES

Adams, M. S., and Neel, J. V. 1967. Children of incest. *Pediatrics* 40:55–62.
Beach, F. A. 1951. Instinctive behavior: Reproductive activities. In *Handbook of experimental psychology*, ed. S. S. Stevens. New York: John Wiley and Sons.
Carpenter, C. R. 1940. A field study in Siam of the behavior and social relations of the gibbon (*Hylobates lar*). *Comparative Psychol. Monographs* 16:1–212.
Carter, C. O. 1967. Risk of offspring of incest. *Lancet* 1:436.
Chivers, D. G., and Aldrich-Blake, F. P. G. 1973. On the genesis of a group of Siamang. *Am. Journal of Physical Anthro.* 38:631–636.
Drickamer, L. C., and Vessey, Stephen. 1973. Group changing in Rhesus monkeys. *Primates* 14:359–368.
Ellefson, J. O. 1968. Territorial behavior in the common white-handed gibbon, *Hylobates lar linn.* In *Primates: Studies in adaptation and variability*, ed. P. C. Jay, pp. 180–200. New York: Holt, Rinehart, and Winston.
Fortes, M. 1949. *The web of kinship among the Tallensi*. Oxford: Oxford University Press.
Freud, S. 1920. *A general introduction to psychoanalysis*, trans. J. Riviere. New York: Liveright.
Gloor, P. 1960. Amygdala. In *Handbook of physiology, neurophysiology II*, ed. J. Field, H. W. Magoun, V. E. Hall, pp. 1395–1420. Washington, D.C.: American Physiological Society.
Heath, R. 1954. *Studies in schizophrenia*. Cambridge: Harvard University Press.
Hrdy, S. B. 1974. Male-male competition and infanticide among the langurs (*Presbytis entellus*) of Abu, Rajasthan. *Folia Primat.* 22:19–57.
Imanishi, K. 1961. The origin of the human family—A primatological approach. *Japanese Journal of Ethnology* 25:119–130.

Itani, J. 1972. A preliminary essay on the relationship between social organization and incest avoidance in non-human primates. In *Primate socialization*, ed. F. E. Poirier, pp. 165–171. New York: Random House.

Kinsey, A. C.; Pomeroy, W. B.; Martin, C. E.; and Gebhard, P. H. 1953. *Sexual behavior in the human female*. Philadelphia: W. R. Saunders.

Kluver, H., and Bucy, P. C. 1937. "Psychic blindness" and other symptoms following bilateral temporal lobectomy in rhesus monkeys. *Am. Journal of Physiology* 119:352–353.

Koford, C. B. 1966. Population changes in rhesus monkeys: Cayo Santiago 1960–64. *Tulane University Studies in Zoology* 13:1–7.

Lawick-Goodall, J. van. 1971. *In the shadow of man*. Boston: Houghton Mifflin.

Lévi-Strauss, C. 1956. The family. In *Man, Culture, and Society*, ed. H. L. Shapiro, pp. 261–286. London: Oxford University Press.

Lindburg, D. G. 1969. Rhesus monkeys: Mating season mobility of adult males. *Science* 166:1176–1178.

Livingstone, F. B. 1969. Genetics, ecology and the origins of incest and exogamy. *Current Anthro.* 10:45–63.

Loy, J., and Loy, K. 1974. Behavior of an all-juvenile group of rhesus monkeys. *Am. Journal of Physical Anthro.* 40:83–97.

MacLean, P. D. 1963. Phylogenesis. In *Expression of the emotions in man*, ed. P. H. Knapp, pp. 16–36. New York: International Universities Press.

Marsden, H. M. 1968. Agonistic behavior of young rhesus monkeys after changes induced in social rank of their mothers. *Animal Behavior* 16:38–44.

Mayr, Ernst. 1970. *Populations, species, and evolution*. Cambridge: Harvard University Press.

Mitchell, G., and Brandt, E. M. 1972. Paternal behavior in primates. In *Primate Socialization*, ed. F. E. Poirier, pp. 173–206. New York: Random House.

Mohnot, S. M. 1971. Some aspects of social changes and infant-killing in the hanuman langur, *Presbytis entellus* (Primates: Cercopithecidae) in Western India. *Mammalia* 35:175–198.

Nishida, T., and Kawanaka, K. 1972. Inter-unit-group relationship among wild chimpanzees of the Mahali Mountains. *Kyoto University African Studies* 7:131–169.

Poirier, F. 1974. Colobine aggression: A review. In *Primate aggression, territoriality, and xenophobia*, ed. R. Holloway, pp. 123–158. New York: Academic Press.

Pribram, K. H. 1961. Implications for systematic studies of behavior. In *Electrical stimulation of the brain*, ed. D. E. Sheer, pp. 563–574. Houston: University of Texas Press.

———. 1967. The limbic systems, efferent control of neural inhibition and behavior. In *Progress in Brain Research*, vol. 27, eds. W. R. Adey, T. Tokizane, pp. 318–336. Amsterdam: Elsevier.

———. 1971. *Languages of the brain: Experimental paradoxes and principles in neuropsychology*. Englewood Cliffs: Prentice-Hall.

Pribram, K. H., and Bagshaw, M. H. 1953. Further analysis of the temporal lobe syndrome utilizing frontotemporal ablations in monkeys. *J. of Comparative Neurology* 99:347–375.

Sade, D. E. 1965. Some aspects of parent-offspring and sibling relations in a group of rhesus monkeys, with a discussion of grooming. *American J. of Phys. Anthro.* 23:1–18.

———. 1967. Determinants of dominance in a group of free-ranging rhesus monkeys. In *Social Communication Among Primates*, ed. S. A. Altmann, chap. 7. Chicago: University of Chicago Press.

———. 1968. Inhibition of son-mother mating among free-ranging rhesus monkeys. *Science and Psychoanalysis* 12:18–38.

Schaller, 1963. *The Mountain gorilla: Ecology and behavior*. Chicago: University of Chicago Press.

Seemanova, E. 1971. A study of children of incestuous matings. *Human Heredity* 21:108–128.

Shepher, J. 1971. Mate selection among second generation kibbutz adolescents and adults: Incest avoidance and negative imprinting. *Archives of Sexual Behavior* 1:293–307.

Sokolov, E. N. 1960. Neuronal models and the orienting reflex. In *The central nervous system and behavior*, ed. M. A. B. Brazier, pp. 187–276. New York: Josiah Macy, Jr., Foundation.

Spiro, M. E. 1954. Is the family universal? *Am. Anthropologist* 56:839–846.

Schwartzbaum, J. S., and Pribram, K. H. 1960. The effects of amygdalectomy in monkeys on transposition along a brightness continuum. *J. Comp. Physiol. Psychol.* 53:396–399.

Sugiyama, Y. 1965. On the social change of hanuman langurs (*Presbytis entellus*) in their natural conditions. *Primates* 6:381–418.

————. 1966. An artificial social change in a hanuman langur troop (*Presbytis entellus*). *Primates* 7:41–72.

Talmon, Y. 1964. Mate selection in collective settlements. *Am. Sociological Review* 29:491–508.

Tokuda, K. 1962. A study on the sexual behavior in the Japanese monkey troop. *Primates* 3:1–40.

Westermarck, E. 1922. *The history of human marriage.* 3 vols. London: Macmillan and Co.

Wolf, A. P. 1966. Childhood association, sexual attraction, and the incest taboo: A Chinese case. *Am. Anthropologist* 68:883–898.

————. 1970. Childhood association and sexual attraction: A further test of the Westermarck hypothesis. *Am. Anthropologist* 72:503–515.

————. 1974. More on childhood association and fertility in Taiwan. Unpublished manuscript.

Yoshiba, K. 1968. Local and intertroop variability in ecology and social behavior of common Indian langurs. In *Primates: Studies in adaptation and variability,* ed. P. C. Jay, pp. 217–242. New York: Holt, Rinehart, and Winston.

12 The Role of Learning Phenomena in the Ontogeny of Exploration and Play

JOHN D. BALDWIN
JANICE I. BALDWIN

■ INTRODUCTION

Exploration and play are unusually complicated behavioral patterns to describe and explain. The motor patterns involved in these activities are extremely varied and are influenced by numerous complex and interacting factors including genetic, physiological, maturational, and environmental variables. An animal's age, sex, past learning experiences, social relationships, physical environment, and personal idiosyncracies all contribute to the causal network that determines the ontogeny of play. Clearly, a multifactor theory of play is needed to account for the complex intertwining of variables that affect exploration and play.

In attempting to locate a central strand of theoretical issues that carry through and tie together much of the data on play in primates, we have selected several themes that center around learning theory, especially the type of learning that is shaped by the reinforcers of sensory stimulation. These themes constantly reappear in dealing with many aspects of exploration and play and serve as a central topic around which to organize other data on physiology, maturation, nutrition, ecology, species adaptations, and so forth. The importance of learning phenomena in exploration and play has already been shown by the numerous studies that have focused on the functions of these behaviors (Lorenz 1956; Welker 1961; Loizos 1967; Eibl-Eibesfeldt 1970; Dolhinow and Bishop 1970). As is apparent from Table 1, most of the adaptive functions that have been postulated to result from exploration and play experience involve learning: for example, the practice of adult skills, learning up-to-date information about the environment, learning one's place in the social structure, learning reproductive skills, and learning how to manipulate objects that could be of potential use. In addition, the primary reinforcers associated with sensory stimulation have been implicated as the major motivational determinants of exploration and play (Schultz 1965; Butler 1965; Ellis 1973). This paper will attempt to demonstrate that numerous other reinforcement and learning phenomena are central to understanding the dynamics of exploration and play in primates and aid in explaining many of the features of these behaviors that have been described in the literature.

The present emphasis on learning is not intended to imply that other causal factors are not influencing the ontogeny of exploration and play. Species genetic, physiological, and maturational factors explain a large portion of the variance in many behaviors, and their role in constraining or biasing the course of learning in different species has been well documented (Breland and Breland 1966; Seligman and Hager 1972; Hinde and Stevenson-Hinde 1973). It is the contention of this paper, however, that learning deserves a central role in explaining the ontogeny of exploration and play. Genetic, physiological, and maturational variables have been woven into the present learning approach in the areas where their impact has been demonstrated (especially in the sections on early development, sexual differences and species differences).

The present theory attempts to specify the natural contingencies of reinforcement that shape the form and frequency of exploration and play throughout the course of development. Whereas many previous theories emphasized that sensory stimulation is important in determining play, and some even pointed out that it is a reinforcer, none began to elaborate in detail the ongoing contingencies of reinforcement that shape the ontogeny of exploration and play activities and how these change throughout the individual's developmental history.

▪ 1. EXPLORATION AND PLAY PROVIDE SENSORY STIMULATION.

To begin with, it is important to recognize that exploration, inquisitiveness, curiosity, manipulation, and play all have much in common as has been pointed out by Beach (1945), Nissen (1951), Lorenz (1956), Bindra (1959), Berlyne (1960), and others. First we will evaluate the features these various activities have in common, then, in later sections, progress to specify the differences among them. Throughout the paper, the words "exploration" and "play" will be used in the broad, general sense to incorporate all inquisitive types of behavior, except where special modifiers—such as "social play"—are employed to specify more restricted usage. Following Welker (1961), we tend to associate exploration with more careful, restrained, and cautious investigation, whereas play (in either its social or nonsocial forms) suggests more unrestrained and active behavior.

One of the first steps in conducting a learning analysis of behavior is to ask: "What consequences follow immediately after the behavior under analysis?" This strategy allows one to either (a) isolate the reinforcers that control the behavior or, (b) determine that the behavior is not under reinforcement control. Exploration, manipulation, and play all tend to share the common trait that they produce, and hence are followed by, sensory stimulation. The sensory stimulation can be in the form of novel events, alterations in the stimulus situation, increased stimulus intensity, or increased stimulus complexity. When an exploring or playing animal hops into new positions, looks at the world upside down from between its legs, or leaps after a playmate through the vines, the player's activities alter its sensory input in ways that produce novel experiences, changes in stimulus intensity, increased stimulus complexity, and so forth. In fact, one of the features of play that most clearly distinguishes it from nonplayful behavior is that play produces novel, varied, and jumbled sequences of activity, whereas nonplayful behavior is more routine, habitual, predictable, and devoid of novel variations (Marler and Hamilton 1966; Loizos 1967; Miller 1973).

It is important to note that the novel, intense, or complex sensory input produced by play can come from both outside and within the body. The proprioceptive or introceptive inputs from twisting and turning the body in new, intense, or complex manners may be as important as sensory inputs from the external environment. Although it is more difficult for the observer to monitor introceptive stimulation, by being sensitive to this type of sensory input one can at least appreciate that many playful activities generate introceptive stimuli important to the player.

The long range implications of internal sensory stimulation for the developing organism have been shown in the research on isolate rhesus monkeys. Sackett (1971) has shown that although isolate monkeys typically display numerous abnormal responses, those isolates that have received electrical stimulation to the brain behave much more like normal animals when tested from 9 to 17 months.

A. DEFINITIONS OF SENSORY STIMULATION.

Although there are numerous definitions of sensory stimulation, most point to the central role of stimulus novelty. Novelty indicates something about both the stimulus and the perceiving organism. First, the more complex, varied, unusual, and rare a stimulus pattern is, the more likely that a randomly chosen individual will find it novel. Second, the longer a given individual experiences a given stimulus, the less likely it is that the stimulus (or similar stimuli) will be perceived as novel. Novelty points to a transaction between physical stimulus properties and the past learning experience of the organism.

On the stimulus side, Fiske and Maddi (1961) have delineated three properties of stimuli that can contribute to the total "impact" of those stimuli on the organism— variety, intensity, and meaningfulness. Stimulus variety or complexity usually enhance the sensory stimulation available from a source by increasing the number of possible novel combinations of patterns within the whole. Variations in color, hue, contrast, pitch, intensity, texture, or movement within a stimulus unit tend to increase the potential novelty. The degree to which a stimulus alerts or arouses an organism also depends on the intensity of the stimulus—low intensities may not easily be detected, high intensities can be overarousing, and the continuum of intermediate intensities can evoke a variety of responses. The meaningfulness or amount of new information that a stimulus carries also enhances the sensory stimulation value of an input. For example, a conditioned stimulus of low intensity and low complexity can induce a high degree of arousal if it is associated with strong positive or negative reinforcers.

On the organism side, the phenomena of habituation, recovery, and learning affect an individual's response to a given stimulus (Welker 1961; Eisenberger 1972). After a novel stimulus is presented several times to an organism, the animal begins to habituate to the stimulus and ceases to find it (or similar stimuli) novel. However, when deprived of the particular stimulus for a sufficient period, the animal recovers from some of the habituation effects and the stimulus partially regains novelty value. Young, inexperienced individuals are likely to find most of the common or easily available stimuli of their environment more novel than older animals do, merely because they have had less chance to habituate to and learn the commonplaceness of these stimuli. However, as individuals grow older, some learn perceptual, cortical, or motor skills that open access to new realms of stimulus qualities that were unavailable to younger or less skilled animals. An older individual may carefully dissect an object and attend to all the components; whereas a younger, less skilled, conspecific might leave the object after a cursory investigation of its exterior (Menzel 1969).

In a very important sense, the world is filled with endless sensory stimulation for those with the skills to locate and extract it. A learning theory of play makes it possible to specify what factors promote or hinder the acquisition of the perceptual, cortical, and motor skills that make the less accessible stimulation available (see sections 6, 7 and 8 below).

■ 2. SENSORY STIMULATION CAN FUNCTION AS A REINFORCER THAT STRENGTHENS EXPLORATORY, MANIPULATIVE AND PLAYFUL BEHAVIORS.

Butler (1958, 1965) has shown that sensory stimulation is an important reinforcer to primates. Rhesus monkeys in a modified Skinner box learned to lever-press when the only consequence was that a window opened and allowed them to view the environment outside the box. Visual input is only one type of sensory stimulation affecting one of the sense modalities. Mason, Hollis, and Sharpe (1962) demonstrated that young chimps would pull an operant lever that opened a window for access to play with a human caretaker. Harlow (1950), Harlow, Harlow, and Meyer (1950) and Harlow and McClearn (1954) have shown that rhesus monkeys would learn to solve puzzles, open latches and manipulate objects merely for the sensory stimulation of doing novel activities. Orangutans, chimpanzees, and other species have also been shown to learn complex tasks with only the reinforcers of manipulation and sensory stimulation (Döhl and Podolczak 1973). Campbell (1972) has demonstrated these effects in squirrel monkeys. Both males and females learned to operate capacitance probes that activated lights of different intensities and colors. Increases in both stimulus intensity and variety were found to be reinforcers—with increases in light intensity and changes in color, the animals responded with increasing frequency. Continuous exposure to the sensory stimulation reinforcers led to satiation within approximately 1 hour; but alternating periods of access (15 minutes per hour) did not lead to an equivalent level of satiation until 8 hours later. Data to be presented in subsequent sections also demonstrate that appropriate levels of sensory stimulation are reinforcers that strengthen other types of exploratory and playful behaviors, and that satiation or habituation operate to weaken the reinforcement effects.

A. MOTIVES FOR PLAY.

Various authors (Buhler 1930; Morris 1962; Dobzhansky 1962; Eibl-Eibesfeldt 1970; Dolhinow and Bishop 1970; Miller 1973; Döhl and Podolczak 1973) have noted that play does not involve extrinsic reinforcers such as food, water, sex, etc., and have concluded that play is *intrinsically rewarding*, is done for the sheer *fun* or *pleasure* of it, or is the consequence of an internal drive. Since reinforcement does produce the phenomenological experiences of "pleasure," "reward," and sometimes even a "drive" to repeat the experience, these other formulations of the motives of play may not be incompatible with the present theory, though the terminology and emphasis often differ considerably.

■ 3. SENSORY STIMULATION THAT BRINGS OR KEEPS AN INDIVIDUAL WITHIN ITS OPTIMAL AROUSAL ZONE IS REINFORCING; BUT OVERSTIMULATION AND UNDERSTIMULATION ARE AVERSIVE AND PUNISHING.

Sensory stimulation from extroceptive or introceptive sources can excite or arouse various neural, hormonal and muscular systems to different levels. Hebb and Thompson (1954), Schneirla (1959), Fiske and Maddi (1961), Schultz (1965), Zuckerman (1969), Ellis (1973), and others have noted that on a continuum from low to high physiological arousal, there is an optimal level of internal excitation for each organism. Excessive levels of sensory input tend to produce overarousal; and overarousal is aversive and hence a negative reinforcer. Too little sensory input leads to suboptimal levels of central arousal, i.e., to boredom; and underarousal is also aversive and hence a negative reinforcer. Between the extremes of overarousal and underarousal is an optimal zone; and optimal arousal is a positive reinforcer. Thus, the organism will tend to avoid or escape places and activities that are associated with overstimulation or understimulation; and it will tend to approach, instigate, or maintain situations that provide optimal levels of arousal. Readily visible examples of these phenomena appear in the exploratory behavior of primates in the laboratory and in their natural habitats. (More subtle phenomena will be elaborated in later sections.)

In the laboratory, the role of optimal arousal is easily visible in the early exploratory activities of the young primate infant. As Mason (1965a, 1968a, 1971), Harlow and Harlow (1965, 1969), and others have shown, the following pattern is common in young primates. The infant vacillates between activities that are arousal-increasing and arousal-reducing as it alternates between exploring the environment and returning to its mother. When the exploring infant is exposed to sensory input above its optimal level, it runs to its mother (or surrogate mother) to cling. The mother is a source of low-level sensory input in the form of warm, soft, familiar sensory input; and clinging makes these stimuli relatively more salient compared with the overstimulating outside environment. After clinging to the arousal-reducing mother for a period, the infant becomes understimulated and will tend to venture away from the low-level sensory input of the mother and out into the environment where the novelty of exploration and play activities provide increased sensory input and arousal-induction. (The mother can also serve as a source of arousal-induction if she plays with or otherwise provides vigorous stimulation for the infant.) When play or exploration becomes overarousing, the infant again flees from the source of overarousing inputs and returns to its mother and clings to her warm, soft, familiar body.

The arousal-reducing properties of the mother provide input that can reduce the infant's arousal level into the optimal zone when the infant has become overly excited. But too much contact with the mother can reduce the infant's arousal level into the suboptimal zone, and after a period of contact the infant will leave the mother to escape aversive understimulation. The cycle of going out to explore and play, then returning to mother, is shown in figure 1. As Fiske and Maddi (1961) point out, an important feature in the timing of the cycles of exploration and play is the inertia in the arousal systems—the infant's behavioral systems (nervous, muscular, endocrine systems) do not instantaneously increase or decrease their activity as soon as there is an increase or decrease in sensory input. The infant may cling to its mother for several seconds or minutes to recover or calm down from an overstimulating bout of exploration. Likewise, the exploring infant may be exposed to a novel environment for several minutes before its arousal level becomes aversively high and the infant returns to its mother. This inertia explains why the arousal decline after "A" and increase after "B" in figure 1 are not instantaneous. (The arousal curve in this figure is

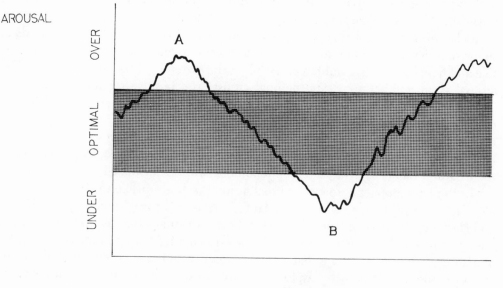

FIGURE 1. *After becoming overaroused during exploration or play (A), the infant returns to its mother for arousal-reducing contact. After a period of arousal-reduction, which may or may not take the infant below its optimal zone (B), the infant returns to exploration or play as arousal-inducing activities. Thus, periods of arousal-induction and arousal-reduction tend to alternate.*

not shown as a smooth line in order to indicate that one would expect numerous perturbations from various internal and external input channels.)

In the wild, young primates of many species show the same patterns of using the mother as a "base of explorations" from which they investigate the world. When the troop is quiet—such as during a general rest period—the infant leaves its low-arousal mother and explores the environment. However, as the environmental inputs cause increases in general arousal (which are often indicated by increased motor activity, vocalization, defecation, urination, or general agitation), the infant becomes over-aroused and returns to its mother for arousal-reduction. This pattern is vivid in the field descriptions of langurs (Jay 1965), chimpanzees (van Lawick-Goodall 1967a, 1967b), gorillas (Schaller 1963), baboons (Hall and DeVore 1965), squirrel monkeys (Baldwin 1969), howler monkeys (Baldwin and Baldwin 1973b), and others. The pattern has also been reported for human children (Ainsworth 1964; Rhinegold and Eckerman 1970).

The optimal level of sensory stimulation is most graphically conspicuous at the age when the young infant is actively moving back and forth between its arousal-reducing mother and the arousal-increasing environment. However, optimal levels of

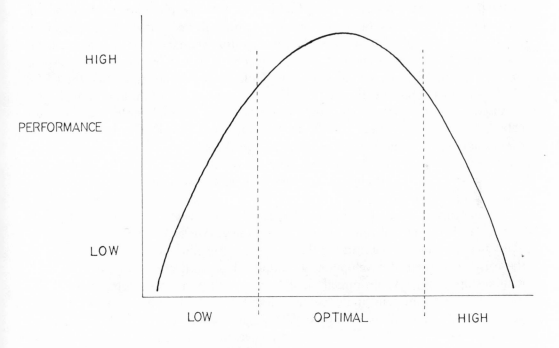

FIGURE 2. *Underarousal tends to produce grogginess, lethargy, or sleepiness which retard performance. Overarousal tends to produce anxiety, CNS overload, or distractions that interfere with performance. Optimal arousal is most compatible with efficient performance.*

sensory input are reinforcing before and after this age, as will be demonstrated in later sections. Hebb (1955), Leuba (1955), Welker (1956a, 1956b, 1961, 1971), Bindra (1959), Schneirla (1959), Fiske and Maddi (1961), Schultz (1965), Hinde (1966), Zuckerman (1969), and others have demonstrated this type of relationship in numerous ways. For example, Hebb (1955) and Ellis (1973) have demonstrated the relationship between arousal levels and optimal performance as an inverted U (Figure 2). Too little input is conducive to sleep, grogginess, lethargy, and boredom, and too much input is highly disruptive, anxiety-evoking, and distracting. An intermediate level of arousal is conducive to efficient learning and performance of behaviors.

More recently it has been determined that different subsets of an animal's behavioral repertoire may have different optimal levels, and that there are diurnal variations in optimal arousal preferences related to physiological changes induced by sleep-wakefulness cycles, hunger-satiation cycles, etc. (Fiske and Maddi 1961; Zubek 1969; Hebb 1972; Ellis 1973). Also, as discussed later, there are many idiosyncratic differences among the animals within a species in the optimal zone for a given behavior.

A. MEASUREMENT PROBLEMS.

At the present stage of research, it is difficult to measure the physiological correlates of arousal or determine the optimal arousal zone for a given animal doing a specified behavior. General arousal can often be inferred from such external cues as vocalizations, muscle tension, response latency, anxiety urination or defecation, nervous movements, or other behavioral indicators. Duffy (1957, 1962) argues that the EEG, GSR, and EMG provide the best physiological measures of arousal, though cardiovascular and temperature measures are useful, too. Neurophysiological data indicate that the reticular system and diffuse thalamic nuclei are involved in monitoring the arousal level of the central nervous system (Welker 1961; Schultz 1965; Ellis 1973). Presumably, measurements based on the brain's own monitoring mechanisms would provide the most useful determinations of an animal's optimal zone or current arousal level.

In this paper, however, it is less important to know *what* an individual animal's arousal level and optimal zones are, than it is to acknowledge that optimal stimulation plays a central role in determining the dynamics of exploration and play activity. In the early or theoretical development of empirical generalizations, one can establish relationships (e.g., *x* varies directly or inversely with *y*) or make comparative statements (e.g., older animals find *x* less arousal-inducing than younger animals) without precise quantification of the variables. Working on a theoretical level, it will suffice to say that rough-and-tumble play is likely to be overarousing for a young nursing infant monkey, but optimally arousing for an independent juvenile. For example, Mason (1968a, 1971) has made significant progress toward explaining early mother-infant relations by utilizing the simple dichotomy of identifying arousal-inducing versus arousal-reducing behaviors and environments. Basic patterns can be sketched out and the theoretical implications evaluated even before the methodology for precise measurement is developed.

Also, the idea of "precise measurement" may obscure the fact that all measurements fall on a continuum of relative precision. Although we are far from having good measures of current arousal and optimal zones, that does not mean that no useful measures can be made. In the field and in the laboratory, one often sees indicators that permit the application of arousal theory. Overarousal is frequently indicated by distinctive vocalizations, by anxiety urination or defecation, by quick or nervous movements, by a retreat to familiar areas or social companions, and so forth. Underarousal is indicated by stimulus seeking, by stereotyped behaviors, by restlessness, by attempts to leave familiar areas or social companions, etc. Also, the degree of novelty or familiarity of a stimulus for a given animal can be inferred from data on the extent to which the individual has had prior experience with similar stimuli. Sophisticated methodologies are being developed for measuring the physiological components of arousal. The EEG, GSR, and EMG are widely used measures of physiological arousal (Duffy 1957, 1962; Zubek 1969; Zuckerman 1969; Schultz 1965). Candland and Nagy (1969), Candland and Mason (1969), Mason, Hill and Thomsen (1971), and others have shown that heart rate is a good indicator of arousal. J. W. Mason (1968), Hill et al. (1973), and Sackett et al. (1973), have shown the usefulness of plasma cortisol levels in obtaining independent measures of arousal level. Meyer and Knobil (1967) and Brown et al. (1971) have shown that growth hormone titer also varies with stress and arousal level, though it reflects different mechanisms from those stimulating plasma cortisol release.

Clearly, many of the problems of measurement still lie ahead; but as more correlations are made between overtly visible behavior and physiological factors, the more easily applicable sensory stimulation theory will become. Skinner (1969) and Andrew (1972) have forcefully argued that because of the methodological problems of making measurements on physiological states, it is advisable to specify motivational states in terms of overtly visible antecedent conditions. That is, hunger can be specified by the number of hours of food deprivation or the organism's ratio of present body weight to free feeding weight. At present, these measures can be obtained more easily and precisely than physiological measures of hunger. An important area of development for sensory stimulation theory is the specification of overtly visible antecedent conditions that determine the nature of exploration and play activities. This will require extensive empirical study. In the meantime, we believe that sensitization to sensory stimulation and arousal theory is a major step in helping to utilize the presently available indicators of stimulation and arousal.

■ 4. EXPLORATION AND PLAY EMERGE FROM EARLY REFLEX-LIKE BEHAVIORS AS THESE BECOME SHAPED INTO MORE COMPLEX, OPERANT STIMULUS-SEEKING BEHAVIORS VIA THE REINFORCERS OF OPTIMAL SENSORY STIMULATION.

The laboratory research by Butler (1958, 1965) has shown how a particular operant pattern of exploratory behavior emerges in adults. When the monkeys were placed in a closed compartment with a lever that opened a window looking out onto a sensory-rich environment, the animals learned the lever-press response in the same manner that other types of operant behaviors are learned when other reinforcers are used. At first the lever was pressed only by accident and the window opened as an operant consequence. Several repetitions of this experience were necessary before the animals began to learn the habit of regularly pressing the lever. Once the response of lever-pressing for sensory input was well established, the sensory reinforcer could be used to train other, more complex, discriminations and responses.

There are differences in the ways an infant learns exploration and play behavior compared with the patterns displayed by the older animals in Butler's experiment. Most neonatal behaviors are highly reflex-like, based more on phylogenetically "wired-in" patterns than on prior learning experience. In the neonate, shaping by optimal arousal reinforcement begins at a more basic level than in Butler's experiment.

From our observations on howler and squirrel monkeys, we suspect that operant conditioning of early exploratory behaviors begins in the first few days of life.[1] Before the conditioning of exploration begins, the infant's basic clinging, crawling, rooting, sucking, and other reflex-like[2] behaviors are probably responsible for many of the infant's first movements. When these reflex-like behaviors are followed by the reinforcers of optimal sensory stimulation, they begin to be strengthened and shaped into explorative responses. Through the process of shaping, new behavioral patterns begin to emerge, the exact form of which will depend on the contingencies of reinforcement in each individual's own environment. The behavior of infant squirrel monkeys demonstrates these processes. Schustermann and Sjoberg (1969) have shown that newborn squirrel monkeys have numerous reflex-like responses, which make them actively behaving animals from the beginning of life. The head-lifting orientation response, climbing response, persistent grasping, and early prehensile use of the tail are among the primitive infant responses that are present in the first days of life. There is natural variance in these primitive behaviors, and some of this variance is followed by reinforcement while other variations are followed by aversive consequences. This differential reinforcement leads to the beginning of operant conditioning.

One example of how infant exploration is shaped can be seen in the processes by which early crawling and clasping behaviors are modified by operant feedback. First, if an infant begins to crawl on its mother's body during a troop rest period, the crawling will probably not lead to aversive consequences. The mother is stationary and the infant (with a strong clinging response) is unlikely to fall off during its slow movements. The movements produce gentle activity and various tactile and proprioceptive experiences as the infant's body moves over the mother's body. Because these sensory experiences are novel for the infant, it is feasible to assume that they bring sensory stimulation reinforcers. (Section 5 will further explain why these early behaviors tend to be optimally arousing for the young infant.) Thus, crawling during troop rest periods is reinforced and gains greater habit strength. Naturally, crawling is enhanced by other factors such as the infant's increasing physical strength during day-to-day maturational development and the reinforcers of reaching the nipples and nursing. The point we wish to make is that early reflex-like behaviors become modified by reinforcers, and sensory stimulation reinforcers are among those operative. Because the infant's crawling behaviors are reinforced during troop rest periods, the stimuli associated with troop rest become S^D's (discriminative stimuli) that begin to acquire stimulus control properties to evoke crawling during future periods of troop rest. For example, the S^D's that come to evoke learned crawling responses might include (a) the quietness and immobility of the mother, (b) general inactivity of nearby animals, and (c) low noise level.

In contrast, the infant's crawling response will become inhibited in other S^D contexts, such as troop travel or presence of an aggressive animal. When an infant begins to crawl on the mother's body while the troop is beginning to move from one area to the next, the infant is very likely to be caught in an awkward position—such as riding backwards or clinging precariously to the mother's hind legs—as the mother begins to move. After several experiences of this nature, the infant learns inhibitions about crawling during certain stimulus conditions. The infant will tend to discriminate on cues—such as the mother's restlessness, certain vocalizations, the movement of other animals—that correlate with the onset of troop progression; and these cues become S^D's which control response inhibitions from the infant.

Therefore, from the very beginning of life, the reinforcers of novelty operate in shaping early reflex-like behaviors into exploration in certain S^D contexts. The negative reinforcers of overarousal and hard knocks inhibit explorative behaviors in other contexts.

One set of reflex-like behaviors—which includes alerting, directing attention, and startling—may play an especially important role in initiating the learning of exploration and play. These reflexes are triggered by a sudden stimulus onset or the appearance of novel stimuli and, hence, they are especially likely to expose the infant to sensory stimulation. In contexts associated with optimal levels of input, alerting and orienting responses are very likely to be strengthened and elaborated into more complex sequences of stimulus-seeking. However, it must be noted that any of a complex combination of early reflex-like behaviors can provide the initial responses from which exploration and play behaviors will be shaped. Crawling, hand-flexing, rooting, mouthing, etc. can all lead to novel input in some S^D contexts and hence become reinforced into early explorative patterns, i.e., into behaviors which tend to be strengthened under the control of sensory stimulation rather than food, water, thermoregulation, or other reinforcers.

When a reflex-like behavior such as crawling is modified by milk reinforcers during sucking or by the aversive consequences of being hit by a thorny twig, the casual observer may be inclined to utilize purposive jargon[3] and say that the infant has learned to crawl "in order to get to the nipple," or that he has learned not to crawl during troop movement "in order to avoid getting hurt by branches." The learned responses appear to be functional, "purposeful," and "goal directed" when closely associated with conspicuous positive reinforcers such as food, water, sex, and thermoregulation or with negative reinforcers such as cuts, stings, falls, cold, and hunger. The perception of sensory stimulation is a less conspicuous reinforcer and often has been overlooked as the factor that shapes exploration and play. However, when the infant starts crawling about on the mother's body, feeling her hairs, and moving its hands through her fur, most observers are likely to label it exploration (or even playfulness). Because exploration and play have often been called "intrinsically motivated," we suspect that most observers often use the words exploration and play when a behavior is under the control of sensory stimulation reinforcers rather than the more conspicuous positive and negative reinforcers mentioned above.

Thus, the infant's early reflex-like crawling behaviors begin to differentiate into several new habitual behaviors, under the control of different S^D's. After several learning experiences, hunger S^D's begin to control crawling to the nipples; troop movements begin to control inhibition of crawling; and troop quiet periods begin to control tactile and locomotor exploration.

A second example of the role of reinforcement in the ontogeny of exploration can be seen in visual orientation behavior. The infant's first eye-opening responses are the consequence of numerous prenatal influences, and the early gazes often look blank, untargeted, slow, and uncoordinated. During early experience, the eye-opening and looking responses become shaped by reinforcers and come under stimulus control of environmental cues; the reflex-like qualities begin to disappear and directed orientation becomes apparent.[4] If the infant's looking behavior is reinforced by finding food in the presence of certain S^D's, then the animal will develop stronger habit strength for "looking for food" in similar S^D contexts. The behavior takes on orientation, directedness, coordination with the environment, and apparent "purpose" as it becomes shaped by differential reinforcement. Food reinforcers shape "looking for food." Thermoregulatory reinforcers shape "looking for and approaching sources of warmth" (or coolness). And the reinforcers of sensory stimulation shape the behaviors that are typically identified as exploration, inquisitiveness, curiosity, and play. If an infant's early looking behavior is followed by the reinforcers of novel sensory input, which is very frequently the case for neonates, then the looking behaviors will become shaped by operant conditioning into strong habits. Simultaneously the responses come under finer S^D control for orientation to novel stimuli.

Exploration can involve any of the sense modalities. Thus there is explorative touching, moving, smelling, mouthing, tasting, looking, and listening. Any of these behaviors which increase the likelihood of experiencing optimal sensory stimulation will become reinforced under the categories of exploratory or playful behavior. If explorative or playful behaviors are reinforced by no other reinforcers except sensory stimulation they represent "pure" cases of stimulus-seeking behaviors. If a certain behavioral pattern also results in the discovery of a tasty morsel of food, the responses become a complex mixture of "pure" exploration and exploration in the service of foraging (Welker 1971). When other reinforcers become involved, one can have exploration in the service of finding sex partners, water, optimal temperatures, etc. We find it useful to conceptualize stimulus-seeking behaviors on the following continuum (see Figure 3): (1) *pure* exploration and *pure* play at one end, controlled by sensory stimulation reinforcers alone; (2) *normal* exploration and play covering the adjacent half of the continuum, predominantly, but not completely, controlled by sensory stimulation reinforcers; (3) *mixed* exploration and play covering the next 30%

or 40% of the continuum, with sensory stimulation being dominated by other reinforcers; (4) exploration in the service of food, sex, water, thermoregulation, and so forth, covering the remaining portion of the continuum; and (5) nonexplorative and nonplayful behaviors at the second pole. (Sex play would be an example of mixed play—containing more sexual elements than normal play but less than exploration and play in the service of sexual interaction.) The locations of the transition points on the continuum are offered as tentative suggestions. Although we have been talking in terms of "pure" exploration up to this point, it is important to realize that pure cases of stimulus-seeking behavior with only one controlling reinforcer are probably less common than "normal" exploration and play. Much of our discussion will emphasize the role of sensory stimulation reinforcers in controlling exploration and play because (1) much of normal exploration and play occurs near the pure stimulus-seeking end of the continuum, and (2) it is easier to describe the pure type than to constantly add caveats about the possible admixture of other reinforcers. We hope the reader will bear these facts in mind and realize that multiple reinforcers are operating in most cases of exploration and play. As we progress to later sections, it will become clear that multiple reinforcers occur in much of exploration and play and that these are important in explaining the transition from youthful playfulness to nonplayful adult behavioral patterns.

Naturally, exploration and play can lead to overarousal and the aversive consequences of sensory overload. When this occurs, exploration is temporarily inhibited and avoidance responses are reinforced. When an infant becomes overaroused, it can retreat to its mother or avoid input in other ways. Visual input, for example, can be terminated by closing the eyes.

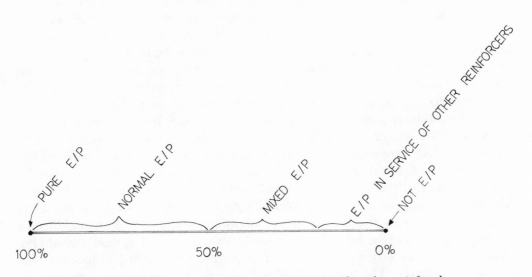

FIGURE 3. *There is a continuum of behaviors from those reinforced only by sensory stimulation (left end) to those reinforced by the other primary reinforcers (right end). (E/P = exploration and/or play.)*

To summarize this section, those behaviors that are learned due to sensory stimulation reinforcers are usually labeled exploration or play. In contrast, behaviors that are reinforced by access to food, sex, and warmth tend to be identified as nonexplorative, "serious,"[5] or "purposive" behaviors that function in acquiring the relevant goal or reinforcer. And, behaviors that are reinforced by the escape or avoidance of aversive stimuli such as stings, cuts, jabs, and cold also tend to be identified not as explorative, but as directed to serving immediate bodily functions. It is only when sensory stimulation is considered as important and common a reinforcer as food, thermoregulation, pain, and so forth that it becomes clear that exploration and play are as "serious" behavioral patterns as food-getting, avoiding thorns, and other nonexplorative responses.

■ 5. THE INDIVIDUAL'S OPTIMAL SENSORY INPUT LEVEL CHANGES WITH AGE.

For the young infant, the whole world is new and arousal-inducing. On one hand, this novelty reinforces practically any exploratory behaviors that the infant does. Crawling leads to new tactile and proprioceptive stimulation. Looking leads to new sights. Manipulation produces a variety of sights, sounds, and feelings. However, the extreme novelty of the world can easily make the young infant overaroused, moving its arousal level into the aversive zone in the upper part of figure 1. This reinforces retreat to mother and clinging. As Mason (1965a, 1968a, 1971) has pointed out, the arousal-reducing properties of the mother are probably the major maternal qualities that reinforce the early infant's attraction and attachment to her. Harlow and Harlow (1965) showed that contact comfort was much preferred to milk when an infant monkey had a choice between a terrycloth surrogate without milk and a wire surrogate with milk. The warm, soft, passive mother provides low-intensity stimulation and thermoregulatory reinforcers that are effective in relaxing the infant when it becomes overaroused by excessive sensory stimulation. The reinforcing experience of contact comfort, arousal-reduction, and thermoregulation all strengthen the young infant's attachment to its mother and help it escape overarousal. Hopf (1970) and Kaplan (1974) found similar patterns in hand-raised and surrogate-reared squirrel monkeys.

As the infant engages in repeated bouts of exploration, it gains experience with the sensory world and begins to find it less novel. Repeated exposure to specific sensory stimuli leads to habituation and satiation with those stimuli and thus decreases the amount of arousal-induction those stimuli produce in the organism (Hebb 1949; Montgomery 1954; Inhelder 1955; Welker 1961; Eisenberger 1972). As the infant gains familiarity with the sensory inputs of its immediate environment, the net novelty decreases and the infant becomes less easily overaroused by those stimuli. For example, in nature very young infants often go through a period of touching and manipulating leaves. The young infant, while clinging to its mother, reaches awkwardly toward leaves near its mother, and the first times it touches them it experiences novel inputs different from feeling the fur of the mother's body. The novelty is arousal-increasing and brings sensory stimulation reinforcers. Since leaves provide low-intensity and noncomplex stimulation, there is little chance that they will over-arouse the infant. Novelty is reinforcing; and during the early phases of life, leaf-touching tends to become a common exploratory behavior in the young monkey's repertoire. However, the leaves lose their novelty after repeated experience; and as the novelty reinforcers begin to decline, the leaf-touching habit begins to extinguish. Through such exploration the riding infant exhausts its sources of nearby stimuli in its environment. However, novelty still exists in the world only a few centimeters away from the mother's body; and the infant is reinforced for turning its attention to these unexplored and, hence, still novel stimuli. If its muscular and central nervous systems are not sufficiently developed the infant is likely to be inhibited from leaving the mother's body by aversive experiences resulting from clumsy behaviors. In this way maturational factors serve to limit premature exploration and play. However, if the infant is sufficiently physiologically developed, it is reinforced for looking farther, reaching farther, and even dismounting from the mother's body. As its locomotor skills increase and hand-eye coordination improves, the infant can venture away from its mother to explore novel stimuli previously beyond its access.

The seduction out into the wider world is a consequence of the reinforcers of the untapped novelty to be discovered there. With time, the infant no longer focuses its exploration toward low levels of sensory input—such as touching leaves or watching butterflies—and it climbs off the mother's body to spend progressively more time traveling on its own. At first, the bouts of exploration away from the mother are short, since the overwhelming novelty of the world quickly overstimulates the infant causing the infant to be reinforced for returning to its mother for arousal-reduction. Through repeated learning experiences the infant is shaped into habits of alternating between bouts of arousal-inducing exploration and bouts of arousal-reducing contact with its mother. As it gains familiarity with the world and habituates to many stimuli, the infant is less rapidly brought into its overarousal zone during exploration and hence it returns to its mother less often. It is reinforced for going farther afield by locating new sources of stimulating activities like running, manipulating, and social play. As will be explained in greater detail in section 7, social play develops naturally from early exploration when certain environmental (and genetic) conditions are present. Social play tends to be more arousing than touching leaves or climbing through the branches; hence, prolonged bouts of social play are avoided until the infant has habituated to somewhat higher levels of input after exploring the inanimate environment. When social play begins, it tends to occur at low arousal levels, and/or for short bouts, then progresses to high activity forms only when the maturational constraints have been overcome and the animals have habituated to higher levels of input. In those situations where the mother and infant play together it is clear that the mother can serve as an arousal-increasing as well as arousal-reducing stimulus for the infant.

Figure 4 shows the general trend in the awake infant's preference for optimal sensory input, averaging out the hourly and daily fluctuations (due to diurnal cycles, fatigue, differences in current behavior patterns, etc.) in order to focus on broader developmental patterns. The neonate finds a short exposure to passive looking and touching to be optimal or even overarousing. But as it habituates to these stimuli and

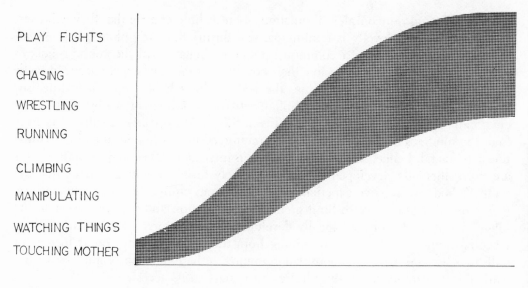

PLAY FIGHTS

CHASING

WRESTLING

RUNNING

CLIMBING

MANIPULATING

WATCHING THINGS

TOUCHING MOTHER

AGE

FIGURE 4. *An infant's optimal input level (stippled area) rises as the infant habituates to more active and complex behaviors. Activities that are overstimulating for a young infant can become understimulating for an older infant.*

they lose their impact, the infant is reinforced for probing into more stimulating activities, i.e., activities of higher novelty, variability, intensity, and complexity. The rising curve is also a consequence of maturational factors, i.e., the infant's developing physical strength and coordination which permit it to move through the environment without excessive aversive experiences from falls, hard knocks, or exhaustion.

Clearly, as the infant is reinforced for exploring its environment, it becomes exposed to opportunities to learn many skills, discriminations, and bits of information about its world. This learning is often adaptive, though maladaptive learning is possible (see section 11).

■ **6. THE SHAPE OF AN INDIVIDUAL'S OPTIMAL INPUT CURVE WILL DEPEND ON THE LEVEL OF SENSORY STIMULATION IN ITS ENVIRONMENT, THE AVAILABILITY OF AROUSAL-REDUCING STIMULI (SUCH AS THE MOTHER), THE AMOUNT OF LEISURE TIME AVAILABLE FOR EXPLORATION, AND THE POSITIVE AND NEGATIVE REINFORCERS ASSOCIATED WITH STIMULUS-SEEKING BEHAVIORS.**

The particular activities and stimuli that a given infant will explore depend in part on (1) what potential sensory stimulation is available in its given social and nonsocial environment; (2) the availability of arousal-reducing experiences with a mother or mother-substitute; (3) the amount of leisure time in which the individual is not occupied by foraging, resting, thermoregulating, fleeing, and other important activities; and (4) the history of positive and negative reinforcers contingently related to exploration and play. The following paragraphs will demonstrate various aspects of these four principles as they appear in the laboratory and the field.

Numerous laboratory experiments have shown that infants raised in sensory-rich environments tend to learn more exploration skills and to habituate to higher levels of sensory input than infants raised in sensory impoverished environments (Jensen et al. 1967; Mason 1968a; Harlow and Harlow 1969; Sackett 1971, 1972; Rosenblum 1971). Primates raised in sensory isolation or deprivation exhaust the readily available sensory stimulation of their environment during early exploratory activity and do not have the chance to habituate to higher levels of stimulation. Once they have exhausted the novelty of their limited environments, there are no longer any sensory stimulation reinforcers to strengthen exploration and play. Thus, the animals cease exploring and habituate to relatively low levels of input which become their normal (or optimal) input levels. As shown in figure 5, the optimal sensory input level of deprived animals is lower than the optimal zone of animals raised in more stimulating environments. When these deprivation-reared animals are exposed to high levels of input (i.e., exposed to conspecifics or complex stimuli), they quickly become over-aroused and are likely to respond inappropriately, either huddling in a corner with fear or striking out in blind attack (Harlow and Harlow 1965, 1969).

Animals raised in sensory-rich environments have more to explore than those

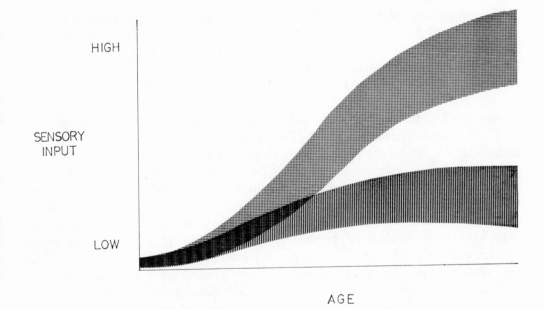

FIGURE 5. *The optimal level of sensory input is higher when an individual is raised in a sensory-rich environment (upper curve) rather than in a sensory-deprived environment (lower curve).*

raised in limited environments, and they develop stronger exploration habits because there are many experiences that bring the reinforcers of novelty which strengthen exploration behaviors. With rapid and active exploration, they quickly habituate to high levels of sensory input. On the other hand, if an animal is raised in an *extremely* active and arousal-inducing environment, explorativeness can be inhibited by too frequent overarousal. Using pigtail macaques, Rosenblum (1971) found that mothers that punish their infants more than normal have infants which show above average rates of clinging, presumably because the punishment is arousal-inducing and exacerbates the need for arousal-reduction. Rosenblum hypothesized that primates raised in excessively stimulating environments would not explore as much as animals in moderately stimulating environments—the excessive stimulation would make them overaroused, which in turn would cause them to stay close to mother rather than to explore. Laboratory studies have also shown that when there is no mother or surrogate mother in the infant's environment, exploration is inhibited. King (1966) and Bronson (1968) suggest that the absence of a mother can also retard an infant's habituation to higher levels of novelty by depriving it of an important source of arousal-reduction (i.e., contact comfort with the mother) which can counteract its arousal-increasing experiences during exploration. An infant with no source of arousal-reduction or security will become excessively inhibited about exploring, since it cannot easily escape overarousal once overstimulated. Although a primary function of the mother is to provide arousal-reduction, it must be stressed that at times the mother can serve as a source of arousal-induction—for example, during rapid locomotion, play, and weaning.

The importance of leisure for stimulus-seeking activities has been shown in laboratory studies. Reynolds et al. (in preparation) and Baldwin and Baldwin (1976) found that play declines to a very low level when animals are forced to spend long periods of time foraging; to the degree that foraging is a monotonous task, it will deprive the animals of variety and retard their habituation to high levels of varied sensory input.

The effect of contingent reinforcement on stimulus-seeking activities has been demonstrated in laboratory studies such as Butler's (1958, 1965), discussed above.

Field studies have demonstrated different aspects of the four factors determining an individual's arousal curve. In the wild there are differences in the sensory deprivation or richness of environments. Several factors can affect the sensory richness of natural environments. The structure of the physical environment can be monotonous or highly complex and thus affect the reinforcers for exploration. The complexity of the social milieu tends to influence social play—the more social variety there is, the more likely the individual will be to develop strong social play habits. Small troops with few peers of similar age deprive a young monkey of many of the opportunities for exploring social play that young monkeys in large troops have. For example, gibbons (Ellefson 1967, 1968), orangutans (Horr 1972, personal communication; MacKinnon 1971), and *Callicebus* (Mason 1968b) live in such small units that young animals have few possible play partners and thus they are deprived of the opportunity to explore the type of activities that social play permits. When orangutans are placed in situations where several young animals live together, such as in zoo enclosures, they do engage in social play (Aldis, in press). Likewise, squirrel monkeys in small troops have few possible play partners and they play less than squirrel monkeys in large troops (Baldwin and Baldwin 1971).

In the wild, few observations have been made on infants deprived of their mothers. One exception is van Lawick-Goodall's (1968, 1973) report on three young chimpanzees who lost their mothers; in two cases the infants ceased exploring and playing and became progressively more like laboratory isolates. The one infant who was adopted by an older and experienced sister continued to develop relatively normally.

Natural environments do not always provide primates with the leisure time to explore and play. The reinforcers associated with hunger, thirst, thermoregulation, escape from danger, and other high-priority survival activities appear to pre-empt sensory stimulation reinforcers in harsh environments. For example, when food is scarce, exploration and play become very infrequent. Hall (1963a) observed chacma baboons that had become marooned on a food-poor island, and his report indicates that play was nonexistent. Loy (1970) found a 94% decrease in play when a troop of rhesus monkeys experienced a 22-day period of starvation. Two troops of squirrel monkeys at Barqueta, Panama, showed a total absence of social play under food restriction conditions that forced them to forage 95% of each day (Baldwin and Baldwin 1972, 1973a). This does not mean that nonplaying primates can not habituate to moderate or high levels of sensory stimulation. During foraging, troop travel, and troop alarm, animals may learn many skills and habituate to high sensory input level activities without engaging in extended periods of exploration or social play. The data (Baldwin and Baldwin 1973a) on squirrel monkeys indicate that an individual can adjust socially and be fully habituated to the input level of its environment even without social play experience.

Positive and negative reinforcement from sources other than sensory stimulation can also affect an individual's rate of exploration and habituation to novelty. In an earlier study (Baldwin 1969) on squirrel monkeys, it was found that the infants with overprotective mothers were slower to explore the environment and to enter into active social play than their peers. The overprotective mothers frequently alarmed and retrieved their infants during exploration in a fear-inducing manner; and these aversive experiences tended to inhibit the infants' explorative behaviors and, hence, to retard the learning of those skills needed for independence. Limited coping skills put an animal at a relative disadvantage in exploration and play—with limited skills, an individual is more likely to fall, be hurt in play, become fatigued, or generally receive more aversive consequences than a skilled animal. These additional aversive consequences also inhibit exploration and thus retard habituation to higher levels of novelty. During the 4 months after the protective squirrel monkey mothers stopped hindering the exploration of their infants, the infants were still less bold and less explorative than average infants in the same environment.

Other factors could conceivably inhibit exploration and play. An infant's level of exploratory skills might be retarded by physical defects in central or peripheral behavioral systems (Berkson 1973; Fedigan and Fedigan, this volume). Punishments from other troop members besides the mother could also inhibit exploration. In rhesus monkeys, where there exists a hierarchy among the adult females, the offspring of low-rank mothers receive negative reinforcers from numerous other troop members and their low-rank mothers cannot defend or protect them. One would expect these punishments to inhibit exploration and play, especially if they were contingently associated with those behaviors.

■ 7. THE APPARENT STAGES OF EXPLORATION AND PLAY RESULT FROM CHANGES IN OPTIMAL STIMULUS INPUT LEVELS, GIVEN THE APPROPRIATE MATURATIONAL DEVELOPMENT.

It has often been noted that primates typically go through several "stages" of development in exploration and play behavior. Harlow and Harlow (1965) discriminated the following stages of rhesus play: (1) reflex stage; (2) exploration stage; (3) interactive play; (3a) rough-and-tumble; (3b) approach-withdrawal or noncontact play; and (4) aggressive play, to which one might add (5) end of play. Ploog et al. (1967) and Baldwin (1969) have noted similar trends in squirrel monkeys. For squirrel monkeys that have the leisure time to play, the following stages have been noted: (a) nonsocial exploration; (b) early social exploration; (c) contact play and wrestling; (d) distance play or chasing; (e) play fights or aggressive play; and (f) end of play. We will describe how animals progress through a full series of stages as one would expect in sensory-rich, play conducive environments, although the data in sections 6 and 12 indicate that the full process can be limited by several factors, all stages need not appear, and limited experience does not necessarily create behavioral abnormalities. An auxiliary goal of section 7 is to discriminate more finely among the different kinds of exploration and play activities.

Describing play as changing through "stages" is not intended to imply that there are discrete time periods associated with each stage or that in any given stage young animals do only one type of activity (as would be shown in figure 6A). The stages are more appropriately thought of as highly overlapping and merely reflecting trends in behavioral development (Hansen 1966). As shown in figure 6B, the young animal at any age x is likely to do behaviors from several stages, though the activities associated with one stage tend to dominate its behavioral repertoire.

As was explained in section 4, the earliest explorative behaviors (stage a) are shaped from reflex-like activities. Given that the world is filled with novelty for the neonate, the newborn is probably easily overaroused; yet it would be expected that inert, noncomplex objects (like leaves or mother's fur) would create less overarousal than intense or complex stimuli (such as those associated with contacting active, rowdy, or aggressive older animals). However, as the infant habituates to low levels of sensory stimulation and becomes physically stronger and better coordinated, it slowly broadens its sphere of exploration to include more arousal-inducing activities as is shown in figure 4. Exploring the physical environment—such as twigs, leaves, flowers, and the fur on mother's body (stage a)—eventually becomes underarousing and the infant begins to seek out the more variable aspects of its environment, which leads

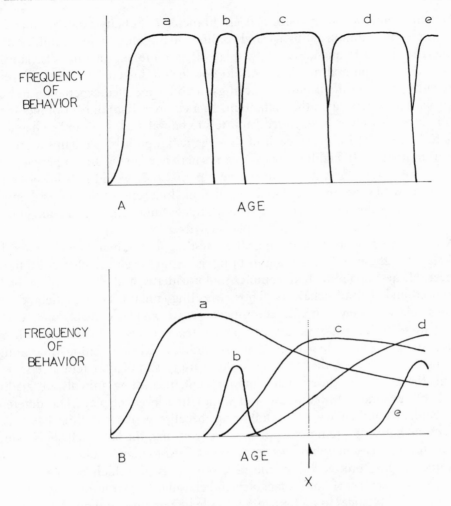

FIGURE 6. *Dividing the ontogeny of play into several stages is not intended to imply that the stages occupy discrete, nonoverlapping time periods (as in A); rather, behaviors from each stage may occur with differing frequencies at various points in development (as in B). Some stages overshadow others in frequency and early stages can outlast later stages.*

it to explore the active, unpredictable stimuli generated by responsive social peers (stage b). For example, when two squirrel monkeys 8 weeks of age first begin to explore touching each other (stage b), there are numerous signs that the complex, novel experience brings them close to the overarousal zone given their level of development. They are cautious and attentive when first touching each other, often making approach-withdrawal movements with the hands as if hesitant to explore the novel stimulus; and their initial interactions are brief. After several such interactions, habituation effects become visible—touching another active infant becomes less over-arousing, and the number of withdrawal movements decreases. Eventually, full body contact and gentle wrestling (stage c) begin to increase. Solitary exploration and play (stage a) are still common at this age and continue to be common for months, but they decline in relative frequency as social play offers more stimulating alternative behaviors. Exploration and manipulation do not decline to zero, however. Players do temporarily satiate on their current social games after repeated experience and other nonsocial inputs may catch their attention. In a complex natural environment, there are numerous insects to be explored, squirrels to be watched, birds to be chased, tree holes to be reached into, and each of these activities probably contains a moderate amount of novelty. In addition, players can experience fatigue after a period of high activity, and during the subsequent recovery period they would find lower arousal-inducing activities to provide optimal stimulation. In a sensory-rich natural environment, solitary exploration and manipulation can continue long after a young monkey has reached advanced levels of social play activity.

After players have habituated to gentle wrestling and to being touched and handled (stage c), these activities cease to bring as many novelty reinforcers. There is, however, always variability in all complex motor patterns, and play is among the least stereotyped and rigid of behaviors. Thus, wrestling contains both gentle and rough episodes, old and new patterns, continued versus sporadic contact, and so forth. Among these varied patterns, the new and unexperienced activities—which often include more vigorous behaviors such as pulling, twisting, biting, and rougher wrestling—are the ones that produce more sensory stimulation reinforcers. Thus, through the process of differential reinforcement, the young animals are gradually shaped into generating more active and stimulating levels of play. The differential reinforcement contingency is the following: because gentle wrestling has become familiar and hence less reinforcing than active wrestling, the latter activity is followed by more novelty reinforcers and becomes conditioned to higher habit strength. Eventually the shaping process brings the players to stage d, which involves the high activity level behaviors of noncontact play and chasing. Higher levels of sensory input created by rapid changes in motor patterns, agile locomotion, and quick coordination with another player are more reinforcing than previous levels.

The escalation to rougher and more vigorous activity takes on new qualities when play activity begins to involve painful consequences. As players become physically stronger and escalate to more intense play, biting can occasionally lead to sharp nips; rough wrestling can lead to hard knocks; rapid chases can lead to falls. As play activity begins to include these aversive experiences, the frequency of avoidance, withdrawal, and distance maintenance gradually increases. Two older players may paw at each other at arm's length, but hesitate to come closer together. In the field it is usually difficult to distinguish when approach-avoidance play grades into play fights or aggressive play (stage e); but at some point the observer may begin to notice that the shrieks, threats, chases, and biting look more like fights than fun. The amount of play fighting or aggressive play that occurs in a given troop is doubtless affected by many factors such as species genetic traits, social stress, crowding, or food deprivation. For example, truly aggressive play may be more likely to occur in situations where the animals cannot easily escape from or terminate these interactions as is the case in laboratories or crowded troops. Because play fighting and especially very aggressive play are potentially aversive, one would expect these activities to be brief when escape or discrimination learning are easy. Symons (1973) observed play fights but no cases of truly aggressive play in the semifree-ranging troops of La Cueva Island (off Puerto Rico).

When rough play escalates to the painful level, the animals learn to discriminate between aversive, aggressive play (which they avoid) and controlled play (which they still seek out). Eventually, the level of play activity tends to decline because the aggressive play is often painful (and hence punishing) and the novelty of controlled rough play is exhausted. However, several factors can retard the eventual decline in play and hence prolong the stages of controlled play (stages c and d). First, stimulus recovery effects (Welker 1961) lengthen the period in which novelty reinforcement can be obtained from a repertoire of play activities. Animals with a long history of play experience have explored and discovered a large number of games. After each game has been discovered and repeated to the point of habituation, the reinforcement for that game declines and it tends to be repeated less frequently. During the period in which the game is not repeated, there is a recovery effect and if the game is rediscovered at a later period, it will have more novelty effect (and reinforcement value) than it did at the point of maximum habituation. Older players, with a large repertoire of games, are in the position to rediscover many old games even after they have ceased to discover new games. The recovery effect extends the length of time that novelty and reinforcement are associated with play, and should be most pronounced in individuals with large repertoires of play activities. Second, once players have arrived at high levels of optimal sensory stimulation needs, few alternative activities besides play can begin to provide the high levels of activity to which the animals have habituated. Passive adult activities such as resting, grooming, leisurely foraging, and merely looking around do not provide the high level of input.[6] Even without many novelty effects, play is the best alternative for avoiding underarousal and hence is the most reinforced activity until the individual eventually habituates to lower input levels. Third, play habits may be exceptionally strong and resistant to extinction in animals with a long history of play experience. Early exploration and play provide high ratios of reinforcement since there is so much novelty to be discovered. With increased experience, the easily discovered novelty declines and the animals move toward more "stretched out" schedules, receiving intermittent reinforcement rather than continuous reinforcement. Intermittent schedules produce strong habits that are very resistant to extinction (Reynolds 1968). Thus, even when there are absolutely no more novelty reinforcers for play (i.e., during extinction conditions), animals with histories of intermittent reinforcement for play show strong play habits that resist extinction for long periods. The overall effect of these three factors is that older players may cease to discover new games, cease to show novelty and variety in their play activities, and yet continue to engage in high levels of play for a long period before the frequency of play declines. In squirrel monkeys with the leisure time to play, the play fight activities

become inhibited before gentler play is fully extinguished (i.e., stage e terminates
before stages c and d). Subadult males often wrestle together and chase after the age
when play fights have ceased to occur. When one considers the contingencies of
reinforcement, it is understandable why an "earlier stage" might outlive a "later
stage." During troop rest periods, juveniles with high optimal activity levels would
receive more reinforcers for engaging in wrestling play (with its moderate level of
stimulation) than for no play (with aversive underarousal), or for aggressive play (with
the negative reinforcers of pain).[7] Therefore, among the various behaviors in their
repertoires, wrestling play brings more sensory stimulation than not engaging in play
and less pain than play fights.

This is one way in which natural contingencies of reinforcement can shape in
what Suomi and Harlow (1971) call the "control of aggression." As is pointed out
elsewhere, the Barqueta study (in which there was controlled aggression and no play)
demonstrates that there are other natural contingencies of reinforcement that can lead
to controlled aggression (Baldwin and Baldwin 1973a). The decline of play, stage f, is
discussed in section 8.

A. CHANGES WITHIN STAGES.

It must be stressed that the concept of "stages" of play development does not imply
static or fixed behavioral categories. The present emphasis on shaping via the contingent reinforcement of sensory stimulation stresses that within each of the early stages
there is constant molding and modification of behavior. These changes give the
impression that young players are frequently "discovering new games." As yesterday's
games are repeated today, the accidental variations on the old games stand in contrast
as relatively more novel and reinforcing than the old games themselves. The following personal observations from a laboratory study on squirrel monkeys illustrate how a
simple nonsocial game progresses through successive changes as novel variations
appear.

A persistent game in 2-year-old Corwin's repertoire consisted of hiding food
pellets under gratings in the cage, then attempting to recover them. When a food
pellet was lost, Corwin went to the cage floor, picked up a new pellet and returned
to the hiding areas. While climbing up to higher parts of the cage, he occasionally
dropped a pellet and it went bouncing around the cage. This made possible a new
game of chasing the pellets. However, the pellets fell to the floor and became inert
so rapidly that there was not much chance for prolonged chasing. One day the
pellet dropping response occurred while Corwin was making a vigorous upward
jump, and the pellet flew upward and bounced around the upper parts of the cage
before falling to the floor. More prolonged games of chasing and catching now
became possible, and the habit of releasing pellets on upward jumps became
stronger. Within two weeks Corwin was becoming skilled at hurling pellets up and
catching them in midair with both hands. By the end of the study he was able to
catch the pellets in his mouth.

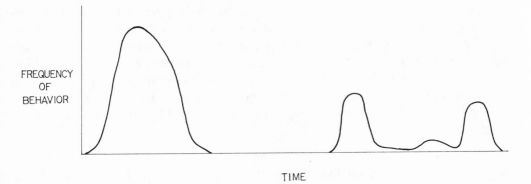

FREQUENCY
OF
BEHAVIOR

TIME

FIGURE 7. *The history of a game includes the discovery period (left)
and possible later rediscoveries (right).*

Thus, yesterday's games are shaped into new patterns; and even within a given stage of play development, the animal may discover countless different games and variations. Social games offer more potential variability than nonsocial ones because of the greater combinational possibilities that are available when an individual interacts with another player as opposed to an inanimate object. Which series of games the individual is shaped through depends upon the activities available in its environment, the accidental variances it happens to make, and the reinforcements that follow. The process is based partly on adventitious contingencies—many behaviors can produce sensory stimulation reinforcers, and each player will tend to receive reinforcers for different accidental combinations of activities. In addition to the new games, old games are rediscovered from time to time after recovery periods in which they gain arousal-increasing potential, compared with prior periods when the players were maximally habituated to those games. Considering the opposing effects of habituation and recovery, one would expect to find that some games took a course like that shown in figure 7—a peak of activity associated with discovery followed by occasional rediscovery periods. The contingencies that produce rediscovery peaks are similar to those that create play rebound effects after periods of play deprivation (Baldwin and Baldwin 1976).

B. CAVEATS ABOUT STAGES.

Although numerous students of behavioral development have found a useful heuristic device in dividing behavioral ontogeny into stages, it is imperative that stages not be reified. In some cases, attempts to delineate stages may obscure the continuous nature of development. In other cases, postulating certain stages for a given species may carry implications that development can follow only one genetically preprogrammed pattern; whereas it is clear that play stages are shaped by an interplay of physiological and environmental causes. Because each individual of a given species grows up in a unique social context, no two animals experience exactly the same models, or contingencies of reinforcement, and, hence, no two individuals will develop through precisely the same behavioral patterns. Stages point to general trends in the maturational and learning process; but detailed observation must stress the effects of each individual's specific interactions with its unique environment. Finally, data on chimpanzees (van Lawick-Goodall 1968, 1971) and observations on humans indicate that clear stages are less easy to identify in these species, because of the enormous behavioral complexity and idiosyncratic development that animals of these species show.

■ **8. A DECLINE IN OPTIMAL AROUSAL LEVEL AND PLAY IS TYPICAL IN THE MATURING ANIMAL.**

The tendency for adults to become nonplayful and sedentary is common in many primate species. Adults often appear to be reinforced for sitting, exchanging social stroking, and avoiding being pestered by active, playful, younger animals. In one sense, this is adaptive in that it produces passive adult females which are well suited for the role of providing a source of low sensory stimulation and arousal-reduction for their young infants. Kummer (1971), Rowell (1972), and others have pointed out that it is functional for adults to be the conservative members of a primate society who retain cultural traditions with little change; whereas the young can explore, innovate, and take risks since they are more expendable according to bioenergetic calculations.

There appear to be several factors that are involved in the decline of play during juvenile and/or adult ages. Several of these are related to a decline in optimal activity (Figure 8). First, the "ceiling of readily available sensory stimulation" in any given environment strongly influences how many months or years of exploration and play an individual engages in. In sensory-impoverished environments, animals can only explore for a relatively short period of time until they have "found it all out," i.e., until they have exhausted the readily available sensory stimulation by becoming familiar with their surroundings. In addition, exploration and play behaviors cannot gain much habit strength in deprived environments where there is little sensory stimulation reinforcement and, thus, these behaviors would be likely to extinguish quickly. On the other hand, in sensory-rich environments the ceiling of readily available sensory stimulation is high and an animal can explore for long periods without exhausting the novelty. After repeated reinforcement, especially with intermittent schedules, exploration and play would become very resistant to extinction. Finally, when an environment is chronically overstimulating, exploration and play would be inhibited due to aversive overarousal (Rosenblum 1971).

In even the most play-conducive of environments, however, the ceiling of readily available sensory stimulation is usually reached after a period of youthful exploration and play, and thereafter there is little more novelty reinforcement to be had. After this point, exploration and play begin to extinguish for lack of reinforcement, although several effects may make play relatively resistant to rapid extinction (as was explained at the end of section 7). The most effortful and/or pain-inducing behaviors are likely

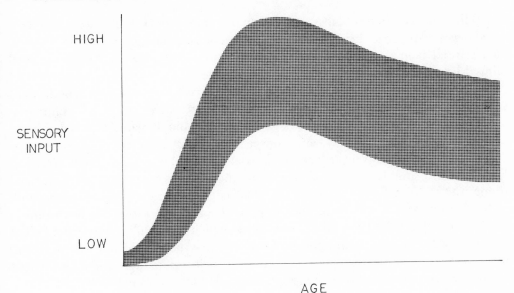

FIGURE 8. *A decline in optimal arousal level is typical during the years after peak exploration and play activity.*

to be among those that drop out of an individual's repertoire early. Activities associated with pain (such as aggressive play can be) become inhibited due to the contingent aversive consequences. Effortful behaviors have also been shown to have aversive properties (Keller and Schoenfeld 1950) which function to inhibit those behaviors once the novelty associated with them is exhausted. It is not uncommon to find that when older juveniles and adults play, they do not engage in effortful activities such as chasing, swimming, running, and climbing (Simonds 1965; Symons 1973). It is the "cheap behaviors," or low effort behaviors (Breland and Breland 1966) that are the last to extinguish.

As exploration and play behaviors become less frequent, the once playful individual begins to spend more of its time in sedentary activities, and it begins to habituate to lower and lower levels of sensory input. As high activity, stimulus-seeking behaviors extinguish, the individual begins to come under a relatively stronger influence of other reinforcers and it begins to develop stronger habit strength for behaviors such as resting, sunning, grooming, foraging, seeking sexual interactions, etc. (The precise pattern of adult behavior will depend on species, sex, ecology, nutrition, reinforcement, and other variables.) After an individual's optimal input levels begin to decline and stronger habits of quiet behavior develop, the individual may begin to withdraw from, or threaten, active, playful youngsters. Their playful rowdiness and constant bounding about may now be above the adult's descending optimal zone, and their unpredictable play antics often disrupt the adult's less playful activities. Both of these situations are aversive to the adult, and this can condition threats or avoidance depending on the circumstances. Adult females often retreat from rowdy juveniles, which again aids in producing a quiet environment for their younger infants. (As will be discussed below, maturing females usually show a steeper decline in optimal arousal level than their more active male age mates.)

Even in the same forest, different individuals have access to different amounts of sensory stimulation. Van Lawick-Goodall (1967b, 1971) describes a young chimpanzee, Gilka, whose mother avoided other chimpanzees and therefore deprived Gilka of a chance to explore and play with conspecifics. Gilka used to spend long hours "doodling," hanging by one arm and looking at the world while she turned in circles. Finally, Gilka found a young baboon with whom to play, and the two developed strong mutual attractions and play habits that lasted for several months. Compared with other chimpanzees living in the same forest, Gilka was exposed to fewer opportunities to explore social play with conspecifics, which clearly illustrates that different individuals that live in a common environment may be shaped into very different patterns of exploration and play. The contingencies of reinforcement that shape exploration and play must often be specified in detail to explain individual differences among animals in the same general environment.

Second, an individual's level of behavioral and "cognitive" skills affects how much sensory stimulation it can locate or produce in a given environment, and how long into adulthood it can continue to do so.[8] The amount of sensory stimulation readily available depends not only on the sensory-richness of the environment, but on the explorers' skills. Individuals who are inhibited, fearful, inexperienced, or clumsy may not be able to locate as much explorable novelty as uninhibited, bold, innovative, or dexterous animals. For example, infants of overprotective mothers or low-status mothers often have limited skills and more inhibitions; thus, they may reach their ceiling of readily available sensory stimulation earlier than animals more skilled and uninhibited about exploration and play. However, quiet animals may possess skills other than those that create stimulation by boldness or force. Different types of skills are suited to locating different types of sensory stimulation. For example, female Japanese macaques are less active and are less easily distracted from quiet activities than males; and it is the females who excel at many of the exploration and innovation tests the Japanese researchers have applied in the field (Tsumori 1967).

Third, the availability of a mother or caretaker for arousal-reducing contact is an important factor that prevents early termination of exploration and play (section 6).

Fourth, a young animal may encounter various aversive consequences associated with stimulus-seeking or high activity levels, and this contingent relationship would inhibit its explorative activities and hasten the decline to quiet adulthood. The aversive consequences of aggressive, rowdy play have already been mentioned as one reason why high levels of activity become inhibited. In addition, mothers often threaten or avoid their older infants or juveniles when the latter are very active around the mothers. Other adults may threaten, avoid, withdraw grooming, and so forth when young animals do not calm down or behave at a quieter level. The reinforcement contingencies that aid in producing calm behavior can be seen in this typical observation on squirrel monkeys during the cool season in a subtropical environment.

February 5, 1967. 9:00: There are clusters of animals sunning in various places on the east side of the forest. Adult females rest quietly together, cuddled in close body contact. Infants and juveniles seem unable to sit together quietly during sunning: not a minute goes by that one does not turn around to put its hands on another, climb over other animals, or grab at its neighbor. When a juvenile enters a cluster of sunning adult females, the adults will remain if the juvenile is quiet; but if the juvenile is restless or playful, the adults either threaten with "hand on" or leave the area. (personal observation)

These contingencies reinforce passivity with social cuddling and thermoregulation; and playful behaviors are inhibited by threats or withdrawal of positive reinforcement. During the winter months the juveniles (especially the females) were noticeably shaped by these contingencies and eventually learned to behave in a quiet (nonplayful) manner with adult females in sunning clusters.

In situations where females have a dominance hierarchy, being the offspring of a low-ranking mother may expose a youngster to negative reinforcers that offspring of high-ranking mothers do not experience (Koford 1963; Sade 1967; Kaufmann 1967). Careful observations on the ontogeny of play in the infants of high- and low-ranking mothers would provide data on the relative importance of social reinforcement in shaping stimulus-seeking behaviors. If play is significantly affected by social reinforcers, one would expect that the negative reinforcers from which the low-ranking mother cannot protect her offspring would be likely to occur when the offspring is active and, hence, salient; and one would expect these contingencies to inhibit active behaviors such as exploration and play.

Fifth, health, physical strength, and maturational factors may play important roles in determining how high an individual's optimal sensory stimulation curve can rise and how late into adulthood it will remain high. Disease, parasites, malnutrition, and old age all take their toll on physical strength and endurance. Strong, coordinated animals can create more noise, breakage, opened objects, and gymnastic variations than weak animals; they can stay in play fights longer without being hurt; they can roam out further from the troop to discover a larger range of novelty in the outside world. Since strength is one of the factors that often aids an animal in gaining social status, strength may free an animal from some of the negative reinforcers and inhibitions that low-status individuals experience.

Sixth, other positive reinforcers besides those of discovering novel sensory inputs may strengthen the animal's habits for spending its time in behaviors other than play. As the maturing individual learns improved skills at attaining sexual or food reinforcers, the increased success (and hence increased frequency of reinforcement) may strengthen habits related with these reinforcers. For example, when a female gives birth a new range of reinforcing activities becomes available to her, and she is likely to focus increasing attention to mothering behaviors at the cost of other activities, including play.

There are other factors that affect the decline in optimal stimulation with age, such as sex differences, which are discussed in the following section. Because so many factors can affect the rise and decline in an animal's optimal stimulation zone, it should be clear that there is much room for individual differences in the ontogeny of exploration and play activities.

■ 9. SEX DIFFERENCES IN PLAY AND EXPLORATION.

Male primates of many species tend to spend more time in active exploration and play than females. Figure 9 shows data on squirrel monkeys in a sensory-rich environment (Baldwin 1969) and reflects the fact that female squirrel monkeys withdraw from active play earlier in life than males. The two curves probably serve as general indicators of the relative levels of optimal sensory stimulation for the two sexes at each age. By the early juvenile age, females begin to prefer quieter types of activities than males, and by the late juvenile period, females are well settled into the quiet life typical of adult females. Similar sex trends in play have been found in langurs (Jay 1965; Sugiyama 1965; Poirier 1970), vervets (Lancaster 1971), bonnet macaques (Simonds 1965, 1974), rhesus macaques (Harlow and Harlow 1965; Symons 1973),

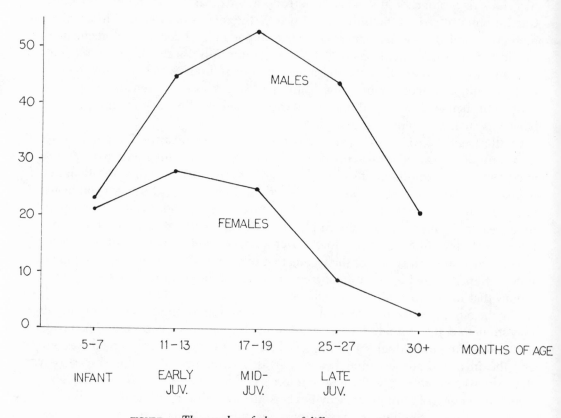

FIGURE 9. *The number of players of different age and sex in randomly sampled social play groups in a study on squirrel monkeys (Baldwin 1969).*

hamadryas baboons (Kummer 1968, 1971), savanna baboons (Hall and DeVore 1965), and other primates. Several factors may be involved in this trend.

First, it has been shown that females of several species (e.g., rats, hamsters, and rhesus monkeys) that are exposed to androgens during a critical period of early development become masculinized as pseudohermaphrodites (Phoenix, Goy, and Resko 1968; Goy and Resko 1972; Goy and Phoenix 1972). These genetically female, but partially masculinized, animals tend to develop play patterns that are approximately halfway between the typical male and typical female patterns. They engage in rough-and-tumble play, and they mount other animals more than normal females, but less than normal males. It is certainly feasible that in primates, prenatal exposure to male hormones affects various neurophysiological structures and biases the central nervous system toward a more activity-seeking, male-like condition. Goy and Resko (1972) and Goy and Phoenix (1972) suggest that prenatal androgens may produce changes in the reinforcement systems of the individual to make active, assertive, male-like behaviors more reinforcing to males and to masculinized pseudohermaphrodites. Behavioral studies have confirmed that "important biological differences may exist *in infancy* that reflect differential requirements by males and females for early life stimulation" (Sackett 1972).

Second, males tend to be physically larger and stronger than females. Even before sexual maturity, male rhesus monkeys tend to gain weight faster than females, although the differences are not statistically significant (Dawes 1968). The data on squirrel monkeys from several laboratories consistently show the same trend in this species (Long and Cooper 1968; Fryar and Hupp 1970, personal communication; Kaplan 1974; Hopf, personal communication). As was discussed in the previous section, additional size, weight, and strength can give an individual an advantage in successfully exploring and playing. The larger, stronger animal may be able to evoke more sensory stimulation from the environment during exploration, and it may be more successful in doing social play without getting hurt or receiving social punishment than smaller, weaker animals. Under these contingencies, strong animals would receive more novelty reinforcers and fewer aversive consequences, hence, they would develop stronger habits of active exploration and play. Thus male size and strength alone may explain a portion of the difference between the frequency and duration of play between the sexes.

Either of the first two factors would increase the likelihood that females experience less novelty and more aversive contacts during exploration or social play than males. Females of many species withdraw from rowdy play with males and they have often been observed to form quiet, all-female play groups separate from the males (Jay 1965; Fady 1969; Baldwin 1969). These aversive contingencies may be responsible in part for shaping the quiet, withdrawn, gentle activities of females and the tendency for females to orient to object manipulation play rather than the social contact play typical of males. Apparently, as the females become inhibited from participating in high arousal social play, they learn to attend to the lower, but safer, arousal level of manipulatory exploration (Tsumori 1967).

Third, as young females withdraw from rowdy male play, they may discover "play-mothering." Young females of many species show interest in small infants, e.g., langurs (Jay 1965); chimpanzees (van Lawick-Goodall 1967a), vervets (Lancaster 1971), and squirrel monkeys (Baldwin 1969). The young females doubtless find the infants a source of novelty. Juveniles sometimes engage in the gentle game of "trying to get the infant away from the mother," which can be a complex, interesting, and novel activity that may last for many minutes at a time as the young female explores various activities for distracting the mother and gaining access to the infant. Once the young female has contact with the infant, their interactions will tend to be nonaversive for the young female since she is stronger than the infant. If the female plays quietly, she may experience many mild but novel play patterns with the infant, and she is unlikely to be hurt by it. However, if she plays too roughly, the infant is likely to give distress calls and the mother may threaten the juvenile and/or remove the source of reinforcement from her. These contingencies tend to shape restrained mother-like behavior during "play-mothering." Most observers agree that young females quickly learn a variety of mother skills during these play experiences, and that these skills are potentially of value once the young female has her own first infant. Trial-and-error experiences while handling the infant, punitive feedback from the mother for mistreating the infant, and perhaps social modeling of the mother's behavior toward the infant can all contribute to the play-mother learning experience. The point of the present discussion is that play-mothering reinforces quiet, low arousal activities in adolescent females and accentuates sex differences in play activities. As was mentioned above (section 8 [4]), once juvenile females become reinforced for being near adult females, adults put extra contingencies on them to enforce quiet resting together and other similar behaviors.

There is at least a fourth factor that affects male-female differences in play. Females of many primate species tend to reach social-sexual maturity at an earlier chronological age than males. This is an additional factor that hastens the termination of play in females. With the birth of the first infant, the female is put under new social contingencies that allow less play and reinforce maternal activities by providing an optimal level of sensory stimulation. Also, the young adult female has less time and metabolic energy to devote to play once she has become a mother. Males, on the other hand, may have several additional years of adolescence or subadult status with ample time to play and explore once their other needs are met. However, these additional years are not necessarily easy for the subadult males, since the transition to adulthood can be fraught with problems. For example, van Lawick-Goodall (1967b) notes that young male chimpanzees face many difficulties in establishing themselves in the social system of the adult males. Subadult male Japanese macaques go through a period of living outside the troop as peripheral members, and some individuals never gain re-entry into their troop (Imanishi 1957; Kawai 1964; Nishida 1966). In some environments, subadult squirrel monkeys also have great difficulties becoming integrated members of the adult male sector of their troops (DuMond 1968; Baldwin 1969; Baldwin and Baldwin 1972). If several males are peripheral at the same time they may play together, explore extensively, venture far away from the troop, and habituate to very high levels of arousal, novelty, danger, etc. These extra experiences that females do not have may be very important in shaping the more venturesome and bold behavior typical of the males of many species. By adulthood, male and female squirrel monkeys have developed very different responses to danger and sources of highly arousing stimuli—in cases of extreme alarm, the adult males come to investigate the sources of danger while the remainder of their troop retreats in quiet (Baldwin 1968). In various species the adult males who are in their prime assume the "troop protector" role in times of danger (Hall and DeVore 1965; Baldwin 1968; Rowell 1974); this may be in part because they are the troop members who are least likely to find highly novel or dangerous stimuli to be overarousing.[9]

Other factors could also operate to produce sex differences in play and thus affect the differential socialization of males and females. For example, in hamadryas baboons, young females are herded into a male's harem and greatly deprived of the opportunity to play (Kummer 1968, 1971). Any contingency imposed by adults, group structure, or the physical environment that affects males and females differently can contribute to sex differences in play activity. Observational learning may also be operative, as it doubtless is in humans (Bandura and Walters 1963).

A. DIFFERENCES IN QUALITY.

Some of the most conspicuous sex differences in play in nonhuman primates seem to relate to stimulus intensity and quantities of novelty, with the males habituating to higher levels than the females. The females' lower position on the stimulus quantity variable does not imply that females fall lower than males on measures of other qualities. Female monkeys outperform males in many tasks requiring patience, prolonged attention, and object exploration (Tsumori 1967). Mothers learn many fine discriminations about how to handle infants that males do not learn. If one applied an information flow model to the analysis of sex differences, it might be found that in their quiet ways the females were processing and discriminating on more stimulus inputs than is obvious from their less active behavior. Much of our present data on sex differences in exploration and play are biased by the fact that exploration and play activities are more intense and conspicuous in males than in females. Future research into the nature of female stimulus-seeking behaviors should reveal more about their stimulus-processing activities.

Concerning sex differences in human behavior, it appears that beyond the edge that males have in brute strength, there are few constraints on exploratory and play behaviors that are not traceable to the social learning contingencies of the culture and that are not modifiable via altered learning contingencies (Bandura and Walters 1963).

■ 10. EXTRAVAGANT AND EXAGGERATED BEHAVIORS BECOME ORDERLY AND EFFICIENT DURING PLAY.

Various studies on play have noted that some of the most typical features of play are that the motor patterns are often extravagant, uneconomical, exaggerated, disorganized, clumsy, fragmented, and repetitious (e.g., Marler and Hamilton 1966; Loizos 1967; Dolhinow and Bishop 1970; Miller 1973). Play sequences are often a jumble of numerous disrupted activities following one after another in apparently random or inappropriate order, unlike the orderly and efficient behavior of adults. At first glance, the wasted energy of play may seem terribly inefficient. This observation led Loizos (1967) to question whether such uneconomical behavior is, in fact, functional. In order to resolve the inefficiency problem, Loizos (ibid.: 211) postulated that perhaps the function of play is to provide the young with imprinting or learning experience about "which species it belongs to." The research on precocial birds has shown that the more effort a hatchling expends in following its mother, the stronger it imprints on her (Hess 1959, 1964). If imprinting were assumed to occur in primates, the wasteful, repetitious, and uneconomical nature of play could be expected to strengthen the imprint. Although several authors have suggested that imprinting may occur in primates, there are no empirical data that unequivocally demonstrate this.[10] Dolhinow and Bishop (1970: 170) cope with the "uneconomical" appearance of play by suggesting that perhaps, to the contrary, play may be "very efficient." These authors suggest that many adaptive behaviors develop during play, and the long-term benefits of play may well outweigh the short-term costs of apparent inefficiency or extravagance.

A learning theory analysis of play takes a different approach to the question of the extravagance, disorganization, and randomness of play activity. Before investigating whether this extravagance is functional, adaptive, or efficient, we will ask: What are the contingencies of reinforcement and the constraints on learning which are operating to make early play activities appear exaggerated, and which make later play activities become progressively more orderly and functional? The questions of adaptiveness will be dealt with in section 11 of this analysis.

The uneconomical and disrupted appearance of infant exploration and play can be analyzed on at least two levels: (1) single, short units of behavior and (2) longer sequences of several behaviors. The remainder of this section will be devoted to discussing these two levels.

First, single units of behavior—such as the simple movements of reaching for a target stimulus—often are clumsy and uncoordinated in the young animal. Hand-eye coordination is poor in most young primates, as can be seen when watching a young infant groping clumsily to touch a leaf that is fluttering right in front of its nose. Such poor coordination—leading to apparent exaggeration, inefficiency, and "unnecessary" repetitions—is doubtless a consequence of many factors. The central nervous system is not completely developed at birth, and the absence of functional, fully myelinated circuitry certainly handicaps complex, cross-modal associative activity (Flechsig 1920, 1927; Gibson 1970; Chevalier-Skolnikoff 1971, 1973, this volume). Weak muscles, under poor central control, also produce instability and faltering movements not present in older animals. The lack of learning experience is complexly interwoven with both of these factors. Single units of behavior which adults have practiced thousands of times will be under much more precise stimulus control

(S^D control) in adults than in young infants. Weak S^D control, by itself, creates the appearance of clumsy, disorganized activity as behaviors repeatedly fall far from the target stimulus or fail to be guided by the control stimuli that coordinate the actions of practiced animals.

Because the purpose of this paper is to explicate the role of learning in the ontogeny of exploration and play, we will look in greater detail at the learning rather than the maturational factors. From the vast number of single-unit responses shown by young animals, we will deal with one example—the reaching or grabbing response of infant monkeys—in order to demonstrate how behavioral units develop from inefficient to efficient levels. Our observations on early exploration and play in howler monkeys and squirrel monkeys suggest the following generalizations. The earliest reaching responses are modified by the contingent positive or negative reinforcers that occur, depending on whether the infant has reached for a leaf, a wasp, a butterfly, or another young monkey. After repeated trial-and-error learning, the young infant is shaped by differential reinforcement to grasp leaves, withdraw from wasps, and grab rapidly for butterflies. Later it learns to touch peers in yet another set of ways. Each environmental stimulus gains S^D control over separate response patterns learned via operant shaping (and perhaps via observational learning). Thus an early unit of behavior, like the reaching response, is shaped into many variations under the control of different stimulus conditions. Reaching responses can be shaped by a variety of positive and negative reinforcers such as sensory novelty, food getting, painful stings, and threats from intolerant adults. Some reaching responses—for example, squirrel monkeys grabbing at and catching flying insects—may take many months to be shaped into efficient adult-like patterns. At first the grabs may appear to be clumsy or ill-targeted, but the differential reinforcement of catches versus misses brings the behavior under subtle stimulus control of many environmental cues. Early grabbing at insects is probably reinforced by novelty, but after several successful catches the behaviors begin to come under the additional control of food reinforcers. By adulthood, squirrel monkeys have learned such coordinated and efficient insect-grabbing skills as the following: the monkey looks upward and inspects the bottom side of sunlit leaves above it for shadows of insects, and it approaches insects from below the leaf for a rapid snatch from underneath. Thus the clumsiness of early behavioral units can progress to very complex skills as the behaviors are reinforced and practiced repeatedly; and naturally the behaviors of the young are inefficient, extravagant, and fragmented, in contrast with adult sophisticated skills.

The reinforcement schedules also determine whether the behavioral units will be repeated and practiced. The positive and negative reinforcers that follow the behaviors control the frequency of the behaviors evoked in response to various stimuli. The early response classes like grabbing behaviors, for example, are often under contingencies dominated by sensory stimulation reinforcers. Early reaches for butterflies may rarely lead to a catch and hence to food reinforcement; but the game of hopping from branch to branch following the elusive butterfly provides the reinforcers of novelty and variety. Thus, the infant's early, clumsy behaviors such as grabbing at butterflies may be reinforced by novelty for a long period before they become skillful enough to be effective in producing food reinforcers and, hence, before the monkey associates them with food rather than fun.

Many nonsocial responses follow this trend of beginning as poorly coordinated, exploratory, and playful behaviors done by infants for novelty reinforcers, then progressing to be well practiced, skillful responses under the control of other reinforcement contingencies such as food getting, thermoregulation or avoidance of pain.

Numerous social responses appear to be shaped from clumsy to polished in much the same way as nonsocial responses, except that they depend more heavily on social contingencies. Early playful grabbing behaviors are modified into different types of social grabbing as they begin to be shaped by the specific contingencies involved in sex play, aggressive interactions, and so forth.

As an example, we will focus on social grabbing done by male squirrel monkeys in a sensory-rich environment to demonstrate the transition from the play reinforcers to sex reinforcers (Baldwin 1969). During play, 9- to 12-month-old squirrel monkeys grabbed at each other in many ways including wrestling, fighting, holding, and swatting. However, as the females began to withdraw from rougher play between 12 to 16 months, the contingencies of reinforcement for males' grabbing behaviors underwent the following changes. If males were gentle when reaching for and holding females, the females would play; but if the males were rough, the females would threaten or withdraw. Reinforcers from genital stimulation began to enter the picture, too—young females were the only play partners that juvenile males could find who would engage in relatively lengthy bouts of sex play with them, including mounting and thrusting. Thus, for the juvenile males, genital stimulation and sexual reinforcement were most likely to be prolonged when their grasping, holding, and sexual playing was done with the females and was done in a gentle manner. During the following years, the juvenile and subadult males learned conspicuously different grabbing and holding responses for interacting with females than they did for contacting other males. By four years of age, the subadult male squirrel monkeys had complex repertoires of rough play grabbing behaviors that they used with other males; but they approached young females gently, followed them slowly, and reached to mount them sexually, only after the females showed signs of permitting contact. By the subadult stage, genital stimulation had probably replaced more general sensory stimulation as the dominant reinforcer in shaping and maintaining heterosexual contact; and the playful and varied aspects of early sex play had declined significantly. In a subsequent study (Baldwin and Baldwin 1972) on squirrel monkeys at Barqueta, Panama, where no social play was observed, neither the subadult males nor the adult males showed such slow, cautious, and gentle approaches to females before copulations. They tended to grab onto females in a much more direct, blunt, and brief manner.

For the young animal, the reinforcers of novelty strengthen many behaviors and lead to their repeated practice. As the individual matures, the reinforcers of food, sex, avoidance of pain, etc. take on increasing importance, especially as the ceiling of readily available sensory stimulation is reached and the novelty reinforcers for exploration and play are exhausted. It is these contingencies that explain why there is a shift from normal play, to mixed play, to well practiced, nonplayful adult behavior (from left to right in Figure 3). If there were numerous opportunities for play in youth, the animal may bring extra variety and sophistication to its adult behaviors that animals with less play experience and practice do not have.

Second, we will turn to the level of analysis that focuses on longer sequences of behaviors, rather than on single units of behavior. Various writers have noted that the sequencing of behaviors during exploration, and especially during social play, often seems unordered, random, and repetitious (Loizos 1967; Dolhinow and Bishop 1970; Bekoff 1972). For example, an active play sequence may contain many apparently randomly ordered units: animal A chases animal B to the end of a branch, then they turn and B chases A back in the other direction; A stops in a matted vine where B catches up and they wrestle; at one point B's head is against A's genitalia, and A does pelvic thrusting at B; there is a relaxed pause with no motion, then A leaps up in the air, and B starts to chase A into higher parts of the tree, and so forth. Often it is difficult and discouraging to attempt to record play sequences in the field because so many combinations of behaviors are possible, and at times little order emerges.

The apparent randomness may be, in part, a consequence of the predominant role of novelty reinforcers in shaping the "games" of play. Strange, bizarre and varied recombinations of behavioral elements provide more novelty for the players than rigid, unvarying combinations do. The players are receiving differential reinforcement for creativity—the more creative their behaviors are, the more novelty reinforcers they receive. Thus, behavioral sequences that appear at first to be fragmented, uneconomical, extravagant, random, or inconsistent may actually be favored by the contingencies, at least in the period before the players have reached the ceiling of readily available sensory stimulation. Routine, economical, and orderly sequences of behavior may be the most efficient in attaining the other reinforcers, but they rarely optimize sensory stimulation reinforcers. Pryor et al. (1969) have demonstrated that dolphins can learn to generate novel and creative responses without prompting or guidance from human trainers merely by arranging the contingencies of reinforcement such that only innovative responses are reinforced. It would be valuable to extend such research to primates.

In addition to the above mentioned contingency of reinforcement—creative behaviors produce sensory stimulation reinforcers—there is a social contingency. The relaxed, exaggerated behaviors of one animal doubtless communicate its readiness to play to other animals (Bekoff 1972; Miller 1973). To the degree that the other animals discriminate on the behaviors of the first animal, they are more likely to respond with play or mild rebuff than with serious threats or lunges. Their differential response pattern creates a social contingency that reinforces the exaggerated behaviors of the first animal. The metacommunicative qualities of a playful approach coordinate players, minimize "misunderstandings," and often serve to elicit play from a previously nonplaying individual, all of which are reinforcing.

The maturational trend from random, innovative, varied sequences of behavior toward sequences of greater order and pattern is in part related to changes in the contingencies of reinforcement. During early exploration and play, young primates are under the control of other reinforcers besides novelty: genital stimulation occurs during early play; food is obtained during the exploration of certain parts of the environment; pain and the avoidance of pain become important shapers of behavior once play becomes rough-and-tumble. As young animals are repeatedly shaped by these positive and negative reinforcers, and as novelty reinforcers are exhausted, the contingencies favoring variety and creativity give way to those favoring efficiency. For example, during the play of the young male squirrel monkey, thrusting occurs at a low frequency and in a variety of different contexts. Randomly oriented thrusting is not as likely to be followed by genital reinforcers as thrusting at certain animals in specific positions. As animals begin to reach the ceiling of readily available sensory stimulation in the area of sexual play, the novelty reinforcers for randomized thrusting decline, and thrusting comes increasingly under the control of genital stimulation reinforcers which tend to shape performance of highly ordered patterns of behavior instead of varied, random, and novel patterns. When a subadult squirrel monkey approaches a juvenile female for thrusting play, the closer that his sequence of behavior matches the pattern of quiet approach, leisurely timing, gentle grasping, and cautious mounting, the more likely he will be to receive the reinforcers of a prolonged thrusting interaction. The differential reinforcement contingencies related to food getting, pain avoidance, smelling sexual odors, thermoregulation and so forth also tend to shape long coordinated sequences of behavior rather than randomized sequences. Order emerges from chaos. The maturing animal tends to become a creature of habit—less "creative" and less innovative—as novelty ceases to be readily available and as his habits come increasingly under the control of the contingencies that reinforce skill and efficiency. The social reinforcers resulting from the metacommunicative properties of exaggerated behavior may, however, maintain certain playful gestures in an animal's repertoire even after the more creative play behaviors have become significantly less frequent.

Another perspective on the randomness of longer play sequences is gained by focusing on the phenomenon of stimulus control. Whenever new behaviors are being learned, they tend at first to be under poor S^D control, but to come under stronger control of more precisely discriminated S^D's with practice. At first, a young animal may not discriminate which stimuli are associated with reinforcement, and thus it may generate a behavioral sequence under "inappropriate" stimulus conditions. An adult may discriminate on many cues that lead to an efficient sequence of behaviors such as A, B, C, then D; but the inexperienced younger animal may fail to discriminate on many of the cues and produce a larger number of different sequences that deviate from the A-B-C-D pattern. It is trial-and-error experience that shapes better discriminations and stronger S^D control of the "appropriate stimuli."[11] The randomness and "inappropriateness" of behavior disappears as finer discriminations are learned.

In summary, the randomness and creative combinations of play sequences should be most conspicuous in young players before reaching the ceiling of readily available sensory stimulation. After the novelty reinforcers become less available and other reinforcers bring behavioral sequences into more effective S^D control, play takes on a more stereotyped appearance. As was explained at the end of section 7, play may continue at high frequencies for months or years after an animal reaches its ceiling of readily available sensory stimulation. The fact that younger players are under contingencies favoring creativity, whereas older players are on extinction, produces visible differences in the play patterns of younger and older players. Symons's (1973) slow motion film analysis of rhesus play revealed that, compared with younger animals, "the play of older animals is slower, less vigorous, contains fewer moves, is more stylized, and contains little or no chasing."

A. STEREOTYPED OR VARIABLE.

The strong focus on the variability of play in section 10 and elsewhere is not intended to obscure the facts that there are species-typical behavioral patterns in play, and that animals of a given species may repeat similar games many times before abandoning them.

In spite of the variability of play within a species, a given species may show a range or cluster of common play forms that are distinctive for that species and that may differ markedly from the behaviors of other species. One might describe the range of play behaviors seen in a given species as the species-typical variance in play, using the term "variance" to indicate that the species-genetic potential permits the learning or development of behaviors anywhere within the range of variance, though the environment controls which subset of the variance will appear in any given animal's repertoire. Although the concept of species-typical behavior is often presented in association with descriptions of highly stereotyped or ritualized behaviors (as in ethological descriptions), it is also the case that even among highly variable behavioral patterns clear species differences can be identified. For example, the social play of howlers and squirrel monkeys is quite different—howlers are usually slow and gentle players who paw softly at each other and rarely engage in chasing play, but squirrel monkeys frequently engage in high-activity play with rapid transitions from one behavioral pattern to another. Considering that howlers of all ages tend to move rather slowly, it is easy to see how their quieter style of play relates to their generally lower activity level. Each species—with different perceptual apparatus, different limb and motor structures, different central nervous and hormonal systems—is likely to learn different clusters of behavior under the general contingencies of sensory-stimulation reinforcement outlined in this paper.

When describing the play behavior of a given species, there is a tendency to list a small and finite number of discrete behaviors that players engage in. This approach is useful in comparing species, but it obscures the important variability inherent in play. Most observers agree that play is one of the most variable types of behaviors that primates generate; thus it is important to realize that semantic categories should not limit our attempts to describe the variability. For example, data on wrestling play may lead one to observe that young players engage in wrestling for months on end. This description might give the impression that precisely the same behavior patterns are being repeated whenever the animals wrestle. However, close observations indicate that there are an enormous number of variations in wrestling behaviors and that the several sense modalities through which these variations are perceived allow for a very large number of novel sensory experiences. Before individuals reach the ceiling of readily available sensory stimulation, the motor patterns of wrestling change frequently and new games are visible week-by-week if not day-by-day. Tactile, olfactory, oral, and proprioceptive feedback converge in numerous combinations as the players squirm into countless positions, kicking out with limbs, twisting, mouthing, etc. Finally, because there are two or more individuals involved, the "next move" is highly unpredictable, therefore unexpectable, and, hence, surprising and arousal-inducing to each individual. In sum, the words used to describe a species play repertoire can fail to convey the variability of play activities.

■ 11. EXPLORATION AND PLAY CAN LEAD TO ADAPTIVE AND MALADAPTIVE CONSEQUENCES.

Numerous theorists have argued that exploration and play are adaptive and functional behaviors, both for the individual and the species (e.g., Loizos 1967; Dolhinow and Bishop 1970; Suomi and Harlow 1971; Bekoff 1972; Poirier 1972). Many of the functions that have been postulated for exploration and play are listed in table 1. They focus on the beneficial aspects that these behaviors have for both physiological development and for increasing the sphere of learning experience. First, exploration and play tend to involve young individuals in active behaviors that exercise the skeletal muscles, cardiovascular system, and so forth. Second, exploration and play tend to lead individuals into varied experience rather than monotony, and varied experience has been shown to have beneficial consequences for the development of the central nervous system. Without adequate sensory input, nerve cells atrophy or fail to develop normally (Riesen 1961, 1965; Levitsky and Barnes 1972; Volkman and Greenough 1972). Thus, the fullest development of the peripheral and central behavioral systems benefit from exploration, play, and other sources of varied experience. Third, as most of the entries in table 1 indicate, a wide variety of learning experience can accrue during exploration and play. Doubtless, the list of beneficial types of learning will expand as play is further studied. Any given individual may not be exposed to all of the beneficial experiences listed in table 1, and young animals can acquire many of the adaptive benefits from the list via other activities besides exploration and play (see section 12). It would be necessary to study the whole range of an individual's developmental experiences in order to measure the degree to which exploration or play are involved in the acquisition of the items from table 1.

As Beach (1945), Müller-Schwarze (1971), Welker (1971), and others have noted, many of the speculations concerning the degree to which play actually contributes to the development of adaptive behaviors are based on very limited evidence. More data are needed to evaluate the amount of adaptive behavior that can develop without exploration and/or play and the other kinds of experiences which can produce the adaptive learning listed in table 1.

Furthermore, exploration and play can have maladaptive consequences, and

TABLE 1 / A LISTING OF THE COMMON FUNCTIONS THAT HAVE BEEN ATTRIBUTED TO EXPLORATION AND PLAY. ALTHOUGH SEVERAL CATEGORIES OVERLAP EACH OTHER, THEY HAVE BEEN KEPT SEPARATE IN ORDER TO RETAIN THE VIEWS OF THE AUTHORS.

FUNCTION	REFERENCE
1. Providing physical exercise.	Brownlee (1954), Dobzhansky (1962), Ewer (1968).
2. Providing sensory input which stimulates development of the central nervous system.	Riesen (1961, 1965), Levitsky and Barnes (1972), Volkman and Greenough (1972).
3. Developing perceptual skills and latent learning.	Welker (1961).
4. Sampling and discovering diversified information about environment.	Washburn and Hamburg (1965), Loizos (1967).
5. Discovering how the environment responds to attack, shaking, biting, pulling, etc.	Lorenz (1956).
6. Providing familiarity with objects that may facilitate tool use.	Birch (1945), Schiller (1957), Eibl-Eibesfeldt (1967).
7. Learning what species the individual belongs to.	Loizos (1967), Poirier (1972).
8. Developing social perception.	Fedigan (1972).
9. Developing motor skills.	Southwick et al. (1965), Dolhinow and Bishop (1970), Poirier (1970).
10. Facilitating adaptive innovation and learning in new environments.	Tsumori (1967), Fedigan (1972).
11. Developing predator defenses.	Lancaster (1971), Symons (1973).
12. Desensitizing early fears of the environment.	Baldwin (1969).
13. Overcoming early helplessness by developing mastery and competence.	White (1961), Dolhinow and Bishop (1970).
14. Practicing adult behaviors.	Pycraft (1912), Hansen (1962), Washburn and Hamburg (1965), Loizos (1967), Ewer (1968), Dolhinow and Bishop (1970), Suomi and Harlow (1971).
15. Facilitating initiation and integration into troop structure.	Southwick et al. (1965), Rosenblum and Lowe (1971), Poirier and Smith (1974).
16. Developing social bonds.	Carpenter (1934), Jay (1965), Southwick et al. (1965), Suomi and Harlow (1971), Poirier (1972).
17. Facilitating normal personality development.	Harlow and Harlow (1969), Poirier (1972).
18. Learning communication skills.	Mason (1965a), Dolhinow (1971), Jolly (1972), Poirier and Smith (1974).
19. Learning troop culture.	Itani (1958), Tsumori (1967), Baldwin (1969).
20. Developing sex roles.	Harlow and Harlow (1965), Baldwin (1969).
21. Developing reproductive skills.	Eibl-Eibesfeldt (1967), Dolhinow and Bishop (1970).
22. Learning dominant and subordinate roles.	Jay (1965).
23. Establishing dominance relations.	Carpenter (1934), Harlow and Harlow (1965), Hall (1965), Dolhinow and Bishop (1970).
24. Learning controlled aggression.	Dolhinow and Bishop (1970), Suomi and Harlow (1971).
25. Working out aggression.	Dolhinow (1971).
26. Learning limitations of self-assertiveness.	Poirier (1972).
27. Overcoming anxiety.	Dolhinow (1971).
28. Increasing behavioral flexibility.	Miller (1973).
29. Learning maternal skills during play-mothering.	Lancaster (1971).
30. Play-mothering increases the chances that abandoned infants will be adopted.	Lancaster (1971).

these deserve attention too, since most of the current play literature focuses on the adaptive consequences. Exploration and play can lead individuals into dangerous or risky situations. Glickman and Morrison (1969) have shown that in populations of mice, the most explorative individuals were the most likely to be eaten by owls. It is possible that explorative or playful primates are exposed to slightly higher risks by being more conspicuous and vulnerable to predators. The research on predation in primates is at present limited, but Berger (1972) found that in olive baboons, mortality was highest in the juvenile male age-sex class which had the greatest "vulnerability to predation because of their exploratory behavior, and being driven from the troop by the dominant males." Teleki (1973) also found that young primates were the primary target of the predatory behavior of the chimpanzees at the Gombe Stream Reserve. In addition, more explorative and playful animals may be exposed to higher risks of having dangerous falls, becoming poisoned, getting lost, or experiencing other harmful consequences. As every human parent knows, exploring and playing children are capable of exposing themselves to countless dangerous situations.

In addition to danger and harmful consequences, playing animals may be exposed to maladaptive learning experiences. A growing body of literature (Stretch et al. 1968; Kelleher and Morse 1968; McKearney 1969; Byrd 1972) has shown that numerous environmental contingencies can lead to maladaptive conditioning. This is effectively demonstrated in the laboratory. For example, although squirrel monkeys normally respond to electric shock as a negative reinforcer, there are various schedules of reinforcement that will condition individuals to lever-press for contingent shock (Byrd 1972). One variation on these experiments consists of the following phases (McKearney 1969). First, the animal learns to avoid shocks by doing the operant behavior of pressing a response key. After the response is learned, a second concurrent contingency is introduced—the first response after each 10-minute interval is followed by a response-contingent shock. The monkey continues to key-press to avoid the shocks (as it learned to in phase one), and it habituates to the occasional contingent shocks of phase two. During the third phase, the original shock avoidance schedule is eliminated, and the monkey is only influenced by the second contingency (10-minute fixed-interval shock presentation). Throughout the remainder of the experiment, the animal would not receive another shock if it never made a subsequent key-press response. However, the early learning strongly conditions animals to respond rather than to abstain, and all the experimental animals become locked into the maladaptive habit of responding (as if to avoid shocks) when, in fact, responding only produces shock. Response rates vary from 0.4–0.8 per second on the F-I 10-minute schedule and accelerate to 0.9–1.7 per second when the fixed-interval shock presentation is shortened to 1 minute.

This type of maladaptive learning has been demonstrated in various species, including humans (Weiner 1965, 1969). Many psychiatric and behavioral problems in humans may result from maladaptive learning histories (Szasz 1961; Weiner 1965). The foundations for maladaptive learning are laid when an individual overlearns one habitual behavior pattern as its dominant response to certain S^D's; then, if the environment changes in a way that is difficult to discriminate, but that makes the old response become inappropriate, the animal will have a strong tendency to maintain its old—but now maladaptive—response. When primatologists become more attuned to studying learning contingencies in the natural environment, it may be possible to determine the commonness of maladaptive learning in natural populations.

In the field it is usually difficult to be certain what is adaptive and what is maladaptive for the individual or the species. As Rowell (1972) points out, many statements about the adaptiveness of a given primate behavior are mostly speculative, since there is presently very little direct evidence available with which to evaluate them. Welker (1971) takes a more extreme position and states that it would be so difficult to evaluate many of the speculations about the adaptiveness of play that their status as untestable hypotheses makes them virtually unscientific.

The difficulty of evaluating the adaptiveness or maladaptiveness of a given behavior in a complex natural environment can be seen in the following example. During the Barqueta study (Baldwin and Baldwin, in preparation), juvenile howling monkeys often followed their troops up to wide gaps that were beyond their skills to cross. If two or three adults jumped immediately in front of a juvenile, the juvenile was very likely to jump, too. Juveniles experienced some serious falls at these wide gaps and often became separated from their troops. Four out of the six solitary howlers at Barqueta were lone juveniles. Our data suggest that some of these solitary juveniles may have become separated from their troops after making dangerous jumps and becoming too injured to maintain contact. This raises the question of whether it is adaptive or maladaptive for juvenile howlers to imitate the jumping responses of older and stronger animals—a case could be made for either position. On one hand, the juveniles' behavior fits the maladaptive learning model. In general, the tendency to imitate the behavior of social models is strengthened by prior reinforcement associated with imitation (Baer, Peterson and Sherman 1967; Danson and Creed 1970, 1971). During infancy, howlers doubtless receive numerous reinforcers for imitating their mothers' activities in the areas of feeding and selecting arboreal pathways. As the infant matures to the juvenile status, the mother ceases carrying it over wide gaps. At this point, when it contributes to inducing the juvenile to jump wider gaps than it can cross successfully, the tendency to imitate travel patterns becomes maladaptive. Thus, the imitation reflects a pattern that was adaptive and strengthened under the old contingencies, but becomes maladaptive once the contingencies change. A counter argument would stress that perhaps it is adaptive for an animal to experience some contingent aversive consequences (painful falls) in order to learn to discriminate when to imitate, and when not to imitate, the jumping behavior of others. Also, a few "bad jumps" might facilitate the learning of more accurate jumping skills. Nevertheless, consequences for the juveniles who become isolated from their troops can be serious—two of the four lone juveniles at Barqueta had suffered handicapping injuries, and two skeletons of juveniles were found at a spot where isolates often fled to avoid contacting strange troops.

The question of adaptiveness or maladaptiveness can also be viewed from the perspective of natural selection: Is the population benefited by removing the isolated animals from the gene pool? There are many possible factors one could analyze—

such as their health, intelligence, and physical strength—in search of calculations relevant to the ratio of adaptive to maladaptive consequences.

A. THE REASON FOR EMPHASIZING MALADAPTIVE CONSEQUENCES.

Almost all students of exploration and play have agreed that the adaptive consequences of these behaviors are numerous. In fact there are very few references in the literature that mention maladaptive consequences. Section 11 devoted space to both adaptive and maladaptive consequences in order to emphasize the necessity of studying both types and the continuum between them. It seems very likely that the majority of play experiences are close to the adaptive end of the continuum, and our discussion of maladaptive consequences was not meant to imply otherwise.

B. FUNCTIONAL AND MECHANISTIC EXPLANATIONS.

Functional explanations tend to focus on the evolutionary causes of behavioral adaptation, whereas mechanistic explanations focus on the physiological and psychological mechanisms that underlie the current changes and variations in behavior.

Exploration and play have doubtless been shaped by evolutionary contingencies. Adaptive exercise and learning clearly outweigh maladaptive consequences for most playing primates. Thus, through natural selection, play has evolved to have the potential to carry several important functional roles in primate socialization. The causal mechanisms that have evolved to activate or motivate play appear to be the ones that establish optimal levels of sensory stimulation as a primary reinforcer, since this reinforcer plays an important role in strengthening exploration and play activities. Sensory deprivation, social punishment, fatigue, injury, and other aversive stimuli operate in inhibiting play. During the developmental history of each individual, the contingencies of positive and negative reinforcement shape much of the ontogeny of play.

Thus the phylogenetic origins of play are explained by the contingencies of natural selection which favor functional and adaptive traits. The ontogeny of play is explained by specifying the interactions of the physiological organism with the unique contingencies of its environment.

■ 12. A GENERALLY ADAPTIVE REPERTOIRE OF BEHAVIOR MAY DEVELOP EVEN WITHOUT SOCIAL PLAY.

In chimpanzees, squirrel monkeys, and other species, where social play has been observed to occur at various frequencies in different individuals, there is the opportunity to determine some of the contributions and long-term consequences of play. The fact that young squirrel monkeys engage in social play for 3 hours per day in some environments and o hours in other environments demonstrates that there is no single "species-typical" level of playfulness (Baldwin and Baldwin 1972, 1973a). Apparently, with increasing sensory richness, peer group size, food abundance, and leisure, there will be higher frequencies of play. In some environments social play is apparently not crucial to the survival of the species. In the Barqueta study, where no social play was observed, the squirrel monkeys had a repertoire of social behaviors that led to stable troop integration and social organization. The 85% reproductive rate was normal for the species;[12] the frequency of aggression appeared to be average or lower than average; there was no sign of behavioral pathology or increased vulnerability to predation. Compared with other troops, the only major difference was that social interactions at Barqueta were relatively infrequent and less elaborate.

In environments where squirrel monkeys have ample opportunity to play, the frequency of play among the young can rise to 3 hours or more per day; and adults tend to interact more frequently and complexly than adults in nonplaying troops. Previous analyses (Baldwin and Baldwin 1973a) of the behavior of squirrel monkeys in different environments has led us to the view that social play per se is not essential to species survival. If play is present, the animals may learn larger repertoires of social interactions and may develop stronger habit strength for spending spare time in social activities. However, without play, life can continue in a simpler, less social style, but in a manner compatible with survival.

Squirrel monkeys are not the only species in which cases of little or no social play have been observed. Social play is not as universal in primates as some authors imply. Several primate studies which have reported little or no social play have described social units which are apparently adaptive and normal and in which there is no social pathology or lack of troop organization. MacKinnon (1971) and Horr (1972, personal communication) have reported that wild orangutans (*Pongo*) rarely engage in social play, and yet they mature to be normal adults; that is, they are typical for their species in the sense that they appear to be relatively well adapted to their natural ecologies. Schaller (1963) concluded that, "on the whole, gorillas are not playful," and commented that "Frequently several days went by without my observing a single instance of this behavior. . . ." (Moynihan (1964, 1970) mentioned that social play was very rare in tamarins (*Saguinus*) and nonexistent in night monkeys (*Aotus*). Petter and Petter (1967) indicated that aye-ayes (*Daubentonia*) probably played very rarely. Other studies have reported situations in which social play occurred but was infrequent: Poirier (1972) remarked on the infrequency of social play in Nilgiri langurs (*Presbytis*), which engaged in only 1 sequence of play per 6.9 hours of observation; and Richard (1970) recorded less than 12 seconds per hour of all social interactions including play in howling monkeys (*Alouatta palliata*). Ellefson (1967) indicated that many gibbons (*Hylobates lar*) experience very little social play due to the small group size and large discrepancies in body size, strength, and behavioral capacities between siblings.

Given the difficulties in recording the duration of social play and that each observer may have different standards for evaluating whether play was infrequent or not, it is at present difficult to compare the actual frequency of play from different studies. However, the fact that little or no social play has been observed in various species from prosimians to great apes suggests several things. Foraging, daily travel, and nonsocial exploration may provide sufficient exercise for muscular development, adequate sensory input for neural development, and numerous experiences contributing to learning and socialization. The mother-infant social relationship may provide the young animal with sufficient positive social experience to establish an attachment with conspecifics. The above-mentioned nonplaying primates do not show "uncontrolled aggression," "personality derangement," or other pathological conditions, even though several theories of play have argued that play is necessary to prevent the development of such abnormal behavior (Carpenter 1965; Loizos 1967; Suomi and Harlow 1971). The fact is that the primates with little or no social play often display many of the items from table 1. Thus there must be other means by which apparently adaptive and normal behavior can develop—play does not have a monopoly.

Perhaps the laboratory isolation experiments (along with cage and zoo studies) have unduly biased the primate literature toward the position that social deprivation—especially the absence of play—leads to abnormal behavior. Rhesus monkeys that are deprived of social interaction with their mothers and peers develop into very maladapted, nonsocial adults (Harlow and Harlow 1965, 1969). But when infant isolates are given 18 minutes per day of experience with peer group play, virtually all the isolation effects are avoided and the monkeys mature into relatively normal laboratory adults (Harlow and Harlow 1969). These data, in conjunction with the more recent experiments which show the effectiveness of play with a therapist monkey for overcoming isolation effects (Harlow and Suomi 1971), clearly indicate that social play can have powerful beneficial effects for preventing isolation abnormalities. But they do not demonstrate that the absence of social play, per se, is detrimental. Laboratory deprivation experiments not only deprive the infant of play, they often deprive it of its mother, social models, sensory input, troop relationships, and so forth; and the removal of so many beneficial experiences contributes to abnormal development. Meier (1965) raised infant rhesus monkeys so that they were deprived of physical contact and social play but were given close visual and vocal contact—the gross abnormalities usually associated with total isolation did not appear. Mason (1968a) raised isolates on stationary and moving surrogate mothers and found that infants benefiting from the extra sensory stimulation provided by their randomly roaming robot mothers showed few signs of abnormality. The highly aberrant behaviors associated with total isolation should not be confused with the effects of limited social play.

From the functional point of view, it would seem most dysfunctional if primates could not survive without social play. Many wild populations may go through a period of "lean years" when food is scarce and hunger replaces social play. Primates have long been typified as adaptable, flexible, and intelligent, and the ability to survive without social play would certainly appear to be functional.

The degree to which the frequency of social play varies in most primate species is not well known. Comparative studies on selected species, each of which is observed in several different ecological settings, are only beginning to be conducted. The preliminary data do show, however, that there is a wide range of variance in the behavior of many species across different environmental settings (Rowell 1966; Frisch 1968; Gartlan and Brain 1968; Jay 1968; Eisenberg et al 1972; Ransom and Rowell 1972; Southwick 1972; Paterson 1973). Since social play has been observed to vary in frequency in chacma baboons (Hall 1963a), rhesus monkeys (Loy 1970; Southwick

1972), and squirrel monkeys (Baldwin and Baldwin 1973a), it is reasonable to assume that social play will be among those behaviors that can have variable frequencies in different environments. However, given the problems in defining, describing, and timing play and the different amounts of attention various researchers focus on play activities, it is difficult to compare most of the play data from the literature that is presently available.

■ 13. THERE ARE OTHER DEVELOPMENTAL ALTERNATIVES BESIDES SOCIAL PLAY THAT CAN LEAD TO "NORMAL" ADULT BEHAVIOR.

Given that much of primate behavior shows variability across different habitats, the concept of "normal" adult behavior ceases to mean fixed, stereotyped, or species-specific and takes on a much looser quality. A species that can survive in a variety of different environments will have members that are exposed to a range of different developmental and socialization experiences. For example, the data (Baldwin and Baldwin 1973a) on squirrel monkeys indicate that adults that are raised with a history of social play tend to have a larger repertoire of social behaviors under subtler stimulus control, and to have stronger habit strength for social interaction. This is "normal" for them. Adults which have had little social play experience have a less social pattern of life, and that is "normal" for them. This perspective encourages the student of primate development to focus on a variety of developmental pathways for normal (and abnormal) behavior and to realize that different pathways can overlap in contributing to the formation of a specific behavioral pattern. Most developmental pathways can be analyzed as involving the following processes: (1) behavioral development that does not involve learning, (2) learning that does not involve social play, or (3) learning that involves social play.

First, there are doubtlessly numerous behaviors that develop without play or any other learning experience. Given adequate nutrition, sensory input, and health, the endocrine and central nervous systems are genetically programmed to produce a variety of behavioral mechanisms that can function without practice or learning. The ethologists (e.g., Eibl-Eibesfeldt 1970) have provided numerous examples of this. The production and (to a lesser degree) the perception of many primate vocalizations appear to fit this model (Marler 1965; Winter 1969; Winter et al. 1973). The "play face" or open-mouth grin typical of playing primates of many species (Bolwig 1964; van Hooff 1967; van Lawick-Goodall 1967a; Chevalier-Skolnikoff 1973) appears to be a ritualized, unlearned signal often associated with play. Sexual behaviors are highly influenced by biological determinants (Goy and Resko 1972; Harlow et al. 1972; Phoenix et al. 1973); and both male and female rhesus monkeys raised in isolation cages that permit only auditory and vocal contact show very normal copulatory and maternal behaviors when tested at adulthood (Meier 1965). Basic locomotor coordination develops with a minimum of feedback, as is shown in studies on monkeys whose limbs are deafferentated at birth (Berman and Berman 1973). Thus, it is likely that part of the behavioral repertoire of monkeys which never have the opportunity to play develops as a consequence of genetically programmed structuring of the nervous, endocrine and other behavioral systems. These behaviors might be called the "species-typical minimal behavioral endowment," noting that different individuals may express different subsets of their species-typical endowment depending on environmental conditions.

Species-typical responses are dependent on the environment in at least two ways. (1) As Lehrman (1953, 1970), Schneirla (1956), Kuo (1967), and other epigenetic behaviorists have pointed out, environmental factors such as nutrition, sensory input, temperature, and numerous feedback loops will affect and modify the development of species-typical behaviors even when no learning processes are involved. Thus, any given species-typical response may be expected to show a range of variance across different environments. (2) Given that the basic requirements for nutrition, sensory stimulation, etc. are present to guarantee the healthy development of the "machinery of behavior" (i.e., the nervous, endocrine, and musculature systems), only a subset of the potentially available species-typical behaviors will be elicited from the organism, depending on which stimuli are present in its environment. Thus, for example, if there is no play in a certain environment, play presentation or the play face might never by elicited even though all of the behavioral machinery necessary for their production was adequately assembled.

Second, the behavioral equipment of most primates includes a large cerebral cortex capable of learning many complex associations about the environment. The second process of behavioral development includes the types of learning available even when there is no social play. This does not exclude nonsocial exploration and manipulation from which it is likely no wild primate is ever completely deprived. Even without social play, growing monkeys are exposed to numerous learning contingencies in which varied responses are conditioned according to each individual's unique experiences. A monkey can learn a great deal about its nonsocial environment while foraging, inspecting a wasp nest, climbing through a spiny palm tree, coming across a sleeping squirrel, etc. Likewise, a young monkey can experience a great deal of social learning during mother-infant interactions, trial-and-error encounters with other troop members, and imitation of other animals.

A large number of field studies—more than can be reviewed here—contain data on learning that involves no social play. Hall (1968) outlines several types of learning situations that do not involve play and which lead to important behavioral adaptations and social structuring. Most field and laboratory studies on mother-infant interactions indicate that mother-infant accommodation depends heavily on learning experiences which rarely include social play in most species. Observational learning is an important means by which feeding habits, nest-building skills, travel patterns, and so forth are passed on from mother to offspring in chimpanzees (van Lawick-Goodall 1973) and orangutans (Horr 1972). Observational learning plays a central role in the diffusion of cultural innovations and skills in Japanese macaques (Kawai 1965).

It is especially important to ask which types of learning contigencies, other than those involving social play, lead to the development of the behaviors listed as adaptive consequences of social play in table 1. In the Barqueta study on squirrel monkeys that did not play, it was possible to identify several learning contingencies that were operative in shaping controlled aggression, sexual behaviors, and individual spacing. These learning contingencies are described in detail elsewhere (Baldwin and Baldwin 1972; 1973a); however, one subset of these data (dealing with aggression) is included here in order to provide the reader with a specific example of natural learning contingencies not associated with social play.

At Barqueta, most agonistic encounters terminated when one participant fled from the conflict area. Submissive postures appeared to quell minor incidents, but rapid flight through the canopy was the most common response to more actively agonistic encounters. In the laboratory, where there is not space to flee 200 meters through the trees, aggression may be more difficult to avoid than in the wild. From laboratory studies, Suomi and Harlow (1971) theorize that social play is essential for monkeys to learn to cope with and "control" aggressive responses. The data from Barqueta suggest a different interpretation. Because wild animals can flee through the trees to escape or avoid agonistic interactions, they can learn "controlled" aggression (even without playing), although this might not be easily learned in confined laboratory environments.

In addition, observations on chasing sequences at Barqueta suggest that the animals learned when and how to avoid aggressive contacts through trial-and-error operant conditioning and possibly through observational learning. For example there were differential reinforcement consequences associated with directing agonistic behaviors toward larger versus smaller animals. If a juvenile threatened an adult male, the latter was likely to deliver a contingent negative reinforcer such as a loud vocalization, branch shake, lunge, or chase. (A vigorous branch shake or lunge by a large animal could knock a small animal off balance or catapult it off the branch. Thus these threats actually had a punishing effect on the threatened animal. Loud vocalizations appear to be conditioned negative reinforcers since the threat vocalizations of males who do not follow up their vocalizations with stronger aversive consequences have little effect.) Not surprisingly, younger and smaller animals showed inhibitions and hesitance about directing agonistic behaviors toward older, larger animals.

There were other contingencies also operating that had differential effects on infants, adults, males, females, and peripheral animals. For example, one might expect that adult males could show unbridled aggression, since there were no individuals in the troop who were larger or stronger to deliver negative reinforcers and inhibit them. However, adult males often encountered aversive consequences after being aggressive. A noisy, aggressive male usually attracted the attention of nearby animals; and when several juveniles and adult females were close to the scene, they would tend to coalesce against the male and chase him. Naturally, these contingencies would reinforce coalition formation among adult females and young and would inhibit adult males about showing aggression near these groups. It usually took a coalition of two or three adult females along with some young to succeed in chasing a male away. When the coalition was smaller, the interaction tended to look like a "semichase" because two or three females or young would slowly follow or pursue an adult male rather than actively chase him. If the coalition moved too far from the remainder of the troop or if it lost a member, the male usually became less inhibited and was likely to threaten his pursuers. Upon being threatened, a pursuer that did not have coalition support usually fled. In these circumstances, nearby onlooking juveniles and adult females also usually fled or retreated. It was clear that juveniles and adult females learned the limits to which they could harass the adult males. A coalition of two or three animals might semichase a male from a distance of 0.5 to 2.0 meters, but if the coalition became smaller, the pursuing animals began to keep larger distances of 1.5 to 4.0 meters and then to retreat. There were numerous opportunities for learning via observation, such as when an onlooking monkey saw a fellow coalition member get threatened for coming too close to an adult male, but it was difficult to evaluate the relative importance of observational learning versus shaping.

These examples only begin to describe the complex contingencies of reinforcement that related to aggressive behavior in a small troop with infrequent social interactions. However they do demonstrate that without any social play experience, troop members were under contingencies conducive for learning complex sets of instigations, inhibitions, and discriminations related to aggressive behavior. At Barqueta, learning without social play also affected troop cohesion, sexual behavior, individual distances, and probably other behaviors also (Baldwin and Baldwin 1973a).

In order to establish precisely how much can be learned without social play it will be necessary to locate and study more troops of primates in which social play is absent or infrequent, or to create such conditions in the laboratory.

The third avenue of behavioral development that can affect an animal's behavioral repertoire is the experience of social play. For aye-ayes, squirrel monkeys, night monkeys, orangutans, gibbons, and probably other species, it is not necessary for social play to be a part of behavioral development; however when individuals do have the opportunity to engage in social play, numerous experiences are opened that permit a new realm of complex social development. Social play creates varied and often complex contingencies that may condition subtle discriminations (perceptual skills) and behavioral sequences. A major difference between squirrel monkeys that have played extensively and those that have not is that the play-experienced animals generally have larger behavioral repertoires, more social activities, and a tendency to spend more time per day engaged in social interactions.

Because social play provides a variety of social experiences, it might be possible to assume that animals with a history of play are better adapted, smarter, or more flexible than animals without play experience. "The more play, the better." To some degree this is probably true. However, exceptions are possible—there may be maladaptive consequences of play experience. In addition to the kinds of maladaptive behaviors listed in section 11, there may be specific limitations to the assumption that "the more play the better." For example, there may be a point of maximum efficiency beyond which there are diminishing returns for additional metabolic energy expended in social play. Also, it might be the case that learning social habits is adaptive up to a point, but that additional social learning experience could produce hypersocial individuals that focus more attention on social interaction than is optimal. It is conceivable that a hypersocial primate would not spend as much time as a less social individual attending to exploration and learning about its nonsocial environment, and this could be a disadvantage for adapting to dangers or changes in the environment. As explained in section 11, it is difficult to evaluate the relative adaptiveness of different behavioral repertoires; thus, at present, it is not possible to estimate the optimal amounts and types of play.

■ 14. GENETIC FACTORS HAVE COMPLEX INFLUENCES ON THE SPECIES DIFFERENCES IN EXPLORATION AND PLAY.

Exploration and play are uncommon in the lower vertebrates and tend to be found mostly in the mammals, especially the carnivores and primates (Welker 1961; Glickman and Sroges 1966). However, at present, it is difficult to assess the role of species genetics in determining the form and frequency of play in a given animal. Members of a given species can spend significantly different amounts of time engaged in social play (depending on local environmental conditions) as has been shown in chacma baboons (Hall 1963a; Hall and DeVore 1965), rhesus monkeys (Loy 1970; Southwick 1972), and squirrel monkeys (Baldwin and Baldwin 1974). These and other studies suggest that a species' genetics establishes a wide potential range of play activities and that environmental conditions determine how play will develop in any given ecological setting. Because there have been only a limited number of studies that have collected comparative data on a selected species living in a variety of ecological conditions, it is difficult to assess the range of play behaviors that most species may show in different environments. Even as field studies on a species in different environments do become available, it may be problematic to compare the frequency and form of play observed because the operational definitions of play used by various researchers are difficult to formulate or standardize. A second way of measuring a species potential for exploration and play is to study animals in a standardized environment in either zoos or laboratories. However, not all species adapt or perform well in a given man-made test environment (Zimmerman and Torrey 1965; Jolly 1966; Jarosch 1968). Minor design errors can strongly handicap the performance of certain species and thus invalidate the measure of their competence in the standardized test environment (Rumbaugh and McCormack 1967). Emotional factors and docility can produce significant differences in the way that very closely related species respond to the same test situation (Green 1972). Bernstein (1971) measured the frequency of several behaviors including social play in nine primate taxa, but no unambiguous pattern emerged in comparing the data on play. Bernstein qualified these findings by pointing out that play is affected by the age-sex composition of the groups and other factors. Even if group size, age-sex composition, environmental complexity, and daily regimen were all carefully standardized, it is not clear that any *single* standardization would yield unbiased measures of the species potential for different levels of play in different environments. Also, laboratory and zoo studies tend to be difficult to interpret since many facts about the origin, age, past conditioning, and health of the test animals are not known or controlled for. Since any of these factors can affect play activities significantly, a large amount of variance in the data usually cannot be accounted for.

Despite the numerous problems in estimating the range of a species' potential for exploration and play, several studies have measured the frequency of play and described the common play patterns seen in different species. The common strategy of interpreting these data consists of (a) combining the material from zoo, laboratory and field research; (b) looking for correlations between exploration, play, and ecological adaptations; and (c) trying to locate functional or logical relationships. This method indicates that carnivores and primates tend to play more than most other mammals and that play becomes less common as one descends the phylogenetic scale (Welker 1961; Aldis, in press). Predators and opportunistic species tend to be more playful than herbivores, and large-brained species more playful than small-brained species (Glickman and Sroges 1966). Species that are adapted to living in variable ecological niches and capitalizing on flexible behavioral styles (the "neophiles") are more playful than conservative species which are adapted to specialized behaviors in rigid niches (Morris 1964). Also, animals that show neoteny tend to play more than less neotenous species (Mason 1968a; Bekoff 1972).

These generalizations point to broad patterns in the data, but they are filled with exceptions and they often fail to indicate that there can be significant within-species differences in exploration and play. Also, while they emphasize the broad patterns visible among several orders of vertebrates, they are not very useful at present in predicting variances within an order, such as primates. For example, *Lemur catta*, a prosimian, has been observed to play more frequently than some of the great apes (Jolly 1966). Yet *Lemur fulvus*, a closely related species, shows significantly less play (Sussman 1973, this vol.). The complexity within the primate order is great. As the study of play becomes a more central topic in primate socialization and more precise data become available about the variability in the frequency and form of play within different species, we can expect exciting new breakthroughs in the analysis of the phylogenetic determinants of play.

▪ CONCLUSION

Crook (1970) has emphasized that most aspects of primate behavior are so complex that only multifactor theories can begin to integrate the relevant data. Exploration and play are no exception. Of the many factors influencing exploration and play, genetics, physiology, ecology, nutrition, and learning all have important roles. In searching for a central theme around which to organize the present primatological data on play, we have found learning theory to be the most useful. Around this central theme we have introduced genetic, physiological, and maturational variables whenever appropriate. The following paragraphs present an overview of the present learning theory of play.

Sensory stimulation affects the general arousal level of an organism in ways that can be positive or negative reinforcers. Both extroceptive and introceptive sensory stimulation derived from exploration and play generate positive or negative reinforcers that shape the frequency and form of these responses. The more novel or complex a stimulus is, the more it is likely to induce higher levels of general arousal. Satiation or habituation to a stimulus decreases the effect of that stimulus in inducing increased arousal levels, but removal of the stimulus permits recovery effects that partially counteract the effects of satiation and habituation.

Because the consummatory act of perceiving sensory input is less salient than the acts of feeding, drinking, copulating, etc., the importance of the reinforcement properties of sensory stimulation have not been widely recognized. The absence of tangible extrinsic reinforcers such as food, water, or sex partners has led many theorists to conclude that play is intrinsically reinforcing, self-motivated, or done for pure pleasure. The present theory is not incompatible with these interpretations—positive reinforcement gives rise to the phenomenological experiences of motivation and pleasure, and the low salience of the consummatory act of perceiving stimuli gives rise to the intrinsic appearance of stimulation-related reinforcers.

Both overstimulation and understimulation are aversive. Positive reinforcement due to sensory stimulation occurs only at intermediate levels. These optimal levels of stimulation change during the life of the individual, usually following a rapidly ascending, then a slowly descending curve. The height, slope, and shape of the curve depend on numerous overlapping factors, including age, sex, environmental complexity, availability of a mother or surrogate, leisure time for play, troop social composition, and contingent reinforcement associated with play. Because the young infant finds low-stimulus input levels to be optimal, contact with the mother (a source of warm, soft, familiar stimuli) is highly reinforcing and conducive to the conditioning of positive emotional responses to the mother. As the infant habituates to higher levels of sensory input, it is reinforced for venturing away from its mother and exploring widening subsets of its environment. A dynamic relationship emerges as the infant explores away from its mother for arousal-inducing stimulation and returns to her for arousal-reducing contact and comfort.

The earliest exploratory activities are shaped from reflex-like responses, including alerting responses elicited by stimulus change and all the other neonatal behaviors that produce varied sensory input. As the individual matures, the topography of its exploratory and playful behaviors will depend on the unique contingencies of reinforcement and social models to which it is exposed. There are general patterns that emerge as individuals explore, and these allow one to sketch basic "stages" of development in exploration and play. The natural contingencies of reinforcement for primates with peers and leisure time to play often lead an individual through stages such as these: presocial exploration, early social exploration, contact play, rough-and-tumble play, distance play and chasing, play fighting, decline in play, and end of play. Numerous contingencies are discussed that produce such patterns and deviations away from them. Because each individual is shaped by unique contingencies, the ontogeny of stimulus-seeking behaviors is not rigid or invariant, hence caveats are presented about the utilization of stage conceptualizations.

Other trends in the ontogeny of play have to do with the apparent "purpose" or "seriousness" of the behaviors seen. Early exploration and play often approach the end of a continuum one might call "pure" stimulus-seeking—the motivation is almost exclusively due to the reinforcers of sensory stimulation. With time there is, however, a trend toward mixed motive play. Sensory reinforcement loses its power as the individual exhausts the readily available sensory stimulation in its environment and ceases to find as much novelty there. Other reinforcers begin to shape play as sex, food, fights, and other conditioning experiences are encountered during play. Thus, early play for relatively pure sensory reinforcement appears to be done for the "pure fun of it" (i.e., intrinsic, sensory reinforcement), but later, play may begin to include longer sequences of sex-related or food-oriented behaviors. This mixed-motive play represents a transition to adult behavior where playful behaviors tend to become fully extinguished. Some writers have typified adult behaviors as "serious" or "purposeful" due to the fact that they are usually efficient and goal directed (i.e., oriented to visible, external reinforcers). In contrast, infant behavior may appear unserious or done for no specific goals other than the joy of it. Our analysis, which places sensory reinforcers on the same level of importance with other reinforcers, leads one to the conclusion that play is no less serious and purposeful than adult behavior, and hence those adjectives do not differentiate between play and adult behavior. The fact that much early playful behavior is so chaotic, disorganized, and uneconomical in appearance is also explained in part by certain characteristics of sensory stimulation reinforcers.

A major factor in determining the decline of playfulness typically associated with adolescence or maturity is the ceiling of readily available sensory stimulation. The ceiling is determined by both the amount of complex stimuli in the environment and the individual's skills, strength, and leisure time for discovering the stimulation potentially available to it. The greater the environmental complexity and the individual's skills, strength, and leisure time, the higher the individual's ceiling of input will be, and greater amount of stimulus-seeking behaviors will occur before the sensory stimulation is exhausted. These, and other factors, determine how late into life an individual will explore and play. Even after the ceiling is reached, play may continue for a long period before extinguishing, especially in individuals with strong habit strength for play or who have been on intermittent reinforcement schedules. Males typically play later into adolesence or adulthood than females, and at least four factors influence the differences that are seen in the play of males and females.

Although exploration and play are basically adaptive and provide the individual with much beneficial experience, maladaptive consequences can occur during stimulus-seeking activities. The exploring young animal may be exposed to higher chances of predation, poisoning, falls from the trees, separation from the troop, and maladaptive learning. Although much of the primate literature on play stresses the adaptive nature of play, it is suggested that the consequences of much stimulus-seeking behavior lie on a continuum between adaptive and maladaptive, though predominantly near the adaptive pole. In addition, the assumption made by some writers—that social play is essential to normal and adaptive functioning—is questioned. Data are presented that demonstrate that many of the beneficial consequences usually attributed to social play can develop in the absence of this activity. Several field studies have reported an absence or very low level of social play in species from prosimians to great apes. There are numerous possible routes to normal and adaptive social development, and because of the redundancy of the socialization system, individuals can develop functional behavioral patterns without being exposed to all the possible socialization experiences.

The effects of species genetics on exploration and play are difficult to establish. As the number of laboratory and field studies on a given species in a variety of environments increases, it will become possible to specify the interaction of genetic and environmental factors that shape the topography of play in different species.

Although the present theory is derived primarily for the nonhuman primates, aspects of it are applicable to other species. Exploration and play in humans is clearly more complex than the present theory can account for, yet there is evidence that stimulus-seeking activities of humans have much in common with those of the other primates.

■ NOTES

1. It is possible that early manipulative and exploratory behaviors may be conditioned prenatally, and we do not intend to preclude this possibility on an a priori basis.

2. We agree with Kuo (1967) and others who contend that the concept of "reflex" is misleading when discussing many complex neonatal behaviors, since early reflex-like behaviors are usually the consequence of complex developmental processes. The term "reflex-like" is used to label early infant behaviors without purporting to explain them.

3. Although the lay observer may attribute "purpose" to an animal that has been conditioned to do certain behaviors (such as looking for food), the conditioning process does not imply the existence of purpose in either a teleological or higher mental insightfulness sense. Skinner (1969) has shown how lay observers will consistently impute purpose, insight, volition, and planned strategies to simple conditioned responses as those behaviors become shaped into efficient, highly organized patterns. Purpose is in the eyes of the beholder.

4. As mentioned elsewhere in this paper physiological maturation is occurring concommitantly and contributes to the development of coordinated, targeted responses.

5. Several early writers have commented that play can be identified by its "nonserious" qualities (e.g., Heckhausen 1964; Marler and Hamilton 1966). More recent writers have taken the opposite view and emphasized that play must be seen as a "serious" part of the development of an individual's life (e.g., Dolhinow and Bishop 1970; Bekoff 1972). By considering play as a behavior controlled by the reinforcers of sensory stimulation, we would emphasize that play is as "serious" as any other acquisitory response and just as "enjoyable" as any other behavior controlled by positive reinforcers.

6. As mentioned in section 3, individuals show diurnal variations in their optimal input zones, and a playful animal may show strong stimulus-seeking habits during only certain periods of the day. Thus, it may be only during these periods that a juvenile avoids low activity behaviors and engages in play.

7. Naturally, there can be variance among individuals in the timing of when rough, aggressive play ceases. For example, one would expect that the largest or most dominant animals would be the last to extinguish or be inhibited from aggressive play.

8. The level of "cognitive" skills of an animal depends on the number of discriminations it can make; the more stimuli that it can discriminate, the more novelty it can find in a given environment. These discrimination skills vary with species cortical capacity, age, and history of learning (Menzel 1969).

9. When a danger context contains aversive stimuli other than those produced by overarousal, males may be the first to flee rather than serve as troop protectors (Rowell 1972). Utilizing Hebb's (1972) dichotomy that a stimulus can have both cue properties and arousal-controlling properties, it is clear that the presence of a special cue stimulus (such as a gun or a dangerous predator) could cue flight behavior that is independent of the overall sensory stimulation level of the stimulus context.

10. Among the authors who suggest that imprinting and critical periods may occur in primates are Loizos (1967), Hinde (1972), and Salzen (1968). However, these authors only present data from bird research and argue by analogy that imprinting in critical periods could be occurring in primates. Harlow, Gluck, and Suomi (1972) conclude from their extensive research on rhesus monkeys that critical periods and imprinting do not occur in infant social development. Their data show that what was once considered irreversible damage done by early isolation can be altered by six months of play therapy. Mason and Kenney (1974) obtained data on infant attachment behavior that are incompatible with theories of imprinting and critical periods in primate development. Hoffman and Ratner (1973) have presented data that may resolve some of the apparent conflict between the imprinting and reinforcement models of attachment.

11. Some responses may be more easily conditioned than others. Research on rhesus monkeys shows how quickly correct copulatory behavior is learned (Michael and Wilson 1973), which demonstrates that the genetic "constraints on learning" (Hinde and Stevenson-Hinde 1973) or "preparedness to learn" (Seligman and Hager 1972) can significantly influence the rate of learning of different types of responses.

12. Observations in several natural and seminatural environments—on the llanos of Colombia (Thorington 1968; Baldwin and Baldwin, personal observation); on Santa Sophia Island near Leticia; Colombia (Bailey, personal communication)—indicate that approximately 85% of the adult females in a squirrel monkey troop bear 1 infant per year.

■ REFERENCES

Ainsworth, M. D. 1964. Patterns of attachment behavior shown by the infant in interaction with his mother. *Merill-Palmer Q.* 10:51–88.

Aldis, O. forthcoming. *The function of play in animals and man*. New York: Academic Press.

Andrew, R. J. 1972. The information potentially available in mammal displays. In *Non-verbal communication*, ed. R. A. Hinde, pp. 179–204. New York: Cambridge University Press.

Baer, D. M.; Peterson, R. F.; and Sherman, J. A. 1967. The development of imitation by reinforcing behavioral similarity to a model. *J. Exp. Anal. Behav.* 10:405–416.

Baldwin, J. D. 1968. The social behavior of adult male squirrel monkeys (*Saimiri sciureus*) in a seminatural environment. *Folia Primat.* 9:281–314.

――――. 1969. The ontogeny of social behavior of adult male squirrel monkeys (*Saimiri sciureus*) in a seminatural environment. *Folia Primat.* 11:35–79.

Baldwin, J. D., and Baldwin, J. I. 1971. Squirrel monkeys (*Saimiri*) in natural habitats in Panama, Colombia, Brazil and Peru. *Primates* 12:45–61.

――――. 1972. The ecology and behavior of squirrel monkeys (*Saimiri oerstedi*) in a natural forest in western Panama. *Folia Primat.* 18:161–184.

――――. 1973a. The role of play in social organization: comparative observations of squirrel monkeys (*Saimiri*). *Primates* 14:369–381.

――――. 1973b. Interactions between adult female and infant howling monkeys (*Alouatta palliata*). *Folia Primat.* 20:27–71.

――――. 1974. Exploration and social play in squirrel monkeys (*Saimiri*). *Amer. Zool.* 14:303–314.

――――. 1976. Effects of food ecology on social play: A laboratory simulation. *Z. Tierpsychol.* 40:1–14.

――――. in prep. Mother-juvenile interactions in howling monkeys (*Alouatta palliata*) in southwestern Panama.

Bandura, A., and Walters, R. H. 1963. *Social learning and personality development*. New York: Holt, Rinehart and Winston.

Beach, F. A. 1945. Current concepts of play in animals. *Amer. Nat.* 79:523–541.

Bekoff, M. 1972. The development of social interaction, play and metacommunication in mammals: An ethological perspective. *Q. Rev. Biol.* 47:412–434.

――――. 1974. Social play and play-soliciting by infant canids. *Amer. Zool.* 14:323–340.

Berger, M. E. 1972. Population structure of olive baboons (*Papio anubis* J. P. Fischer) in the Laikipia district of Kenya. *E. Afr. Wildl. J.* 10:159–164.

Berkson, G. 1973. Social responses to abnormal infant monkeys. *Amer. J. Phys. Anthrop.* 38:583–586.

Berlyne, D. E. 1960. *Conflict, arousal, and curiosity.* New York: McGraw-Hill.

Berman, A. J., and Berman, D. 1973. Fetal deafferentation: The ontogenesis of movement in the absence of peripheral sensory feedback. *Exp. Neurol.* 38:170–176.

Bernstein, I. S. 1971. Activity profiles of primate groups. In *Behavior of nonhuman primates* vol. 3, eds. A. M. Schrier and F. Stollnitz, pp. 69–106. New York: Academic Press.

Bindra, D. 1959. *Motivation: A systematic reinterpretation.* New York: Ronald Press.

Birch, H. G. 1945. The relation of previous experience to insightful problem-solving. *J. Comp. Physiol. Psychol.* 38:367–383.

Bolwig, N. 1964. Facial expression in primates with remarks on a parallel development in certain carnivores (a preliminary report on work in progress). *Behaviour* 22:167–193.

Breland, K., and Breland M. 1966. *Animal behavior.* New York: Macmillan.

Bronson, G. W. 1968. The development of fear in man and other animals. *Child. Dev.* 39:409–431.

Brown, G. M.; Schalch, D. S.; and Reichlin, S. 1971. Patterns of growth hormone and cortisol response to psychological stress in the squirrel monkey. *Endocrinology* 88:956–963.

Brownlee, A. 1954. Play in domestic cattle in Britain: An analysis of its nature. *Br. Vet. J.* 110:48–68.

Buhler, K. 1930. *The mental development of the child.* New York: Harcourt Brace.

Butler, R. A. 1958. The differential effect of visual and auditory incentives on the performance of monkeys. *Amer. J. Psychol.* 71:591–593.

———. 1965. Investigative behavior. In *Behavior of nonhuman primates* vol. 2, eds. M. Schrier, H. F. Harlow, and F. Stollnitz, pp. 463–493. New York: Academic Press.

Byrd, L. D. 1972. Responding in the squirrel monkey under second-order schedules of shock delivery. *J. Exp. Anal. Behav.* 18:155–167.

Campbell, H. J. 1972. Peripheral self-stimulation as a reward in fish, reptile and mammal. *Physiol. Behav.* 8:637–640.

Candland, D. K., and Mason, W. A. 1969. Infant monkey heart rate: Habituation and effects of social substitutes. *Dev. Psychobiol.* 1:254–256.

Candland, D. K., and Nagy, Z. M. 1969. The open field: Some comparative data. *Ann. N. Y. Acad. Sci.* 159:831–851.

Carpenter, C. R. 1934. A field study of the behavior and social relations of howling monkeys. *Comp. Psychol. Monogr.* 10:1–168.

———. 1965. The howlers of Barro Colorado Island. In *Primate behavior: Field studies on monkeys and apes,* ed. I. DeVore, pp. 250–291. New York: Holt, Rinehart and Winston.

Chevalier-Skolnikoff, S. 1971. The ontogeny of communication in *Macaca speciosa.* Ph.D. dissertation, University of California, Berkeley.

———. 1973. *The ontogeny of communication in the stumptail macaque, Macaca arctoides, Bibl. Primat.* 2. Basel: Karger.

Crook, J. H. 1970. Introduction—Social behaviour and ethology. In *Social behaviour in birds and mammals,* ed. J. H. Crook, pp. xxi–xl. New York: Academic Press.

Danson, C., and Creed, T. 1970. Rate of response as a visual social stimulus. *J. Exp. Anal. Behav.* 13:233–242.

———. 1971. Successive reversals of a visual social stimulus. *Psychon. Sci.* 22:283–285.

Dawes, G. S. 1968. *Foetal and neonatal physiology.* Chicago: Year Book Medical Publishers.

Dobzhansky, T. 1962. *Mankind evolving.* New Haven: Yale University Press.

Döhl, J., and Podolczak, D. 1973. Versuche zur Manipulierfreudigkeit von zwei jungen Orang-Utans (*Pongo pygmaeus*) im Frankfurter Zoo. *Zool. Gart.* 43:81–94.

Dolhinow, P. J. 1971. At play in the fields. *Nat. Hist.* Special Supplement, pp. 66–71.

Dolhinow, P. J., and Bishop, N. 1970. The development of motor skills and social relationships among primates through play. In *Minnesota symposia on child psychology,* vol. 4, ed. J. P. Hill, pp. 141–198. Minneapolis: University Minnesota Press.

Duffy, E. 1957. The psychological significance of the concept of "arousal" or "activation." *Psychol. Rev.* 64:265–275.

———. 1962. *Activation and behavior.* New York: Wiley.

DuMond, F. V. 1968. The squirrel monkey in a seminatural environment. In *The squirrel monkey*, eds. L. A. Rosenblum and R. W. Cooper, pp. 87–145. New York: Academic Press.

Eibl-Eibesfeldt, I. 1967. Concepts of ethology and their significance in the study of human behavior. In *Early behavior*, eds. H. W. Stevenson, E. H. Hess, and H. L. Rheingold, pp. 127–146. New York: Wiley.

———. 1970. *Ethology: The biology of behavior.* New York: Holt, Rinehart and Winston.

Eisenberg, J. F.; Muckenhirn, N.; and Rudran, R. 1972. The relation between ecology and social structure in primates. *Science* 176:863–874.

Eisenberger, R. 1972. Explanation of rewards that do not reduce tissue needs. *Psychol. Bull.* 77:319–339.

Ellefson, J. O. 1967. A natural history of the gibbon of the Malay Peninsula. Ph.D. dissertation, University of California, Berkeley.

———. 1968. Territorial behavior in the common white-handed gibbon, *Hylobates lar*. In *Primates: Studies in adaptation and variability*, ed. P. C. Jay, pp. 180–199. New York: Holt, Rinehart and Winston.

Ellis, M. J. 1973. *Why people play.* Englewood Cliffs: Prentice-Hall.

Ewer, R. F. 1968. *Ethology of mammals.* New York: Plenum.

Fady, J.-C. 1969. Les jeux sociaux: Le compagnon de jeux chez les jeunes. Observations chez *Macaca irus*. *Folia. Primat.* 11:134–143.

Fedigan, L. 1972. Social and solitary play in a colony of vervet monkeys (*Cercopithecus aethiops*). *Primates* 13:347–364.

Fiske, D. W., and Maddi, S. R. 1961. *Functions of varied experience.* Homewood: Dorsey.

Flechsig, P. 1920. *Anatomie des menschlichen Gehirns und Ruckenmarks auf myelogenetischer Grundlage.* Leipzig: Thieme.

———. 1927. *Meine myelogenetische Hiernlehre.* Berlin: Springer.

Frisch, J. E. 1968. Individual behavior and intertroop variability in Japanese macaques. In *Primates: Studies in adaptation and variability*, ed. P. C. Jay, pp. 243–252. New York: Holt, Rinehart and Winston.

Fryar, A. A., and Hupp, E. W. 1970. Blood values in infant and growing squirrel monkeys, *Saimiri sciureus*. *Tex. J. Sci.* 21:343.

Gartlan, J. S., and Brain, C. K. 1968. Ecology and social variability in *Cercopithecus aethiops* and *C. mitis*. In *Primates: Studies in adaptation and variability*, ed. P. C. Jay, pp. 253–292. New York: Holt, Rinehart and Winston.

Gibson, K. R. 1970. Sequence of myelinization in the brain of *Macaca mulatta*. Ph.D. dissertation, University of California, Berkeley.

Glickman, S. E., and Morrison, B. J. 1969. Some behavioral and neural correlates of predation susceptibility in mice. *Communications in Behav. Biol.* 4:261–267.

Glickman, S. E., and Sroges, R. W. 1966. Curiosity in zoo animals. *Behaviour* 26:151–188.

Goy, R. W., and Phoenix, C. H. 1972. The effects of testosterone propionate administered before birth on the development of behavior in genetic female rhesus monkeys. In *Steroid hormones and brain function*, eds. C. Sawyer and R. Gorski, pp. 193–200. Berkeley: University of California Press.

Goy, R. W., and Resko, J. A. 1972. Gonadal hormones and behavior of normal and pseudohermaphroditic nonhuman female primates. In *Recent progress in hormone research*, vol. 28, ed. E. B. Astwood, pp. 707–733. New York: Academic Press.

Green, P. C. 1972. Temperamental variation as a determinant of learning efficiency in cercopithecoid monkeys. *Primates* 13:35–41.

Hall, K. R. L. 1963a. Variations in the ecology of the chacma baboon (*Papio ursinus*). *Symp. Zool. Soc. Lond.* 10:1–28.

———. 1963b. Observational learning in monkeys and apes. *Br. J. Psychol.*, 54:201–226.

———. 1965. Behaviour and ecology of the wild patas monkey, *Erythrocebus patas*, in Uganda. *J. Zool.* 148:15–87.

———. 1968. Social learning in monkeys. In *Primates: Studies in adaptation and variability*, ed. P. C. Jay, pp. 383–397. New York: Holt, Rinehart and Winston.

Hall, K. R. L., and DeVore, I. 1965. Baboon social behavior. In *Primate behavior: Field studies of monkeys and apes*, ed. I. DeVore, pp. 53–110. New York: Holt, Rinehart and Winston.

Hansen, E. W. 1962. The development of maternal and infant behavior in the rhesus monkey. Ph.D. dissertation, University of Wisconsin.

———. 1966. The development of maternal and infant behavior in the rhesus monkey. *Behaviour* 27:107–149.

Harlow, H. F. 1950. Learning and satiation of response in intrinsically motivated complex puzzle performance by monkeys. *J. Comp. Physiol. Psychol.* 43:289–294.

Harlow, H. F.; Gluck, J. P.; and Suomi, S. J. 1972. Generalization of behavioral data between nonhuman and human animals. *Amer. Psychol.* 27:709–716.

Harlow, H. F., and Harlow, M. K. 1965. The affectional systems. In *Behavior of nonhuman primates*, vol. 2, eds. A. M. Schrier, H. F. Harlow, and F. Stollnitz, pp. 287–334. New York: Academic Press.

———. 1969. Effects of various mother-infant relationships on rhesus monkey behaviors. In *Determinants of infant behaviour*, vol. 4, ed. B. M. Foss, pp. 15–35. London: Methuen.

Harlow, H. F.; Harlow, M. K.; Hansen, E. W.; and Suomi, S. J. 1972. Infantile sexuality in monkeys. *Arch. Sex. Behav.* 2:1–7.

Harlow, H. F.; Harlow, M. K.; and Meyer, D. R. 1950. Learning motivated by a manipulation drive. *J. Exp. Psychol.* 40:228–234.

Harlow, H. F., and McClearn, G. E. 1954. Object discrimination learned by monkeys on the basis of manipulation motives. *J. Comp. Physiol. Psychol.* 47:73–76.

Harlow, H. F., and Suomi, S. J. 1971. Social recovery by isolation-reared monkeys. *Proc. Natl. Acad. Sci.* 68:1534–1538.

Hebb, D. O. 1949. *The organization of behavior.* New York: Wiley.

———. 1955. Drives and the CNS (conceptual nervous system). *Psychol. Rev.* 62:243–254.

———. 1966. *A textbook of psychology.* 2d ed. Philadelphia: Saunders.

———. 1972. *Textbook of psychology.* 3d ed. Philadelphia: Saunders.

Hebb, D. O., and Thompson, W. R. 1954. The social significance of animal studies. In *Handbook of social psychology*, ed. G. Lindzey, pp. 551–562. Cambridge: Addison-Wesley.

Heckhausen, H. 1964. Entwurf einer Psychologie des Spielens. *Psychol. Forsch.* 27:225–243.

Hess, E. H. 1959. Imprinting. *Science* 130:133–141.

———. 1964. Imprinting in birds. *Science* 146:1128–1139.

Hill, S. D.; McCormack, S. A.; and Mason, W. A. 1973. Effects of artificial mothers and visual experience on adrenal responsiveness of infant monkeys. *Dev. Psychobiol.* 6:421–429.

Hinde, R. A. 1966. *Animal behavior: A synthesis of ethology and comparative psychology.* New York: McGraw-Hill.

———. 1972. Social behaviour and its development in subhuman primates. *Condon Lectures*, Eugene: Oregon State System of Higher Education.

Hinde, R. A., and Stevenson-Hinde, J. 1973. *Constraints on learning: Limitations and predispositions.* New York: Academic Press.

Hoffman, H. S., and Ratner, A. M. 1973. A reinforcement model of imprinting: Implications for socialization in monkeys and men. *Psychol. Rev.* 80:527–544.

Hopf, S. 1970. Report on a hand-reared squirrel monkey (*Saimiri sciureus*). *Z. Tierpsychol.* 27:610–621.

———. 1972. Sozialpsychologische Untersuchungen zur Verhaltensentwicklung des Totenkopfaffens. Ph.D. dissertation, Philipps-Universität, Marburg/Lahn.

Horr, D. A. 1972. The Borneo orang-utan. *Borneo Research Bull.* 4:46–50.

Imanishi, K. 1957. Social behavior in Japanese monkeys *Macaca fuscata*. *Psychologia: Int. J. Psych. Orient* 1:47–54.

Inhelder, E. 1955. Zur Psychologie einiger Verhaltensweisen-besonders des Spiels—von Zootieren. Z. *Tierpsychol.* 12:88–144.

Itani, J. 1958. On the acquisition and propagation of new habits in the natural group of the Japanese monkeys at Takasakiyama. *Primates* 1:84–98.

Jarosch, E. 1968. Untersuchungen über die Einflüsse der sozialen Organisation der Totenkopfäffchen (*Saimiri sciureus*) auf das instrumentelle Lernen bzw. die selbstgewählte Reihenfolge in den Versuchen. *Folia Primat.* 9:135–153.

Jay, P. C. 1965. The common langur of north India. In *Primate behavior: Field studies of monkeys and apes*, ed. I. DeVore, pp. 197–249. New York: Holt, Rinehart and Winston.

―――. 1968. Studies of variability in species behavior: Comments. In *Primates: Studies in adaptation and variability*, ed. P. C. Jay, pp. 173–179. New York: Holt, Rinehart and Winston.

Jensen, G. D.; Bobbit, R. A.; and Gordon, B. N. 1967. The development of mutual independence in mother-infant pigtailed monkeys, *Macaca nemestrina*. In *Social communication among primates*, ed. S. A. Altmann, pp. 43–53. Chicago: University of Chicago Press.

―――. 1968. Effects of environment on the relationship between mother and infant pigtailed monkeys (*Macaca nemestrina*). *J. Comp. Physiol. Psychol.* 66:259–263.

―――. 1969. Patterns and sequences of hitting behavior in mother and infant monkeys (*Macaca nemestrina*). *J. Psychiatr. Res.* 7:55–61.

Jolly, A. 1966. Lemur social behavior and primate intelligence. *Science* 153:501–506.

―――. 1972. *The evolution of primate behavior*. New York: Macmillan.

Kaplan, J. 1974. Growth and behavior of surrogate-reared squirrel monkeys. *Dev. Psychobiol.* 7:7–13.

Kaufmann, J. H. 1967. Social relations of adult males in a free-ranging band of rhesus monkeys. In *Social communication among primates*, ed. S. A. Altmann, pp. 73–98. Chicago: University of Chicago Press.

Kawai, M. 1964. *Ecology of Japanese monkeys*. Tokyo: Kawade-Shobo.

―――. 1965. Newly acquired precultural behavior of the natural troop of Japanese monkeys on Koshima islet. *Primates* 6:1–30.

Kelleher, R. T., and Morse, W. H. 1968. Schedules using noxious stimuli. III. Responding maintained with response-produced electric shocks. *J. Exp. Anal. Behav.* 11:819–838.

Keller, F. S., and Schoenfeld, W. N. 1950. *Principles of psychology*. New York: Appleton-Century-Crofts.

King, D. L. 1966. A review and interpretation of some aspects of the infant-mother relationship in mammals and birds. *Psychol. Bull.* 65:143–155.

Koford, C. B. 1963. Rank of mothers and sons in bands of rhesus monkeys. *Science* 141:356–357.

Kummer, H. 1967. Tripartite relations in hamadryas baboons. In *Social communication among primates*, ed. S. A. Altmann, pp. 63–71. Chicago: University of Chicago Press.

―――. 1968. *Social organization of hamadryas baboons*. Chicago: University of Chicago Press.

―――. 1971. *Primate societies*. Chicago: Aldine-Atherton.

Kuo, Z. 1967. *The dynamics of behavior development: An epigenetic view*. New York: Random House.

Lancaster, J. B. 1971. Play-mothering: The relations between juvenile females and young infants among free-ranging vervet monkeys (*Cercopithecus aethiops*). *Folia Primat.* 15:161–182.

Lawick-Goodall, J. van. 1967a. Mother-offspring relationships in free-ranging chimpanzees. In *Primate ethology*, ed. D. Morris, pp. 287–346. London: Weidenfeld and Nicolson.

―――. 1967b. *My friends the wild chimpanzees*. Washington: National Geographic Society.

―――. 1968. The behavior of free-living chimpanzees in the Gombe Stream area. *Anim. Behav. Monogr.* 1:161–311.

―――. 1971. *In the shadow of man*. Boston: Houghton-Mifflin.

―――. 1973. Behavior of chimpanzees in their natural habitat. *Amer. J. Psychiatry* 130:1–12.

Lehrman, D. S. 1953. A critique of Konrad Lorenz's theory of instinctive behavior. *Q. Rev. Biol.* 28:337–363.

———. 1970. Semantic and conceptual issues in the nature-nurture problem. In *Development and evolution of behavior: Essays in memory of T. C. Schneirla*, eds. L. R. Aronson, E. Tobach, D. S. Lehrman, and J. S. Rosenblatt, pp. 17–52. San Francisco: W. H. Freeman.

Leuba, C. 1955. Toward some integration of learning theories: The concept of optimal stimulation. *Psychol. Rep.* 1:27–33.

Levitsky, D. A., and Barnes, R. H. 1972. Nutritional and environmental interactions in the behavioral development of the rat: Long term effects. *Science* 176:68–71.

Loizos, C. 1967. Play behaviour in higher primates: A review. In *Primate ethology*, ed. D. Morris, pp. 176–218. London: Weidenfeld and Nicolson.

Long, J. O., and Cooper, R. W. 1968. Physical growth and dental eruption in captive-bred squirrel monkeys, *Saimiri sciureus* (Leticia, Colombia). In *The squirrel monkey*, eds. L. A. Rosenblum and R. W. Cooper, pp. 193–205. New York: Academic Press.

Lorenz, K. Z. 1956. Play and vacuum activities. In *L'Instinct dans le comportement des animaux et de l'homme*, pp. 633–638. Foundation Singer-Polignac. Paris: Masson et Cie Editeurs.

Loy, J. 1970. Behavioral responses of free-ranging rhesus monkeys to food shortage. *Amer. J. Phys. Anthrop.* 33:263–272.

MacKinnon, J. 1971. The orang-utan in Sabah today. *Oryx* 11:141–191.

Marler, P. 1965. Communication in monkeys and apes. In *Primate behavior: Field studies of monkeys and apes*, ed. I. DeVore, pp. 544–584. New York: Holt, Rinehart and Winston.

Marler, P., and Hamilton, W. J. 1966. *Mechanisms of animal behavior*. New York: Wiley.

Mason, J. W. 1968. A review of psychoendocrine research on the pituitary-adrenal cortical systems. *Psychosom. Med.* 30:576–585.

Mason, W. A. 1965a. The social development of monkeys and apes. In *Primate behavior: Field studies of monkeys and apes*, ed. I. DeVore, pp. 514–543. New York: Holt, Rinehart and Winston.

———. 1965b. Determinants of social behavior in young chimpanzees. In *Behavior of nonhuman primates*, vol. 2, eds. A. M. Schrier, H. F. Harlow, and F. Stollnitz, pp. 335–364. New York: Academic Press.

———. 1968a. Early social deprivation in the nonhuman primates: Implications for human behavior. In *Environmental influences*, ed. D. Glass, pp. 70–101. New York: Russell Sage Foundation.

———. 1968b. Use of space by *Callicebus* groups. In *Primates: Studies in adaptation and variability*, ed. P. C. Jay, pp. 200–216. New York: Holt, Rinehart and Winston.

———. 1971. Motivational factors in psychosocial development. In *Nebr. symp. motiv.*, eds. W. J. Arnold and M. M. Page, pp. 35–67. Lincoln: University of Nebraska Press.

Mason, W. A.; Hill, S. D.; and Thomsen, C. E. 1971. Perceptual factors in the development of filial attachment. *Proc. 3rd Int. Congr. Primat.*, vol. 3, ed. H. Kummer, pp. 125–133. Basel: Karger.

Mason, W. A.; Hollis, J. H.; and Sharpe, L. G. 1962. Differential responses of chimpanzees to social stimulation. *J. Comp. Physiol. Psychol.* 55:1105–1110.

Mason, W. A., and Kenney, M. D. 1974. Redirection of filial attachments in rhesus monkeys: Dogs as mother surrogates. *Science* 183:1209–1211.

McKearney, J. W. 1969. Fixed-interval schedules of electric shock presentation: Extinction and recovery of performance under different shock intensities and fixed-interval durations. *J. Exp. Anal. Behav.* 12:301–313.

Meier, G. W. 1965. Other data on the effects of social isolation during rearing upon adult reproductive behavior in the rhesus monkey (*Macaca mulatta*). *Anim. Behav.* 13:228–231.

Menzel, E. W., Jr. 1969. Chimpanzee utilization of space and responsiveness to objects: Age differences and comparison with macaques. *Proc. 2nd Int. Congr. Primat.*, vol. 1, ed. C. R. Carpenter, pp. 72–80. Basel: Karger.

Meyer, V., and Knobil, E. 1967. Growth hormone secretion in the unanesthetized rhesus monkey in response to noxious stimuli. *Endocrinology* 80:163–171.

Michael, R. P., and Wilson, M. 1973. Changes in the sexual behavior of male rhesus monkeys (M. mulatta) at puberty. Folia Primat. 19:384–403.

Miller, S. 1973. Ends, means, and galumphing: Some leitmotifs of play. Amer. Anthrop. 75:87–98.

Montgomery, K. C. 1954. The role of the exploratory drive in learning. J. Comp. Physiol. Psychol. 48:254–260.

Morris, D. 1962. The biology of art. London: Methuen.

———. 1964. The response of animals to a restricted environment. Symp. Zool. Soc. Lond. 13:99–118.

Moynihan, M. 1964. Some behavior patterns of platyrrhine monkeys: I. The night monkey (Aotus trivirgatus). Smithson. Misc. Coll. 146:1–84.

———. 1970. Some behavior patterns of platyrrhine monkeys: II. Saguinus geoffroyi and some other tamarins. Smithson. Contrib. Zool. 28:1–77.

Müller-Schwarze, D. 1971. Ludic behavior in young mammals. In Brain development and behavior, eds. M. B. Sterman, D. J. McGinty, and A. M. Adinolfi, pp. 229–249. New York: Academic Press.

Nishida, T. 1966. A sociological study of solitary male monkeys. Primates 7:141–204.

Nissen, H. W. 1951. Phylogenetic comparison. In Handbook of experimental psychology, ed. S. S. Stevens, pp. 347–386. New York: Wiley.

Paterson, J. D. 1973. Ecologically different patterns of aggressive and sexual behavior in two troops of Ugandan baboons, Papio anubis. Amer. J. Phys. Anthropol. 38:641–647.

Petter, J.-J., and Petter, A. 1967. The aye-aye of Madagascar. In Social communication among primates, ed. S. A. Altmann, pp. 195–205. Chicago: University of Chicago Press.

Phoenix, C. H.; Goy, R. W.; and Resko, J. A. 1968. Psychosexual differentiation as a function of androgenic stimulation. In Perspectives in reproduction and sexual behavior ed. M. Diamond, pp. 33–49. Bloomington: Indiana University Press.

Phoenix, C. H.; Slob, A. K.; and Goy, R. W. 1973. Effects of castration and replacement therapy on sexual behavior of adult male rhesuses. J. Comp. Physiol. Psychol. 84:472–481.

Ploog, D.; Hopf, S.; and Winter, P. 1967. Ontogenese des Verhaltens von Totenkopfaffen (Saimiri sciureus). Psychol. Forsch. 31:1–41.

Poirier, F. E. 1970. Nilgiri langur ecology and social behavior. In Primate behavior: Developments in field and laboratory research, ed. L. A. Rosenblum, pp. 251–383. New York: Academic Press.

———. 1972. Introduction to Primate socialization, ed. F. E. Poirier, pp. 3–28. New York: Random House.

Poirier, F. E., and Smith, E. O. 1974. Socializing functions of primate play. Amer. Zool. 14:275–287.

Pryor, K.; Haag, R.; and O'Reilly, J. 1969. The creative porpoise: Training for novel behavior. J. Exp. Anal. Behav. 12:653–661.

Pycraft, W. P. 1912. The infancy of animals. London: Hutchinson.

Ransom, T. W., and Rowell, T. E. 1972. Early social development of feral baboons. In Primate socialization, ed. F. E. Poirier, pp. 105–144. New York: Random House.

Reynolds, G. 1968. A primer of operant conditioning. Glenview: Scott Foresman.

Reynolds, P. C.; Oakley, F. B.; and Noble, M. in preparation. Evidence for a rebound effect after social play deprivation in rhesus monkeys.

Rheingold, H. L., and Eckerman, C. O. 1970. The infant separates himself from his mother. Science 168:78–90.

Richard, A. 1970. A comparative study of the activity patterns and behavior of Alouatta villosa and Ateles geoffroyi. Folia Primat. 12:241–263.

Riesen, A. H. 1961. Stimulation as a requirement for growth and function in behavioral development. In Functions of varied experience, eds. D. W. Fiske and S. R. Maddi, pp. 57–80. Homewood: Dorsey.

———. 1965. Effects of early deprivation of photic stimulation. In The biosocial basis of mental retardation, eds. S. F. Osler and R. E. Cooke, pp. 61–85. Baltimore: Johns Hopkins Press.

Rosenblum, L. A. 1971. The ontogeny of mother-infant relationships in macaques. In Ontogeny of vertebrate behavior, ed. H. Moltz, pp. 315–367. New York: Academic Press.

Rosenblum, L. A., and Lowe, A. 1971. The influence of familiarity during rearing on subsequent partner preferences in squirrel monkeys. *Psychon. Sci.* 23:35–37.

Rowell, T. E. 1966. Forest living baboons in Uganda. *J. Zool. Lond.* 147:344–364.

———. 1972. *The social behaviour of monkeys.* Baltimore: Penguin.

———. 1974. Contrasting adult male roles in different species of nonhuman primates. *Arch. Sex. Behav.* 3:143–149.

Rumbaugh, D. M., and McCormack, C. 1967. The learning skills of primates: A comparative study of the apes and monkeys. In *Progress in primatology.* eds. D. Starck, R. Schneider, and H. J. Kuhn, pp. 289–306. Stuttgart: Gustav Fischer.

Sackett, G. P. 1971. Isolation rearing in monkeys: Diffuse and specific effects on later behavior. *Colloq. Int. Centr. Nat. Rech. Sci.* 198:61–110.

———. 1972. Exploratory behavior of rhesus monkeys as a function of rearing experiences and sex. *Dev. Psychol.* 6:260–270.

Sackett, G. P.; Bowman, R. E.; Meyer, J. S.; Tripp, R. L.; and Grady, S. S. 1973. Adrenocortical and behavioral reactions by differentially raised rhesus monkeys. *Physiol. Psychol.* 1:209–212.

Sade, D. S. 1967. Determinants of dominance in a group of free-ranging rhesus monkeys. In *Social communication among primates,* ed. S. A. Altmann, pp. 99–114. Chicago: University of Chicago Press.

Salzen, E. 1968. Imprinting in birds and primates. *Behaviour* 28:232–254.

Schaller, G. B. 1963. *The mountain gorilla.* Chicago: University of Chicago Press.

Schiller, P. H. 1957. Innate motor action as a basis of learning: Manipulative patterns in the chimpanzee. In *Instinctive behavior,* ed. C. Schiller, pp. 264–287. New York: International Universities Press.

Schneirla, T. C. 1956. The interrelationships of the "innate" and the "acquired" in instinctive behavior. In *L'Instinct dans le comportement des animaux et de l'homme,* ed. P.-P. Grassé, pp. 387–452. Paris: Masson.

———. 1959. An evolutionary and developmental theory of biphasic processes underlying approach and withdrawal. In *Nebr. symp. motiv.* ed. M. R. Jones, pp. 1–42. Lincoln: University of Nebraska Press.

Schultz, D. D. 1965. *Sensory restriction: Effects of behavior.* New York: Academic Press.

Schusterman, R. J., and Sjoberg, A. 1969. Early behavior patterns of squirrel monkeys (*Saimiri sciureus*). *Proc. 2nd Int. Congr. Primat.* vol. 1, ed. C. R. Carpenter, pp. 194–203. Basel: Karger.

Seligman, M., and Hager, J. 1972. *The biological boundaries of learning.* New York: Appleton-Century-Crofts.

Simonds, P. E. 1965. The bonnet macaque in south India. In *Primate behavior: Field Studies of monkeys and apes,* ed. I. DeVore, pp. 175–196. New York: Holt, Rinehart and Winston.

———. 1974. Sex differences in bonnet macaque networks and social structure. *Arch. Sex. Behav.* 3:151–166.

Skinner, B. F. 1969. *Contingencies of reinforcement.* New York: Appleton-Century-Crofts.

Southwick, C. H. 1972. *Aggression among nonhuman primates.* An Addison-Wesley Module in Anthropology, no. 23. Reading: Addison-Wesley.

Southwick, C. H.; Beg, M. A.; and Siddiqi, H. R. 1965. Rhesus monkeys in north India. In *Primate behavior: Field studies of monkeys and apes,* ed. I. DeVore, pp. 111–159. New York: Holt, Rinehart and Winston.

Stretch, R.; Orloff, E. R.; and Dalrymple, S. D. 1968. Maintenance of responding by fixed-interval schedule of electric shock presentation in squirrel monkeys. *Science* 162:583–586.

Sugiyama, Y. 1965. Behavioral development and social structure in two troops of hanuman langurs (*Presbytis entellus*). *Primates* 6:213–247.

Suomi, S. J., and Harlow, H. F. 1971. Monkeys at play. *Nat. Hist.* Special Supplement, pp. 72–75.

Sussman, R. 1973. Socialization, social structure, and ecology of two sympatric species of *Lemur.* Paper read at meetings of the American Association Physical Anthropologists.

Symons, D. A. 1973. Aggressive play in a free-ranging group of rhesus monkeys (*Macaca mulatta*). Ph.D. dissertation, University of California, Berkeley.

Szasz, T. S. 1961. *The myth of mental illness*. New York: Hoeber-Harper.

Teleki, G. 1973. *The predatory behavior of wild chimpanzees*. Lewisberg: Bucknell University Press.

Thorington, R. W., Jr. 1968. Observations of squirrel monkeys in a Colombian forest. In *The squirrel monkey*, eds. L. A. Rosenblum and R. W. Cooper, pp. 69–85. New York: Academic Press.

Tsumori, A. 1967. Newly acquired behavior and social interactions of Japanese monkeys. In *Social communication among primates*, ed. S. A. Altmann, pp. 207–219. Chicago: University of Chicago Press.

Van Hooff, J. 1967. The facial displays of the catarrhine monkeys and apes. In *Primate ethology*, ed. D. Morris, pp. 7–68. London: Weidenfeld and Nicolson.

Volkman, F. R., and Greenough, W. T. 1972. Rearing compexity affects branching of dendrites in the visual cortex of the rat. *Science* 176:1445–1447.

Washburn, S. L., and Hamburg, D. A. 1965. The implications of primate research. In *Primate behavior: Field studies of monkeys and apes*, ed. I. DeVore, pp. 607–622. New York: Holt, Rinehart and Winston.

Weiner, H. 1965. Conditioning history and maladaptive human operant behavior. *Psychol. Rep.* 17:935–942.

———. 1969. Controlling human fixed-interval performance. *J. Exp. Anal. Behav.* 12:349–373.

Welker, W. I. 1956a. Effects of age and experience on play and exploration of young chimpanzees. *J. Comp. Physiol. Psychol.* 49:223–226.

———. 1956b. Some determinants of play and exploration in chimpanzees. *J. Comp. Physiol. Psychol.* 49:84–89.

———. 1961. An analysis of exploratory and play behavior in animals. In *Functions of varied experience*, eds. D. W. Fiske and S. R. Maddi, pp. 175–226. Chicago: Dorsey.

———. 1971. Ontogeny of play and exploratory behaviors: A definition of problems and a search for new conceptual solutions. In *The ontogeny of vertebrate behavior*, ed. H. Moltz, pp. 171–228. New York: Academic Press.

White, R. W. 1961. Motivation reconsidered: The concept of competence. In *Functions of varied experience*, eds. D. W. Fiske and S. R. Maddi, pp. 278–325. Homewood: Dorsey.

Winter, P. 1969. Dialects in squirrel monkeys: Vocalization of the Roman arch type. *Folia Primat.* 10:216–229.

Winter, P.; Handley, P.; Ploog, D.; and Schott, D. 1973. Ontogeny of squirrel monkey calls under normal conditions and under acoustic isolation. *Behaviour* 47:230–239.

Zimmerman, R. R., and Torrey, C. C. 1965. Ontogeny of learning. In *Behavior of nonhuman primates*, vol. 1, eds. A. M. Schrier, H. F. Harlow, and F. Stollnitz, pp. 405–447. New York: Academic Press.

Zubek, J. P. 1969. Physiological and biochemical effects. In *Sensory deprivation: Fifteen years of research*, ed. J. P. Zubek, pp. 254–288. New York: Appleton-Century-Crofts.

Zuckerman, M. 1969. Theoretical formulations: I. In *Sensory deprivation: Fifteen years of research*, ed. J. P. Zubek, pp. 407–432. New York: Appleton-Century-Crofts.

13 | The Effects of Social Isolation on Sexual Behavior in Macaca fascicularis[1]

THOMAS J. TESTA
DAVID MACK

- **INTRODUCTION**

The isolation experiment is a valuable method for assessing the effects of the socialization process on behavior. It has been profitably employed by many researchers such as Harlow (1961), Harlow and Harlow (1962), Harlow and Zimmerman (1959) and Mason (1968). Yet before the effects of socialization can be properly determined using the isolation experiment, it is necessary to fully describe the components of *normal* socialized behavior. Harlow and Mason as well as others have attempted to gather descriptive data on the behavior of fully socialized individuals. However, new methods of behavioral observation, recording, and analysis have recently been perfected and their application to the isolation experiment should reveal new data on the effects of socialization on behavior more clearly.

The two studies presented in this paper represent such analyses. The first study examines the sexual behavior of ferally-reared (wild-born-and-reared) male-female dyads of crab-eating macaques (*M. fascicularis*). It includes: (1) a descriptive analysis of motor patterns, (i.e., topographical analysis), which should reveal the *form* of certain species-typical sexual responses; (2) frequency analyses, which should reveal the relative probabilities of certain responses, which in turn reveals something about response *thresholds* and *situational* cues; and (3) sequential analyses, which should reveal the nature of eliciting stimuli (i.e., the more specific relationships or interdependencies which exist between the responses of one member and the other). Such analyses should provide a more comprehensive description of the end-product of primate socialization than has been made previously.

The basis of the second study is to analyze the mechanisms which permit such socialization by examining isolation-reared animals in similar situations. In order to accomplish this, topographical, frequency, and sequential analyses of isolation-reared primate social interactions are again performed. Such analyses should permit a determination of the *locus* of the effect of the socialization process on sexual behavior; that is, whether the socialization process determines the *form* of species-typical sexual responses, the *threshold* of such responses, or the prevailing eliciting *contingencies* or interdependencies which determine the sequences of such responses.

In attempting to specify the mechanisms which permit socialization, we might assume that any relationship which exists between two responses, either within or between individuals, is based on the development during phylogeny of a reflex (i.e., a stimulus-response dependency which results from the direct transmission of that stimulus to the muscles which innervate that response and which owes its existence to some physiological substrate whose structure is solely determined by the genetic makeup of the individual) or the formation during ontogeny of a classical and/or instrumental association (i.e., some stimulus-response dependency which requires exposure during the individual's life-time to some natural interdependency among events). One can assess these particular contributions of phylogeny and ontogeny by examining the behavioral relationships in feral- and isolation-reared animals. If feral- and isolation-reared animals do not reveal similar sequential dependencies it can be argued that conditioning regimes (i.e., exposure to those natural interdependencies among events which result in classical and instrumental associations) must be experienced by the to-be-socialized individual if such sequences are to develop.

However, processes other than conditioning may also be involved in socialization. For instance, Mason (1968) has reported on the effects of isolation on macaque infants raised with mothers made of terry cloth, which were moved on a random schedule. Mason found that such a form of stimulation is, by itself, an effective determinant of later social behavior. Thus, response-contingent experiences (i.e., conditioning) are not the only kinds of experiences that can affect later social behavior. Similarly, Sackett (1965) and Melzach (1968) have suggested that socialization exposes the infant of a particular species to a level of stimulation which is typical for that species. Isolation prevents adaptation to this optimal level of stimulation and presumably results in a hypersensitivity which is not conducive to social interaction. Thus, changes in the form and frequency of sequential dependencies of social responses which result from isolation are not necessarily merely the result of changes in the social conditioning regime.

Finally hormones, as well as the other forms of stimulation, exert their effect on social behavior and their interaction with feral- and isolation-rearing experiences will reveal more information concerning mechanisms of socialization. For instance, we might ask if the effect of some hormonal state influences receptivity and/or attractiveness (see Michael 1971) because it modifies the attractiveness of certain conspecific stimuli for either the male or female, or because it lowers the threshold of certain responses on the part of the female thus providing especially strong stimuli for the male, or because it lowers the threshold of certain stimuli emanating from the male. A detailed analysis of dyadic interactions of ferally-reared *M. fascicularis* and of feral-isolate dyads should be able to distinguish among these alternatives. In later sections the relationship between hormonal state and the frequency of certain behaviors is examined. Such an analysis is equivalent to a sequential analysis, but is conducted on a different level. Instead of examining the probability of certain response sequences, the relationship of ovulatory cycle to the frequency of some response is examined. One can then assess the role played by socialization in developing this contingency (if one is shown to exist) by comparing feral- and isolation-reared *Macaca fascicularis.*

■ SEXUAL BEHAVIOR IN FERALLY-REARED SUBJECTS

CONDITIONS OF THE STUDY

The subjects were housed in a climate-controlled colony room in separate metal cages which measured 81.2 cm high × 60.9 cm wide × 55.8 cm deep. Two ferally-born female and two ferally-born male crab-eating monkeys were assigned to two heterosexual dyads. All feral animals were adults. Each animal was placed alone in the observation cage for a single one-hour session to permit acclimatization. Each dyad was then placed in the observation cage for a single hour to allow for habituation on the following day. Thereafter each pair was observed for one-half hour every four days for a total of 23 observation periods which spanned three menstrual cycles.

The observation cage measured 150 cm long × 122 cm high × 67 cm wide. The front wall of the cage was made of clear plexiglass. The cage was placed along one wall of an experimental room, and a television camera was mounted on a tripod in front of the opposite wall. A plaster-board partition concealed the observer who sat in front of a table with a closed-circuit television monitor, an Esterline-Angus event recorder, and a panel of event buttons in front of him. The observer could open and close the guillotine doors on the observation cage from this position and could view the behavior of the dyad via the T.V. monitor. An ethogram developed at the Wisconsin Regional Primate Center for use with *M. mulatta* was used as a basis for the construction of the present ethogram. Each of 19 discrete behaviors which appeared to be most common was initially assigned to one of the buttons on the panel. Each button when pressed would activate one of the 20 pens on the Esterline-Angus recorder. Preliminary observation revealed that the 19 behaviors which were chosen were indeed the most common. In fact, although any other observed behavior was noted directly on the recorder chart, other behaviors rarely occurred more than once or twice.

THE ETHOGRAM

The following list describes each behavior which was recorded:

1. *Male withdrawal:* Any movement of the male which increased the distance between two members.

2. *Female climb:* This behavior was recorded whenever the female was completely off the floor of the cage but in contact with the walls or ceiling.

3. *Female look back:* This behavior was recorded whenever the female turned and looked at the male over her shoulder during a present and/or mount.

4. *Female withdrawal:* Any movement of the female which increased the distance between the two members.

5. *Female aggressive:* This behavior was recorded whenever the female behaved agonistically either by approaching or withdrawing with a start, grabbing, biting, etc.

6. *Female approach:* Any movement of the female which decreased the distance between the members.

7. *Female genital examination:* This behavior was recorded if the female touched, groomed, or sniffed the male's genital region.

8. *Female presents:* This behavior was described by Luttge (1971) as consisting of one animal standing on all fours and orienting the exposed perineal region toward another animal.

9. *Female positions:* This behavior was recorded whenever the female backed into the male, posterior first.

10. *Female reach back:* This behavior was recorded whenever the female reached back with one hand towards the male during a present and/or mount.

11. *Female grooms the male:* Grooming behavior was recorded whenever the female touched, plucked, or brushed the hair of the male. (No instances of the male grooming the female were observed.)

12. *Female tooth-chatters:* This behavior is described by van Hoof (1972) as a combination of the jaw movements of smacking and baring of the teeth. (No instances of male tooth-chattering were recorded.)

13. *Male head-bob:* This behavior is described by Michael and Zumpe (1970) as a rapid raising of the head relative to the shoulders accompanied by an extension or retraction of the chin.

14. *Male approach:* Any movement of the male member of the dyad which reduced the distance between the two members.

15. *Male aggressive:* This behavior was recorded whenever the male behaved agonistically either by approaching with a start, grabbing, biting, etc.

16. *Male genital examination:* This behavior was recorded if the male touched, sniffed, nuzzled, or stared at the female's genital region.

17. *Male positions:* This behavior was recorded whenever the male placed his hands on the pelvic or rump area of the female and/or pulled her in an upward direction towards him.

18. *Male mount:* This behavior is described by Kanagawa et al. (1972) as occurring whenever the male grasped the female at the waist with his hands and gripped her in the ankle region with his foot, so that he was completely off the ground.

19. *Male intromission and pelvic thrust:* This behavior was recorded whenever a mount was followed by insertion of the male's penis into the female's vagina and followed by thrusting movements of the male's pelvic area.

In addition to the behaviors being recorded during each pairing session, the female's menstrual condition was also observed. That is, the number of days which had transpired since the last menstruation as well as the color and degree of swelling of the perineal area was noted.

The following descriptive analysis will be divided into several sections: (a) general description: a brief description of the behavioral sequences which were observed; (b) formal analysis: a description of the motor patterns involved in each behavior; (c) frequency analysis: a description of the relative frequency of important behaviors and a correlation of frequency with menstrual state; and (d) a sequential analysis: the transitional probabilities that behavior X was followed by behavior Y.

GENERAL DESCRIPTION

The ferally-reared dyads of *Macaca fascicularis* reacted predictably to the experimental situation. When the doors to the observation cage were raised, each member of the pair typically rushed toward each other. The female would usually turn and present to the male as soon as each reached the center of the cage. The female would often look over her shoulder and tooth chatter at the male during such presentations. The male in turn would quickly examine the female's genital area and then mount her. Pelvic thrusts often followed briefly, but the female usually did not stand still for long periods of time during these early mounts. She would usually begin to withdraw from the male who would either follow or withdraw himself. After a brief period of time, grooming would follow which was initiated by a male or female approach. Grooming periods would then alternate with periods of sexual interaction for the remainder of the session.

These interactions were typical. If both members were standing or sitting close to each other the male would often withdraw and then approach. During the return he would attempt to genitally examine the female either visually and/or manually. If she presented at this point the male would either mount her directly or genitally examine her first. If she did not present, the male would quickly move his head so as to gain a direct view of the female's face and eyes. When her gaze was caught, the male would quickly move into a position appropriate for the genital examination. This back and forth gesture (stare at the genitals—stare at the face) would be repeated. The female would either react by presenting or would withdraw. In either case the male could then attempt a mount. The head bob response of the male was very similar to these back and forth staring gestures and may in fact be a ritualized derivative of this response.

ANALYSIS OF MOTOR PATTERNS

Several descriptions of rhesus monkey sexual behavior have been reported. It is of interest to determine the degree of similarity which exists among the motor patterns which comprise certain behaviors for *Macaca mulatta* and *M. fascicularis*.

All dyadic interactions were recorded on video tape. Analyses of motor patterns were accomplished by reviewing these video tapes and stopping the action at appropriate points. At each of these points, a sheet of acetate was taped to the monitor screen and the outline of each member of the pair was traced on the acetate with a felt marker. These analyses revealed that sexual behavior in *Macaca fascicularis* is very similar to the sexual behavior in *Macaca mulatta* in terms of the motor patterns. For instance, in both species the male usually first grasps the female at the waist if she is sitting, and pulls her in an upward direction, thereby "positioning" her. Once the female is standing the male may sniff, lick, or merely visually examine the female's perineal area. In both species the female stands on all fours and directs the exposed

TABLE 1 / OVERALL PROBABILITY OF EACH BEHAVIOR FOR FERALLY REARED MACACA FASCICULARIS.

Behavior:	1	2	3	4	5	6	7	8	9	10	11	12	13	14	15	16	17	18	19
Probability for Pair A:	.04	.01	.04	.06	*	.10	.02	.15	*	.01	.09	.03	.07	.01	*	.15	.06	.06	.19
Probability for Pair B;	.08	.01	*	.06	.01	.13	*	.04	*	.01	.15	.01	.02	.04	*	.15	.13	.07	.06
Mean Probability:	.04	.01	.02	.06	*	.11	.01	.09	*	.01	.12	.02	.04	.02	*	.15	.09	.06	.06

* probability less than .01

perineal region toward the male; he grasps the female at the waist with his hands and grips her ankle region with his feet, so that he is completely off the ground. Michael and Zumpe (1970) described the head-duck and head-bob behavior of the rhesus as a rapid lowering of the head relative to the shoulders accompanied by an extension of the chin, as well as the inverse of this movement. Again, a similar pattern was observed in *Macaca fascicularis*. Finally, analyses of motor patterns also demonstrated that the female look-back and reach-back response during a mount sequence was very similar in both species.

FREQUENCY ANALYSIS

Frequency analyses were conducted for each pair separately and then for both pairs combined. Each dyad was paired for one one-half hour every four days throughout three menstrual cycles for a total of 23 sessions. The frequency of occurrence for each response was recorded and an activity measure was determined by computing the total frequency of all behaviors and dividing this by the total number of sessions, 23. A total of 3,072 behaviors was reported for pair A. Thus, the average activity score for this pair was 133.5. The average activity score for pair B was 1484/23 = 64.5. The relative frequency for each behavior was determined by dividing the total frequency of each

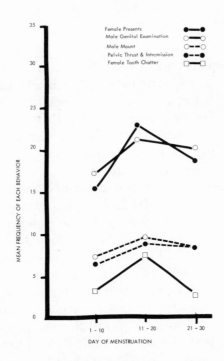

FIGURE 1. *Mean frequency of presenting, male genital examination, male mounting, male pelvic thrust and intromission, and female tooth-chatter as a function of day of menstruation.*

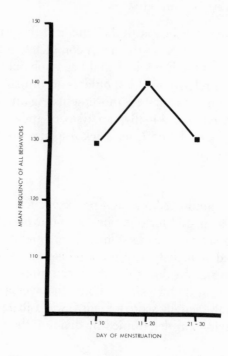

FIGURE 2. *Mean frequency of all behaviors as a function of day of menstruation.*

response by the total frequency of all behaviors across 23 sessions. Table 1 reports these relative frequencies or probabilities of each behavior for pairs A and B and for the overall probability of both pairs combined. Since the score for each response is weighted by the total activity measure of each pair, these scores can be meaningfully averaged in order to obtain an estimate of the overall probability of each behavior which is likely to occur in a male-female interaction. Given these baseline probabilities it is possible to examine deviations from these scores as a function of menstrual state, i.e., does the probability of behavior X increase, decrease, or remain unchanged as the menstrual cycle advances? Such an analysis should give the least biased measure of the interaction of menstrual state with behavioral propensities. In order to illustrate this point, consider some of the more traditional measures which have been reported. Figure 1 graphs the mean frequency of several behaviors (presenting, genital examination, mounts, pelvic thrust and intromission and female tooth chatter) as a function of day of menses. Such a frequency analysis reveals a clear peak in the frequency of these behaviors for days of menses 11 through 20. However, the relation of overall activity ($\bar{X}$ total number of behaviors) to day of menses shown in figure 2 reveals a similar peak. Hence it is difficult to determine whether the peak which appears for certain sexual behaviors is in fact due to an increase in these behaviors or whether or not this increase is an artifact of an overall increment in activity. The relative frequency measure, however, is not subject to this confounding because it is weighted by the overall activity measure. Figure 3 graphs these prob-

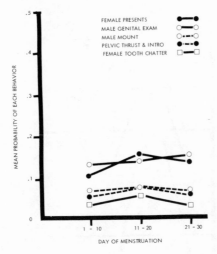

FIGURE 3. *Mean probability of presenting, male genital examination, male mounting, male pelvic thrust and intromission, and female tooth-chatter as a function of day of menstruation.*

abilities as a function of the day of menses. A peak in the probability of these behaviors is maintained. However, analysis of variance revealed no significant differences for any relative frequency measure as a function of day of menses with the exception of the female tooth chattering response ($F = 3.17$; df $= 5$, 17; $p < .05$). Given this shift in overall activity as a function of day of menses, it is reasonable to assume that the influence of the hormonal state on sexual behavior is diffuse. Rather than influencing threshold levels for specific behaviors, the effect of the hormonal condition may simply be an increase in general activity which in turn increases the probability that conditions will arise which are appropriate for the subsequent stimulation of a "sexual" reaction. This explanation must be viewed with caution, however, because the overall activity measures are based on both male and female behaviors. A closer examination of the frequency score is essential. It is possible that, for example, the overall activity of the male does not increase and that his behaviors do show a true peak as a function of cycle; this would indicate an increase in female attractiveness. It is equally possible that the female's activity level is not generally elevated and that her propensity to perform specific sexual behaviors is, in fact, elevated during certain phases of the cycle. In order to assess these possibilities, the overall activity measures were separated into male behaviors and female behaviors. For pair A, the average activity level of the female was 71.22 and of the male, 62.35. For pair B, the average activity of the female was 28.43 and of the male, 36.09. Figure 4 graphs the changes in mean activity level for each animal in each dyad as a function of the day of menses. The straight horizontal lines correspond to the overall mean activity level for each animal. It can be seen that in each case the mean activity level recorded for days 6 through 15 of menses is above the overall activity level while the mean activity level

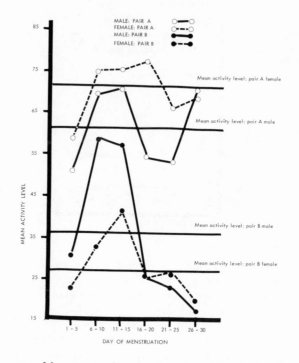

FIGURE 4. *Mean activity level for the male and female of each dyad as a function of day of menstruation.*

recorded during the remaining phases of menses is clearly below the average activity level. Hence, it can be concluded that a peak in activity (i.e., the mean frequency of all behaviors) exists for both male and female members of the pair.

SEQUENTIAL ANALYSIS

As stated in the previous section, the demonstration that certain behaviors, either male or female, increase in frequency at certain phases of the menstrual cycle implies a hormonal influence. However, it tells us little or nothing about what mechanisms underlie sexual behavior and how these mechanisms are affected by changes in the hormonal state. For instance, as mentioned previously, it is possible that during certain phases of the ovulatory cycle, the female's hormonal state acts to lower her threshold in such a way that she will respond more readily to stimuli provided by the male (increased female receptivity). This should be reflected in a change of sequential probabilities. That is, the probability of certain female responses following male responses should change as a function of cycle. Conversely, the female may provide certain stimuli to the male which result in increased male sexual activity. Michael and Kaverne (1968) have shown that a female pheromone which is communicated olfactorily tends to increase male sexual behavior. Hence, nonbehavioral stimuli (odor, color, or size of the perineal area) may enhance the effectiveness of female behavioral stimuli. In such cases we would expect to observe an increase in the probability of certain male behaviors following female behaviors (increased female attractiveness).

Another objective of the present analysis is to record those sequential probabilities which exist between certain responses which could permit conditioning to take place, i.e., which would permit the formation of an association between these responses. Rescorla (1967) argued that an association cannot form between two events unless there is a contingency between them; that is, the probability of event two occurring, given that event one has occurred, must be different from the probability of event two occurring, given that event one did not occur. Unless this condition is satisfied a relationship does not exist between the events and an association cannot be expected to form. Hence, sequential probabilities are examined in order to detect contingencies or relationships between pairs of responses. If, in fact, the socialization process involves exposure to conspecific behavior which is reliably related to ongoing events in the environment, then an examination of those relationships existing in the fully socialized adult should yield some information concerning the nature of the conditioning regime which the to-be-socialized infant may experience. For this reason a sequential analysis was performed for all sessions. A matrix designed after Altmann (1965) was constructed and conditional probabilities were calculated. The probability that behavior X followed, given that behavior Y occurred, was calculated. Graphs were then constructed which plotted the joint probability of behavior X, given behavior Y, and behavior X, given no behavior Y. As described by Rescorla (1967) the condition of independence between two events is met when $p(B/A) = p(B/\bar{A})$. If, however, the probability of B occurring is substantially changed when A has occurred as compared with the probability of B when A has not occurred, then A and B are dependent to some extent. Figure 5 presents the joint probabilities of selected pairs of behavior in the form of a contingency space. The abscissa is the probability that behavior B will follow, given that behavior A has occurred [$p(B/A)$]. The ordinate gives the probability that behavior B will follow, given behavior A has *not* occurred [$p(B/\bar{A})$]. Joint probabilities which lie on or near the 45 degree line reveal behaviors which are close to being independent. Joint probabilities which are above or below this line display some dependence. For instance, if the joint probability of two behaviors, A and B, fell above the 45 degree line of the contingency space, this would mean that behavior B has a greater probability of occurring if A has not occurred than if A did occur. Thus, an inhibitory association would develop between the two behaviors. If, on the other hand, the joint probability of behaviors A and B fell below the 45 degree line this would mean that B is *more* likely to occur when A has just occurred than if A has not occurred. According to Rescorla (1967), associations of the type necessary for conditioning to take place are formed under these circumstances. As the graphs in figure 5 indicate, the circumstances necessary for the formation of associations are present for several behaviors. For instance, the occurrence of a male pelvic thrust could readily be associated with (i.e., become conditioned to) a male mount. This is because the pelvic thrust has a very low probability of occurring (about $p = .01$) unless a mount has preceded (which boosts the probability to about $p = .96$). The occurrence of a male genital examination could also readily be associated with either a female present or a male positioning. Under "normal" circumstances one of these behaviors can be "expected" to follow a male genital examination. Knowledge of such existing relationships or contingencies not only permits the investigator to precisely distinguish normal from abnormal social interactions (i.e., to identify the locus of the effect of isolation), but also permits the investigator to identify the social conditioning regime experienced by the to-be-socialized individual. Thus, exposure to adult interactions or participation in such interactions should lead to the formation of associations between these responses.

Associations could also develop between the male approach and the female present, male genital examination, and male positioning. Similarly, the conditions for the formation of an association are also met for female presentations and either a male genital examination or a male positioning. Regardless of the mechanism which causes a male genital examination to be followed by a female present, the fact that a present does consistently follow an examination means that the male, for example, is constantly exposed to this relationship. Thus, an association should form between the two behaviors and genital examination can become an instrumental response which will lead to presentation.

This use of sequential analysis permits a more detailed examination of the various

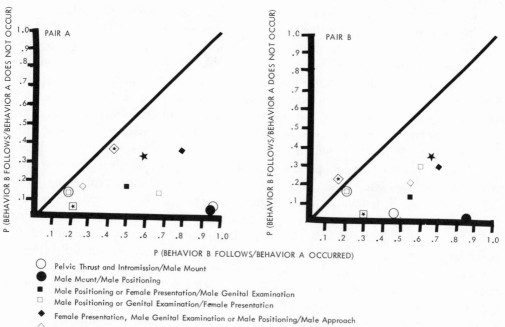

FIGURE 5. *Joint probabilities of each dyad of ferally-born subjects.*

roles played by the responses which we call sexual behavior. This becomes clear in the Craig-Lorenz Schema of natural behavior patterns. In this schema certain parts of a behavior sequence are considered to be consummatory acts. It is these acts which permit the final release of the action-specific drive which motivated the response sequence in the first place. However, the entire sequence of events leading up to the point of drive dissipation can involve a large number of behaviors which are not consummatory. These responses which precede the consummatory act are assumed to function in such a way as to bring the animal into the most ideal stimulus conditions for the occurrence of the consummatory action; they are called appetitive behaviors. General appetitive behaviors bring the animal into a situation in which a more specific appetitive behavior can be activated. This continues to occur until the conditions are appropriate for the release of the consummatory action, at which point all behavior, appetitive and consummatory, come to a temporary end. The Craig-Lorenz Schema thus assumes that appetitive or instrumental responses can be classified as general and specific. The general responses bring the organism into a situation which permits the occurrence of more specific appetitive responses and so on. It should be noted, therefore, that the probability of a general appetitive response leading to a consummatory act should be low because a specific appetitive response must intervene. Similarly, the general appetitive responses may or may not lead successfully to specific appetitive responses. This means that the transitional probability of a general

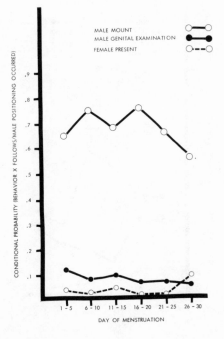

FIGURE 6. *Conditional probability of a male mount, male genital examination, and female present following a male positioning as a function of day of menstruation.*

appetitive response being followed by a specific appetitive response will be lower than the probability of a specific appetitive response being followed by a consummatory response. Hence, casual observation as well as detailed sequential analysis should reveal a trend in sexual behavior whereby the responses of each member of a dyad appear to vary randomly as they fluctuate from one behavior to another in the beginning parts of the behavioral sequence. As the sequence advances, the behaviors become more and more specific and lead with greater probability from one response to the next, until, as the consummatory response is reached, one behavior leads almost with certainty to the next. According to the Craig-Lorenz Schema almost all consummatory acts (eating, drinking, sexual interactions) are characterized by this targeting in on the consummatory act via appetitive behaviors.

An examination of figure 5 indicates that, if a mount and pelvic thrust with intromission is considered the culmination or consummatory part of the sequence of behaviors leading to copulation, then a male genital examination, a female presentation, and/or a male positioning response are the specific instrumental, operant, or appetitive responses leading to this consummatory response. This is indicated by the high probability of transition from one of these behaviors to the consummatory act. Experience with conspecifics should lead to the eventual association of these pre-copulatory behaviors with the act of copulation since the contingencies under seminatural circumstances are such that the conditions for association are met. Simi-

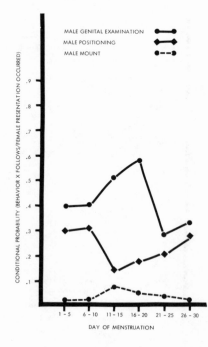

FIGURE 7. *Conditional probability of a male mount, male genital examination, and male positioning following a female presentation as a function of day of menstruation.*

larly, if male genital examination, male positioning, and female presentation are the specific precopulatory appetitive responses, then the approach, the female tooth-chatter, and the male head bob are the general appetitive behaviors which lead to these more specific precopulatory appetitive behaviors. The data recorded reveal that sequential probabilities are such that an association could very well form between these responses. However, as discussed previously, the presence of the conditions necessary for association in no way implies that associative formation is necessary for proper sexual behavior.

Whether or not these contingencies between certain responses are the result of acquired associations or elicited species-typical motor patterns, the identification of certain behavioral sequences permits a more detailed analysis of the relationship of the menstrual cycle to sexual behavior. Rather than determining the relationship between cycle and sexual-response frequencies, it is possible to examine the relationship between changes in the degree of dependence between two behaviors and the phase of the menstrual cycle. Such an analysis can, in turn, shed some light on the question of female receptivity and attractiveness as well as questions concerning response thresholds. Figure 6 graphs the conditional probabilities of several behavior pairs as a function of day of menses. Since male positioning is considered to be a more specific appetitive response with a high probability of leading directly to the mount and intromission response, the probability of this sequence may very well remain

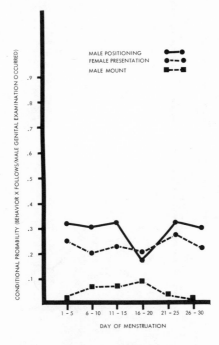

FIGURE 8. *Conditional probability of a male mount, male positioning, and female present following a male genital examination as a function of day of menstruation.*

consistently high. Figure 6 reveals that given a positioning response, the probability of a mount response remains high and does not fluctuate appreciably. Perhaps the less specific appetitive behaviors which sometimes lead to the positioning response tend to vary in their probability of being followed by a positioning response as a function of menstrual state. Female presentations and male genital examinations were considered behaviors of this sort. As Figures 7 and 8 reveal, the frequency of positioning responses following either presentations or genital examinations remains stable or even decreases as the mid-cycle ovulation period is approached. However, the frequency of genital examinations which follow female presents increases. This may be taken to indicate that the increased frequency of presentations and genital examinations which were recorded as a function of menses occurred because of a greater tendency for female presentations to be followed by male genital examinations. However, it should be noted that while the probability of a male genital examination following a female presentation increases toward mid-cycle, the probability of a female presentation following a male genital examination remains unchanged. This may be taken as an indication that the male may find the female more attractive at these times, but there is no evidence that the female is more responsive or receptive to the male, given attempts to genitally examine her.

Perhaps the less specific appetitive responses are more variable as a function of

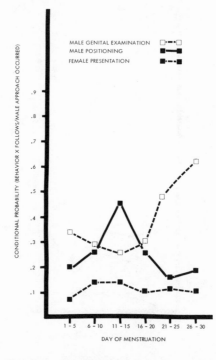

FIGURE 9. *Conditional probability of a male positioning, male genital examination, and female present following a male approach as a function of day of menstruation.*

cycle. Approaches, male head-bobs and female tooth-chatters were considered responses which could lead to the more obvious precopulatory behaviors. Figure 9 shows that the probability of male positioning responses following a male approach increases at mid-cycle. This is another bit of evidence which indicates that female attractiveness increases with cycle. The decrease in the probability of male genital examinations following male approaches implies that approaches by the male were more often followed directly by positioning responses at mid-cycle with the intervening genital examination being omitted.

An examination of figures 10 and 11 provides the first indication of an effect of menstrual cycle on female receptivity. However, rather than affecting the probability that certain female precopulatory behaviors will follow male precopulatory behaviors, the only effect of menstrual cycle on female behavior appears to be restricted to facial courting gestures, most notably the female tooth-chatter. Figure 10 indicates that the female is more likely to respond to the male's head-bob courting gesture by returning a tooth-chatter at mid-cycle than at any other time. This finding correlates with the mid-cycle increase in frequency of female tooth-chatters reported in the previous section. Figure 11 shows that given that a female tooth-chatter has occurred, there is a higher probability that a male positioning response will follow at mid-cycle than at any other time. One could thus tentatively assume that increases in male mounts and positioning are a function of increased female receptivity to the head-bob courtship

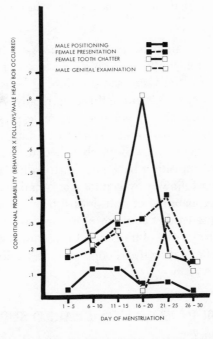

FIGURE 10. *Conditional probability of male positioning, male genital examination, female present, and female tooth-chatter following a male head-bob as a function of day of menstruation.*

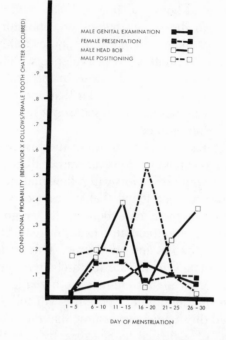

FIGURE 11. *Conditional probability of a male positioning, male genital examination, male head-bob, and female present following a female tooth-chatter as a function of day of menstruation.*

response which results in an increase in female tooth-chattering and, in turn, an increase in positioning. Such tooth-chattering may occur in the process of female presentation and account for the increased positioning behavior. Analysis of triplet sequences would be necessary to determine this empirically. At any rate, the probability of a transition from precopulatory behaviors (female presentations, male genital examinations) to copulatory behaviors appears to remain constant across days of menses. Precopulatory behaviors may themselves become more frequent because of an increased probability of transition back and forth within this group of responses, i.e., of presentations being followed by genital examinations and vice versa. Rather than a change in the probability of transitions from precopulatory behavior to copulatory behavior, the increased frequency of mounts and intromission seems to be a result of the increased probability of transition from courting behaviors (approaches, head-bobs, and tooth-chatters) directly to copulatory behaviors (i.e., the positioning response).

■ SEXUAL BEHAVIOR IN ISOLATION-REARED SUBJECTS

Given the above analysis of sexual interactions in ferally-reared monkeys, we can begin to assess the effects of isolation on these social responses. For example, we might examine the topographical characteristics (i.e., motor patterns) of the responses which are observed in isolation-reared subjects. At such a point we can ask whether or not some of the typical behaviors which appeared in feral animals also appear in the isolation subjects. To the extent that some "social response" is observed, and to the extent that such a response is not aberrant from those observed in ferally-reared animals, we can assume that such behaviors do not require contingent or noncontingent stimulation which is peculiar to the normal social environment. Such behaviors can be assumed to arise naturally from an interaction of the genetic structure of the individual with the environment of isolation. Because such environments are devoid of social stimulation, we can accept that they do not require the normal socialization process for their occurrence.

Only one observation of some response which is performed by an isolation-reared individual proves that such a response does not require social stimulation during ontogeny in order to develop the proper motor patterns for its performance. However, the frequency of that response should reveal something about the conditions which are necessary for its occurrence. Similarly, the relationship of frequency to day of menses should reveal something about the mechanisms which underlie the occurrence of that response. And finally, the use of sequential analysis should provide some information concerning the organization of such behaviors.

CONDITIONS OF THE STUDY

Isolation-reared animals were housed in the same colony room as ferally-born subjects. These subjects were purchased from the University of Chicago where they were born and raised under conditions of social restriction. All isolate animals were removed from their mothers at birth, hand-reared and housed in individual wire cages for one year thereafter. These cages prevented visual and physical contact with conspecifics but permitted auditory interaction. Four male and four female isolation-reared crab-eating macaques were used as subjects. Pair A of the feral subjects displayed the highest incidence of sexual behavior and were therefore employed as the other member of each dyad. In other words, each of the four male isolation-reared subjects was paired every four days with the female from feral pair A, and each of the four female isolation-reared subjects was paired with the male of pair A every four days. Each pair was observed for one-half hour throughout one menstrual cycle for a total of ten sessions.

THE ETHOGRAM

The 19 behaviors described previously were used for these observations. However, pilot observations revealed that other behaviors occurred for the feral-isolation pairs which were not observed in the previous dyads. These behaviors were included in the expanded ethogram. They were:

20. *Male fear grimace:* Although neither feral male displayed this facial gesture during pairings with a feral female, it was often observed in the isolation monkeys. It is described by van Hoof (1972) as fully retracted mouth corners and lips (so that an appreciable part of the gums is bared), closed or only slightly opened mouth.

21. *Male climbing:* Neither male displayed this response with feral females. It was observed in the feral-isolation pairs.

22. *Male grooms self:* Although not observed in the feral dyads, it was observed in the feral-isolation pairs.

23. *Female grooms self:* Although not observed in the feral dyads, females in the feral-isolation pairs were observed grooming themselves.

24. *Male self-directed aggression:* This behavior was recorded on those occasions when the male subject engaged in agonistic or attack responses which were directed at a part of his own body.

25. *Female fear grimace:* This behavior was recorded whenever the female was observed making this facial gesture.

Other behaviors were observed which were peculiar to the isolation-reared subjects. However, such actions could be readily recorded by using one or a combination of appropriate behavioral categories which already existed. For instance, pacing could be recorded as a series of approaches and withdrawals.

GENERAL DESCRIPTION

The isolation-reared members of the dyads reacted similarly to the experimental situation when compared with ferally-reared *Macaca fascicularis*, with some important exceptions. When the guillotine doors were raised, each member of the pair rushed toward one another in a manner similar to that observed in feral pairs. However, isolation-reared females typically fear grimaced at the approaching ferally-reared male. The isolation-reared females rarely presented but rather turned and quickly withdrew. The feral male would then attempt to mount. Isolation females under such circumstances would withdraw further and, if successful, would begin to climb the walls of the observation cage. The male might then genitally examine the retreating female or simply withdraw himself. If the feral male managed to grasp or position the isolation-reared female, she would usually fear grimace and struggle away; successful mounts were rare for this reason. After several initial pairings the feral male no longer attempted such mounts in the beginning of the session. He might approach and genitally examine a standing female and, if she did not withdraw, he might attempt to mount. However, even the most cooperative isolation-reared females would start to withdraw at this point.

Isolation-reared males would also initially "rush" toward the ferally-reared female in the typical manner. However, when such males reached the female who was presenting by this time, they might "crawl over" the female or grab her by the tail and attempt to "climb up" her rear quarters. Later, as the sessions continued, the isolation-reared males would continue to rush toward the female when the doors were opened, but the female quickly began to withdraw from such approaches and would spend more of her time climbing than usual. As time advanced within a particular session, the isolation-reared males would approach the female often. The female typically withdrew at these times. However, isolation males were persistent in their attempts to get close to the female. When they were successful they might actually position the female (touching or pulling up on her rump). If the female did present, the male might genitally examine her in the typical manner. However, if the isolation-reared males attempted to mount, the behavior was sufficiently abnormal to cause the female to start and/or withdraw. These abnormal mounts might involve the males grasping the female's hindquarters, awkwardly attempting to climb up her rear legs. Although those responses were poorly organized and usually frightened the female, they did resemble rudimentary components of the leg grasp response typical of normal male mounting behavior (see next section).

ANALYSIS OF MOTOR PATTERNS

The records for each pair of subjects were examined, and responses which were recorded as having occurred at least once in all pairs were noted. The isolation-reared female-ferally-reared male pairs revealed that all such females had performed the following behaviors at least once: female climbing, female withdrawal, female approach, female present, and female tooth-chatter. One or more pairs never displayed

TABLE 2 / OVERALL PROBABILITY OF EACH BEHAVIOR FOR ISOLATION-REARED MACACA FASCICULARIS.

	BEHAVIOR:	1	2	3	4	5	6	7	8	9	10	11	12
Group FI	Pair A:	.09	.16	*	.18	*	.10	0	.05	0	*	0	*
	Pair B:	.26	.01	0	.10	*	.20	.01	*	*	0	0	.01
	Pair C:	.09	.01	0	.31	*	.34	*	*	0	0	0	.06
	Pair D:	.07	.07	0	.32	0	.30	0	.01	0	0	.01	.01
	Mean:	.13	.06	0	.23	*	.24	*	.02	*	*	*	.02
Group MI	Pair A:	.17	.06	*	.17	.02	.04	0	.03	0	0	*	.04
	Pair B:	.42	.01	*	.04	.01	.02	0	.01	0	0	0	.03
	Pair C:	.14	.03	.01	.13	.01	.09	.01	.07	*	*	.01	.02
	Pair D:	.08	.10	0	.28	.11	.03	0	.02	0	0	0	.02
	Mean:	.21	.05	*	.15	.04	.05	*	.03	*	*	*	.03

	BEHAVIOR:	13	14	15	16	17	18	19	20	21	22	23	24
Group FI	Pair A:	0	.20	0	.19	*	*	*	0	.02	*	0	0
	Pair B:	.01	.21	*	.04	*	0	0	0	.10	.02	0	0
	Pair C:	.02	.09	*	.05	0	0	0	0	.03	.01	*	0
	Pair D:	*	.11	0	.06	*	0	0	0	.02	*	0	0
	Mean:	*	.15	*	.09	*	*	*	0	.04	*	*	0
Group MI	Pair A:	0	.30	*	.04	0	0	0	.01	.06	0	.01	.03
	Pair B:	*	.43	*	.01	0	0	0	.01	*	*	*	*
	Pair C:	.04	.23	*	.08	.08	.02	*	0	*	0	.01	0
	Pair D:	*	.31	*	.01	0	0	0	*	.03	0	0	0
	Mean:	*	.32	*	.03	.02	*	*	*	.02	*	*	*

female look back, female aggression, female grimace, female positioning, female reach back, or female grooms the male. Viewing of the video tapes revealed that the presenting response of the isolate females was similar to the present of the feral females in most ways. However, one female (number 6) displayed some deviation from the normal pattern. In this case, although the female might present normally with tail raised, etc., she failed to stand still during such behavior. "Walking presents" were common in this case.

The isolation-reared male-ferally-reared female pairs revealed that all such males had performed the following behaviors at least once: male withdrawal, male approach, male aggression, male genital examination, and male climbing. One or more pairs never displayed male head-bob, male positioning, male mount, male intromission and pelvic thrust, and male grooms self. Viewing of the video tapes revealed that all behaviors which occurred consistently in all males were similar in form to those observed in the feral males. One male which had attempted to mount on several occasions displayed atypical mounts (orienting to the side of the female) and other similar abnormalities. For instance, several attempted mounts consisted of this male grabbing the female's tail and attempting to climb up her rear quarters. It is interesting to note that nearly all attempted mounts included the leg grasp response.

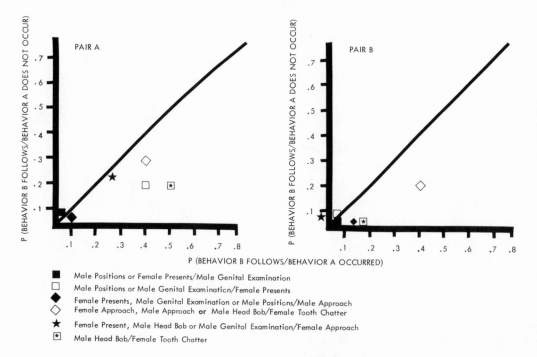

■   Male Positions or Female Presents/Male Genital Examination
□   Male Positions or Male Genital Examination/Female Presents
◆   Female Presents, Male Genital Examination or Male Positions/Male Approach
◇   Female Approach, Male Approach or Male Head Bob/Female Tooth Chatter
★   Female Present, Male Head Bob or Male Genital Examination/Female Approach
⊡   Male Head Bob/Female Tooth Chatter

FEMALE ISOLATES

FIGURE 12. *Joint probabilities for Female-Isolate (FI) subjects, pair A and B.*

FREQUENCY ANALYSIS

Hereafter each isolation pair will be designated as follows: female isolates paired with the feral male will be referred to as FI; male isolates which were paired with feral female will be referred to as MI. Activity scores were computed for these groups in the same way that such scores were computed for the feral pairs. That is, the total frequency of all behaviors was computed for each pair and divided by the total number of sessions, in this case ten. For pair A of the FI group the activity measure was 91.2; for pair B of that group it was 67.1. The activity measures for pairs C and D respectively were 209.7 and 268.5. The activity measures for the MI group, pairs A and B respectively, were 195.1 and 303.1. For pairs C and D these scores were 197.2 and 247.6. Again the relative frequency or probability of each behavior was determined by dividing the total frequency of each behavior by the total frequency of all behaviors. These probabilities are presented in table 2. If one assumes that the entire behavioral repertoire of isolate and feral *Macaca fascicularis* in such situations is fully specified by the present 24-behavior ethogram, then if all 24 behaviors were equally probable we would expect that each behavioral event would have a probability of .04. A large deviation from this score would therefore imply that such a behavior was more probable than others under the particular circumstances. The present frequency

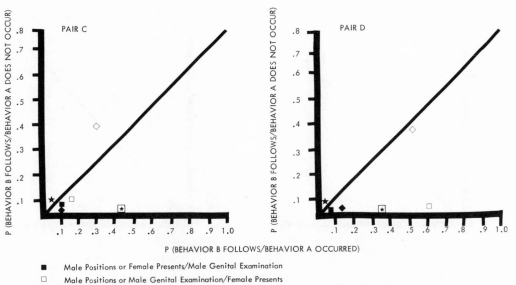

P (BEHAVIOR B FOLLOWS/BEHAVIOR A OCCURRED)

■ Male Positions or Female Presents/Male Genital Examination
□ Male Positions or Male Genital Examination/Female Presents
▣ Female Presents, Male Genital Examination or Male Positions/Male Approach
◇ Female Approach, Male Approach or Male Head Bob/Female Tooth Chatter
★ Female Present, Male Head Bob, Male Genital Examination/Female Approach
◆ Male Head Bob/Female Tooth Chatter

FEMALE ISOLATES

FIGURE 13. *Joint probability for Female-Isolate (FI) subjects, pair C and* D.

analysis revealed that male and female approaches and withdrawal were the most probable responses for all pairs and that the male genital examination was highly probable for the feral male when paired with the isolate females. All other behaviors occurred very infrequently. Activity scores were computed for each pair across days of menses (average total number of responses for each five-day block of the menstrual cycle). No indication of a peak or any consistent change as a function of menstrual cycle was found. Because either member of the pair might have obscured a peak in the activity of the other member, similar measures were taken for each pair member. Again, no consistent change in the overall activity level was noted. The probability of each behavior was also determined as a function of day of menses. None of the 24 behaviors revealed a consistent change in probability as a function of cycle state.

SEQUENTIAL ANALYSIS

Matrices similar to those described for the feral dyads were constructed for feral-isolate pairs. Conditional probabilities were determined for each possible pair of behaviors for each dyad. Figures 12 through 15 give the joint probabilities of certain behaviors for each group for those cases in which probabilities exceeded .100. As indicated in these "contingency spaces," fewer behaviors displayed a strong contingency or relationship with other behaviors. No contingency was revealed for mounting and/or

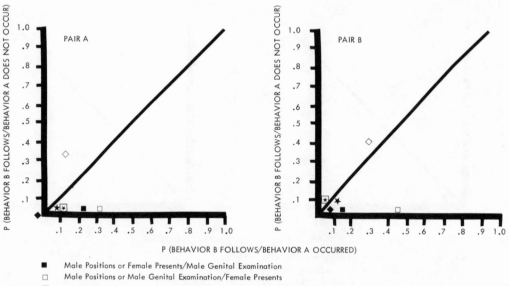

■	Male Positions or Female Presents/Male Genital Examination
□	Male Positions or Male Genital Examination/Female Presents
⊡	Female Presents, Male Genital Examination or Male Positions/Male Approach
◇	Female Approach, Male Approach or Male Head Bob/Female Tooth Chatter
★	Female Present, Male Head Bob or Male Genital Examination/Female Approach
◆	Male Head Bob/Female Tooth Chatter

MALE ISOLATES

FIGURE 14. *Joint probabilities for Male-Isolate (MI) subjects, pair A and* B.

positioning responses since the behaviors never or rarely occurred. For those behaviors which did occur, contingencies were less obvious.

Those contingencies which did exist were examined as a function of menses. Unlike the ferally-reared subjects, no consistent change in any contingency was observed as a function of day of menses. However, the conditional probability of a male genital examination given a male approach did reveal a peak for the feral male-isolate female pairs but not for the feral female-isolate male pairs. This finding is expected because the male is a ferally-born monkey which displayed an increase in the frequency of genital examinations and conditional probability of genital examination given an approach during mid-cycle when paired with a feral female. However, the increase in the likelihood of this response as a function of menses in both feral and feral-isolate pairs implies that female behavioral factors do not account for this effect of menstruation on male behavior. The data give no evidence that isolate females react differently at any phase of menstruation, either in terms of the probability or frequency of a response or in terms of sequential or conditional probabilities. If this is correct we can assume that an increase in male genital examinations following male approaches in the feral male-isolate female groups is not the result of behavioral

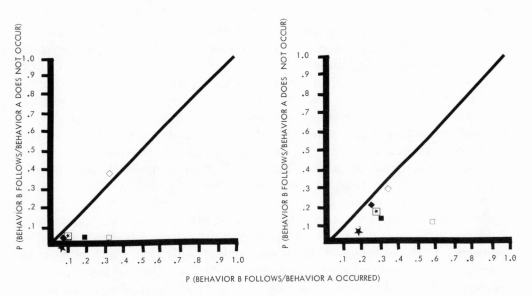

P (BEHAVIOR B FOLLOWS/BEHAVIOR A OCCURRED)

■ Male Positions or Female Presents/Male Genital Examination

□ Male Positions or Male Genital Examination/Female Presents

◆ Female Presents, Male Genital Examination or Male Positions/Male Approach

◇ Female Approach, Male Approach or Male Head Bob/Female Tooth Chatter

✶ Female Present, Male Head Bob or Male Genital Examination/Female Approach

★ Male Head Bob/Female Tooth Chatter

MALE ISOLATES

FIGURE 15. *Joint probabilities for Male-Isolate (MI) subjects, pair C and D.*

changes. Perhaps the pheromone reported by Michael et al. (1968) is responsible for this effect. However, given the presence of such a pheromone we are left to question its lack of effect on isolate-male behavior. Again, the data show no evidence of a change in frequency or conditional probabilities of male behaviors as a function of menses. This implies that the effect of female pheromones on male behavior requires experience with social stimulation. Perhaps the pheromone acts as a cue which provides information concerning the likely reaction of the female to male sexual overtures. A social conditioning regime would therefore be required in order to make this pheromone effective. Experimentation similar to the Michael et al. (1968) study using anosmic and intact isolate males would verify this possibility.

■ CONCLUSION

The data reported in these studies can be summarized as follows. (1) An increase in the frequency of certain courting and precopulatory behaviors during the middle portion of the menstrual cycle was recorded for ferally-reared heterosexual dyads of *Macaca fascicularis*. (2) This increase was not maintained, however, when frequency scores were converted to probability scores, indicating that the "peak" was an artifact of increased general activity. (3) Sequential analysis revealed three classes of sexual behavior: copulatory responses which consisted of the mount, intromission, and pelvic thrusts; precopulatory responses which consisted of the positioning response, genital examination of the female by the male, and the female present; and courting responses which consisted of the male and female approach, the male head-bob, and the female tooth-chatter. This classification is based on transitional probabilities (i.e., the probability that a transition from one response to the next will take place) which indicated that precopulatory responses had a high probability of being followed by copulatory responses, while courting responses had a lower probability of being followed by precopulatory responses. This classification is similar to the appetitive response-consummatory response distinction, discussed by Craig and Lorenz, which assumes that certain specific appetitive behaviors lead, with a high probability, to conditions which permit the consummatory response (in this case copulation). Such a distinction usually assumes that less specific appetitive responses (in this case courting approaches, tooth-chattering and head-bobbing) lead, with a lower probability, to the more specific appetitive responses. (4) An examination of sequential probabilities as a function of day of menses revealed that while the probability of precopulatory responses leading to copulation remained constant as ovulation approached, the probability of courting responses leading directly to copulatory behavior (e.g., positioning) increased during the middle portion of the cycle. (5) Isolation-reared subjects displayed a capacity to perform nearly all the behaviors observed in ferally-reared subjects. In certain cases, however, the form of the response itself was abnormal. For instance, mount attempts were often improperly oriented toward the female, and/or all components of the response were improperly integrated. Similarly, while female presentations were observed, these responses were often incomplete, usually occurring when the subject was walking. (6) Frequency analysis revealed a low probability

of many courting and precopulatory behavior. (7) No consistent change in the probability of any response was recorded as a function of menses. (8) Sequential analysis
revealed that while similar responses were interdependent in both isolate-feral pairs
and feral pairs, the degree of dependence was less obvious in the isolate-feral dyads.
(9) The sequential dependencies did not vary consistently as a function of day of
menses. This last fact is of special interest. It implies that those changes which do
increase sexual activity during mid-cycle owe their effect to previous social experience. If pheromones act to enhance female attractiveness it is possible that this
attraction is due to a conditioning of the pheromone to certain behavioral contingencies. It does not appear to act innately.

Given these findings what can be deduced about the role of socialization in
primate sexual behavior? It will be recalled that one objective of the comparison of
sexual behavior in feral- and isolation-reared *Macaca fascicularis* was to identify the
locus of the effect of socialization. By locus we mean one of three precise characteristics of social behavior: (1) the form of some social response; (2) the frequency of
occurrence of that response; or (3) the organization of that response within the entire
social repertoire. As these studies suggest, all three characteristics of sexual responses
seem to be strongly affected by socialization. Isolation-reared primates displayed differences in each aspect of behavior. The form of certain sexual responses was modified, the frequency of certain responses was lowered, and sequential dependencies
were different.

Hence, these studies allow us to draw some tentative and limited conclusions.
First, we can describe normal socio-sexual interactions in detail. The high degree of
consistency which is present across interactions implies that such social responses
function as communicative signals which transmit a great deal of information. Secondly, by comparison of feral- and isolation-reared individuals we can conclude that
social stimulation functions to modify the form, frequency, and organization (i.e.,
sequences) of sexual responses. However, we can make no definitive statements
regarding the mechanism by which social stimulation accomplishes these
modifications in behavior because too many unknowns remain. We may be able to
specify the relevant classes of social stimulation (i.e., the proper level of stimulation,
the proper types of noncontingent stimulation, and the proper contingencies among
the different stimuli). We may also be able to specify the various effects of social
stimulation (on the form, frequency, and sequences of responses). However, we do
not know what mechanisms are activated by each form of social stimulation, nor
which mechanisms determine each aspect of social responding. This will require
further research which attempts to selectively control stimulation and selectively affect
social responses.

Yet it should be noted that previous researchers have suggested what these
mechanisms of socialization might be like and how they might interact. For instance,
genetic mechanisms probably play a role in determining the nature of social responses. The reader will recall the definition of a reflex offered in the introduction;
the evolutionary processes of variation and selection will bring about species which
display certain stimulus-response dependencies (i.e., the individual reacts reliably and

predictably to certain stimuli resulting in discernible sequences of behavior). Such reflexes depend on genetic mechanisms for their occurrence. Thus, in the social context certain social stimuli will "elicit" responses in the individual. These responses may or may not elicit reactions in other individuals via the same genetic mechanisms causing further elicitation of other responses in the individual and so determining contingent sequences of interaction.

However, the individual raised in isolation maintains an intact phylogenetic heritage and the genetic mechanisms which such a heritage entails. Therefore, the response aberrations which result from isolation-rearing imply that normal social behavior is not solely dependent on genetic mechanisms. For this reason social experience, and the learning which results from such experience, is frequently offered as a major determinant of social behavior. In this way learning mechanisms become crucial to the socialization process. Social stimuli are reacted to by the individual in a random manner; successful or *reinforced* responses are repeated and preserved in the social repertoire, while unsuccessful or punished responses are abandoned. Such initially rewarded behaviors can be "liberated" from their original contexts by contiguity conditioning and secondary reinforcement.

However, the present studies revealed that those individuals deprived of social learning experience were capable of limited social responding. Mounting attempts were observed and easily recognized as such, as were presents and genital examination. It is therefore unlikely that social behavior results solely from learning experiences. It is for these reasons that suggestions of researchers such as Schneirla (1965) and Lehrman (1970) concerning the epigenetic view of behavior are so compelling. This view assumes that the organism develops in a state in which experiential and genetic factors constantly interact to determine behavior. Several attempts to specify this sort of interaction as it occurs in the socialization process have appeared in the literature. For example, Anthoney (1968) has suggested that the greeting response of *Papio cynocephalus* develops during ontogeny from a species-specific sucking response. The author suggests that the lip-smacking greeting response develops from the species-typical motor pattern used in sucking at the mother's nipple—the genetic component. In later life, however, other female baboons lip-smack when approaching the infant who returns the gesture. The approaching female then lifts the infant and nuzzles the genital area. If the infant is a male, this usually results in penile erection. The infant may also lip-smack the female's nipple. If the female is in estrous she may present to the infant and lip-smack over one shoulder. The infant then mounts, thrusts, and often achieves intromission. Anthoney suggests that although the lip-smack response had a strong genetic component, it later comes to be used in other social interactions (e.g., sexual greeting) due to the associations which have developed through experience—the learning component. Burton (1972) describes an interaction of learning and genetic mechanisms in the ontogeny of the chatter response in *Macaca sylvana*. Lancaster (1972) describes another form of interaction in the development of female maternal behavior. Rather than a genetic response predisposition, Lancaster suggests that female juvenile vervet monkeys are strongly attracted to infants—the genetic component. When in contact with or in possession of such an infant, the juvenile female's responses are rather arbitrary and often clumsy. However, selective punishment of inappropriate mothering responses occurs because the infant's distress calls brings the mother who may threaten and/or remove the infant—the learned component.

In these, as well as other examples, the interaction of genetic and learned components takes the following form. Some behavior occurs "naturally," either because the response has been "elicited" or because some stimulus is naturally attractive or pleasant. Within the context of socialization this eliciting stimulus is either some behavior pattern or some characteristic of a conspecific. These genetic species-typical reactions on the part of the individual are then accompanied by pleasant or negative emotional states (either caused by the individual's behavior directly or by the reaction of the recipient). Such behaviors are thus modified in accordance with the principles of learning. Similarly, these very important reactions of the conspecific are themselves accompanied by gestures which can become associated with these reactions and thus serve as communicative signals. In this way socialization appears to be a process by which simple S-R transactions become merged through maturation and experience into higher-order transactions (see Welker 1971).

Similarly, reports on the learned acquisition of instrumental responses (i.e., behaviors which result in some reward) under laboratory conditions have focused on the related issue of biological response predispositions. Staddon and Simmelhag (1971), Segal (1972), Shettleworth (1972), Seligman (1970), and others have attempted to clarify the role which response innateness plays in the instrumental acquisition of certain behaviors. For example, Staddon and Simmelhag as well as Segal have indicated that instrumental responses are probably not acquired simply by reinforcing random responses. Instead they suggest that when certain reinforcers are present in some situation these stimuli *naturally* evoke or elicit certain responses—the genetic component. Instrumental contingencies (i.e., the presence or absence of reward following a response) then act to modify these crude natural reactions until at some point the final response may only resemble the original elicited behavior. It is suggested that many of the responses observed in primate sexual interactions have a natural or innate component. However, social experience is required to maximize the tendency to elicit these responses in reaction to certain social stimuli and to minimize their occurrence in the presence of other inappropriate stimuli; hence, the effect of social isolation on the *organization* of primate sexual responses. This interpretation is reasonable because, as stated, sequential dependencies appear to exist among the same behaviors in both feral- and isolate-reared *Macaca fascicularis*, although the degree of dependence is less obvious in isolate-reared subjects. This implies that a biological predisposition to respond "appropriately" to conspecific behavior exists without social experience. However, without such socialization these natural response tendencies do not become appropriately modified.

This form of interaction between the genetic and learning mechanisms seems to be the most attractive theoretical explanation as to how the socialization process proceeds. The reader should be aware that such an explanation leaves us with a pair of interacting mechanisms which are clearly capable of determining the sequences of responses which will occur. Reflexes and learned responses are identified by virtue of their consistent relationship with other stimuli. Thus the modified sequential characteristics of isolate social behavior are readily explained. In this way both the data and conclusions presented here are completely amenable to statements of researchers such as Alexander and Harlow (1965), Mason (1968), Senko (1966), or Menzel (1964). Their suggestions that social deprivation may leave a subject's ability to perform some behavior intact and only affect the organization of that behavior (species-typical motor patterns occurring out of sequence or at inappropriate times) are reasonable in view of the nature of the socialization mechanisms which are offered here. It is assumed that response *tendencies* are naturally elicited. However, the lack of experience in which certain responses are consistently elicited in the presence of certain stimuli (as might occur during socialization) could quite conceivably prevent the fine tuning of these reflexes which is necessary for the proper organization of social behavior.

However, how then do we explain the modified motor response patterns and the modified frequency characteristics also observed in isolation-reared individuals? Such effects of isolation do not at first glance appear to be the consequence of a learning-genetic mechanism which determines sequences. Yet reports of aberrations in the motor patterns of certain species-typical responses (immature versus mature mounting patterns), as observed in these studies and as reported by Goy and Goldfoot (1974), are in fact understandable in terms of such a socialization mechanism. If our level of analysis is appropriately modified we can explain *species-typical* response aberrations in more molecular terms. Although some behavior may not occur intact in the isolation-reared animal we may observe aspects or remnants of it. Again, a species predisposition to perform these motor patterns must exist. However, experience in which these patterns vary with social stimulation may be essential to their modification via learning and conditioning. Therefore the assertion that response predispositions arise phylogenetically and are modified experientially to occur (1) at proper times and (2) with proper organization of major sequences, in no way contradicts the finding that improper behavioral organization as well as aberrant response forms result from isolation-rearing. Both aspects of behavior entail response sequencing.

Similarly, the reader will recall that frequency (as well as hormonal) characteristics may also be interpreted as aspects of response sequencing. The frequency of some response is easily modified by general situational cues. Hence, the frequency characteristic of a response is also the probability that this response will occur given these general situational cues. In this way the scope provided by the learning-genetic mechanism makes this explanation extremely appealing.

■ ACKNOWLEDGMENTS

The authors would like to thank Drs. S. Carter-Porges, L. Klein, and E. Banks for their support and advice.

■ REFERENCES

Alexander, B. K., and Harlow, H. F. 1965. Social behavior of juvenile rhesus monkeys subjected to different rearing conditions during the first six months of life. *Zool. Jahrb. Physiol.* 71:489–508.

Altmann, S. A. 1965. Sociobiology of the rhesus monkey II: Stochastics of communication. *J. Theor. Biol.* 8:490–522.

Burton, F. D. 1972. The integration of biology and behavior in the socialization of *Macaca sylvana* of Gibraltar. In *Primate socialization*, ed. F. E. Poirier, pp. 29–62. New York: Random House.

Goy, R. W., and Goldfoot, D. A. 1974. Experiential and hormonal factors influencing development of sexual behavior in the male rhesus monkey. In *The neurosciences: Third study program*, ed. F. O. Schmitt, pp. 571–581. New York: Rockefeller University Press.

Harlow, H. F. 1961. The development of affectional patterns in infant monkeys. In *Determinants of infant behaviour*, ed. B. M. Foss, pp. 75–97. New York: Wiley.

Harlow, H. F., and Harlow, M. K. 1962. Social deprivation in monkeys. *Scientific American* 207:137–146.

―――. 1966. Learning to love. *American Scientist* 54:244–270.

Harlow, H. F., and Zimmerman, R. R. 1959. Affectional responses in the infant monkey. *Science* 130:421–432.

Hooff, J. A. R. A. M., van. 1972. A comparative approach to the phylogeny of laughter and smiling. In *Non-verbal communication*, ed. R. A. Hinde. Cambridge: Cambridge University Press.

Kanagawa, H.; Hafez, E. S. E.; Newar, M. M.; and Jaszczak, S. J. 1972. Patterns of sexual behavior and anatomy of copulatory organs in macaques. *Z. Tierpsychol.* 31:449–460.

Lancaster, J. B. 1972. Play-mothering: The relations between juvenile females and young infants among free-ranging vervet monkeys. In *Primate socialization*, ed. F. E. Poirier, pp. 83–104. New York: Random House.

Lehrman, D. S. 1970. Semantic and conceptual issues in the nature-nurture problem. In *Development and evolution of behavior*, eds. L. Aronson, E. Tobach, D. S. Lehrman, and J. Rosenblatt, pp. 17–52. San Francisco: Freeman.

Luttge, W. G. 1971. Role of gonadal hormones in the sexual behavior of the rhesus monkey and human: A literature survey. *Arch. Sex. Beh.* 1:61–88.

Mason, W. A. 1968. Early social deprivation in the non-human primates: Implications for human behavior. In *Environmental influences, biology and behavior series*, ed. D. C. Glass, pp. 70–101. New York: Rockefeller University Press.

Melzack, R. 1968. Early experience: A neuropsychological approach to heredity-environment interactions. In *Early experience and behavior*, eds. G. Newton and S. Levine, pp. 65–82. Springfield: Charles C Thomas.

Menzel, E. W., Jr. 1964. Patterns of responsiveness in chimpanzees reared through infancy under environmental restriction. *Psychol. Forsch.* 27:337–365.

Michael, R. P. 1971. Determinants of primate reproductive behavior. Paper read at *Symposium on the use of non-human primates for research on problems of human reproduction.* Sukhumi, USSR.

Michael, R. P., and Keverne, E. B. 1968. Pheromones in the communication of sexual status in primates. *Nature* 218:746–749.

Michael, R. P., and Zumpe, D. 1970. Sexual initiating behavior by female rhesus monkeys (*Macaca mulatta*) under laboratory conditions. *Behaviour* 36:168–185.

Rescorla, R. A. 1967. Pavlovian conditioning and its proper control procedures. *Psychol. Rev.* 74:71–79.

Sackett, G. P. 1965. Effects of rearing conditions upon the behavior of rhesus monkeys (*Macaca mulatta*). *Child Development* 36:855–868.

Schnierla, T. C. 1965. Aspects of stimulation and organizing in approach/withdrawal processes underlying vertebrate behavioral development. In *Advances in the study of behavior*, eds. D. S. Lehrman, R. A. Hinde, and E. Shaw, pp. 1–74. New York: Academic Press.

Segal, E. F. 1972. Induction and the provenance of operants. In *Reinforcement: Behavioral analyses*, eds. R. M. Gilbert and J. R. Millenson, pp. 1–34. New York: Academic Press.

Seligman, M. E. P. 1970. On the generality of the laws of learning. *Psychol. Rev.* 77:406–418.

Senko, M. G. 1966. Effects of early, intermediate and late experiences upon adult macaque sexual behavior. Master's thesis, University of Wisconsin, Madison.

Shettleworth, S. 1972. Constraints on learning. In *Advances in the study of behavior*, 4 eds. D. S. Lehrman, R. A. Hinde, and E. Shaw, pp. 1–68. New York: Academic Press.

Staddon, J. E. R., and Simmelhag, V. L. 1971. The superstition experiment: A reexamination of its implications for the principles of adaptive behavior. *Psychol. Rev.* 78:3–43.

Welker, W. I. 1971. Ontogeny of play and exploratory behaviors: A definition of problems and a search for new conceptual solutions. In *The ontogeny of vertebrate behavior*, ed. H. Moltz. New York: Academic Press.

14 Foot Clasp Mounting in the Prepubertal Rhesus Monkey: Social and Hormonal Influences

KIM WALLEN, CRAIG BIELERT, JEFFERSON SLIMP

- **INTRODUCTION**

All primates studied have been found to exhibit a complex set of motor patterns and social behaviors which comprise their sexual repertoire. Courtship, female soliciting behaviors, and male and female copulatory behaviors are all part of most primates' sexual behavior. The copulatory behavior of the adult rhesus monkey (*Macaca mulatta*) is characterized by a series of mounts by the male during which the female partner stands with fore- and hindfeet flat on the ground as the male grasps her back legs with both of his hindfeet ("foot clasp mount"—see Figure 1). During the mount the male achieves vaginal intromission and, after several mounts with intromission, ejaculates. The foot clasp mount has been reported by both laboratory and field investigators as the copulatory mount and is apparently an obligatory element of successful copulation (Carpenter 1942; Altmann 1962; Michael and Saayman 1967).

The display of the foot clasp mount begins very early in the young rhesus male's life and its execution is hard to distinguish from that of the adult (see Figure 2), although it is rarely if ever accompanied by intromission or ejaculation before 2½ or 3 years of age. By 9 months to 1 year of age, the predominant mount type in male rhesus observed on Cayo Santiago, Puerto Rico, is the foot clasp mount (Wallen, unpublished observations). A similar finding was reported for 1-year-old males in a captive troop of Japanese macaques, a species which has an adult copulatory pattern very similar to the rhesus (Hanby and Brown 1974). These findings demonstrate that foot clasp mounting behavior clearly occurs before the onset of testicular activity associated with puberty.

The critical importance of the foot clasp mount to the adult copulatory pattern, as well as evidence that its early display is a good predictor of subsequent adult copulatory success (Goy and Goldfoot 1974), make it an appropriate behavioral end point for the study of sexual behavior in the prepubertal male rhesus. In this chapter we will describe both hormonal and social factors which influence the display of the foot clasp mount in rhesus monkeys raised in the laboratory.

Young monkeys of both sexes raised under socially restricted conditions in this and other laboratories display some form of mounting behavior within the first year of

FIGURE 1. *The adult rhesus foot clasp copulatory mount.*

life (Goy 1966; Harlow and Lauersdorf 1974). Foot clasp mounts account for a small proportion of mounts seen at this time and are usually only displayed by males. Not all males display foot clasp mounts and some do so only at extremely low frequencies (Goy and Goldfoot 1974; Goy, Wallen, and Goldfoot 1974). Females limit their mounting mainly to improperly oriented mounts with pelvic thrusting and rarely display foot clasp mounts (Goy 1970; Harlow and Lauersdorf 1974). The sexual dimorphism in the display of foot clasp mounting, and the great variability in display among comparably aged males, led to investigations of possible endocrine contributions to prepubertal mounting behavior.

It has been known for many years from studies in other species (for reviews see Lisk 1974; Phoenix 1973; Young 1961) that the full adult male copulatory pattern is dependent upon secretions from the testes. Data concerning testicular influences on the foot clasp mounting of young adult and fully mature male rhesus have recently become available. They show that castration of young postpubertal males results in a marked reduction in mount frequency (Michael, Wilson, and Plant 1973; Michael and Wilson 1974). In contrast, castrated fully adult males continue to mount at preoperative levels for as long as 1 year post-castration (Phoenix, Slob, and Goy 1973), but intromission frequency and ejaculations are markedly reduced (Michael et al. 1973; Michael and Wilson 1974).

Furthermore, studies of intact free-ranging male rhesus have demonstrated a contribution of the testes to mounting behavior as a result of the seasonal testicular cycle. During the nonbreeding season, testes become reduced in size, the output of androgen lowers, spermatogenesis ceases, and mounting behavior declines markedly

FIGURE 2. A *foot clasp mount as displayed by a one-year-old male rhesus.*

(Conaway and Sade 1965; Sade 1964; Wilson and Vessey 1968; Vandenbergh 1973; Robinson et al. 1975). With the onset of breeding season and the associated increase in female receptivity, the males' testes recrudesce and mounting behavior increases. Although there is likely a female contribution to this decline and reinstatement of mounting behavior (Vandenbergh 1973; Gordon and Bernstein 1973), it also seems clear that there is a definite testicular contribution to this cyclical fluctuation of mounting behavior. This evidence from both laboratory and field suggests that the integrity of the testis is necessary for the adult rhesus copulatory pattern and can have a marked influence on mounting behavior as well.

Previous investigators (Phoenix, Goy, and Young 1967; Goy 1968; Phoenix 1973) have examined the possible testicular contribution to foot clasp mounting in prepubertal males by castrating them at birth or 3 months of age and observing their behavior in heterosexual peer groups during the first 4 years of life. Phoenix (1973) reported no differences in the frequency of mounting between castrates and intact males except during the second year of life. However, the data presented by Phoenix were not restricted to foot clasp mounting and his intact and castrate male subjects did not receive uniform social experience. We have reanalyzed the data, taking these factors into consideration, and can find no difference between intact and castrate males at any time prepubertally in the frequency of foot clasp mounting. Castrates show the same variability as do intact males in their display of foot clasp mounts, with some never displaying any, and others showing high frequencies of mounts. Both types of males, as a group, show a higher frequency of foot clasp mounts than do comparably reared females. The differences between males and females as well as differences among males cannot be accounted for by a postnatal-prepubertal testicular contribution. In contrast to the adult, it appears that the testis contributes little if anything to the prepubertal display of foot clasp mounting.

If the differences in mounting behavior observed between males and females cannot be accounted for by postnatal endocrine influences, possibly they are the result of prenatal endocrine influences. Ample evidence in many species exists for the masculinizing influence of testicular androgen when present during a critical period of development (Phoenix et al. 1959; Goy, Bridson, and Young 1964; Barraclough and Gorski 1962; Harris and Levine 1965; Grady, Phoenix and Young 1965; Eaton 1970; Goy 1970a, 1970b). In the genetic male rhesus, it is the activity of the testes during this prenatal period which normally produces the hormones necessary for complete virilization (Resko 1970). Prenatal exposure of the genetic female to exogenous androgens results in the formation of a phallus and scrotum with closure of the vaginal orifice externally and the internal formation of a prostate and seminal vesicles. The androgen treatment, however, does not interfere with the formation of functional ovaries (Goy and Resko 1972), fallopian tubes, or uterus (Wells and van Wagenan 1954). For this reason these females have been termed "pseudohermaphrodites."

Work on rodents has shown that exposing genetic females to testosterone early in development decreases the predisposition to display female responses and increases the display of masculine behaviors (for reviews see Davidson and Levine 1972; Gorski 1971; Whalen 1968; Young 1964; Goy and Goldfoot 1973). Prenatal exposure to androgen in the rhesus results in the display of foot clasp mounting at frequencies which are not significantly different from comparably reared males, but which are significantly higher than those displayed by comparably reared genetic females (Goy 1966, 1969, 1970a, 1970b; Goy and Phoenix 1972; Goy and Resko 1972; Goy and Goldfoot 1974). These pseudohermaphroditic female rhesus show the same variability in the display of foot clasp mounting exhibited by males. Again, some never display any foot clasp mounts, whereas others do so at high frequencies. Prenatal androgen exposure therefore seems to account well for the observed differences in foot clasp mounting between normal males and females, but it does not apparently account for the degree of variability observed across all animals.

Primates, in contrast to many other socially living animals, exhibit a long period of dependence and development prior to the onset of puberty and an additional period of extended adolescence prior to fertile reproductive activity. It is during this prepubertal period that much of what has been described as socialization occurs (Poirier 1972).

It is known that limited and specific social interaction during this period can have a profound effect upon eventual adult copulatory behavior (Mason 1960; Harlow et al. 1966; Missakian 1969; Goy and Goldfoot 1974). The effects of social restriction can also affect the type of mounting displayed early in life, so markedly, in fact, that one author was led to conclude erroneously that the "double foot-clasp of the female's legs . . . is not an infant male posture" (Harlow and Lauersdorf, 1974: 350). This was in reference to animals 1 year of age which had been raised with access only to other peers. Probably the variability in the occurrence of foot clasp mounting observed in other studies (Goy 1970a; Goy and Goldfoot 1974; Goy, Wallen, and Goldfoot 1974) involving 1-year-old animals is due to the specific social environment in which they were reared. The remainder of this chapter will be directed towards an analysis of the factors within a social environment which influence the display of foot clasp mounting. We will mainly concern ourselves with the amount and type of social experience an individual male is given during the first 3 years of life. Some of the data reported here are the result of a reanalysis of the raw data presented in previous studies (Goy 1970a; Goy and Goldfoot 1974; Goy, Wallen, and Goldfoot 1974); other data are previously unreported and will serve to clarify aspects of social influences on foot clasp mounting.

For all of the studies being presented the following behavioral definitions were used:

1. *Abortive mounts*: Mounts which do not contain a foot clasp and in which the mounter's pelvic area is not properly oriented toward the partner's genitalia. These mounts may be standing, sitting, or lying, but they must always be accompanied by pelvic thrusting.

2. *No foot clasp mounts*: A flat-footed stance in which the mounting animal's pelvic area is properly oriented toward the partner's genitalia and the hands are placed on the partner's back, waist, or hips. No back foot clasping occurs in this mount, but there is pelvic thrusting (Figure 3).

3. *Single or double foot clasp mounts*: A properly oriented mount resembling the adult copulatory posture previously described, which includes pelvic thrusting. Although the double foot clasp mount is the most common adult copulatory mount,

FIGURE 3. A *no-foot clasp*, or *immature*, *mount as displayed by a one-year-old male rhesus.*

some males use the single foot clasp mount during copulation; for this reason, no distinction was made in this scoring system between the single and double foot clasp mount.

The social behaviors recorded were as follows:

1. *Aggression:* A vigorous bite with or without headshaking given to another animal.

2. *Fear grimace:* A stereotyped submissive behavior in which the lips are fully retracted to expose the teeth.

3. *Rough-and-tumble play:* A vigorous form of wrestling play with whole body involvement.

4. *Grooming:* Stereotyped spreading and picking of a partner's fur.

■ **SOCIAL FACTORS INFLUENCING THE EXPRESSION OF FOOT CLASP MOUNTING**

The first social environment in which foot clasp mounting was studied is one in which infants are separated from their mothers at 90 days of age and placed in individual cages. For 20 weeks they are observed for 30 minutes per day, 5 days per week, in heterosexual groups of 4–6 animals in a playroom situation (see both "peer-reared conditions" in Figure 4). After the formation of the group the composition remained constant, barring the death of one of the group members, thus limiting an individual's social interaction to its group. For this first study, data from 22 males and 16 females are reported on.

During social observation runs, a checklist of sexual and social behaviors is employed. Sexual behaviors are scored for all animals in the group for the entire ½-hour test period. Concurrently, selected social behaviors are scored utilizing a "shifting focal animal" technique in which the behaviors initiated by a predetermined animal are scored regardless of the recipient for a period of 5 minutes. Each animal within the group becomes the focal animal once during a daily observation period.

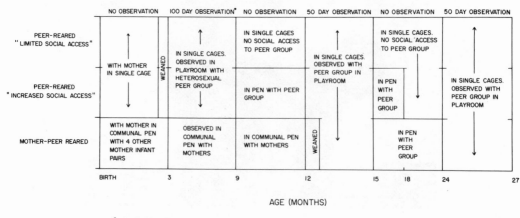

FIGURE 4. *Rearing conditions used in the study of social influences on foot clasp mounting.*

PEER-REARED: FIRST YEAR OF LIFE

During the first observation period, 21 of the 22 males studied displayed 5 or more abortive and/or no foot clasp mounts. The great majority displayed many more than 5 (mean for abortive mounts per male per 100 days of observation was 62.1 mounts, and the mean for no foot clasp mounts per male per 100 days of observation was 63.2 mounts). The occurrence of foot clasp mounts was quite different with only 14 of the 22 males ever displaying a foot clasp mount. Of these 14, only three displayed more than five foot clasp mounts during 100 days of observation (mean for males displaying foot clasp mounts was 4.4 mounts per male per 100 days of observation).

The groups in which all of these males, both mounters and nonmounters alike, were reared were characterized by high levels of rough play, high frequencies of aggression, and rare incidences of grooming. By using a fear grimace matrix in which the occurrence for an individual of this facial expression (Figure 5), frequently associated with submission (Hinde and Rowell 1962), is plotted against the animal to which it is displayed, it is possible to define a linear dominance hierarchy within these small heterosexual peer groups (Schjelderup-Ebbe 1935). Usually the Alpha or most

FIGURE 5. An example of the "fear grimace" which is used as a measure of submission in social interactions between rhesus monkeys.

dominant animal was distinguished from other group members only by the fact that he fear grimaced to no other animal and all animals fear grimaced to him, but in some groups the Alpha became an apparent "bully" that dominated the other animals through excessive threatening and aggression. It appeared in these groups that there was a relationship between the presence of a "bully" and a lowered frequency of the mature mount type (Goldfoot, unpublished data). The low frequencies of grooming observed in all groups suggest social environments in which close associations between group animals are rare. The high levels of aggression also seemed likely to have a negative effect on the development of stable affiliative relationships among group members. The social characteristics of the groups during the first year of life led us to hypothesize that although some males were able to develop and display the foot clasp mount, the great majority of males were being inhibited from expressing this particular mount pattern by social factors within the groups. It was hypothesized that the relatively short period of daily group exposure, although sufficient for the expression of dominance interactions, was too brief for the development of positive social relationships.

PEER-REARED: SECOND YEAR OF LIFE

To test the hypothesis that the amount of time which animals had to interact was critical to the formation of a social environment in which foot clasp mounting could be expressed, the following study was undertaken.

When the original peer-reared groups had reached approximately 9 months of age, 4 groups containing 12 males and 6 females were allowed 3 months of continuous living experience within their own groups. The males in these groups, referred to as "increased social access" males (Figure 3) are identical to the animals previously described as "socialized" by Goy, Wallen, and Goldfoot (1974). These 12 increased social access males were compared to 14 comparable aged males from 5 other peer-reared groups which were housed in individual cages for the 3-month period and allowed no social interaction with other monkeys. These will be referred to as "limited social access" males. At the end of the 3-month period when all subjects were approximately 1 year of age, animals from both conditions were housed in individual cages and a 50-day period of 30-minute daily test sessions was carried out (Figure 3).

There were very clear differences between the groups reared under different conditions in terms of both the frequency and proportion of males displaying foot clasp mounts. Eleven of the 12 increased social access males displayed foot clasp mounts, which is significantly greater than the 5 of 14 limited social access males displaying foot clasp mounts. In males with increased social experience, the foot clasp mount accounted for 80% of all mounts displayed, whereas for limited social access males it accounted for less than 5% of total mounts (Figure 6). Those males with increased social experience which displayed the foot clasp mount also did so at a significantly higher frequency than males in the socially limited groups (48.4 mounts per responding increased social access male as opposed to 5.0 mounts for the responding limited social access males). The limited social access males displayed significantly higher frequencies of no foot clasp mounts than did males with increased social access (28.4

mounts per male responding as opposed to 7.3 mounts per responder). However, there were no differences between the two populations in the proportion of males displaying either this less mature form of mounting or abortive mounts. The differences, then, in the proportion of males displaying a given mount type, were limited solely to the foot clasp form of the mount.

We were interested in whether the observed differences in the proportion of males displaying foot clasp mounts and the increased frequency of display could be related to qualitative changes in social interactions. The amount of previous social contact allowed the different groups appeared not to affect the frequency of rough play behavior. Both populations displayed extremely high levels of play and there were no significant differences between the limited social access males and those which received increased social access. It seemed possible that the key factor might be the

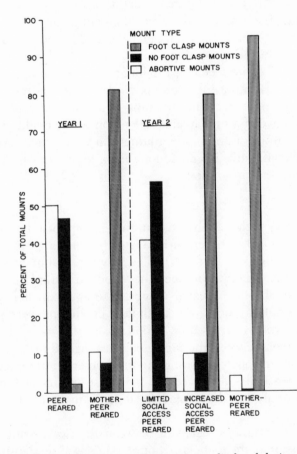

FIGURE 6. *The distribution of mount types displayed during the first and second year of life by peer-reared males receiving differential amounts of social contact and by males reared with their mothers in communal pens during the first year of life.*

amount of agonistic behavior occurring among the animals within these groups. One appropriate measure of agonistic behavior is the amount of aggression individual males within the groups initiated or received. Surprisingly, there were no differences either in the frequency or the proportion of males initiating or receiving aggression. A large majority of males in both rearing conditions received aggression at fairly high frequencies (mean number of aggressions received per increased social access male is equal to 12.87 versus 10.42 for the limited social access males). Likewise, fear grimacing showed no differences between the two conditions either in frequency or the proportion of males displaying it.

The receipt of grooming from another animal within the male's group, however, did show a marked difference between the two experimental conditions. A significantly larger proportion of increased social access males than limited social access males received grooming (9 of 12, as opposed to 4 of 14), and those which were groomed received a significantly greater number of grooming bouts (9.1 for increased social access males versus 2.0 for limited social access males).

Marler (1965: 544) has described the effect of grooming on the social environment of many primates as "a general relaxation of tension and the adoption by the recipient of postures that invite further grooming... an activity... that is incompatible with agonistic behavior...." More recently, Lindburg (1973) has proposed that grooming promotes cohesiveness within a social group. In this light, the results of the present study suggest that the more frequent grooming observed after a period of increased social access is indicative of a fundamental alteration in the quality of social interaction within the peer groups from an environment in which high levels of aggressive encounters with little affiliative interaction are seen, to one in which both aggressive and affiliative interactions are present.

PEER REARED: THIRD YEAR OF LIFE

At the end of the 50-day period of observation, the 4 groups of animals which received the increased social contact were returned to large pens for a 3-month period during their second year of life (see Figure 3). The other 5 groups were again kept in individual cages for the 9 months between the second and third observation periods. When all animals in the heterosexual groups were approximately 2 years of age, another 50-day observation period was instituted; conditions were similar to those used in the previous 50-day observation period.

The differences seen between the increased social access males and socially limited males became more apparent during the third observation period. All 12 of the increased social access males displayed the foot clasp mount as their predominant mount type, whereas only 4 of the limited social access males displayed the foot clasp mount.

The difference in mounting behavior between the two rearing conditions was limited to the expression of the foot clasp mount. There was no statistically significant difference in the proportion of males showing at least one form of mounting behavior (12 out of 12 for the increased social access males versus 11 out of 14 for the limited social access males) or in the total number of mounts displayed of all types (1,140 as opposed to 865 for the limited social access males). What was striking was the distribution of the mounts displayed. Only 3% of the mounts displayed by the increased social access males were abortive and 4% were no foot clasp mounts, while 93% were foot clasp mounts. In contrast, abortive mounts by the limited social access males accounted for 36% of their total mounts, no foot clasp mounts accounted for 54% of the mounts, and foot clasp mounts represented only 10% of the total mounts displayed. The differences between these two populations of animals cannot be accounted for simply by an across-the-board inhibition of mounting, but rather, it is suggested that there were selective behavioral and perhaps social differences, possibly due to the partner's behavior, which resulted in the preferential display of an immature and reproductively dysfunctional form of mounting behavior in the limited social access males.

As was the case in the second observation period, there were no differences between rearing conditions with respect to the proportion of animals displaying or the frequency of rough play, fear grimacing, or aggression. There was, however, a striking and highly significant difference in the frequency of grooming bouts received by the increased social access males. Whereas limited social access males received an average of only 4 grooming bouts during the entire 50-day observation period, the socialized males received an average of 32. Although the proportion of males receiving any grooming bouts was not significantly different between the two groups, there was a marked difference in the number of males receiving more than a "token" groom or two during the observation period. All 12 of the socialized males received more than 5 bouts of grooming (with only 3 receiving fewer than 20), whereas only 2 of the 14 limited social access males received more than 5 (and none of these males ever received 20 bouts of grooming). This marked social difference reinforces the notion that increased social exposure allowed the development of affiliative relationships between the individual members of the social group, resulting in more time spent in nonhostile, close contact, such as grooming, possibly needed for the accommodation of the foot clasp mount.

PEER-REARED: AFTER THE THIRD YEAR OF LIFE

After the third period of observation, males in all groups were housed in individual cages. One group of males which previously had 2 periods of extended social contact was housed in an outdoor pen for 1 month during their fourth year. Subsequent to that period they were housed individually and only had contact with other monkeys during several 50-day observation periods over the next 2–3 years.

One group of 4 limited social access males was housed together for 5 months during their fourth year of life. Aside from this period of continuous contact, their only contact with other monkeys was during 50-day observation periods during their fourth and fifth year of life.

When males reared under both conditions had passed through puberty and reached between 4 and 6 years of age, they were tested with appropriately receptive females. As has been published previously (Goy and Goldfoot 1974), 6 of 7 males which were reared with extended periods of group living displayed intromissions during the period of observations. The 4 males reared without a period of group living during the first 2 years of life did not display any intromissions during the test.[1] The differences between the two rearing conditions in the percentage of males displaying intromissions was significant and was unrelated to the age of the male at the time of testing.

Even though the males which had experienced group living early in their developmental history did achieve intromissions when tested in adulthood, they were far from robust copulators. Of the 7 males achieving intromission, only 2 actually ejaculated. Commonly, 85% of the wild-caught males brought into the laboratory will ejaculate with an appropriately receptive female (Bielert 1974). It seems clear that although extended group living increased the copulatory ability of males so reared, they were still far from comparable to feral males in their copulatory ability.

MOTHER-PEER-REARED: FIRST YEAR OF LIFE

The period of group living possibly had occurred too late in the development of the males described above. It had been seen that the 5 months of continuous living experience given during the fourth year of life to males not previously having any continuous group experience had no observable effect on their ability to display intromission, as none were observed to intromit when tested in adulthood. It also seemed likely that limiting experience to peers only may not have provided a rich enough social environment for the full development of successful copulators. For these reasons we formed heterosexual groups of 5 or 6 infants with their mothers which lived together continuously in large indoor pens from the time the infants were 2 months of age ("mother-peer-reared"—Figure 3). At 3 months of age a series of 100 days of ½–hour–per–day observations in the home pen began. To date, 24 males distributed among 12 heterosexual groups have been observed through the first year of life.

Since the animals within the groups are free to interact for 24 hours per day, our method of sampling behavior for ½ hour per day can only record a part of the total frequency for the display of any given behavior. In the peer-rearing condition previously described, animals were allowed to interact only during the ½ hour observation period, so the frequencies obtained in that condition were an accurate estimate of the actual frequency of occurrence. This methodological difference between males reared during the first year of life in mother-peer groups and males reared with other peers only, precludes direct comparison of the frequency data obtained under both rearing conditions. Qualitative comparisons are valid, however, and provide striking contrasts between the 2 rearing conditions.

Of the 24 males reared with mothers and peers during the first year of life, 22 displayed at least 1 foot clasp mount during the 100-day observation period (mean per male per 100 days was 46.8 mounts). This is in marked contrast to the peer-reared males in which only 14 of the 22 males studied ever displayed a foot clasp mount. Even though the frequency of foot clasp mounts displayed by mother-peer-reared males probably only represents a portion of the total frequency due to the sampling technique used, it is still significantly greater than the frequency of foot clasp mounts displayed by peer-reared animals. In addition to being displayed at high frequencies, the foot clasp mount was the predominant mount type, accounting for nearly 82% of all mounts displayed by mother-peer-reared males (Figure 6). This is in marked contrast to peer-reared males, in which it accounted for less than 3% of the total mounts displayed. Not only were foot clasp mounts the predominant mount type, but they were also displayed more consistently throughout the period of observation. A comparison between mother-peer and peer-reared males of the number of 10-day observation blocks in which a foot clasp mount was observed reveals that the mother-peer-reared males displayed foot clasp mounts during significantly more observational blocks than did peer-reared males (a mean of 6.5 blocks per mother-peer-reared male displaying foot clasp mounts as opposed to 2.4 blocks per peer-reared male).

In addition to the differences in the character of the mounting behavior displayed, there were marked differences in the quality of the social interactions observed. Fear grimacing to other infants, which was quite common in the all-peer social group, was seen very infrequently in the mother-peer groups. In fact, 13 of the 24 males observed were never seen to fear grimace to another infant. This is significantly more than the 5 out of 22 peer-reared males that similarly showed an absence of fear grimacing. It is not likely that the lack of infant-to-infant fear grimacing is an artifact of the observation techniques, as all of the mother-peer-reared males were observed to fear grimace to a mother during the 100 days. It seems more reasonable that the lack of fear grimacing to other infants is indicative of a different social environment in which infants do not have to compete with each other for positions in a dominance hierarchy. This impression is further borne out by the lack of infant-infant aggression observed in mother-peer groups. Only 3 males of the 24 studied were seen to aggress another infant, whereas all 22 of the peer-reared males aggressed other infants. The low occurrence of aggression that did occur in mother-peer groups was directed to the females in the groups, as none of the 24 males were ever seen to receive aggression. This is in contrast to the peer groups where 58.3% of the aggression was directed towards other males and 14 of the 22 males studied received aggression at some point during the 100-day observation period. The less intense agonistic interactions between infants in the mother-peer situation was not reflected in the occurrence of grooming, however. Males in either rearing condition were groomed by other infants only infrequently, and there was no significant difference between the rearing conditions in the proportion of males receiving grooming, with 12 of the 22 peer-reared males being groomed by other infants as opposed to 5 of the 24 mother-peer-reared males. In no case did any animal receive more than 5 grooming bouts.

The difference in the apparent social atmosphere had no observable effect on the play behavior in the 2 rearing conditions, as 100% of the males, regardless of their rearing environment, engaged in rough play.

MOTHER-PEER-REARED: SECOND YEAR OF LIFE

When the animals in the mother-peer pens had reached 1 year of age, they were separated from their mothers and put in individual cages. Following separation, the groups were observed in a play room similar to the one used for the peer-reared groups for ½ hour a day. After a 2-week acclimatization period, the daily observations were recorded for a period of 50 days. This observation procedure duplicates that used for the peer-reared males and allows quantitative as well as qualitative comparisons to be made.

Twenty-three of the 24 males reared with their mothers and peers displayed foot clasp mounts at some time during the 50-day observation period and at significantly higher frequencies than either the increased social access or limited social access males (a mean of 191.5 mounts per 50 days per mother-peer male as opposed to means of 48.4 and 5.0 mounts, respectively, for the peer-rearing condition). As can be seen in figure 6, the mother-peer-reared males displayed foot clasp mounts almost exclusively, a mount type distribution similar to that for increased social access males, but more pronounced.

Animals reared under the two different conditions continued to display differences in their social behavior in this second run. Although the proportion of mother-peer-reared males initiating aggressive interactions was not significantly different, the frequency with which they engaged in aggression was significantly lower than either the increased social access or limited social access peers, with the mother-peer males initiating aggression 3.9 times on the average per 50 days as opposed to 18.8 and 16.8 for the two types of peer-reared males.

Fear grimacing, on the other hand, no longer differentiated the two rearing conditions; without the mothers present, the animals in the heterosexual groups formed linear dominance hierarchies and fear grimaced at frequencies not significantly different from either of the types of peer-reared males, a mean frequency of 10.8 fear grimaces per male for the peer-rearing conditions as compared with 12.8 fear grimaces per male for the mother-peer-reared males.

The grooming behavior observed among the mother-peer-reared males during this observation period without their mothers was very similar to that observed for the increased social access males. Eighteen of the 24 mother-peer-reared males received grooming bouts with an average of 10.3 bouts of grooming per male. This is comparable to the 9.1 grooming bouts received on average by increased social access males.

The play behavior exhibited by the mother-peer males was significantly different from that seen in either peer-only rearing condition. Although there was no difference in the proportion of males engaging in play, the frequency with which mother-peer males engaged in rough play, an average of 6.1 play episodes per 10-day observation block, was markedly lower. The peer-reared males, in contrast, engaged in 19.1 episodes. It has been previously suggested (Goy, Wallen, and Goldfoot 1974) that this difference in play behavior may be an indication that the mother-peer males are suffering from the depressive effects of the separation from their mothers, which has been shown in other laboratories to have the effect of decreasing play activities (Hinde and Spencer-Booth 1971). An alternative explanation can be reached from data comparing socially restricted males with socially experienced males which had as one result a reduction in the frequency of play behaviors exhibited by the socially experienced males (Mason 1960; Anderson and Mason 1974). It is possible, therefore, that this reduction in the frequency of play observed in the mother-peer-reared males may result from increased social experience. The initially drastic reduction in play frequencies, and its subsequent increase during the observation period previously reported (Goy, Wallen, and Goldfoot 1974) for animals reared in this manner, argues strongly that the initial decrease is due to the maternal separation and that play resumes after infants have recovered somewhat from the initial trauma of separation. It is of interest that although rough play appears to have been affected by the trauma of separation, the mounting behavior is much less obviously affected.

MOTHER-PEER-REARED: THIRD YEAR OF LIFE

After the second observation period, all mother-peer-reared animals were housed together in their respective groups in large double cages for 9 months. At the end of this period, the animals within the groups were housed individually and a third 50-day period of ½-hr-per-day observation in a playroom was undertaken (see Figure 3). At this point, only 14 of the 24 males being studied had reached 3 years of age and these comprise the sample on which we now report.

All except one of the mother-peer-reared males continued in their third year to display foot clasp mounts at very high frequencies almost exclusively. Ninety-seven percent of all mounts displayed were foot clasp mounts with a mean frequency of 130.7 mounts per male per 50 days of observation. Although not significant, this is higher than the 88.8 mounts per male observed for the increased social access males during their third year of life.

The mother-peer-reared males continued to show very low levels of aggression, which now differed significantly from those of the increased social access males. Seven of the 14 mother-peer males engaged in aggressive interactions, but at a mean frequency of only 2 episodes in 50 days of observation. Both the proportion of males engaging in aggression and the frequency with which it was displayed differs significantly from both peer-rearing conditions, with 22 of the 26 peer-reared males engaging in aggression with an average frequency of 16.4 bouts per male per 50 days of observation. There was no difference between the limited social access and increased social access males either in the proportion of males engaging in aggression or the frequency with which they did so.

The impression of the relaxed atmosphere within these mother-peer groups was reinforced by the high frequency of grooming observed. Twelve of the 14 males received grooming bouts with an average frequency of 9.7 bouts per 50 days of observation. This frequency, while substantial, is significantly lower than that seen in increased social access groups during their third observation period. This difference in frequency does not necessarily imply that the increased social access groups are more stable or relaxed, but may be an artifact of the manner in which we record grooming behavior. Our data provide information about the frequency of grooming bouts, but cannot offer insight into the duration of these bouts, which has been shown to increase with increased social experience (Mason 1960). When comparing groups with relatively high frequencies, it is necessary to know the duration of grooming bouts before meaningful conclusions can be reached using differences in frequency. In light of this, it would seem that the meaningful consideration in this case is that the mother-peer males exhibit substantial grooming frequencies and very low occurrences of aggression. These factors are compatible with an interpretation of increased affiliative bonding and decreased agonistic behavior.

The play behavior displayed by the mother-peer-reared males was not significantly different from that of the peer-reared males with almost identical mean frequencies per male per 10-day observation block (13.2 episodes per mother-peer male as compared with 13.7 episodes per peer-reared male). This supports our earlier deduction that the low level of play observed during the second observation period was due to separation trauma.

MOTHER-PEER-REARED: AFTER THE THIRD YEAR OF LIFE

At the end of the 50-day observation period all animals were again housed in groups in large double cages for 9 months. At the end of the 9-month period, animals were housed in individual cages and a 50-day play room observation period was initiated. At the time of this writing, data on only 6 males are available, therefore no detailed analysis will be presented. An effect of the rearing is already evident, though, as three of the 6 males have ejaculated with females in the daily observation period; this already is a higher proportion of males ejaculating than was observed among postpubertal increased social access males. It remains to be seen whether or not the copulatory behavior of these males is generalizable to partners other than the ones with which they grew up, and such studies are presently being undertaken.

■ DISCUSSION

Copulatory success in the male rhesus appears dependent upon at least three factors. The male must be motivated to copulate, be socially competent so he will recognize when sexual advances are likely to meet with success, and be able to execute the species-specific copulatory pattern. The present studies demonstrate that alterations in a male's early social environment can have a marked effect on his ability to perform the critical copulatory pattern. The deficiency observed in males with limited social access is not in the display of mounting per se and therefore, by inference, in their motivation to mount; it is in their display of a specific reproductively crucial mount type, the foot clasp mount. It should be pointed out, however, that there are cases among the limited social access males in which social deprivation results in an eventual disappearance of all mounting behavior by 2 or 3 years of age. In these cases the possibility exists that early social deprivation may have as one effect the reduction in a male's motivation to mount. Unfortunately, data presently available can provide only limited insight concerning this crucial question. Complete disappearance of mounting behavior is observed in a limited proportion of animals and the more common effect is a reduction in the frequency of occurrence of foot clasp mounts.

This deficiency in the expression of the mature form of the mount without a concomitant decrease in all mounts agrees well with work on the effects of early social deprivation on mounting behavior in the guinea pig (Valenstein, Riss and Young

1955; Gerall 1963), and the dog (Beach 1968). It is of interest that total social isolation was used in these other species to produce deficiencies in the expression of the adult mounting pattern, whereas only a reduction in the amount of social contact is sufficient to produce similar deficiencies in the rhesus. This suggests that a complex social interaction is necessary for the development of the rhesus copulatory pattern and a relatively moderate alteration in the social environment is sufficient to disrupt or inhibit the development of the foot clasp mount.

The foot clasp mount is unique in the young rhesus male's behavioral repertoire in that, unlike aggression, most forms of play, or even an improperly oriented mount, it depends upon appropriate behavioral responses from both animals participating in the behavior. As seen in figure 2, the recipient of a foot clasp mount must assume the proper stance and support the actor's weight. Figure 7 demonstrates that if the recipient fails to cooperate, then the foot clasp mount cannot be properly executed. This illustration also demonstrates that the foot clasping motor pattern can be present in the behavior of animals which display mainly abortive mounts. One is led to conclude that the expression of the foot clasp mount is the product of an interdependent interaction between two animals.

Willingness to hold the accommodating posture for a foot clasp mount is crucial

FIGURE 7. *An abortive mount as displayed by a one-year-old male.*

to the display of the mount and, we feel, also to its development. Failure to accommodate a mount could occur for many reasons. Several possibilities are fear, misreading another individual's intentions, or general social incompetence. We feel that whether or not an individual accommodates another animal's mount depends in part on what the recipient perceives the actor's intentions to be. Although we cannot present any empirical data which directly supports this contention, it seems likely that an animal which fears that it may be aggressed or which interprets any approach as a play initiation will not display the posture necessary to accommodate a foot clasp mount. Unfortunately, an independent measure of the willingness to accommodate a foot clasp mount has not been obtained in the circumstances under which the animals in this study were observed. We therefore are unable to determine which individual is responsible for the observed deficiency in foot clasp mounting. We can only infer that foot clasp mounting occurs in an appropriate social environment which allows for the behavioral development of both the actor and the recipient so that they interact in a manner that does not inhibit the expression of this prototypical copulatory mount. Experiments by Testa and Mack (this volume) on *Macaca fascicularis* support our speculation. Thus the occurrence of the foot clasp mount appears to be the end product of many underlying social variables and therefore seems to be indicative of appropriate social development.

In consideration of the ontogeny of mounting behavior of the rhesus monkey, Harlow (1965) has described a constellation of male and female behaviors which he suggests are basic response patterns for the rhesus from which heterosexual behavior is derived. These behaviors include threat, passivity, withdrawal, rigidity, and play. Harlow has further hypothesized that it is during the display of these behaviors that basic postures occur which increase the likelihood of sexual behavior (Harlow 1965). The data presented in this paper provide little support for this hypothesis, especially with regard to play behavior. There appears to be no positive relationship between the frequency or proportion of males engaging in play and their subsequent proficiency at mounting. Males were observed in the limited social access group which had extremely high frequencies of play and yet were never observed to display a foot clasp mount. It is also of interest that in studies which compared either feral (Mason 1960) or socially experienced (Anderson and Mason 1974) rhesus males with socially restricted males, one consistent finding was a reduction in the frequency of play among socially experienced males. These socially competent males also exhibited more mounting behavior. Such results suggest the possibility that too much as well as too little play may be indicative of a deficient social environment. Although we have observed many males which play but do not mount, we have not observed any that mount but do not play. This suggests that the relationship between mounting and play is not a clear-cut one, and it could be argued that once the mount pattern has differentiated from a basic response pattern—play in this case—there is no longer any relationship between the two behaviors. If this is the case, and our data cannot rule this out, one still needs to explain what is absent from the environment in which play does occur but foot clasp mounting does not.

It has been suggested (Goy, Wallen, and Goldfoot 1974) that there is not one prototypical response pattern missing from the behavioral repertoire of the limited social access males which is present in either the increased social access males or the mother-peer-reared males. It is more likely that it is the lack of the formation of positive social interactions between peers which inhibits the expression of foot clasp mounting. The low frequency of grooming exhibited by limited social access males in this present study has similarities to a situation described by Marler (1965). He has suggested, in an intraspecific context, that old world monkeys with clearly established dominance relations exhibit higher frequencies of grooming than those in which dominance relations are tenuous and undecided. Only with the addition of an extended period of 24 hours per day contact does the frequency of grooming become appreciable, leading to the conclusion that dominance relations become stable when a large amount of time is available during which conflicts can be resolved. The 30 minutes per day for social contact available to the limited social access males may have been insufficient for conflict resolution so that the dominance relations had to be continuously reaffirmed on successive test days, resulting in exaggerated aggressive and submissive interactions. The low occurrence of grooming and high incidence of aggression observed in the limited social access peer groups are suggestive of an environment in which animals lack positive interindividual interactions and must deal with problems of uncertain dominance status.

The opposite situation occurs in the mother-peer-reared groups in which the dominance relations between individuals are clear to a trained observer, yet in many cases it is difficult to assign complete dominance hierarchies using the fear grimace matrix because this behavior occurs so infrequently. Apparently, cues more subtle than the fear grimace or overt aggression are sufficient to convey and reinforce dominance relations in socialized individuals, whereas socially limited animals either use or respond to only the most exaggerated and stereotyped signals. The ability to respond to more subtle social cues appears to be a necessary aspect of the environment in which a complex interactive behavior such as the foot clasp mount can develop. The mother-peer-rearing condition described in this chapter appears to allow the development of this aspect of social competence.

The manner in which mother-peer-reared animals handle dominance conflicts has led us to consider the mother's influence on the development of positive social interactions between peers. Marler and Gordon (1968: 117) have stated "that the growing infant of several species is buffered from more violent social interactions by direct intervention of the mother, and that it gains a great variety of social experiences long before it is called upon to engage in aggressive encounters..."

Intervention by the mother is frequently observed in our mother-peer groups, not only in potentially aggressive encounters, but also in bouts of vigorous play. We suspect that the end result of this maternal intervention is to prevent the infants from learning to fear the approach of another infant. A mother-peer-reared infant, upon being approached by another infant, either retreats to its mother or interacts actively with the approaching infant. Rigidity or passivity of the type described by Harlow

(1965) and purported to be necessary for the development of heterosexual behavior is rarely seen when the mother is present. Its frequent occurrence in the peer-rearing situation is probably a result of the infant having no "safe haven" to which it can run when frightened or bullied by a more dominant animal. The rigidity response is probably a behavioral mechanism by which the peer-reared infant copes with the often overwhelming demands presented by its social environment. In the process of coping with its social environment, the peer-reared infant learns behavioral responses to other monkeys which ultimately interfere with successful copulatory behavior.

While males reared with mothers and peers during the first year of life display high frequencies of foot clasp mounts, females raised in the same rearing condition do not; therefore a sexual dimorphism in foot clasp mounting exists across all social environments studied. Studies of the manner in which the display of foot clasp mounting by prenatally androgen treated females responds to the increased social interaction of the mother-peer environment are valuable to an understanding of the organizing effects of androgen.

Data collected in this laboratory by J. C. Slimp and R. W. Goy demonstrate that the display of foot clasp mounting by pseudohermaphroditic females responds to social environment in a manner similar to that reported for males. When pseudohermaphroditic females are reared in the mother-peer-rearing condition, they display frequencies of foot clasp mounts not different from mother-peer-reared males and higher than peer-reared pseudohermaphrodites. Prenatal exposure of females to androgen results in a predisposition to display foot clasp mounts which is expressed according to the constraints of the social environment. If the social environment results in low mounting frequencies for males, pseudohermaphrodites will be similarly affected. Likewise, if the rearing condition is conducive to high levels of mounting by males as in the mother-peer-reared monkeys, pseudohermaphrodites will also display high levels of mounting.

In contrast to their importance to adult sexual behavior, the testes appear not to have a major influence on the mounting of prepubertal males. Castration of males does not alter the expression of foot clasp mounting; castrates mount at levels similar to those of comparably reared intact males in any rearing condition studied. This indicates that the testes are not necessary for the display of mounting by prepubertal males. Alternatively, neither are they responsible for the deficiencies observed in peer-reared males (Goy and Goldfoot 1974). Furthermore, evidence exists that testosterone is incapable of compensating for socially mediated mounting deficits. Bielert (1974) has demonstrated that testosterone injections administered to 1-year-old peer-reared male rhesus do not alter markedly their mount frequencies within the peer group setting.

In summary, the work presented here indicates that prenatal endocrine factors play a larger role in determining sex-typed behavioral predispositions of the individual in later life than do early postnatal or prepubertal endocrine states. Behavioral patterns can be readily altered and modified by social variables, however, although the extent of experiential influences appears to be limited by the endocrine environment experienced by the individual during prenatal life.

■ ACKNOWLEDGMENTS

This is Publication number 16-017 of the Wisconsin Regional Primate Research Center.

This work was supported by grant RR-00167 to the Wisconsin Regional Primate Research Center from the National Institutes of Health, and grant MH-21312 from the National Institute of Mental Health. Thanks are due to Mary Collins, Jane Cords, Jens Jensen, and Warren Schmidt for their assistance in the collection of these data and to Elizabeth Shirah for her preparation of the manuscript. Special thanks are due to Robert W. Goy and David A. Goldfoot for their support and advice during the preparation of this manuscript.

■ NOTE

1. Two of the males (#4845 and #4850) reported on in Goy and Goldfoot (1974) are not included here as they had received a 1-month period of group living between their first and second year of life. It was felt that this period was too short to include them with males receiving 6 months of group living, but since it may have had some effect on their sexual development, they were not included with the limited social access males.

■ REFERENCES

Anderson, C. O., and Mason, W. A. 1974. Early experience and complexity of social organization in groups of young rhesus monkeys (*Macaca mulatta*). *J. Comp. Physiol. Psychol.* 87:681–690.

Altmann, S. A. 1962. A field study of the sociobiology of rhesus monkeys (*Macaca mulatta*). *Ann. New York Acad. Sci.* 102:338–435.

Barraclough, C. A., and Gorski, R. A. 1962. Studies on mating behavior in the androgen-sterilized female rat in relation to the hypothalamic regulation of sexual behavior. *J. Endocrinol.* 25:175–182.

Beach, F. A. 1968. Coital behavior in dogs. III. Effects of early isolation on mating in males. *Behaviour* 30:217–238.

Bielert, C. F. 1974. The effects of early castration and testosterone propionate treatment on the development and display of behavior patterns by male rhesus monkeys (*Macaca mulatta*). Ph.D. dissertation, Michigan State University.

Carpenter, C. R. 1947. Sexual behavior of free ranging rhesus monkeys (*Macaca mulatta*). I. Specimens, procedures and behavioral characteristics of estrus. II. Periodicity of estrus, homosexuality, autoerotic and non-conformist behavior. *J. Comp. Psychol.* 33:113–142; 143–162.

Conaway, C. H., and Sade, D. S. 1965. The seasonal spermatogenic cycle in free-ranging rhesus monkeys. *Folia Primat.* 3:1–12.

Davidson, J. M., and Levine, S. 1972. Endocrine regulation of behavior. *Ann. Rev. Physiol.* 34:375–408.

Eaton, G. 1970. Effect of a single prepubertal injection of testosterone propionate on adult bisexual behavior of male hamsters castrated at birth. *Endocrinology* 87:934–940.

Gerall, A. A. 1963. An exploratory study of the effect of social isolation variables on the sexual behaviour of male guinea pigs. *Anim. Behav.* 11:274–282.

Gordon, T. P., and Bernstein, I. S. 1973. Seasonal variation in sexual behavior of all-male rhesus troops. *Amer. J. Phys. Anthrop.* 38:221–226.

Gorski, R. A. 1971. Gonadal hormones and the perinatal development of neuroendocrine function. In *Frontiers in neuroendocrinology*, eds. L. Martini and W. F. Ganong, pp. 237–290. New York: Oxford University Press.

Goy, R. W. 1966. Role of androgens in the establishment and regulation of behavioral sex differences in mammals. *J. Anim. Sci.* (Supplement) 25:21–35.

———. 1968. Organizing effects of androgen on the behaviour of rhesus monkeys. In *Endocrinology and human behavior*, ed. R. P. Michael, pp. 12–31. London: Oxford University Press.

———. 1970a. Experimental control of psychosexuality. *Phil. Trans. Roy. Soc. London Ser. B.* 259:149–162.

———. 1970b. Early hormonal influences on the development of sexual and sex-related behavior. In *The neurosciences: Second study program*, ed. F. O. Schmitt, pp. 196–207. New York: Rockefeller University Press.

Goy, R. W.; Bridson, W.; and Young, W. C. 1964. Period of maximal susceptibility of the prenatal female guinea pig to masculinizing actions of testosterone propionate. *J. Comp. Physiol. Psychol.* 57:166–174.

Goy, R. W., and Goldfoot, D. A. 1973. Hormonal influences on sexually dimorphic behavior. In *Handbook of physiology*, vol. II, section 7, ed. R. O. Greep, pp. 169–186. Washington: American Physiological Society.

———. 1974. Experiential and hormonal factors influencing the development of sexual behavior in the male rhesus monkey. In *The neurosciences: Third study program.*, ed. F. O. Schmitt and F. G. Worden, pp. 571–581. Cambridge: MIT Press.

Goy, R. W., and Phoenix, C. H. 1972. The effects of testosterone propionate administered before birth on the development of behavior in genetic female rhesus monkeys. In *Steroid hormones and brain function*, ed. C. Sawyer and R. Gorski, pp. 193–204. Berkeley: University of California Press.

Goy, R. W., and Resko, J. A. 1972. Gonadal hormones and behavior of normal and pseudohermaphroditic female primates. In *Recent progress in hormone research*, vol. 28, ed. E. B. Astwood, pp. 707–733. New York: Academic Press.

Goy, R. W.; Wallen, K.; and Goldfoot, D. A. 1974. Social factors influencing development of mounting behavior in male rhesus monkeys. In *Reproductive behavior*, eds. W. Montagna and W. A. Sadler, pp. 223–247. New York: Plenum.

Grady, K. L.; Phoenix, C. H.; and Young, W. C. 1965. Role of the developing rat testis in differentiation of the neural tissues mediating male behavior. *J. Comp. Physiol.Psychol.* 59:176–182.

Hanby, J. P., and Brown, C. E. 1974. The development of sociosexual behaviour in Japanese macaques (*Macaca fuscata*). *Behaviour* 49:152–196.

Harlow, H. F. 1965. Sexual behavior in the rhesus monkey. In *Sex and behavior*, ed. F. A. Beach, pp. 234–265. New York: John Wiley and Sons.

Harlow, H. F., and Lauersdorf, H. F. 1974. Sex differences in passion and play. *Perspectives in Biology and Medicine* 17:348–361.

Harlow, H. F.; Joslyn, W. D.; Senko, M. G.; and Dopp, A. 1966. Behavioral aspects of reproduction in primates. *J. Anim. Sci* (Supplement) 25:49–67.

Harris, G. W., and Levine, S. 1965. Sexual differentiation of the brain and its experimental control. *J. Physiol.* 181:379–400.

Hinde, R. A., and Rowell, T. E. 1962. Communication by postures and facial expressions in the rhesus monkey (*Macaca mulatta*). *Proc. Zool. Soc. Lond.* 138:1–21.

Hinde, R. A., and Spencer-Booth, Y. 1971. Effects of brief separation from mother on rhesus monkey. *Science* 173:111–118.

Lindburg, D. G. 1973. Grooming behavior as a regulatory of social interactions in rhesus monkeys. In *Behavioral regulators of behavior in primates*, ed. C. R. Carpenter, pp. 124–148. Lewisburg: Bucknell University Press.

Lisk, R. D. 1973. Hormonal regulation of sexual behavior in polyestrous mammals common to the laboratory. In *Handbook of physiology*, vol. II, section 7, ed. R. O. Greep, pp. 223–260. Washington: American Physiological Society.

Marler, P. 1965. Communication in monkeys and apes. In *Primate behavior: Field studies of monkeys and apes*, ed. I. DeVore, pp. 544–584. New York: Holt, Rinehart, and Winston.

Marler, P., and Gordon, A. 1968. The social environment of infant macaques. In *Biology and behavior: Environmental influences*, ed. D. Glass, pp. 113–129. New York: Rockefeller University Press.

Mason, W. A. 1960. The effects of social restriction on the behavior of rhesus monkeys. I. Free social behavior. *J. Comp. Physiol. Psychol.* 53:582–589.

Michael, R. P., and Saayman, G. 1967. Individual differences in the sexual behavior of male rhesus monkeys (*Macaca mulatta*) under laboratory conditions. *Anim. Behav.* 15:460–466.

Michael, R. P., and Wilson, M. 1974. Effects of castration and hormone replacement in fully adult male rhesus monkeys (*Macaca mulatta*). *Endocrinology* 95:150–159.

Michael, R. P.; Wilson, M.; and Plant, T. M. 1973. Sexual behaviour of male primates and the role of testosterone. In *Comparative ecology and behaviour of primates*, eds. R. P. Michael and J. H. Crook, pp. 236–313. New York: Academic Press.

Missakian, E. A. 1969. Reproductive behavior of socially deprived male rhesus monkeys (*Macaca mulatta*). *J. Comp. Physiol. Psychol.* 69:403–407.

Phoenix, C. H. 1973. The role of testosterone in the sexual behavior of laboratory male rhesus. In *Symp. 4th Int. Congr. Primatol.*, vol. 2, pp. 99–122. Basel: Karger.

Phoenix, C. H.; Goy, R. W.; Gerall, A. R.; and Young, W. C. 1959. Organizing action of prenatally administered testosterone propionate on the tissues mediating mating behavior in the female guinea pig. *Endocrinology* 65: 369–382.

Phoenix, C. H.; Goy, R. W.; and Young, W. C. 1967. Sexual behavior: General aspects. In *Neuroendocrinology*, vol. 2, eds. L. Martini and W. Ganong, pp. 163–196. New York: Academic Press.

Phoenix, C. H.; Slob, A. K.; and Goy, R. W. 1973. Effects of castration and replacement therapy on the sexual behavior of adult male rhesus. *J. Comp. Physiol. Psychol.* 184:472–481.

Poirier, F. E. 1972. *Primate socialization*. New York: Random House.

Resko, J. A. 1970. Androgen secretion by the fetal and neonatal rhesus monkey. *Endocrinology* 87:680–687.

Robinson, J. A.; Scheffler, G.; Eisele, S. G.; and Goy, R. W. 1975. Effects of age and season on sexual behavior and plasma testosterone and dihydrotestosterone concentrations of laboratory-housed male rhesus monkeys (*Macaca mulatta*). *Biology of reproduction* 13:203–210.

Sade, D. S. 1964. Seasonal cycle in the size of testes of free-ranging Macaca mulatta. *Folia Primat.* 2:171–180.

Schelderup-Ebbe, T. 1935. Social behavior of birds. In *A Handbook of social psychology*, vol. 2, ed. C. Murchison, pp. 947–972. New York: Russell and Russell.

Valenstein, E. S.; Riss, W.; and Young, W. C. 1955. Experiential and genetic factors in the organization of sexual behavior in male guinea pigs. *J. Comp. Physiol. Psychol.* 48:397–403.

Vandenbergh, J. G. 1973. Environmental influences on breeding in rhesus monkeys. Symp. 4th Int. Congr. Primat., vol. 2, pp. 1–19. Basel: Karger.

Wells, L. J., and van Wagenan, G. 1954. Androgen-induced female pseudohermaphrodites in the monkey (*Macaca mulatta*): Anatomy of the reproductive organs. *Contr. Embryol. Carnegie Instn.* 35:95–106.

Whalen R. E. 1968. Differentiation of the neural mechanisms which control gonadotropin secretion and sexual behavior. In *Perspectives in reproduction and sexual behavior*, ed. M. Diamond, pp. 303–340. Bloomington: Indiana University Press.

Wilson, A. P., and Vessey, S. H. 1968. Behavior of free-ranging castrated rhesus monkeys. *Folia Primat.* 9:1–14.

Young, W. C. 1961. Hormones and mating behavior. In *Sex and internal secretions*, ed. W. C. Young, pp. 1173–1239 Baltimore: Williams and Wilkins.

———. 1964. The hormones and behavior. In *Comparative biochemistry*, vol. 7, eds. M. Florkin and H. S. Mason, pp. 203–251. New York: Academic Press.

15 | Response to Mother and Stranger: A First Step in Socialization

LEONARD A. ROSENBLUM
STEPHANIE ALPERT

- INTRODUCTION

It is axiomatic, as Rowell (1972: 23) has pointed out, that organized social groups "are based on individual recognition, long term association, and a high degree of cooperation." In addition, a number of writers (e.g., Imanishi 1961; Koford 1963; Rosenblum 1971a) in considering the social structure of primate groups have stressed the enduring significance of the mother-infant bond in influencing the course and outcome of socialization. This emphasis has been particularly apparent when researchers have been able to observe the same groups over a number of years. Ransom and Rowell (1972: 126), for example, in considering baboon interactive patterns, indicate that "many of the special relationships in adults, such as enduring preferences for sexual and social partners, may stem directly from familial relationships recognized by the maternal link."

As investigators of human primates have indicated, this maternal link, or what is more commonly referred to as "attachment," encompasses patterns of behavior which: (a) are directed toward a *specific* individual (e.g., the mother, rather than towards a *class* of conspecifics, e.g., adult females); (b) are directed towards these selected individuals during a protracted segment of the infant's life; and (c) must be manifest in a variety of situations and not be specific to any particular setting. It is our hypothesis that the differentiation of the mother from others generally is the first refined social discrimination to occur and serves as a model for subsequent social discriminations. Thus, we view the infant's initial relationship with the mother as significant not only in terms of her protection and nurturance, but we concur with Mason (1965: 351) that, "the endless process of social adjustment begins with the mother, whose behavior is in many ways representative of the larger group. . . ." Hence recognition and preferential response to mother is postulated as a crucial first step in the socialization of individuals within socially organized groups.

It should be noted further that theoreticians concerned with the development of attachment responses in children have emphasized the significance of a changing responsiveness to unfamiliar conspecifics as an accompaniment to the full development of filial attachments (Rosenblum and Alpert 1974). Indeed most workers consider the corollary of attachment to be an increasing tendency towards relative avoidance of and/or behavioral disturbance in the presence of strangers (Ainsworth 1969, 1972; Morgan and Ricciutti 1969; Tennes and Lampl 1964; Yarrow 1972; see also Lewis and Rosenblum 1974). It may be true that in nonhuman primate groups, infants are not frequently exposed to entirely unknown conspecifics. However, depending upon such factors as group size and structure and the mother's status, various individuals of the group will differ in their degree of familiarity to the infant depending upon its opportunity to interact with them. Thus, response to relatively unfamiliar conspecifics in nonhuman primates should also relate to the emergence of specific attachments and provide a dramatic reflection of the developing recognition of individuals which occurs in the course of socialization.

In light of the considerations presented above, a research program was formulated in order to provide some primary experimental data on several important questions. (1) What factors influence the development of visual recognition and preferential response to the mother in macaque infants? At what age do these occur? Are there sex differences in the onset and intensity of this preference? Are there species differences? Does the social milieu within which the infant develops influence the emergence of preference? (2) Do macaque infants manifest any aversive responses to unfamiliar (i.e., strange) conspecifics comparable to those seen in children? If so, at what age do these occur? Are there sex and species differences? How do responses to strangers correlate with emerging selective responses to mother?

Inasmuch as this research program attempted to synthesize various theoretical and experimental views derived from a broad comparative perspective, the authors feel it is appropriate to make explicit the basic assumptions upon which the research program was formulated. The primary assumption was: the attachment process identified in children, with regard to both selective responsiveness to maternal figures as well as relative aversive reactions to unfamiliar conspecifics, is rooted in the evolution of Anthropoidea and therefore will be manifest in macaque development. This basic assumption may be delineated further into several derivative suppositions. (1) There is considerable adaptive significance for the infant, once moving freely in the environment, to recognize, maintain orientation, and be able to move quickly towards the unique caregiver who will consistently and effectively provide protection and sustenance. (2) Such recognition and selective response does not involve a rapid or immediate "imprinting" in the classical Lorenzian sense, but requires some form of gradual learning. (3) Differential response to or relative avoidance of strangers, when it occurs, can only follow appropriate learned discriminations and must, in addition, involve the maturation of appropriate affective response systems. Thus, the ability to differentiate mother from stranger is necessary, but not sufficient, to evoke avoidance of strangers (Schaffer 1966; Rosenblum and Alpert 1974; Hoffman 1974). (4) The rate of development and magnitude of selective responses to mother, as well as the appearance of aversive responses to strangers, involves the confluence of genetic and ontogenetic factors, e.g., species, sex of infant, and conditions of rearing. (5) Relationships and behavior patterns characteristic of ongoing social interactions in the rearing environment will be reflected in essentially parallel fashion in related, moderately arousing test situations, i.e., the selective responses are not situation specific (Ainsworth 1972; Schaffer and Emerson 1964; Cairns 1972).

The development of recognition and preference for the mother and differential response to strange conspecific females was studied in two species of macaques. Pigtail (*Macaca nemestrina*) and bonnet (*M. radiata*) macaques, which have been studied extensively in our laboratory (Rosenblum 1971b), were the subjects of the current research. Since pigtails and bonnets are members of the same genus, the developmental pace and early behaviors of the infants have much in common. Infants are born with strong reflexes, such as rooting and clasping, which facilitate early survival, and infants of both species spend the first several weeks of life in close contact with mother. Following initial excursions from the mother beginning at about two weeks of age, there is a gradual decline in the levels of mother-infant contact. This decline is parallel for the infants of these two species under comparable laboratory conditions (Rosenblum 1971b).

Considerable research has indicated that members of bonnet groups spend prolonged periods in close proximity and contact with one another whereas such behavior between pigtails is relatively rare (Rosenblum et al. 1964). This species difference in aggregative behavior is maintained under a variety of conditions and, most importantly, continues after the birth of infants. Although infants of these two species have much in common, there are major differences in the maternal behavior of bonnet and pigtail mothers toward their infants, some of which appear to relate to the differences in contact patterns. Bonnet mothers allow others of the group to explore, touch, groom, and remain close to their young infants, whereas pigtail mothers only rarely allow others to do so. Pigtail mothers, early in their infant's life, engage in a greater frequency of protective maternal behaviors than do bonnet mothers. Pigtails guard and restrain their infants more than bonnets during the first eight weeks and they retrieve their infants about twice as often throughout the first twelve weeks of the infant's life. As a consequence of all these patterns, bonnet infants have frequent, close interaction with others of the group far more frequently in early life than do their pigtail counterparts. Thus, differences in maternal behavior may be expected to have a profound impact in shaping many of the features of future social interactions of the developing infant. It may be in this way that species-specific preadult and adult social patterns reflect ontogenetic, as well as phylogenetic factors.

■ STUDIES OF GROUP-REARED INFANTS

The data presented below are the result of a series of tests on a total of 30 macaque infants including 7 female and 6 male pigtails, and 10 female and 7 male bonnets. All were reared in conspecific group settings containing several mothers and their infants. All groups lived in relatively large and complex pens of approximately 7 ft. × 7 ft. × 12 ft. (See Rosenblum and Youngstein 1974 for details.) To assess responsiveness to mothers and strangers across the broadest possible age range, all infants available in the laboratory during the last 30 months were tested; thus, various infants began testing at different ages ranging from 2–41 weeks of age. Some were tested only once, others as many as 10 times until they reached 47 weeks of age. Repeated tests were never less than 3–4 weeks apart. Careful assessment of the data has indicated that neither age at the onset of testing, nor the number of repetitions of tests, had any systematic influence on the age-related data obtained. Within the age blocks presented below, 4–7 male and 6–9 female bonnets were tested. For the pigtail testing, 5 females were tested in the first three age blocks and 3 females in the final one; for males, N = 3,5,4, and 2 subjects respectively for the four age blocks presented. All testing was carried out in a chamber 8 feet long × 18 inches wide × 2 feet high (see Figure 1). The experimenter recorded the infant's behavior, with minimal distraction for the infant, by observing through an angled mirror hung above the chamber.

The two end walls of the chamber contained a large one-way vision glass window. No illumination was provided within the chamber itself; however, each stimulus chamber (an open-mesh transport cage) was brightly illuminated by means of a 20-watt fluorescent fixture. These features were critical because with the one-way glass and the differential lighting the infant was easily able to see the mother, the stranger, or an empty stimulus compartment; stimulus animals, however, could not see the infant within the test chamber. Thus, differential response to the infant by the stimulus animals was effectively eliminated. Whereas olfactory discrimination may play a role in early recognition of mother as has been recently suggested for squirrel monkeys (Kaplan, this volume), such cues could not be utilized in the current setting given the glass barrier and frequent alternation of stimulus animals. Similarly, vocalizations by the stimulus animal, although potential cues, were very intermittent, seemed to be difficult to localize because of the construction of the apparatus, and failed to produce any systematic alteration in infant response.

Prior to testing, a mother-infant dyad was captured and the infant removed from

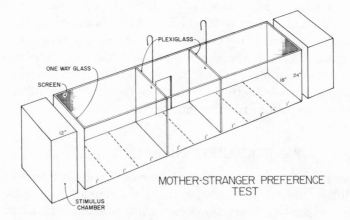

FIGURE 1. *The mother-stranger preference test apparatus. The apparatus had hinged Plexiglas lids covered with fine mesh screening. The floor contained shag carpeting to facilitate locomotion. Observations were made through a mirror hung at about a 45-degree angle, 3 feet above the apparatus (not illustrated).*

the mother while the latter was restrained in a squeeze cage. The mother was then transferred to a stimulus chamber. For pigtail infants, this procedure was slightly modified. Since pigtail mothers, as a result of infant removal, were often violently agitated during testing, whereas strangers generally remained calm, both stimulus animals were mildly tranquilized for testing. Comparisons of response to nontranquilized mothers observed in five infants failed to reveal any effects of tranquilization.

Adult females, from conspecific groups other than the one in which the test infant lived, served as strangers. To start a trial the infant was placed in the start area and allowed to view the stimuli through the two Plexiglas doors. Thirty seconds after the placement of the appropriate stimuli at the ends of the chamber, the Plexiglas doors were raised and the trial begun. Each test day the infant was given seven, 180-second trials. Several minutes elapsed between each trial and during this time the infant was confined once again in the center start area. Each set of trials was begun with an adaptation trial in which the two stimulus chambers were empty. The infant then received two trials in each of three conditions: (1) mother versus empty chamber; (2) stranger versus empty; and (3) mother versus stranger. All conditions were balanced for side on each test day and balanced for sequence across test days, except that the mother-stranger trials always concluded a set of trials on a given day.

When the Plexiglas doors were raised to begin a trial, the observer recorded each entry into, and the total duration of time that the infant spent in each 1-foot segment of the test chamber. However, the primary measure utilized was the average total duration spent in the 1-foot segment closest to the stimulus compartments, i.e., the *choice area*. Except where noted, *t* tests were utilized for statistical evaluations, and two-tailed probability values are presented where appropriate.

MOTHER PREFERENCE IN BONNET MACAQUES

The trials in which mother and stranger were presented simultaneously were run in order to provide a sensitive measure of discrimination of, and preference for, the mother. These preferences were measured in terms of the difference between the average percent of the trial spent in the choice area near the mother and the average percent of the trial spent in the choice area near the stranger (the percent of the trial during which mother was chosen minus the percent of the trial during which the stranger was chosen).

As reflected in figure 2, in the 3–11-week age range bonnet females were spending an average of 25% more time with the mother than the stranger (a *Mean Difference Score* of +25%). This value approached, but did not reach, statistical significance. Males, on the other hand, had a mean difference score of −3.1%. Although females were not showing a clear preference for the mother at this early age, a comparison with the distribution of male scores suggests that females were beginning to show the strong preference which they ultimately manifested later in life. For example, if a minimal difference of 10% between mother and stranger scores is used as a criterion of preference on the part of each animal, four of the five females, but only one of the six males tested in this age range reached this criterion of preference.

As seen in the marked increase in bonnet male and female preference scores

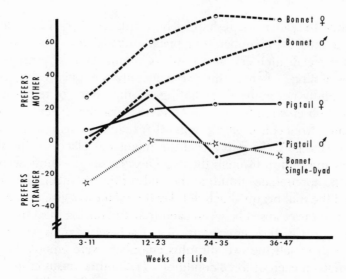

FIGURE 2. *Relative preference in mother-stranger trials shown by group-reared bonnet and pigtail males and females, and single-dyad bonnet infants. Preference was computed by subtracting stranger choice-time from mother choice-time.*

depicted in figure 2, significant preference for mother emerged in both sexes during the 12–23-week period and was sustained in all subsequent age ranges. Females in the 12–23-week age range spent an average of 70.5% of each mother-stranger trial at the mother's end of the chamber and less than 10% of the trial near the stranger, i.e., the females obtained a difference score of 61% (p <.001). (See Figure 3.) Males also showed a significant preference for the mother during this age interval, although the difference between the mother and stranger scores for males was only 32.4%. This difference was significant even though males actually spent less than half the trial in the immediate proximity of the mother, spending a considerable portion of the trial in nonchoice areas. In the 24–35- and 36–47-week age ranges, both males and females increased their significant preference for mother, but females continued to show relatively higher difference scores than males (females $\bar{X}$ = 74.4%; males $\bar{X}$ = 50.4%). Thus in group-reared bonnet infants certain trends emerged. Neither males nor females showed a strong and significant preference for the mother until sometime after the twelfth week of life. Females, however, showed an earlier preference for the mother than did males. Finally, females showed a consistently higher preference for the mother than did males throughout the entire first year.

MOTHER-PREFERENCE IN PIGTAIL MACAQUES

Male and female pigtail infants, as was the case with *male* bonnet infants, gave no indication of preference for their mother during the first 11 weeks of life. The mean differences between mother and stranger scores for females and males were 6.1% and 2.3% respectively. In addition, only two of the five pigtail females and none of the

FIGURE 3. A bonnet female infant, six months of age, in the
mother-choice area in a mother-stranger trial. Note the infant's
close proximity and orientation to her mother. (One-way vision
screen removed for photography.) (Photo by B. Turner.)

three males tested in that age range attained the 10% criterion of preference for the
mother.

In the 12–23-week age range, pigtails of both sexes, like bonnets, showed a
significant preference for the mother. The mean difference score for females was
19.8% (p <.05) and for males was 28.4% (p <.02). Similar to bonnet males, however, pigtail females and males at this age were spending only half the trial in the
choice area next to the mother.

However, in striking contrast to both bonnet males and females, each of whom
showed a continual increase in their preference for mother across the first year, pigtail
preference scores remained the same or decreased with age. Indeed, pigtail male
response to the mother dramatically declined in the second half of the first year. In
fact, they no longer showed any discernible preference for the mother after 24 weeks
of age. Pigtail females also did not show an increase in preference with age but

responded at the same average level in each age block after 12 weeks of age. An analysis of variance of bonnet males and females and of pigtail males and females during 24–47 weeks of life showed that there was a significant species effect ($F = 7.47$; $p < .05$). Therefore, although it is clear that both bonnets and pigtails could discriminate their mothers sometime after the twelfth week of life, pigtails preferred their mothers significantly less than bonnets by the end of the first year ($F = 7.47$; $p < .05$). Moreover, pigtail males after 24 weeks of age showed no preference for their mothers.

RESPONSES TO STRANGERS IN BONNET MACAQUES

Young macaque infants in the first few months of life are either in close contact with the mother or under her close supervision and she regulates virtually all early social interactions of the infant. However, as the infant becomes increasingly independent of her it must develop appropriate responses of its own towards a variety of other conspecifics. As suggested earlier, it is necessary that the infant develop the capacity to respond discriminatively towards these different individuals if socialization is to progress. They may be unfamiliar because the mother avoids them or because they are members of another troop. Although such contact with strangers may not be commonplace, under certain ecological conditions it may occur often enough to make differential responses particularly critical. Hence, we assessed the changing responsiveness of an infant toward strangers as a part of the current research program. The 8-foot-long chamber utilized in this research allowed opportunity for the infant on the "empty" trials ("mother versus empty," "stranger versus empty") to move a considerable distance away from the stimulus animal presented at one end of the chamber.

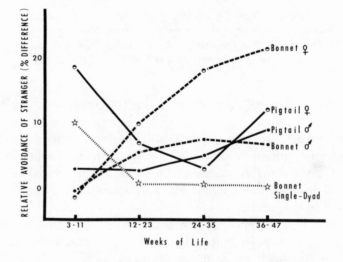

FIGURE 4. *Relative avoidance of stranger shown by group-reared male and female bonnet and pigtail infants and single-dyad bonnet infants. Relative avoidance was computed by subtracting the "empty" choice time in mother-empty trials from the "empty" choice-time in stranger-empty trials.*

We considered the amount of time spent at the empty end of the chamber when mother was present as a measure of basic activity and exploratory interest in that end of the chamber. Therefore, we hypothesized that the difference between the time a given infant spent at the empty end when a stranger was present, and the amount of time spent at the empty end when its mother was present, might provide an indication of that infant's relative avoidance of the stranger. The validity of this measure of avoidance is further supported by the view that exploratory behavior normally is inhibited under conditions of relatively high arousal. Hence, movements to the empty end are less likely to represent exploratory excursions when stranger is present than when mother is present. Direct observations similarly confirmed this view. On tests in which high relative avoidance scores were obtained the stranger-empty trials generally were characterized by an absence of either manual or close visual exploration at the empty end and often involved the subject racing, or even backing, towards

FIGURE 5. *A bonnet female, six months of age, showing avoidance behaviors frequently observed during stranger-empty trials. Note that the infant has moved to the "empty" end, 8 feet from the stranger and is fear-grimacing and screeching. (One-way vision screen removed for photography.) (Photo by B. Turner.)*

the empty end while grimacing or screeching and then maintaining visual fixation on the stranger with tail raised. It should be noted, however, that although the empty trials have been used to assess the degree to which infants at varying ages move more frequently and for longer periods away from strangers as compared to mother, in all infants at all ages the majority of the stranger-empty trials was spent in the proximity of the strange conspecific. This fact suggests that there is a general tendency of young monkeys of these species to seek the company of others when in a strange situation from which an attachment figure is absent. Such attractions may play a role in the general socialization process which ultimately incorporates infants into adult groups. The wariness and relative avoidance discussed below represent the intermittent hesitancy and apprehension of some infants in the presence of a stranger during particular age periods, rather than a continuous or overwhelmingly fearful response to strangers for substantial or sustained portions of the test trials.

From 3–11 weeks of life, both male and female bonnets initially spent almost identical portions of the test period at the empty end of the chamber when either mother or stranger was present. As indicated in figure 4, however, bonnet macaque infants manifested important sex differences after the third month of life in their responsiveness to strangers. For males, the difference between the percent of the trial spent at the empty end when stranger was present and when mother was present, never exceeded 8% at any given age. Females, on the other hand, beginning at 12–23 weeks of life and continuing throughout the rest of the first 47 weeks of life, showed a marked relative avoidance of the strange female when she alone was present (see Figure 5). Females spent significantly more time at the empty end of the chamber when the stranger was present in the last three age intervals tested. These differences were 10%, 19%, and 22% respectively (p <.05; p <.02; p <.05). In fact bonnet females were spending only 1%–2% of the trial at the empty end when mother was present, but more than 10 times this amount of the trial 8 feet away from the stranger.

RESPONSES TO STRANGERS IN PIGTAIL MACAQUES

Both male and female pigtail macaques, in contrast to female bonnets, never showed consistent wariness of strange conspecific adult females. Although in the 3–11-week age range pigtail females on the average did spend more time in the empty end when stranger was present than when mother was present, the scores were too variable to allow us to consider this difference reliable. In all subsequent age blocks, female pigtails gave no indication of a wariness of strangers remotely similar to that of bonnet females. Furthermore, pigtail males showed even less apparent avoidance of strangers than did females throughout the first 47 weeks of life.

The research completed thus far on group-reared male and female bonnet and pigtail infants during the first 47 weeks of life suggests the following initial conclusions. Unequivocal preference for mother is not demonstrated by either bonnet or pigtail infants until after the third month of life. Nonetheless, bonnet females by the third month have begun to show the essential elements of preferential response for mother and a gradual waning of approach to strangers. Bonnet and pigtail infants of both sexes manifest a reliable preference for the mother during the second three months of life. After this age, however, there is a reliable species difference in the magnitude of preference for the mother: as bonnet males and females are showing a continual increase in preference throughout the first year of life, pigtail females maintain relatively low levels of preference and pigtail males show no preference at all. Furthermore, both male and female pigtail infants and bonnet male infants showed no consistent relative avoidance of strange conspecific females. Bonnet females within 1–2 months following the emergence of unequivocal preference for the mother, manifest a relatively strong wariness or avoidance of conspecific strangers. The potential importance of the development of these preference and avoidance patterns can also be understood in terms of some additional experimental data on bonnet mother-infant dyads raised in social isolation.

■ SINGLE-DYAD BONNET INFANTS

In our attempts to differentiate between factors which are primarily genetically determined and those which are socially learned, we examined the development of response to mother and other conspecifics in bonnet infants reared with the mother in isolation. Recent hypotheses regarding selectivity of response in precocial birds in imprinting situations (Hoffman 1974) and mammalian attachment processes (Cairns 1972) have suggested that discrimination and preferential response to mother as well as fear of strangers might be enhanced considerably when infants are raised exclusively with their mothers in isolation. Cairns (ibid.: 58) has suggested that, "Maternal animals in several species isolate themselves and their offspring from others following parturition. Such isolation promotes, among other things, the establishment of individual recognition and mutually supported response patterns... such an outcome seems consistent with the proposal that mutual discrimination develops readily in the course of insulated and exclusive interactions."

Thus far, two male and two female bonnet infants have been raised from birth with only their mothers. These *single-dyad* infants lived in visual isolation from all other animals in pens of approximately 2.5 ft. × 2.5 ft. × 7.0 ft. Each infant was tested at all age points throughout the first 47 weeks (see figure 2). Since the number of animals in each sex was so small, and since there were no indications of any sex differences, the male and female scores were combined for analysis and comparison to other groups.

Testing of the single-dyads revealed response patterns diametrically opposite to those predicted. As indicated in figure 2, single-dyad bonnet infants failed to manifest any overt preferential response to mother in mother-stranger trials throughout the course of the first 47 weeks of life. Considering the four infants together, at no age did the responses to mother differ significantly from those to stranger. For example, in the 24–35-week age range when group-reared bonnet males and females showed peak levels of preferential response to mother, the four single-dyad infants had difference scores ranging from −0.5% to −6.1%. Thus all four infants were actually spending slightly more time with the stranger than the mother. As a consequence of the single-dyad infants' failure to manifest any clear preference for the mother, the difference in responsiveness between single-dyad and group-reared infants was statistically reliable across the full age range (p <.01).

It is worth noting further, that in single-dyad infants the total choice time (time spent in the 1-foot segment next to the stimulus compartments) dropped sharply after the first 6 months of life. They ultimately spent less than half the mother-stranger test period in either choice location. This was in marked contrast to group-reared bonnet infants who spent more than 80% of the test trials in one or another of the choice locations, primarily, of course, in the area immediately adjacent to the mother.

In similar contradiction to prior expectation, the evaluation of the mother-empty and stranger-empty trials indicated that single-dyad infants, both male and female, failed to manifest any degree of wariness or relative avoidance of strangers. As reflected in figure 4, single-dyad infants spent equal amounts of time in the choice area regardless of whether mother or stranger was present. Thus male and female single-dyad infants showed significantly lower avoidance scores than did group-reared bonnets, even when male and female scores were combined (p <.05). Female group-reared infants, however, were largely responsible for this difference.

In summary, then, bonnet infants reared alone with their mothers in the single-dyad situation failed to develop any clear preferential response to mother as compared with stranger. Furthermore, perhaps as a direct corollary of this fact, they failed at any age to manifest any degree of wariness or relative avoidance of conspecific strangers. It is worth noting, in support of this finding, single-dyad squirrel monkey infants, unlike group-reared infants, tested at 6 months of age, failed to manifest any significant preference for mother as compared with stranger (Kaplan and Schusterman 1972).

This lack of differential response to mother and strangers was demonstrated dramatically in a brief home pen observation on one female single-dyad infant, 1 year of age. At the conclusion of a test day, before reuniting the infant with her mother, an unfamiliar adult female was placed in the home pen. The infant was then returned to the pen. The infant immediately ran to the unfamiliar female and repeatedly attempted to make ventral and nipple contact. The stranger threatened, rejected, and punished the infant repeatedly to prevent the infant from making contact with her. After 5 minutes, the mother was returned to the pen. When the mother entered she was preoccupied with the unfamiliar female, and thus it was necessary for the infant to be the initiator in order to achieve contact with the mother. The infant repeatedly looked at her mother and then at the stranger and then back at the mother. The infant then closely explored and sniffed the two females alternately. It was not until the mother retrieved the infant, that the infant was able to make contact with her and was calmed. Thus, the infant's inability to discriminate the mother as seen in the test situation was translated into a type of indiscriminate and potentially maladaptive social behavior in a free response situation.

■ SUMMARY

In bonnet and pigtail macaques reared in complex social groups the mother probably represents the first socializing agent for the infant both in terms of her greater contact with it and also because of her importance for the infant's survival and emotional well-being. Thus, the trends which emerged from the previously described research and outlined below have implications in terms of species-specific social organization as well as class-specific behaviors within a species. In addition, environmental variables can have a profound impact on these two facets of socialization. A brief discussion of more specific aspects of these considerations follows.

First, bonnet, and to a lesser degree, pigtail group-reared infants show unequivocal preference for the mother sometime after the third month of life. The lack of demonstrated preference before that age, in a moderately arousing test situation, implies either an inability or unwillingness on the part of the infant to respond discriminately to the mother when she is not providing differentiating communicative cues. It is suggested that it is not until such discrimination is shown by the infant that socialization may fully progress, since its culmination in a particular role in the group requires differential responses towards various group members.

Second, although bonnet and pigtail infants of both sexes demonstrated preference for the mother after 12 weeks of life, by the forty-seventh week striking species differences emerged. Bonnets showed an increase in preference for the mother as they matured; pigtails showed a decline. In an attempt to speculate on the meaning of this difference it should be noted that the apparatus used allows an assessment of the infant's independent motivation for dyadic contact. Bearing this fact in mind and based on prior work, in these species it is reasonable to suggest that the differences in preference may be the product of the different patterns of bonnet and pigtail maternal behavior (Rosenblum 1971b; see also Rosenblum and Youngstein 1974). In particular, as indicated earlier, bonnet mothers are initially less restrictive and protective than pigtail mothers. In addition, after about 4 months of age pigtail mothers are more rejecting and punitive toward their infants. Thus, throughout the first year, pigtail mothers are more coercive in regulating infant behavior (Rosenblum 1974). Hence, bonnet infants must exhibit more initiative than pigtail infants in promoting and maintaining dyadic proximity and contact, particularly early in life. This different demand on infants of the two species correlates with the differences between bonnet and pigtail social structures within which the infants must assume their developing role. Bonnet group structure compared with pigtails is less clearly hierarchical and one in which relatively fluid patterns of interaction between all group members may occur. In pigtail groups, on the other hand, interactions are more restricted and individual roles are more rigidly defined. Thus, in pigtails, individual behavior is more completely dictated by the total group structure; whereas in bonnets, regulation of behavior appears to reside to a greater degree within each individual. Thus, bonnet infants must ultimately engage in a relatively looser social organization which requires greater social initiative on the part of its members. Pigtails, on the other hand, must be socialized to respond in a relatively fixed social structure which requires a greater adherence to predefined roles on the part of newly maturing animals.

In addition to species differences, important sex differences were found, particularly within the bonnet infants. Bonnet female infants displayed a somewhat earlier and greater preference for the mother and a striking relative avoidance of strange conspecific females. This earlier, stronger, and more enduring bond with the mother may provide the basis upon which female infants of many species are incorporated into the adult female core of the group. Recent hypotheses have suggested the importance of such female subgroupings (Lancaster 1973). The strong avoidance of strangers may be a reflection of the females' hesitancy to establish new relationships, thus strengthening the cohesiveness of the female subgroup, and perhaps it becomes more understandable why females in the wild only rarely are found to leave their natal group. Males, on the other hand, though clearly showing explicit preference for the mother early in life, did not evince such high levels of preference for the mother as females did, and showed a greater readiness to move toward strange conspecifics. Thus for males, the tendency to establish a relatively less intense bond to mother and to show a greater responsiveness to strangers may interact to produce, in many species, the greater potential for peripheralization of males and greater readiness to transfer troop allegiance.

Finally, although our data have suggested certain species and sex differences, the data on single-dyad infants suggest that these differences are neither wholly phylogenetic nor totally dependent upon some sexually differentiated neural substrate. Bonnet single-dyad infants never manifested the high levels of preference for the mother shown by group-reared bonnets or pigtails. In addition, single-dyad infants showed no clear sex differences in their behavior. This suggests that some degree of interaction with other animals, in addition to the mother, during rearing is a prerequisite for the development of differential response to the mother. In addition, the degree of congruence between pigtail group-reared and bonnet single-dyad infants' performance suggests that gradations in the availability of close interactions with nonmother conspecifics will be reflected in the discreteness of infantile discrimination and preference for mother. In this regard, the unusually high levels of affiliative bonnet social interaction to which group-reared infants are exposed may be essential to the patterns of preference and relative avoidance described above and may similarly be critical to the continuance of the characteristic species pattern over successive generations.

■ **ACKNOWLEDGMENTS**

This research was supported by grants #MH 22640 and #MH 15965 from the National Institute of Mental Health Service. The authors wish to express their appreciation to Ms. Barbara Turner for her invaluable assistance.

■ **REFERENCES**

Ainsworth, M. D. S. 1972. Attachment and dependency: A comparison. In *Attachment and dependency*, ed. J. Gewirtz, pp. 97–138. Washington: Winston.
Ainsworth, M. D. S. 1969. Object relations, dependency, and attachment: A theoretical review of the infant-mother relationship. *Child Development* 40:969–1025.
Cairns, R. B. 1972. Attachment and dependency: A psychobiological and social learning synthesis. In *Attachment and dependency*, ed. J. L. Gewirtz pp. 29–80. Washington: Winston.

Hoffman, Howard S. 1974. Fear mediated processes in the context of imprinting. In *Origins of fear*, eds. M. Lewis and L. A. Rosenblum, pp. 25–48. New York: Wiley.

Imanishi, K. 1961. The origin of the human family—A primatological approach. *Jap. Ethnol.* 25:119–130.

Kaplan, J., and Schusterman, R. J. 1972. Social preferences of mother and infant squirrel monkeys following different rearing experiences. *Developmental Psychobiology* 5:53–59.

Koford, C. B. 1963. Ranks of mothers and sons in bonds of rhesus monkeys. *Science* 141:356–357.

Lancaster, J. B. 1973. In praise of the achieving female monkey. *Psychology Today* 7:30–37.

Lewis, M., and Rosenblum, L. A., eds. 1974. *Origins of fear*. New York: Wiley.

Mason, W. A. 1965. The social development of monkeys and apes. In *Primate behavior: Field studies of monkeys and apes*, ed. I. DeVore, pp. 514–543. New York: Holt, Rinehart and Winston.

Morgan, G. A., and Ricciuti, H. N. 1969. Infants responses to strangers during the first year. In *Determinants of infant behavior*, vol. 4, ed. B. M. Foss pp. 253–272 London: Methuen.

Ransom, T. W., and Rowell, T. E. 1972. Early social development of feral baboons. In *Primate socialization*, ed. F. Poirier, pp. 105–144. New York: Random House.

———. 1974. Maternal regulation of infant social interactions. In *Social regulatory mechanisms of primates*, ed. C. R. Carpenter, pp. 195–217. University of Pennsylvania Press.

Rosenblum, L.A. 1971a. Kinship interaction patterns in pigtail and bonnet macaques. *Proc. 3d int. congr. primat.*, vol. 3, pp. 79–84. Basel: S. Karger

———. 1971b. The ontogeny of mother infant behavior in macaques. In *Ontogeny of vertebrate behavior*. ed. H. Moltz, pp. 315–367. New York: Academic Press.

Rosenblum, L. A., and Alpert, S. 1974. Fear of strangers and specificity of attachment in monkeys. In *Origins of fear*. eds. M. Lewis and L. A. Rosenblum, pp. 165–193. New York: Wiley.

Rosenblum, L. A., and Youngstein, K. P. 1974. Developmental changes in compensatory dyadic response in mother and infant monkeys. In *The origins of behavior: The influence of the infant on its caregiver*, eds. M. Lewis and L. A. Rosenblum, pp. 141–161. New York: Wiley.

Rowell, T. E. 1972. *The Social behavior of monkeys*. Baltimore: Penguin.

Schaffer, H. R. 1966. The onset of fear of strangers and the incongruity hypothesis. *Journal of Child Psych. and Psych* 7:95–106.

Schaffer, H. R., and Emerson, P. E. 1964. The development of social attachments in infancy. Monographs of the Society for Research in Child Development, vol. 29, no. 3, serial no. 94.

Tennes, K. H., and Lampl, E. E. 1964. Stranger and separation anxiety in infancy. *Journal of Nervous and Mental Disease* 139:247–254.

Yarrow, L. J. 1972. Attachment and dependency: A developmental perspective. In *Attachment and dependency*, ed. J. Gewirtz, pp. 81–96. Washington: Winston.

16 The Influence of Social Structure on Squirrel Monkey Socialization

LEONARD A. ROSENBLUM
CHRISTOPHER L. COE

- **INTRODUCTION**

Recent reviews of research on primate groups have stressed the importance of ecological and demographic variables as potential modifiers of species-specific group structures (Crook and Gartlan 1966; Strushaker 1969; Eisenberg et al. 1972). However, theories of social organization have not emphasized ontogenetic factors even though laboratory experiments have demonstrated the impact of early experience on the establishment of social affinities and the patterns of social relationships (e.g., Suomi and Harlow 1975). The current authors support the view that the capacity for formative learning allows for flexibility in social organization and, thus, the process of socialization plays a role in the maintenance and evolution of primate social structure (Poirier 1972). Thus, for example, comparisons of isolate-reared versus group-reared juvenile macaques suggest that the early social environment affects the complexity of developing social relations (Anderson and Mason 1974; see also Rosenblum and Alpert, this volume); similarly, the relative exclusivity of early dyadic attachments influences the degree of continued consanguinal association during later development and thereby establishes the kinship basis of group structure in certain species (Rosenblum 1971).

Normal socialization processes must culminate in the ultimate incorporation of young members of the troop into the characteristic adult group structure. Hence, any understanding of these processes must involve a detailed analysis of the structural dynamics which organize adult behavior as a prelude to understanding the progression of infant socialization within that structure. The social organization of squirrel monkeys (*Saimiri sciureus*) represents an excellent model within which this process can be studied.

The social organization of squirrel monkey troops has been classified as intermediate grade in complexity (Crook and Gartlan 1966; Eisenberg et al. 1971) although bisexual groups of up to 300 animals have been reported (Baldwin and Baldwin 1971). A number of important features characterize these *Saimiri* aggregations (paralleled in Cercopithecoidea in *Miopithecus talapoin* [Dixson et al. 1973; Rowell and Dixson 1975]); the adult males travel separately at the periphery of the

479

more numerous and cohesive female group and are organized in an age-graded, multi-male dominance hierarchy. The males tend to avoid other monkeys of different ages and sexes during most of the year (Baldwin 1968, 1971). The spatial and social segregation of the sexes has also been demonstrated in laboratory settings, and the social interaction between the sexes has been described as infrequent and agonistic. This social pattern has been revealed in spatial analyses of dyadic interaction as well as in the evaluation of preferences for social partners (Mason and Epple 1968; Candland et al. 1973).

The field data on *Saimiri* indicate that the females are the social unifiers and leaders of the troop (Baldwin 1971). In addition, adult females are more active than males in initiating those affiliative and agonistic interactions which determine the spatial relations between adult squirrel monkeys (Coe and Rosenblum 1974). Even in the confines of the laboratory the males consistently aggregate only with other adult males and avoid all others, frequently withdrawing from adult females. As a consequence of these male and female behavior patterns, males generally remain spatially segregated from the females of the troop.

This segregation seems to be relatively independent of endogenous hormonal state since the pattern is essentially maintained throughout the year, although the male's disposition for avoidance of adult females changes during the readily demarcated mating season. During this breeding period which occurs at the dry part of the year, males develop a sexually dimorphic "fatting response" which appears to be dependent on testicular androgens (Dumond 1968; Nadler and Rosenblum 1972). As adult females become sexually receptive, males increasingly enter the female part of the troop. Aggression between females and males frequently emerges in association with the males' approaches, and male-male aggression increases as well. In fact, the male's activity at this time has been described as generally disruptive of the social order (Baldwin 1971; Bailey, personal communication) and reflects a transient alteration of the spatial pattern rather than a total shift towards male-female integration. Direct observations of mating behavior in our laboratory indicate that the pattern itself reflects a high degree of intersexual tension since it involves extended chases and false starts before a mount is consummated.

In addition, in studies in our laboratory, adult squirrel monkeys which were gonadectomized after puberty and allowed to live in social groups with intact animals continued to show the pattern of sexual segregation (Coe and Rosenblum 1974). The castrated males and ovariectomized females associated with the spatial clusters of intact adults of their own sex and continued to behave in parallel with them throughout the annual breeding cycle. These results suggest that the fundamental social pattern of segregation and its annual variations are not exclusively dependent on gonadal steroids.

The laboratory studies to be discussed in this section were designed to delineate further the aggregation and dispersion of adult male and female squirrel monkeys. In addition, our objective was to determine the influence of the adult pattern on developing infants (see section II) and the means through which preadult subjects are influenced by, and eventually included in, the adult structure. In Baldwin's study of a semi-free-ranging troop of squirrel monkeys (1971), behavioral distinctions between age groups, such as the peripheralization of subadult males, made it apparent that specifiable juvenile and subadult stages emerged prior to the development of the adult behavior patterns. Bearing this in mind, we studied groups at specific age points, and each group was viewed as a discrete population based on age criteria and containing two subgroups, i.e., males and females. The spatial distribution of each group was used as a basic measure of its internal social organization and also as a variable which could be modified by the presence of, and interaction with, a different sex or age subgroup.

■ THE IMPACT OF ADULTS ON PREADULT SOCIAL STRUCTURE

The three groups of monkeys used in this study each contained 5 males and 5 females, and were classified as adults, subadults, and juveniles. (See Table 1.) All adults were wild-born, of the Peruvian variety, well-acclimated to the laboratory and chosen for

TABLE 1 / DESIGNATION OF STUDY GROUPS AND SUBJECT CHARACTERISTICS

STUDY	DESIGNATION	SUBJECTS	N	GROUP SIZE	WEIGHT (GM)/ ORIGIN	OBSERVATION LENGTH
I	Single-Age: Adults	a) Adult Females	5	10	608/wild	3 weeks (Group
		b) Adult Males	5		835/wild	Formation)
II	Single-Age: Subadults	a) Subadult Females	5	10	654/lab.	4 weeks (Group
		b) Subadult Males	5		680/lab.	Formation)
						1 week (after Study IV)
III	Single-Age: Juveniles	a) Juven. Females	5	10	543/lab.	4 weeks (Group
		b) Juven. Males	5		570/lab.	Formation)
						1 week (after Study) V)
IV	Mixed-Age: Adults + Subadults	Subjects of Studies I + II		20		3 weeks (After Study II)
V	Mixed-Age: Adults + Juveniles	Subjects of Studies I + III		20		3 weeks (After Study III)

similar weights and unfamiliarity with other adults of the same sex. The subadults were laboratory-reared by wild-born mothers within social groups, containing a number of maternal dyads, and were 33–39 months of age when testing began. These subjects were considered subadults based on prior developmental data from our laboratory which indicated puberty for females at about 3 years and for males at about 4 years (Rosenblum 1972). The juveniles were similarly reared and 21–23 months of age at the time of testing. Many subadults were familiar with one another, as were many of the juveniles; however, none of these subjects were familiar with the adults.

   In order to minimize the confounding effect of varying environments, each group of squirrel monkeys was observed in an identical experimental living area which consisted of two pens connected by a tunnel (see Figures 1a and 1b). The arrangement of perches within each pen was identical and designed to facilitate the recording of animal locations and the computation of inter-animal distances. The simultaneous locations of all subjects within a pen were recorded on schematic spatial diagrams. Three diagrams were completed at approximately ten-minute intervals on each observation day, three days per week. The subject locations were numerically coded and computer analyzed to determine the pen location of each subject and the weekly mean distance in centimeters from each type of partner. The inter-animal distances were based only on those occasions when subjects were recorded within the same pen.

   The distribution of animals across the pair of connected pens was also determined

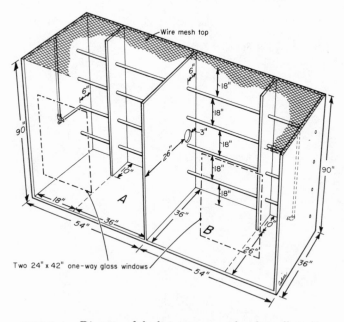

FIGURE 1a. *Diagram of the living areas employed in all studies. Note identical construction of the adjoining pens and the 7-in.-diameter connecting tunnel. For computer analysis, inches were converted to centimeters and all statistical analyses and graphic presentations are in metric units.*

FIGURE 1b. *Two adult males within one pen of the living area and an adult female entering through the connecting tunnel.*

and utilized as an index of aggregation. *Isosexual aggregation* was defined stringently as at least 80% of the males or females within one pen, and the frequency of these occurrences was assessed statistically by comparison to the appropriate Binominal Expansions, i.e., the chance likelihood of such occurrences. *Heterosexual segregation* was similarly defined as the dispersion of at least 80% of the males and 80% of the females in opposite pens, and evaluated in a similar manner.

Additional behavior samplings were obtained in order to provide relevant behavioral data on the patterns of interaction within and between the sex and age groupings. These samples involved 100 seconds of continuous observation of each subject of the group, in random order, during which time the frequency and duration of a variety of predefined behavior patterns were recorded. These observations accompanied the completion of diagrams on each observation day, and each subject was observed for 300 seconds per week. Thus, between 31,000 and 42,000 seconds of data recording were involved in each of the two phases, depending on the number of subjects being observed. Inasmuch as overt aggression was rare, the two opposing behaviors which were most directly pertinent to the observed spacing patterns were (1)

Huddle: a subject sits crouched with its tail curled to its shoulder, while maintaining passive body contact with a partner; and, (2) *Displace:* one subject changes location in immediate response to the approach or aggressive behavior of a partner.

The study consisted of two phases, the single-age groups and the mixed-age groups. During the first phase, the adult group was observed for 3 weeks, and the subadult and juvenile groups were each observed for 4 weeks. In the second phase, the subadult and juvenile groups were each merged separately with the adult group and social interactions were observed for three weeks. In both cases, the adult group was then removed and the subadult or juvenile group observed for one more week in order to assess any residual effects resulting from the experience of living with the adults. Bearing in mind the essential nature of sampling techniques, in which behavioral recordings are made during temporally delimited but representative segments of the total observation time (c.f. "contact-time" in field studies), these studies spanned 19 weeks of observations, approximately 75–100 hours of observation time, and included quantitative samples based on approximately 75,000 seconds of direct recording and nearly 200 spatial location diagrams.

SPATIAL ORGANIZATION OF SINGLE-AGE GROUPS

In the group of ten adult subjects the spatial separation of the sexes developed by the second day after group formation and was maintained throughout the 3 weeks (see Figure 2). During this period the average distance between adults of the same sex was 76 cm, significantly smaller than that between adults of the opposite sex which was 133 cm (Mann-Whitney U, p <.01). Across the 3 weeks, both males and females became increasingly proximal to animals of the same sex but continued to maintain large distances from the opposite sex. Despite the clear-cut separation of the sexes when within the same pen, the tendency to separate did not result in consistent occupancy of opposite pens by the two sexes. The failure of males and females to separate maximally between the two available pens was reflected by insignificant levels of the Aggregation and Segregation Indices. This may have been indicative of the continued motivation of males and females to remain integrated within a sphere of potential interaction. Thus, the male and female subgroups, although spatially separate, were not functioning as discrete social units.

The conclusions drawn from the spatial data were supported by the behavioral observations. Isosexual affinity and proximity were indicated by relatively large within-sex huddling scores. Male-male huddling occurred during 19% of each observation; females huddled together significantly more, i.e., during 45% of the observations. Heterosexual separation, on the other hand, was supported by the total absence of huddling between the sexes. Dispersion of the males and females was also maintained and reinforced by intermittent but consistent female displacement of the males. Each male received more displacements from females than he initiated across the three weeks of observation, and the displaced males usually sustained the spatial separation from the rejecting females.

Neither the subadult group nor the juvenile group manifested the sexual segrega-

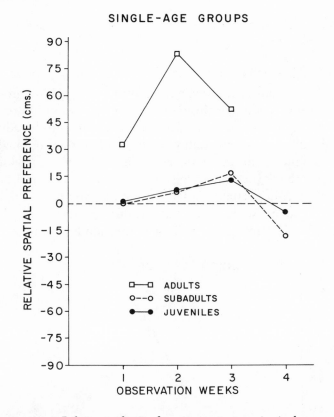

FIGURE 2. *Relative preference for same-sex partners in single-age groups as reflected by differences in average distance scores. Positive scores reflect a greater proximity with same-sex animals.*

tion pattern. As indicated in figure 2, the distances between young monkeys of the same sex and opposite sex were virtually identical. On the average, subadults were 75 cm from monkeys of the same sex and 76 cm from monkeys of the opposite sex. For juveniles, as also indicated in figure 2, the distances between same-sex subjects were 90 cm, different by only 3 cm from the average distance between opposite-sex subjects.

Behavioral observations of the subadult and juvenile groups supported the conclusion from the spatial analyses that no sexual segregation was in evidence. Indeed, positive interaction between the sexes was common. The total time spent in isosexual and heterosexual huddling was similar. However, both types of huddling were infrequent and thus not a good single measure of affiliation. The amounts of hetero- and isosexual displacements were also similar to one another. However, these rather frequent displacements in juveniles and subadults occurred in the context of active social play and did not result in sustained increases in distance between partners. Moreover, unlike adult males who generally were only the recipients of displacements from adult females, preadult males in the course of play frequently approached and displaced their young female peers.

SPATIAL ANALYSIS OF MIXED-AGE GROUPS

Following the observations of the single-age groups, the subadult group was merged with the adult group. At the conclusion of this mixed-age group observation, subadults were removed and the juvenile group was merged with the adults. In each of these mergers the adult monkeys continued to maintain their original social pattern (Figure 3). The same-sex distance for adults averaged 73 cm while opposite-sex distance averaged 138 cm (i.e., an average difference of 55 cm) when in the presence of subadults and juveniles. Huddling levels were also similar to the levels observed when adults were alone. Adult males huddled during 23% of the observations and females during 37%. Heterosexual huddling among adults continued to be very rare.

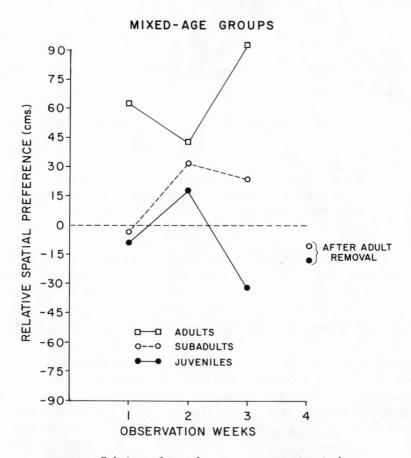

FIGURE 3. *Relative preference for same-sex partners in mixed-age groups as reflected by differences in average distance scores. Positive scores reflect a greater proximity to animals of the same sex and age. Note that the fourth week points for subadults and juveniles are after adult removal.*

In the presence of the adult structure, the subadults responded in a very different manner than did the juveniles. Beginning in the second week of mixed-age grouping, the subadults overcame their initial fear of the adults and increasingly approached them. From this point on, two factors influenced the development of the social structure which ultimately emerged. First, the subadults, primarily the females, frequently approached the adults. Secondly, the adults responded differentially to the approaches of subadults of their own and opposite sex. Adult females were ultimately more tolerant of subadult females than of subadult males. The adult males, though generally avoiding others, appeared more accepting of the subadult males. As a result of those factors there was a significant decrease of 50 cm in the distances between the subadults and adults of the same sex (Friedman test, $p < .01$). There was a correspondingly significant increase of 11 cm in the distance between subadults and adults of opposite sex (Friedman test, $p < .02$) (Figure 4).

The interaction between adults and subadults of the same sex had a dramatic impact on the internal social organization of the subadults. As can be seen in figure 3,

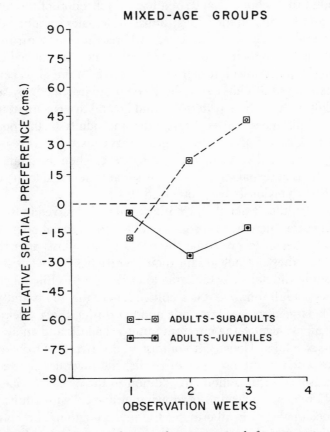

FIGURE 4. *Relative preferences of younger animals for same-sex adults as reflected by differences between same-sex and opposite-sex distance scores. Positive scores reflect a greater proximity with same-sex adults.*

the subadults' spatial pattern underwent a radical and systematic alteration. As the subadults became closer to same-sex adults and further from opposite-sex adults, there was a corresponding increase in the distances maintained between subadults of opposite sex. During the second and third weeks of observation, a sexually segregated subadult distribution emerged. The opposite-sex distances between the subadults were significantly greater than their same-sex distances (Mann-Whitney U, $p < .01$). By the third week, this trend resulted in an average opposite-sex distance of 108 cm as compared to average same-sex distance of 84 cm, i.e., a difference of 24 cm between same-sex and opposite-sex distances.

Behavioral observations supported the finding that social cohesion was developing between subadults and adults of the same sex. Some degree of huddling, increasing over weeks, appeared between subadults and adults of the same sex whereas subadults never huddled with opposite-sex adults. Displacement frequencies between the subadults and adults, however, did not provide a simple explanation for the changing spatial patterns. Subadult females received more displacements from adult females than did subadult males but this appeared to be a reflection of the repeated efforts of subadult females to approach the adults. Subadult females sustained their approaches toward adult females after being displaced, whereas the subadult males rapidly decreased the frequency of their approaches. Adult males, in marked contrast to adult females, did not displace subadults but in fact received more displacements than they initiated. Moreover, adult males were displaced more frequently by subadult females than by subadult males. These patterns of huddle and displacement ultimately led to relatively greater distances between subadults and adults of the opposite sex. The emergence of these mixed-age, same-sex groupings caused the increased sexual segregation among the subadults. As indicated in figure 3, when the adults were removed, the subadult pattern of sexual segregation immediately reverted to the undifferentiated clustering that had previously been observed.

Unlike the subadults, when the juvenile group was observed in the presence of the adult social structure, there was no consistent change in social orientation. Although the juveniles also began to overcome their fear of the adults after the first week of mixed-age testing, they did not sustain their approaches in a consistent manner. In addition, the adults did not show the same tolerance for juveniles of either sex as they had shown for subadult monkeys. As a consequence, the adults and juveniles maintained relatively large distances from one another during the 3 weeks of observation; juveniles were an average of 132 cm from same-sex adults and an average of 121 cm from opposite-sex adults. Hence, in contrast to the significant changes in subadult social structure when in the presence of adults, the juveniles showed no systematic change in their social organization. In addition to these spatial data, the behavioral observations showed no indication of growing affiliation with adults nor changes in the juvenile social pattern. Furthermore, the juvenile group also continued to show an undifferentiated spatial distribution following the removal of the adult group (see Figure 3).

These findings suggest that younger squirrel monkeys initially develop the adult pattern of sexual segregation at least in part through a socialization process involving the redirection of social attachments from age-peers of both sexes to animals of the same sex. The subadults were strongly attracted toward adult monkeys when available, and through a process of differential acceptance by the adults, readily associated themselves with the adult segregated structure. The development of segregation in the brief encounter with adults in this study apparently did not create fundamental changes in the social orientation of subadults towards peers which were sustained after adult removal, although it is likely that longer periods of interaction with adults during development would produce more permanent alterations in the pattern of social affiliations. The fact that juveniles failed to orient in a similar manner in the presence of adults suggests that a certain level of maturation may be necessary in order for this aspect of socialization to occur.

In addition to the finding that age was a critical factor in the readiness for social change, there was a strong indication that males and females also differed in this regard. Subadult females showed a greater readiness to develop interactive and cohesive behavior patterns with same-sex adults and this tendency may be a precursor of the stronger social attachments characteristic of the adult females.

■ THE IMPACT OF ADULT SOCIAL STRUCTURE ON DEVELOPING INFANTS

In order to evaluate the effects of breeding and the presence of infants on the adult segregated pattern, two breeding groups were established in the same experimental pens as reported for the previous study. In addition, the influence of the segregated pattern on developing infants was assessed. Each group initially consisted of 10

TABLE 2 / DESIGNATION OF STUDY GROUPS AND SUBJECT CHARACTERISTICS. NOTE THE WEIGHTS GIVEN FOR THE ADULT MALES IN THE INFANT-DEVELOPMENT GROUP ARE THE WEIGHTS AT GROUP FORMATION SINCE THE MALES UNDERWENT SEASONAL FATTING DURING THE LATER MONTHS OF THIS STUDY.

STUDY	DESIGNATION	SUBJECTS	N	GROUP SIZE	WEIGHT (GM)/ ORIGIN	OBSERVATION LENGTH
I	Breeding	a) Adult Females	16 (8+8)	15	641/wild	3 weeks
	(2 matched groups)	b) Adult Males	6 (3+3)		851/wild	(Group Formation)
		c) OV-X Females	4 (2+2)		536/wild	
		d) Castrate Males	4 (2+2)		737/wild	3 weeks (4 months later)
II	Infant-Development	a) Adult Females	8	28	645/wild	7 months
		b) Adult Males	5		837/wild	
		c) OV-X Females	4		550/wild	
		d) Castrate Males	3		728/wild	
		e) Infant Females	4			
		f) Infant Males	4			

females and 5 males; 2 males and 2 females in each group were gonadectomized (see Table 2). For the purposes of this study the gonadectomized animals were excluded from the analyses. Further reproductive and physiological details regarding these groups will be reported in subsequent publications (see also Coe and Rosenblum 1974).

Initial observations were made when each group was formed and were continued without change until the birth of infants. Mating occurred within four weeks following group formation and resulted in a total of 13 pregnancies in the two groups. Six months later, due to erratic birth success, the two groups were merged in order to form one infant-development group containing 4 male and 4 female infants with their mothers. This merged group also contained 5 intact males, 4 ovariectomized females and 3 castrated males (see Table 2).

Spatial diagrams, identical to those used in the studies described above, were utilized to record the locations of subjects within the experimental pens. Similar procedures were employed for both the separate breeding groups and the subsequently merged infant-development group. These later observations consisted of 6 spatial diagrams per week until the infants were seven months of age. Data from each of these diagrams was computer-analyzed for inter-animal distances and the Segregation and Aggregation Indices defined previously were calculated. In addition, the data from the diagrams in the infant-development group were used to assess mother-infant contact

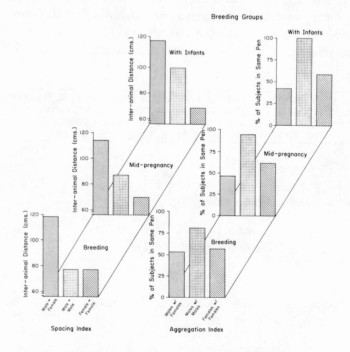

FIGURE 5. Inter-animal distances and aggregative indices for the breeding groups during breeding and mid-pregnancy phases and in the presence of infants.

and distance, the frequency of mother and infant separation between pens, as well as
the frequency of infant proximity (defined as less than 26 cm) to adults of both sexes.
Behavioral data concerning infant interactions with adult males were also recorded.
For this latter purpose, each adult male was observed for 15 minutes per week using a
30-second time-sample check sheet.

Analysis of the spatial diagrams for the breeding and infant-development groups
revealed that these adult subjects manifested the species pattern of sexual segregation.
As indicated in figure 5, the distribution of adults between the two pens was clearly
segregated. Throughout the observations—during breeding, mid-pregnancy, and in
the presence of infants—male aggregative behavior was extremely pronounced
(Binominal Distribution, p <.01). Males were located almost exclusively in one of

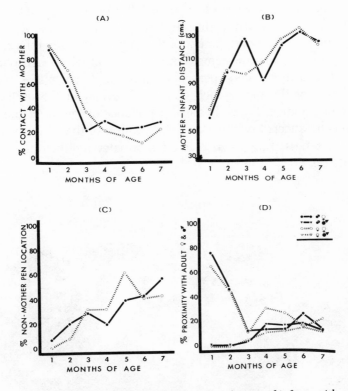

FIGURE 6. Developmental changes in the relations of infants with
their mothers and other adults. Solid line represents infant male
data; and broken line, infant female data.
A. The percent of spatial diagrams on which mothers and infants
   were in contact.
B. The average distance between mothers and infants when apart.
C. The percent of diagrams on which infants were in the opposite
   pen from their mothers.
D. The percent of diagrams on which infants were in proximity to
   adults. Circles indicate scores with non-mother adult females;
   triangles indicate scores with adult males.

the two available pens throughout the study. The number of females in these groups coupled with the generally agonistic behavior of the females towards the males seemed to have been the critical factor in restricting male movement. The females, however, showed greater mobility and consequently were more evenly distributed between the two pens. Due to the greater female movement between pens and the fact that some females regularly were located in the pen which the males habitually occupied, neither the female Aggregation Index nor the overall female-male Segregation Index reached statistically significant levels (see Figure 5). Thus, the distribution of animals between pens reflected the balance between the tendency of males and females to disperse from one another to form isosexual clusters and their continued heterosexual attractions. Within each pen, however, distances between opposite-sex animals were unquestionably greater than distances between same-sex monkeys within every phase of the observations (Mann-Whitney U, p <.01).

It is of interest to examine how this physical environment and the adult segregation pattern influenced the development of the squirrel monkey infants. As reflected in figure 5, the adult males and females rarely came close to each other and often were not even found in the same pen. During the maternal phase, males, in fact, never entered one of the pens which contained several of the mothers and their infants. As a consequence, the early social experience of the developing young may have been radically affected by the spatial distribution of the adults both in terms of the continued closeness of the mothers to other females and the initial inaccessibility of the adult males.

FIGURE 7. *Two adult females huddling between two mothers and their infants. Note the immediate proximity of young infants and familiar females.*

As indicated in figure 6, the development of autonomy in young squirrel monkeys was comparatively rapid. The decreasing percent of the spatial diagrams in which the infants were found in contact with the mother reflected this precocious development (Figure 6A). Contact decreased markedly within the first three months as a result of both the increasing willingness of the mother to allow infant departures and the increasing confidence of the infants. Initially, infant excursions were infrequent but, by three months of age, the infants were spending 80% of the observation time off the mother. In the early excursions the infants did not venture very far but the distances that infants traveled when off the mother increased with age (Figure 6B). In fact, independent infant excursions quadrupled in frequency and doubled in average distance during the course of the first seven months. The increase in exploration of the environment and spatial separation from the mothers was uniquely demonstrated in this environment by the willingness of infants to break visual contact with their mother and enter a different pen (Figure 6C). Initially, neither infants nor their mothers moved into opposite pens from one another, but by seven months of age the infants and their mothers were recorded in opposite pens on an average of 48% of the spatial diagrams.

Observations in this study failed to detect any sex differences between male and

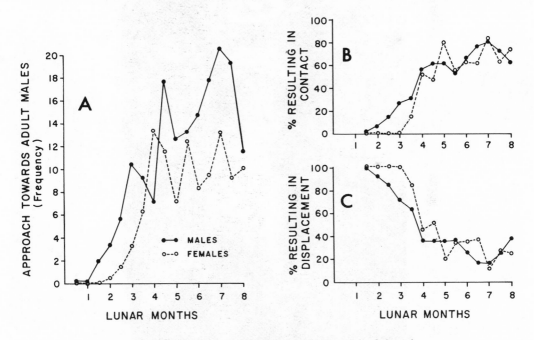

FIGURE 8. *The development of infant interaction with adult males.*
A. *The biweekly frequency of infant approaches towards adult males.*
B. *The percent of infant approaches that resulted in contact with adult males.*
C. *The percent of infant approaches that resulted in displacement of the adult males.*

female infants on the measures of maternal contact and varying degrees of spatial separation from mother. Nonetheless, the adult pattern of sexual segregation had a marked effect on the early experiences of the infants of both sexes. Infants, during the first 2 months of life, primarily associated with their mother and the monkeys which were near to her. Since the mothers were never near adult males, encounters between very young infants and adult males were prevented (Figure 6D). Initially, adult females were almost exclusively the only adults that were observed in proximity with infants (see Figure 7).

However, these high levels of proximity with adult females decreased as the infants developed and as they were spending more time in environmental exploration and play with other infants. At 3 months, when infants were more independent and willing to enter a different pen from their mother, they were spending almost an equal amount of time in proximity with adults of either sex (Figure 6D). Thus, the early

FIGURE 9. *Infant male in contact with an adult male.* (Photo by B. Turner)

social associations of the infants had been restricted by the same-sex preferences of their mothers but this pattern shifted substantially after the third month of life.

As the infants matured, the frequency of their approaches towards adult males significantly increased (F = 15.2, d.f. = 1/6, p <.01) (Figure 8). The development of infant proximity with adult males required a social adjustment both on the part of the infants and the males. Initially, male infants more than female infants approached adult males. However, the response of adult males was to flee rather than to stay near infants of either sex. In fact, almost all of the infant approaches resulted in displacement of the adult males during the first 2 months. Subsequently, the response of the adult males towards infants began to change and infants were allowed to establish proximity and, ultimately, contact. Male infants began to make contact with adult males at 2 months of age, 6 weeks earlier than female infants. However, this sex difference disappeared when the percent of approaches which resulted in contact increased significantly for both sexes (F = 13.1, d.f. = 1/6, p <.02). By 5 months of age, 70% of the infant approaches towards adult males resulted in contact. Male infants approached adult males about once every 19 half-minute intervals of observation and female infants approached them once every thirty intervals (see Figure 9). This sex difference approached, but did not reach, statistical significance because of high within-sex variability.

■ SUMMARY

The groups of adult squirrel monkeys studied in these experiments readily demonstrated the social organization typical of the species. In the groups designated single-age adults and breeding groups, the adults of opposite sexes maintained consistently larger distances between themselves than did adults of the same sex. The spatial analyses were supported by behavioral observations which demonstrated that almost all sustained, affiliative behavior was directed toward same-sex partners. However, the dispersion of male and female adults did not result in a total absence of social interaction between the sexes. In the single-age adult group, which contained 5 males and 5 females, all animals utilized both pens and thus there was a clear potential for encounters between males and females. In the breeding groups, each of which initially contained 10 females and 5 males, the predominance of females resulted in greater antagonism between the sexes and resulted in the males of each group habitually occupying only one of the pens. Yet even in this social setting the females did not choose to remain only in the pen without males and, in fact, intermittently engaged in contact, displacement, food competition, and visual inspection with males. Thus, the females did not maintain a discrete, completely autonomous social group.

Neither subadult nor juvenile squirrel monkeys revealed the adult forms of social and spatial organization. Young males and females readily interacted, and the spatial analyses did not show a discriminative use of space according to sex. However, the readiness to manifest the sexual segregation pattern characteristic of the adults was clearly demonstrated as an age-dependent phenomenon. Subadults, when in the presence of adults, readily gravitated towards same-sex adults and thus the distances between subadults of opposite sexes increased significantly. The juveniles strongly avoided the adults and were not consistently influenced by them.

In these studies, infants that were allowed to develop within the sexually segregated adult structure were affected by the social environment. The initial social experience of infants was primarily within the social milieu of the adult females. The proximity of adult females, the spatial segregation of the sexes, and the avoidance shown by adult males towards infants prevented early interaction between infants and adult males. Nevertheless, the findings of this study suggest that adult males may be an important stimulus for attracting the infants into the larger environment. Infants were strongly motivated to approach adult males as they became increasingly independent. However, adult males showed a marked avoidance of infants initially. In the wild where for most of the year males are located far from females, contact between infants and adult males may be rare and may depend on greater infant initiatives than were required in the current laboratory arrangements.

Male infants were more precocious than females in their encounters with adult males, although no clear sex differences in total occurrence were obtained in this study on any measure. This result is consistent in certain respects with previous findings on squirrel monkeys—that the development of infant independence varies according to environmental setting and the sex of infant (Rosenblum 1974). In small stable groups, each consisting of four mothers and their offspring, no sex differences emerged in the decline of filial attachment and the rise of infant independence. However, male and female infants did differ on these measures when similarly composed groups of age- and sex-matched infants and their mothers were exposed to weekly changes in group membership and the physical environment in which they lived. In this more enriched environmental condition, after 3 months of age, male infants showed a significantly greater readiness than females to leave their mothers.

Thus, in the complex conditions of the wild, after about 3 or more months of age (the approximate age at which infants were beginning to venture into a pen apart from their mother in the current study) male infants may proceed further than female infants towards approaching and interacting with adult males. This sex difference in the potential for adult male-young encounters may encourage and perhaps establish the bases for alliances between animals of the same sex typical of this species.

■ CONCEPTUAL CONSIDERATIONS

A consideration of the data presented above and in the chapter by Rosenblum and Alpert, as well as the relevant material derived from other sources, raises a series of conceptual issues regarding the socialization process and its study. We define socialization as follows: socialization is that process which produces a sustained alteration in the pattern and form of an individual's social relations and which occurs during the course of social interaction. In general, the cumulative consequence of socialization is the intergenerational continuity of species-characteristic behavior patterns. Given this view, and the material currently available, the following considerations and questions require our focused attention in the future.

SOCIALIZATION AS A LIFE-LONG PROCESS

Socialization may not be delimited to a relatively narrow, early period of development but may occur at various junctures during the life span of the individual. The changing pattern of behavior which occurs at puberty, for example, may be as important in determining subsequent social relations as the initial transfer of infant orientation away from the mother and towards peers and other adults. Similarly, modification in role, status, and interaction-networks that accompany the attainment of mature size and reproductive potential must be considered as part of the total socializing experiences of the individual if we are to understand the roles subsequently played by the individual as an adult. It should be clear, however, that this broad view of multiple socialization phases is not incompatible with the view that early socialization events and their outcomes constrain the course of subsequent processes. At this point we can neither fully accept nor completely discard the theoretical view that initial socialization events are the primary determinants of all subsequent social capacities. Indeed, it is among our most urgent needs in this area of research that we examine closely the sequential effects of successive steps in socialization as they unfold during the life span.

AGENTS OF SOCIALIZATION

For socialization to occur, specific individuals of the group must interact with a subject in a manner which coerces or allows the expression of newly emerging behavior patterns. Although the behavior of the infant may be influenced prior to its discrimination of specific individuals, for these agents to exert a more definitive socializing effect, it is critical that the subject in question develop the capacity to differentiate the socializing agent from others in the group. Such discrimination is necessary in order that the subject, ultimately, can respond differently to that agent and perhaps to the class of which that agent is a representative. These agents of socialization are not necessarily restricted to any particular class of individuals. Hence it is critical that our research, even under laboratory conditions, include observations of animals within broadly heterogeneous conspecific groups. Our research should not be limited to observations of restricted age-specific, i.e., peer, groups. The hypothesis that age mates represent a unique source of socializing stimulation, critical to normal social development, is only partially confirmed at present. Thus, as indicated in the study on squirrel monkey socialization, subadult interaction with the segregated adult social structure directly influenced the course of subadult social interaction.

Moreover, the total composition and structure of the social group influences the course of individual socialization by affecting the relative availability of different classes of social partner. Thus, in group-reared bonnet infants, frequent opportunities exist for infant interaction with nonmother adults. Pigtails, on the other hand, lack such opportunities early in life (see chapter 15, this volume). In squirrel monkeys, young infant interactions are largely restricted to nonmother females and only rarely involve interactions with adult males. The importance of the relative availability of social partners raises an additional and very significant conceptual issue. We must be sensitive to the notion that relatively infrequent experiences with particular classes of individuals may have as significant an impact on subsequent behavior as more frequent and perhaps more intimate contact with other individuals. In short, the socializing influence of a given agent may not be a direct corollary of the frequency of agent-subject encounters.

CONFOUNDING VARIABLES IN THE STUDY OF SOCIALIZATION

In the attempt to simplify our understanding of socialization processes, socialization research has tended to focus upon the impact of various factors, social and otherwise, as they influence individuals at various chronological ages. It is critical that we bear in mind, however, that the degree of maturation necessary for receptivity to particular social experiences may not be a consistent correlate of age. Individuals of various species, different sex, and/or those observed within varying environments, may at any age have attained a greater or lesser capacity to incorporate pertinent social information. Hence, if males and females mature at different rates, sex and maturational level will be confounded during early development with regard to their role in socialization. If, in addition, environmental factors influence the respective rates of maturation of the sexes as has been implied in several recent studies (e.g., Rosenblum 1974), the potential for even further confounding is evident.

■ ACKNOWLEDGMENTS

This research was supported by grants #MH 22640 and #MH 15965 from the National Institute of Mental Health Service. The authors wish to express their appreciation to Ms. Barbara Turner for her invaluable assistance.

■ REFERENCES

Anderson, C. O., and Mason, W. A. 1974. Early experience and complexity of social organization in groups of young rhesus monkeys (*Macaca mulatta*). *J. Comp. Physiol. Psych.* 87:681–690.
Baldwin, J. D. 1968. The social behavior of the adult male squirrel monkey in a seminatural environment. *Folia Primat.* 9:281–314.
Baldwin, J. D. 1971. The social organization of a semi-free ranging troop of squirrel monkeys. *Folia Primat.* 14:23–50.
Baldwin, J. D., and Baldwin, S. I. 1971. Squirrel monkeys (*Saimiri*) in natural habitats in Panama, Columbia, Brazil and Peru. *Primates* 12:45–61.
Candland, D. K.; Tyrrell, D. S.; and Wagner, N. M. 1973. Social preference of the squirrel monkey. *Folia Primat.* 19:437–449.
Coe, C. L., and Rosenblum, L. A. 1974. Sexual segregation and its ontogeny in squirrel monkey social structure. *J. Human Evol.* 3:144–154.
Crook, J. H., and Gartlan, J. S. 1966. Evolution of primate societies. *Nature* 210:1200–1203.
Dixson, A. F.; Everitt, B. S.; Herbert, S.; Rugman, S. M.; and Sunton, D. M. 1973. Hormonal determinants of sexual attractiveness and receptivity in rhesus and talapoin monkeys. In *Primate reproductive behavior*, ed. C. H. Phoenix, pp. 30–63. Basel: S. Karger.

Dumond, F. V. 1968. The squirrel monkey in a seminatural environment. In *The squirrel monkey*, eds. L. A. Rosenblum and R. W. Cooper, pp. 88–146. New York: Academic Press.

Eisenberg, J. F.; Muckenhirn, N. A.; and Rudran, R. 1972. The relation between ecology and social structure in primates. *Science* 176:863–874.

Mason, W. A., and Epple, G. 1968. Social organization in experimental groups of *Saimiri* and *Callicebus*. In *Proc. 2d Int. Congress Primat.*, Vol. 1, pp. 59–65. Basel: S. Karger.

Nadler, R. D., and Rosenblum, L. A. 1972. Hormonal regulation of the "fatted" phenomenon in squirrel monkeys. *The Anatomical Record* 173:181–187.

Poirier, F. E. 1972. Introduction to *Primate socialization*, pp. 3–28. New York: Random House.

Rosenblum, L. A. 1971. The ontogeny of mother-infant relations in macaques. In *Ontogeny of vertebrate behavior*, ed. H. Moltz, pp. 315–367. New York: Academic Press.

Rosenblum, L. A. 1971. Kinship interaction patterns in pigtail and bonnet macaques. *Proc. 3rd Int. Congress Primat.*, vol. 3, pp. 79–84. Basel: S. Karger.

Rosenblum, L. A. 1972. Reproduction of the squirrel monkey in the laboratory. In *Breeding primates*. ed. W. I. B. Beveridge, pp. 130–143. Basel: S. Karger.

Rosenblum, L. A. 1974. Sex differences, environmental complexity, and mother-infant relations. *Archives of Sexual Behavior* 3:117–127.

Rowell, T. E., and Dixson, A. F. 1975. Changes in social organization during the breeding season of wild talapoin monkeys. *J. Reprod. Fert.* 43:419–434.

Struhsaker, T. T. 1969. Correlates of ecology and social organization among African Cercopithecines. *Folia Primat.* 11:80–118.

Suomi, S. J., and Harlow, H. F. 1975. The role and reason of peer friendships in rhesus monkeys. In *Peer relations and friendships*, eds. M. Lewis and L. Rosenblum. Origins of behavior, vol. 3, pp. 153–186. New York: Wiley.

Thorington, R. W. 1968. Observations of squirrel monkeys in a Colombian forest. In *The squirrel monkey*, eds. L. A. Rosenblum and R. W. Cooper, pp. 69–87. New York: Academic Press.

17 Some Behavioral Observations of Surrogate- and Mother-Reared Squirrel Monkeys

JOEL KAPLAN

■ INTRODUCTION

It has been only in the last few years that the squirrel monkey (*Saimiri sciureus*) has become increasingly popular for studies dealing with early development, both in the laboratory (e.g., Hopf 1971; Ploog 1969; Rosenblum 1968, 1974; Rosenblum and Lowe 1971) and in seminatural and natural environments (e.g., Baldwin 1969; Baldwin and Baldwin 1971, this volume). Before this almost all of the research that had been conducted in this area with nonhuman primates had utilized the rhesus macaque (*Macaca mulatta*). Although the rhesus is still used extensively, several factors, not the least of which include their high cost and limited supply, have stimulated a great deal of interest in finding other suitable primate models. Moreover, we now know that sufficient variation exists between even closely related species of nonhuman primates—such as the manner in which infants are reared by their mothers and treated by other members of the group—to question the usefulness of generalizations based too heavily on data from only one species. For example, comparisons of pigtail and bonnet monkeys (*Macaca nemestrina* and *Macaca radiata*) have generally indicated that pigtail mothers have a more intense relationship with their infants than do bonnet mothers, displaying a much greater frequency of both protection and rejection types of responses (Rosenblum 1971; Rosenblum and Kaufman 1967). The consequences of such differences are not yet fully apparent, but the finding that pigtail infants are much more disturbed by maternal separation (Rosenblum 1971; Rosenblum and Kaufman 1967) and less responsive to their mothers in preference tests (Rosenblum and Alpert, chapter 15) seems to reflect to some extent these different patterns of maternal care. Results like these imply that responses to other treatments early in development could be influenced by phylogenetic differences, but there is a general lack of comparative data. One interesting observation that has been reported by Berkson et al. (1966) on socially isolated infant marmosets (*Hapale jacchus*) suggests that these rather primitive New World monkeys do not develop the abnormal, stereotyped motor behaviors commonly found in isolation-reared macaques and chimpanzees (Mason 1968; Mason et al. 1968; Mitchell 1970).

The need for making comparisons within the primate order plus relatively recent achievements in squirrel monkey husbandry prompted us to study various aspects of early development in this New World monkey. The present chapter comprises some observations we have made of groups of infants reared with artificial mother substitutes or with real mothers, and is part of a larger program in our laboratory on the effects of early experiences on development. When we began our research in this area we hoped it would help clarify phylogenetic distinctions among primates; since both similarities to, and differences from, previous macaque studies have been found, this approach appears to have been fruitful (Kaplan 1970; 1972).

■ GENERAL ASPECTS OF SQUIRREL MONKEY DEVELOPMENT

Since detailed reports concerning normative aspects of squirrel monkey development have already appeared in the literature (e.g., Hopf 1971; Ploog 1969; Rosenblum 1968), we will only summarize briefly here. A newborn squirrel monkey weighs about 100 grams, one-sixth the weight of the mother. Like most nonhuman primates it has a strong tendency to cling to its mother's fur, normally riding on the mother's back about 90% of the time during the first month of life. As it gets older it gradually spends less and less time on her, so that by the sixth month, over 90% of its time is spent off the mother. The amount of nursing the infant does also subsides over the first six months and is minimal by this age. If a mother allows her infant to continue to nurse, however, it may do so even after it is one year old. Solid foods are sampled toward the end of the second month of life and are consumed in increasing amounts as nursing declines.

At only a few weeks of age the infant first begins to explore the environment while on its mother's back. It reaches out and touches nearby objects which often include other animals. Interactions with peers begin in this manner since mothers often sit next to each other, but these contacts become more intense and involved as infants start to move around by themselves away from their mothers. Although peer contact is initially cautious and exploratory, it rapidly develops into play by the time the infants are two to three months old. After this, the intensity and frequency of play bouts increase until about six months of age and remain at a high level for at least another year.

Even by the end of the first year, when most of an infant's day is spent away from its mother playing with peers, the mother remains the primary source of emotional attachment. The infant often returns to her at moments of heightened emotional excitement and may even continue to sleep on her back overnight. When the mother has her next offspring, which often occurs the following year, the yearling can no longer seek refuge or rest on its mother's body. However, on the basis of a few cases in our laboratory in which young have remained in the same groups as their mother following the birth of a sibling, it was clear that filial ties did not end with the birth of the second baby. Instead, we observed that mothers would continue to aid their older offspring if attempts were made by laboratory personnel to capture it, and that juveniles would often flee to their mother's side under the threat of being captured.

■ BACKGROUND FOR PRESENT STUDIES

Although it has long been recognized that the mother-offspring relationship in primates has a strong influence in shaping development, we now know that much of what the mother contributes can be provided by other sources. Harlow's early studies with rhesus monkeys showed that artificial mothers that provided "contact comfort" also served as a source of emotional security and attachment to infants (Harlow 1958), although infants reared in this manner were abnormal in other respects (Harlow and Harlow 1969; Mason and Berkson 1975; Deutsch and Larsson 1974). However, other studies with rhesus monkeys have shown that alternative rearing practices, such as raising infants only with peers, can produce relatively normal individuals. These differences are probably related to both the quality and quantity of sensory and social stimulation to which infants are exposed. Rhesus infants raised in environments with limited sensory and social stimulation have difficulties in their emotional and social adjustment, the severity of which seems to depend on the nature and degree of deprivation.

The typical artificial mother that has been used in surrogate-rearing experiments with rhesus monkeys (Harlow 1958; Harlow and Suomi 1970; Mason and Berkson 1975) does not provide much in the way of stimulation to the infant. Infants raised on these devices have generally developed abnormal motor and social behaviors even though they may clearly show signs of being strongly attached to their substitute mothers. They clutch themselves, suck on parts of their bodies, rock back and forth, and respond inappropriately in social and sexual situations. Yet when distressed they may seek out and obtain comfort from the familiar surrogate.

FIGURE 1. A—*Three-month-old squirrel monkey on surrogate, and* B—*on real mother.*

In order to see whether similar reactions would also occur in a phylogenetically more primitive primate and at the same time to study the significance of different perceptual factors in the development of attachment (see chapter 7), we raised a group of twelve squirrel monkey infants on surrogates for the first six months of life. The surrogates were made of plexiglas cylinders that approximated the size of a mother's body and were attached to a wall of the home cage. They were heated, provided nourishment from built-in bottles, and were covered with soft acrylic fur material to provide a comfortable surface to which the infants could cling (Kaplan and Russell 1973). Except for some initial training in locating the single nipple that was attached to the one-ounce bottle, and regular bottle changes, no other human assistance was necessary in caring for the infants (Figure 1).

The infants were born in our breeding colony and removed from their natural mothers at approximately one week of age. In training them to locate and nurse on the surrogate's nipple, they were gently forced underneath the surrogate to the nipple's position. If they did not suck voluntarily, the liquid baby formula was forced into their mouths by squeezing the plastic bottle (Kaplan 1974). After a few days of these procedures practically all of the infants learned to nurse by themselves. At approximately two months of age, water-softened monkey biscuits and various fruits were added to their diet.

▪ GENERAL BEHAVIOR OF SURROGATE-REARED SQUIRREL MONKEYS

The growth and general behavior of these infants during the period of surrogate-rearing were very similar to those that had been raised by their own mothers in our laboratory (Kaplan 1972; 1974). The amount of time spent on the surrogate and at the nipple declined at about the same rate as was found earlier for infants raised with mothers under similar living conditions. Like mother-reared infants, they also made low-pitched "purr" sounds prior to nursing and developed the characteristic squirrel monkey genital display and urine wash (see below). Moreover, they developed a strong attachment to their artificial mothers and used them as a source of security in much the same manner as had previously been demonstrated with rhesus monkeys. Strange noises or other disturbances generally caused them to immediately return and cling tightly to their surrogate, and in unfamiliar surroundings away from the home cage, physical contact with the surrogate eliminated distressful vocalizations (Kaplan and Russell 1974). However, unlike any of our mother-reared infants that have never been observed to exhibit any appreciable amount of digit sucking, three of the twelve surrogate-reared infants began to suck their thumbs by the time they were a month old and continued to do so thereafter. This incidence was still much lower than we had found in the past, however, as all of our hand-reared monkeys prior to this study inevitably developed digit-sucking behavior; the major difference in rearing was that in the past our infants were always picked up and fed by hand according to some predetermined schedule, e.g., every four hours. It appeared, therefore, that nonnutritive sucking was greatly reduced by having both the formula and nipple continuously available.

Only one of our surrogate-reared infants exhibited anything that could be distinguished as aberrant behavior. This animal made jerky rubbing movements with his hands and arms over his head and face, generally accompanied by squealing sounds and occasional bouts of self-biting. This appeared to occur most often when the animal was suddenly excited by some form of disturbance but also occurred spontaneously without any apparent provocation. This contrasts markedly with the degree of abnormality found in the surrogate-reared rhesus. Mason and Berkson (1975) reported that nine out of ten rhesus monkeys raised with stationary artificial mothers developed body-rocking as an habitual pattern although this behavior did not occur in infants raised on mechanically driven mobile dummies that provided moving stimulation. The mobile surrogates did not prevent other abnormal responses from developing, however, as monkeys raised on these devices as well as those raised on stationary ones exhibited high frequencies of nonnutritive sucking, self-clasping, and crouching, and were also socially deviant as compared with wild-born monkeys, when tested at about fourteen months of age. Social behaviors were evaluated by pairing surrogate-raised monkeys with wild-born animals that were of similar size and weight. Under these conditions the surrogate-raised monkeys displayed excessive amounts of distress vocalizations, fear grimaces, and withdrawals from social contact, and were also deficient in showing the fully integrated male sexual mounting pattern.

It is unlikely that these differences found between surrogate-reared rhesus and squirrel monkeys were due to differences in rearing procedures. The amount of handling and other forms of stimulation experienced by infants of both of these species in both the Mason and Berkson study and in ours appear to be comparable. The major factor, therefore, would appear to be related to fundamental differences between the two species; a few that are known to exist include differences in maturation rates, mother-infant and group relations, and various aspects of the natural environment. There is, however, a much more basic distinction concerned with the evolution of separate phylogenetic lines of New and Old World species. The New World monkeys, which includes the squirrel monkey, are generally considered to be more primitive and less complex in terms of their structural and behavioral organization than most of the Old World monkeys, which includes the rhesus macaque. The cerebral cortex, for example, is more highly developed in Old World species. What this suggests is that the more primitive monkeys may require less in the way of complex stimulation for normal development and consequently would not be as affected as Old World monkeys by the same degree of deprivation. Of course such a notion is merely speculative and would require a great deal more data from many other species in order to determine its validity. Some support is provided, however, from the work with marmosets mentioned above and from studies on the chimpanzee which, like those on the phylogenetically lower macaque, have shown a variety of abnormal responses developing as a consequence of early isolation conditions (see Mason et al. 1968).

Squirrel monkeys exhibit several kinds of stereotyped behaviors that differ, at least quantitatively, between adults of the two sexes. One such response is the genital display, in which one animal exposes its genitals to another by an outward flexion of the leg (Ploog et al. 1963). Adult males perform this behavior more often than females, especially under conditions of social assertiveness. The response is also basically similar for both sexes, but females do occasionally display by simultaneously spreading both legs outward instead of the more common posture of abducting one leg. This postural dimorphism is seen more clearly in immature animals, and we have studied it to some extent in our own laboratory by testing the response of both infants and adults to their mirror image. Without describing the details of our testing procedures, what we found was that both six-month-old mother- and surrogate-reared females displayed to their mirror image predominantly in the two-legged manner, whereas the majority of displays by adult females were of the one-legged variety. Male infants, on the other hand, displayed in the two-legged fashion much less, regardless of rearing conditions; adult males were never observed to display in the two-legged fashion.

The adults of both sexes also perform a characteristic urine wash to about the same extent, where a few drops of urine are deposited on a hand and rubbed onto the soles of the feet. A third type of stereotyped response consists of chest-kicking with one of the feet. This behavior often accompanies urine-washing, and has only been observed in males (Baldwin 1968).

Of these three responses—genital display, urine-wash, and chest-kick—only the chest-kick was not observed in any of our mother- or surrogate-reared infants. The other two occurred regularly, both prior to weaning (in the home cage with either the surrogate or natural mother and during mirror testing) and after the animals were housed in groups, and did not differ in either quality or quantity on the basis of the different rearing treatments. The reason why chest-kick responses were not observed may be related to the immaturity of our animals. Since only adult males have been observed to respond in this manner, it may be that the behavior is triggered by hormonal changes at puberty.

■ SOCIAL BEHAVIOR OF SURROGATE- AND MOTHER-REARED SQUIRREL MONKEYS

By the time our infants had lived with their surrogates for six months it was clear that this type of rearing was not nearly as damaging as seemed to be the case for the rhesus. However, our data were still rather limited, based mainly on home cage observations. We decided, therefore, to obtain indices of social behavior in our infants for a more complete picture of the effects of their earlier rearing. Several reports had already appeared in the literature on the aberrant social adjustment of rhesus monkeys deprived of early social experiences. Mason and Berkson's (1975) findings with surrogate-reared animals were discussed briefly above, and the results of many other studies dealing with the general topic of early social deprivation have been reviewed in detail by others (e.g., Bronfenbrenner 1968; Mason et al. 1968; Mitchell 1970; Sackett and Ruppenthal 1973).

After the six-month period of surrogate rearing, our infants were housed together as a group and observed over the next fifteen weeks. Each animal was observed for three fifteen-minute sessions each week, in which the total duration and frequency of several individual and interactive behaviors were scored and compared with a separate control group of mother-reared animals that were also weaned and placed together at the age of six months. The surrogate group consisted of eight females and four males, and the mother-reared group of seven females and five males.

The different rearing treatments did not produce any detectable differences in social behavior. Both groups engaged in the same amount and quality of play, were similar with respect to types and degrees of social contact, and were alike with regard to other socially-related responses, e.g., approaches, withdrawals, and aggressive behaviors. There were, however, conspicuous differences in certain socially directed behaviors of males and females, irrespective of prior rearing conditions. The males in each group performed significantly more genital displays and waist grasps (a response in which one animal grasps another around the waist with both arms from the side or the rear) than females in their groups and directed these responses at both males and females alike. These observations, as well as those described above on other stereotypic motor patterns, indicate that at least some responses displayed by *Saimiri* are highly dependent upon genetic "programs" (in certain cases to different degrees in the two sexes) and do not require prior social experience with another animal (either mother or peer) for their expression. This is not to say that experiences cannot influence such responses, for they do develop into rather complex social signals in adulthood which, as far as we know, depend largely on acquired associations with other animals. For example, although males display more often than females at an early age, not until they have learned where they stand with other animals in a group do they concentrate their displays on particular individuals.

After our surrogate- and mother-reared animals had lived in their respective groups for about eight months they were tested to see whether cagemates were preferred to strangers in a free choice situation. It was our feeling that these tests might provide a more sensitive measure of any deviancy in the development of social attachments by the surrogate-reared animals not previously found in our earlier observations of group social behavior. The procedure that was used was one we have used extensively in our laboratory and involves the apparatus shown in figure 2. The apparatus consists simply of four alleys and a central intersection, with cages at the rear of each alley for holding stimulus animals. Tactile cues can be eliminated and olfactory cues minimized by inserting clear plastic windows between the cages and alleys if desired. Preferences are determined by recording the total time spent by a test animal in each of the different choice alleys, usually over a ten-minute period.

In the present series of tests, surrogate- and mother-reared subjects were given several trials in which they were confronted with a cagemate and an animal from the other rearing condition in separate stimulus cages. The two other cages of the ap-

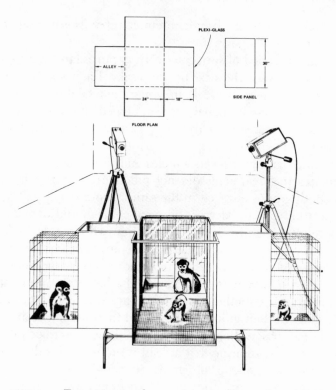

FIGURE 2. *Test apparatus for measuring preferences (cage adjoin-ing 4th alley not shown).*

paratus were always left empty. The two stimulus animals were always of the same sex, although on one-half of the trials they were the same sex as the animal being tested and on the other half of the trials they were of the opposite sex. In addition, to determine whether tactile and possibly olfactory cues played any role in the animals' preferences, we included trials both with and without windows between the cages and alleys. The absence of these windows allowed the test and stimulus animals to achieve close contact through the cage bars.

The results from all of these tests showed that both the surrogate- and mother-reared subjects preferred cagemates to strangers to the same degree (Figure 3). Fur-thermore, alleys that contained stimulus animals were clearly preferred to those that contained only empty cages; about 85% of the total time spent in alleys was spent in those that contained animals. No differences were found for the presence or absence of the window, so close physical contact was apparently not important in the selection of preferred choices. Similarly, the sex of the stimulus animal was not related to preferences. Both same-sex and opposite-sex choices were preferred to the same extent as long as they were from the test animal's own group. This contrasts with the choices of adult squirrel monkeys who have generally been found to prefer members of their own sex (Mason 1971).

The one difference that did stand out, however, for the differently reared groups in

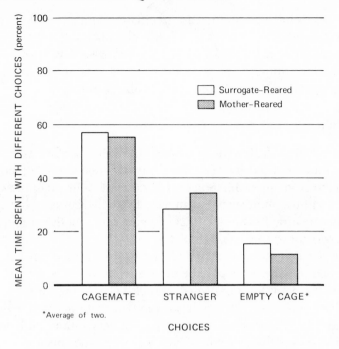

FIGURE 3. *Preferences of surrogate- and mother-reared subjects.*

this test situation actually had nothing to do with their preference behavior. Instead, it was related to differences in general activity levels during the test trials. Although no direct measure of motor activity was recorded, it was possible to evaluate such behavior from the way alley times were scored. No time was recorded when test animals spent less than two seconds in an alley, which was the case whenever there was rapid movement between alleys without any pause to view stimulus animals or empty cages. Since test animals rarely stayed in the central, non-alley, area for extended periods of time, the percentage of total alley time relative to total testing time (ten minutes) was used as an indication of the animals' activity. Using this measure, greater activity would be reflected by lower percentages of alley time. This is exactly what was found after all the test trials were combined, with averages for surrogate- and mother-reared groups being 46.5% and 66.1% respectively, a statistically significant difference using the Mann Whitney U-test. Differences were even greater however during the first few trials. In the first trial, for example, 39% of the surrogate animals' time was spent in alleys, compared with 69% for the mother-reared subjects. As test trials continued the surrogate animals gradually decreased their activity and spent more time in the different alleys. The activity of mother-reared animals, on the other hand, was quite consistent from trial to trial. We are not yet certain what this difference represents but one attractive possibility—especially since differences in activity were not detected in our group-cage observations—is that the novelty of the test situation initially produced higher levels of arousal (Berlyne 1960; Hebb 1955; Malmo 1959) in the surrogate-reared animals. Because of the relatively unstimulating early environment of these animals it seems possible that an increase in environmental stimulation—provided in our case by the novel surroundings of the test chamber and test procedures—might have caused them to be more activated by such changes. This is only speculative, of course, but does have some support from the results and interpretation of other early deprivation-type experiments (Mason 1968; Melzack 1968).

■ CONCLUSIONS

It is obvious that a great deal more work needs to be done before we can establish the extent to which different species react differently to impoverished conditions early in life and the reasons for such differences that do exist. Our preliminary observations suggest that the squirrel monkey is less affected than the rhesus with respect to at least one type of early rearing treatment—that of being raised both without a real mother and without peers for the first six months of life.

Whether other conditions that provide even less stimulation than the surrogate environment used in our study would be more effective in producing abnormalities in squirrel monkeys remains to be answered. It is also clear from the extensive amount of research that has been done with the rhesus monkey that the age at which deprivation is imposed, as well as the length and conditions of deprivation, all have a bearing on the severity of disorders that develop. These variables, too, need to be considered in future studies with other species. We know that our results are far from conclusive. However, they do provide a clue as to what might be expected in comparing species at different phylogenetic levels.

■ ACKNOWLEDGMENTS

This research was supported by USPHS Grant No. HD04905. I wish to thank the following people for their assistance in various aspects of this work: Ann Ball, Barbara Behrens, Ann Lemp, Susan Lingle, Dan McFarland, Nancy Newell, Nancy Nicolson, Robert Pool, and Michael Russell. I also appreciate the helpful comments on the manuscript provided by Dr. W. K. Redican.

■ REFERENCES

Baldwin, J. D. 1968. The social behavior of adult male squirrel monkeys (*Saimiri sciureus*) in a seminatural environment. *Folia Primat.* 9:281–314.
Baldwin, J. D. 1969. The ontogeny of social behavior of squirrel monkeys (*Saimiri sciureus*) in a seminatural environment. *Folia Primat.* 11:35–79.
Baldwin, J. D., and Baldwin, J. I. 1971. Squirrel monkeys (*Saimiri*) in natural habitats in Panama, Colombia, Brazil, and Peru. *Primates* 12:45–61.
Berkson, G.; Goodrich, J.; and Kraft, I. 1966. Abnormal stereotyped movements of marmosets. *Perceptual and Motor Skills* 23:491–498.
Berlyne, D. E. 1960. *Conflict, arousal, curiosity.* New York: McGraw-Hill.
Bronfenbrenner, U. 1968. Early deprivation in mammals: A cross-species analysis. In *Early experience and behavior*, eds. G. Newton and S. Levine, pp. 627–764. Springfield: Charles C Thomas.
Deutsch, J., and Larsson, K. 1974. Model-oriented sexual behavior in surrogate-reared rhesus monkeys. *Brain, Behav. Evol.* 9:157–164.
Harlow, H. F. 1958. The nature of love. *Amer. Psychol.* 13:673–685.

Harlow, H. F., and Harlow, M. K. 1969. Effects of various mother-infant relationships on rhesus monkey behaviors. In *Determinants of infant behaviour*, vol. 4. ed. B. M. Foss, pp. 15–36. London: Methuen.

Harlow, H. F., and Suomi, S. J. 1970. Nature of love—simplified. *Amer. Psychol.* 25:161–168.

Hebb, D. O. 1955. Drives and the c.n.s. (conceptual nervous system). *Psychol. Rev.* 62:243–254.

Hopf, S. 1971. New findings on the ontogeny of social behavior in the squirrel monkey. *Psychiat. Neurol. Neurochir.* 74:21–34.

Kaplan, J. 1970. The effects of separation and reunion on the behavior of mother and infant squirrel monkeys. *Develop. Psychobiol.* 3:43–52.

Kaplan, J. 1972. Differences in the mother-infant relations of squirrel monkeys housed in social and restricted environments. *Develop. Psychobiol.* 5:43–52.

Kaplan, J. 1974. Growth and behavior of surrogate-reared squirrel monkeys. *Develop. Psychobiol.* 7:7–13.

Kaplan J., and Russell, M. 1973. A surrogate for rearing infant squirrel monkeys. *Behav. Res. Meth. and Instru.* 5:379–380.

———. 1974. Olfactory recognition in the infant squirrel monkey. *Develop. Psychobiol.* 7:15–19.

Malmo, R. B. 1959. Activation: A neuropsychological dimension. *Psychol. Rev.* 66:367–386.

Mason, W. A. 1968. Early social deprivation in the nonhuman primates: Implications for human behavior. In *Environmental influences*, ed. D. C. Glass, pp. 70–101. New York: The Rockefeller University Press and Russell Sage Foundation.

Mason, W. A. 1971. Field and laboratory studies of social organization in *Saimiri* and *Callicebus*. In *Primate behavior: Developments in field and laboratory research*, vol. 2. ed. L. A. Rosenblum, pp. 107–137. New York: Academic Press.

Mason, W. A., and Berkson, G. 1975. Effects of maternal mobility on the development of rocking and other behaviors in rhesus monkeys: A study with artificial mothers. *Develop. Psychobiol.* 8:197–211.

Mason, W. A.; Davenport, R. K.; and Menzel, E. W. 1968. Early experience and the social development of rhesus monkeys and chimpanzees. In *Early experience and behavior*, eds. G. Newton and S. Levine, pp. 440–480. Springfield: Charles C Thomas.

Melzack, R. 1968. Early experience: A neuropsychological approach to heredity-environment interactions. In *Early experience and behavior*, eds. G. Newton and S. Levine, pp. 65–82. Springfield: Charles C Thomas.

Mitchell, G. 1970. Abnormal behavior in primates. In *Primate behavior: Developments in field and laboratory research*, vol. 1. ed. L. A. Rosenblum, pp. 195–249. New York: Academic Press.

Ploog, D. W. 1969. Early communication processes in squirrel monkeys. In *Brain and early behavior*, ed. R. J. Robinson, pp. 269–303. New York: Academic Press.

Ploog, D. W.; Blitz, J.; and Ploog, F. 1963. Studies on the social and sexual behavior of the squirrel monkey. *Folia Primat.* 1:29–66.

Rosenblum, L. A. 1968. Mother-infant relations and early behavioral development in the squirrel monkey. In *The squirrel monkey*, eds. L. A. Rosenblum, and R. W. Cooper, pp. 207–233. New York: Academic Press.

———. 1971. The ontogeny of mother-infant relations in macaques. In *Ontogeny of vertebrate behavior*, ed. H. Moltz, pp. 315–367. New York: Academic Press.

———. 1974. Sex differences, environmental complexity, and mother-infant relations. *Archives of Sexual Behavior* 3:117–128.

Rosenblum, L. A., and Kaufman, I. C. 1967. Variations in infant development and response to maternal loss in monkeys. *Am. J. Orthopsychiat.* 38:418–426.

Rosenblum, L. A., and Lowe, A. 1971. The influence of familiarity during rearing on subsequent partner preferences in squirrel monkeys. *Psychon. Sci.* 23:35–37.

Sackett, G. P., and Ruppenthal, G. C. 1973. Development of monkeys after varied experiences during infancy. In *Ethology and development*, ed. S. A. Barnett, pp. 52–87. London: Spastics Intl. Med. Publ.

Part Three

ECOLOGICAL DETERMINANTS OF SOCIALIZATION

18 Socialization, Social Structure, and Ecology of Two Sympatric Species of *Lemur*

ROBERT W. SUSSMAN

■ INTRODUCTION

It is likely that a relationship exists between the habitat preferences of a species and its social structure and methods of socialization. If two closely related species have different ecological preferences and social structures, one would hypothesize that there will also be differences in the socialization of infants and that relationships exist between socialization practices and adult behavior. In some cases, correlates between socialization, ecological adaptations, and social behavior may seem obvious. However, for the most part, they may be extremely obscure and only long-term studies, focused on specific problems, allow us to understand how specific socialization practices help mold individuals into adapted group members.

In this paper I will describe the ecology, social structure, and some aspects of socialization and infant development in two closely related species of *Lemur, Lemur catta* and *Lemur fulvus*. However, since species of *Lemur* give birth seasonally and only during a short period of the year, few data are available yet on infant development and socialization. This paper, therefore, is preliminary and I present it mainly as a possible approach to understanding how different socialization practices may lead to behavioral differences in adults of different species.

The paper is divided into three parts. First, I describe results of an 18-month study of the ecology and social structure of *L. catta* and *L. fulvus* in Madagascar. These two species were studied in forests both where they coexist and where they do not. The second section is a description of the stages of development in the two species and some differences in the way in which infants are reared. Data described in this section are from a number of laboratory and field studies. In the final section I discuss some of the problems involved in attempting to correlate specific aspects of socialization, ecology, and social structure.

Lemuriformes are ideal animals for topical research on the ecology, behavior, and evolutionary history of primates. Isolated on the island of Madagascar since the Eocene, they have undergone a remarkable radiation evolving a wide variety of forms and occupying a number of diverse habitats. Many of these forms show adaptations paralleling those of certain old and new world monkeys. The island of Madagascar is relatively small, and many closely related species can often be found occupying the same forests. By comparing two species occupying the same forest, one can often delineate differences in social behavior and ecological preferences in the two populations.

■ ECOLOGY AND SOCIAL STRUCTURE

The behavior and ecology of two diurnal species of lemur, *Lemur fulvus rufus* and *Lemur catta*, were compared during an 18-month field study in Madagascar. The study was executed in three forests: Antserananomby, Berenty, and Tongobato (Figure 1). At Antserananomby, both species coexist, *L. catta* is found at Berenty, and *L. fulvus rufus* is found at Tongobato. All three forests are primary deciduous forests, with a kily (*Tamarindus indica*) dominated continuous canopy about 7 to 15 m high. Structurally the three forests are quite similar. Intraspecific comparisons of populations at the three forests revealed the degree to which interspecific interactions were directly influencing the behavior of the two species (see Sussman 1972 and 1974 for details, a summary of which follows).

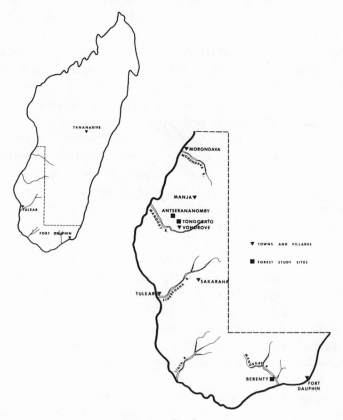

FIGURE 1. *Study sites.*

L. *catta* and L. *fulvus* showed preferences for different habitats, and these preferences did not seem to be influenced by the presence or absence of the other species (Table 1). Groups of L. *fulvus* spent over 70% of their time (during the day) in the continuous canopy and were seen on the ground in less than 2% of the observations. Approximately 85% of group travel was in the canopy, and groups were rarely seen in forest areas which would have necessitated terrestrial locomotion. L. *catta* utilized all of the strata of the forest. For the most part, this species moved and traveled on the ground, rested during the day in low trees, rested at night in the closed canopy, and fed in all of the forest strata. At both study areas L. *catta* spent over 30% of the time on the ground. When groups traveled, approximately 70% of this activity was performed on the ground. The group of L. *catta* at Antserananomby spent 58% of the daylight hours outside the portion of the forest with a continuous canopy, although this represented only 30% of the total area of its range.

The home ranges of groups of L. *fulvus* were very small—approximately 1.0 ha—and overlapped extensively. The borders of the ranges were not defended but, in most cases, groups seemed to maintain spatial separation by means of frequent vocalizations. The groups moved little within a period of a day; day ranges were recorded between 125 and 150 m. The population density was very high at both forests, averaging about 1000/km².

Agonistic group encounters were frequent. Since the population density was very high and groups of L. *fulvus* were often in close proximity, it is likely that these encounters were important in maintaining group coherence. This allowed many groups to utilize resources in close temporal and spatial proximity while group integrity was maintained. Agonistic encounters between members of the same group were infrequent. During such encounters, an animal would simply lunge at or attempt to grab or slap another, and many interactions ended with a brief chase.

The home ranges of groups of L. *catta* were between 6 and 9 ha, and day ranges averaged 950 m.[1] The population density was estimated to be 215/km² at Antserananomby and 250/km² at Berenty. At Berenty in 1963–1964 (Jolly 1966) and at Antserananomby each group of L. *catta* maintained almost exclusive use of its home range, and boundaries of the home ranges of adjacent groups overlapped only slightly. In general, distance was probably maintained between groups by terrier-like barks that were exchanged at various times throughout the day. Intergroup encounters were very rare and, for the most part, when groups noticed each other they both immediately changed directions and moved away. At Berenty in 1970, however, a different pattern of intergroup interactions emerged (Sussman and Richard 1974; Jolly 1972a). In some portions of the reserve, groups shared a large portion of their home ranges. They extensively used the areas of overlap, but were usually found in these areas at different times. Intergroup encounters were more frequent. Although in some of these, groups moved off in opposite directions, many encounters involved agonistic interactions consisting of quick spats, chases, genital and brachial marking, and tail-waving displays between several members of the two groups. The reasons for these different patterns of group interaction are, as yet, not fully understood. Agonistic

		LEMUR CATTA		LEMUR FULVUS RUFUS	
		Berenty	Antserananomby	Antserananomby	Tongobato
1.	Forest level at which the animals spend the greatest percentage of time	Ground (Level 1)	Ground (Level 1)	Closed canopy (Level 4)	Closed canopy (Level 4)
2.	Forest level at which the greatest percentage of travel takes place	Level 1 (71%)	Level 1 (65%)	Level 4 (88%)	Level 4 (83%)
3.	Percentage of time spent on the ground	36	30	0.4	2
4.	Average length of day range	965 m	920 m	125–150 m	—
5.	Average area of home range	6.0 ha	8.8 ha	.75 ha	1.0 ha
6.	Percentage of sightings within the portion of the forest with a closed canopy	—	42%	100%	100%
7.	Population density (per km²)	250	215	1222	900
8.	Percentage of feeding which takes place on the ground	31	28	0	2
9.	Percentage of feeding which takes place in the continuous canopy and emergent layer	12	34	82	81
10.	Number of plant species eaten	24	24	11	8
11.	Percentage of feeding observations in which the animals were seen eating kily leaves	12	11	75	42
12.	Average size of groups	17.50	19.00	9.17	10.20
13.	Range of size of groups	15–20	19	4–15	7–17
14.	Adult sex ratio (♂ : ♀)	1:1.00	1:1.14	1:1.24	1:1.29
15.	Ratio of infants to noninfants	1:3.38	—	—	1:4.66
16.	Ratio of juveniles to adults	—	1:3.75	1:5.11	—

TABLE 1

encounters between members of the same group are frequent in *L. catta*; outside of the mating season there were, on the average, approximately 1–2 per hour (Jolly 1966). The characteristics of these encounters are variable and can range from a stare or slap to a highly ritualized scent-marking display.

Both species live in multi-male groups with an appropriate 1:1 sex ratio. However, groups of *L. catta* were about twice the size of those of *L. fulvus*, averaging 18.8 (N =

8) and 9.5 (N = 17) respectively. The average composition of the groups censused *after the birth season* was as follows:

	Adult male	Adult female	Juvenile	Infant	Total
L. catta (N = 2)	5.5	5.5	2.5	4.0	17.5
L. fulvus (N = 5)	3.4	4.4	0.6	1.8	10.2

In groups of *L. catta* there is a well-defined dominance hierarchy (at least in relation to the outcome of agonistic encounters); there is no noticeable hierarchy in groups of *L. fulvus*. The hierarchical structure in groups of *L. catta* facilitates the division of the large group into subgroups. As a group moves from place to place, the females, juveniles, and dominant males usually move together and ahead of the subordinate males. The subordinate males also tend to feed and rest together, often being joined by some of the older juveniles. Thus the *L. catta* group is frequently divided into subgroups which separate while moving, foraging, or resting. Groups of *L. fulvus*, on the other hand, are very cohesive and regular subgroups were not seen.

The diets of the two species, even where they were studied together, differed radically. The diet of both species was composed of leaves, shoots, flowers, and fruit. but the proportion of these items and the number of species eaten was quite different. The diverse vertical and horizontal ranging pattern of *L. catta* was associated with a varied diet. A group of *L. catta* would usually utilize one part of its home range for about 2–3 days and then change to another part. In a period of 7–10 days the group visited most of its total range. The constant surveillance of a relatively large home range allowed groups of *L. catta* to exploit a number of different plants which had a patchy distribution over a wide area—trees that were blossoming or that had fruit could be located by the group and utilized. In this way the group of *L. catta* at Antserananomby exploited food sources in the brush and scrub vegetation that were not available to *L. fulvus*, in addition to some resources in the portion of the forest with a continuous canopy that were found within the home ranges of only a few groups of *L. fulvus*. Thus, during the dry season at Antserananomby, *L. catta* had access to and ate considerably more species of plants as well as more fruit than *L. fulvus*. *L. catta* was seen feeding on 24 species of plants in both forests and 8 species made up approximately 70% of the diet.

The diet of *L. fulvus* was also composed of leaves, shoots, flowers, and fruit, but consisted of only a few species of plants. In the dry season at Antserananomby, approximately 75% of the observed feeding was done on mature kily leaves. Groups supplemented a diet of kily—leaves, fruit, and flowers—with those species of trees which happened to be within their small home ranges. At both Tongobato and Antserananomby, only 3 species of plants made up over 80% of the diet, with kily accounting for 50% and 75% of the diet respectively.

The major differences between the two species in the three forests are summarized in table 1. The data indicate that the differences in habitat preferences were not the result of the interaction between the two species, but of adaptations to basically different environmental conditions. *L. catta* is the only diurnal species of lemur inhabiting the dry brush and scrub vegetation of the south and southwest of Madagascar and utilizing the dry, rocky, and mountainous areas in the southern portion of the central plateau where only patches of deciduous forest remain. The ranging and foraging pattern of *L. catta* may be related to its ability to cope with these semiarid environments in which the resources are sparse and unevenly distributed. *L. fulvus*, on the other hand, may be more able to efficiently exploit forests in which the resources are abundant and more uniformly distributed.

■ INFANT DEVELOPMENT AND SOCIALIZATION

Lemur catta and *Lemur fulvus* are closely related species with different methods of exploiting the environment and different social structures. Correlated with this, as one would expect, are a number of differences in the stages of development and infant socialization.

As in all species of *Lemur*, mating and births in *L. fulvus* and *L. catta* are seasonal and restricted to a brief period of the year.[2] The dates of the birth season may vary in different localities in Madagascar. In the south, at Berenty, *L. catta* gave birth to infants between about August 20 and September 1 in 1963, 1969, and 1970 (Jolly 1966; Klopfer 1972; Jolly n.d.), whereas in the southwest, near the Mangoky River, in 1970, neither *L. catta* nor *L. fulvus* gave birth to infants until about September 15. Petter's (1962) data indicate that the onset of the birth season in *L. fulvus* may occur even later (mid-October) in the northern part of the island.

The gestation period in *L. catta* and *L. fulvus* is from 120 to 136 days (Petter-Rousseaux 1964; Evans and Goy 1968; Buettner-Janusch 1970). At birth the infants cling to the ventral surface of their mothers. In *L. catta* the infant's body axis is parallel to the mother's. In all other species of *Lemur* the infant is carried with its body axis perpendicular to the body axis of the mother.[3] The body proportions of *L. catta* and *L. fulvus* are essentially the same, and the reasons for the sagittal orientation of the infant's body in *L. catta* are not known. However, it is likely to be related to locomotor and postural differences which may exist between *L. catta* and other species of *Lemur* due to the high percentage of time that *L. catta* spends on the ground. However, detailed film analyses are needed before these differences can be fully understood.

Developmental stages and maternal care in *L. fulvus* and *L. catta* have been studied in the laboratory by Petter-Rousseaux (1962) and Klopfer (Klopfer and Klopfer 1970; Klopfer 1972; Klopfer 1974) and in the field by Jolly (1966, n.d.) and myself. The stages of infant development in these two species cannot easily be divided into discrete units, but are more accurately seen as a series of continuous changes with some degree of individual variation. The major events which occur in both *L. catta* and *L. fulvus* are: the infant clings to the ventral surface of the mother; the infant remains on the back of the mother; the infant begins investigations away from its mother; the infant becomes more and more independent of the mother. During this final stage the infant begins to feed on solid food and is finally weaned. Also, as the infant is becoming independent of its mother, it begins to interact more frequently with other group members.

LEMUR CATTA

Infant *Lemur catta* are far more precocious than infant *Lemur fulvus*. Within three days of birth, *L. catta* infants move actively about on the mother's body. At this age, *L. fulvus* just alternate between the nipple and lap of the mother. The infant *L. catta* is groomed frequently by its mother and by other adult females (especially other mothers) and juveniles. Preliminary observations (Klopfer and Klopfer 1970) in the laboratory indicate that adult males may be less tolerated by the mother than females and less likely to be allowed to groom the young infant. *L. catta* infants regularly ride on the mother's back about a week after birth (Figure 2), and before 2 weeks of age they may venture onto the back of another female or juvenile.

At 2 weeks of age, the infant may attempt some brief, independent exploration, but does not frequently leave its mother until about a month after birth. In laboratory observations, Klopfer (1972) found that the amount of time infants were *not* in contact with their mothers increased from less than 10% in the third week after birth to over 30% by the sixth or seventh week. Also, at about 4–6 weeks old, the infant may taste solid foods and actively engage in play with juveniles and other infants. Mothers

FIGURE 2. *Two-month-old* Lemur catta *infant riding on the back of mother.* (Photograph by A. Schilling)

often exchange or kidnap infants during this period and reciprocal nursing is frequent (Klopfer 1974; Jolly n.d.). In many instances the infant is carried by a female other than its mother for several minutes. This behavior, of course, is similar to what is referred to as "aunting" in other primates and probably functions in much the same way (see Jolly 1972b and Hinde 1974). However, the extent and details of infant sharing in *L. catta* have not been studied.

Although males do tolerate the infants if they tease or jump on them, there is generally little interaction between males and infants at this time. However, in 1970, Jolly (n.d.) observed a male adopt an abandoned 30–40-day-old infant. The male carried the infant for 1½ days until it disappeared, presumably dead of starvation.

After 2 months of age the infants rapidly become more independent. It is during this stage of infant development that most of my field observations were made (beginning November 9 in the south). At about 2–2½ months old, the infant was still carried by its mother when the troop moved. However, whenever she stopped to feed or rest, the infant would get off to play, climb about in the bushes or trees, explore the environment, or taste various foods. It is during this period that infants were first observed to chew solid foods (in the laboratory this was first observed on the fifty-sixth day, as reported by Klopfer and Klopfer in 1970). The mothers frequently left their infants and began moving off without them, but would quickly retrieve the infant if it cried. There were 4 infants in each of the two groups studied at Berenty. The mothers often rested close to each other in the trees or on the ground, and the infants would spend most of the rest periods playing together.

Play groups soon included the juveniles, and by the time the infants were 3 months old, nursery or play groups formed regularly during afternoon rest periods. Between November 19 and November 26 (when the infants were about 3 months old) 15 play groups were observed; all but one of these involved more than one young animal. These play groups always occurred on the ground. Since there were 6 and 7 juveniles and infants in the two groups studied, the play groups were often quite large. The young animals would generally play in the center of the group, surrounded by their mothers and other adults who formed a circle around them, while some adults rested in the trees above (Figure 3). The young animals wrestled and chased each other, often drawing the mothers or other adults (male, as well as female) into the play for short periods. In some instances, most of the animals in the group would enter into the play. The mother still protected the infant during this period, chasing juveniles that played too roughly with her infant.

Between 2½ and 3 months of age, infants began to travel with the group independently of their mothers, especially when the group moved on the ground. The infants usually moved, fed, and sunned together, often with the juveniles. During this period, I first observed an infant (presumably a male) follow an adult male and clumsily imitate the ritualized movements of palmar marking, and a mother chase her infant when it attempted to nurse, although nursing was still frequent. In the laboratory, Klopfer and Klopfer (1970) first observed both of these events earlier—when infants were 8 and 10 weeks old respectively. However, the ages at which particular events first occur are usually quite variable. On the other hand, in comparing the frequencies of certain behaviors of infants in the field and in captivity, Klopfer

FIGURE 3. *Drawing taken from field notes indicating positions of members of a group of* Lemur catta *during afternoon rest session. Play groups invariably developed during these sessions. (Drawing by Barbara Galler)*

(1974) states, "with respect to the specific indices of development which we used, the captive and wild animals appeared strikingly similar."

After about 14 weeks of age, both in the laboratory and in the field, the infants are clearly more a part of a juvenile-infant subgroup than with the mother, although weaning is not complete.

LEMUR FULVUS

The *Lemur fulvus* infant does not begin to move about on the mother until after 2 weeks of age. Klopfer and Klopfer (1970) found that in the laboratory many mothers do not allow other animals to approach the young infant. In the wild, the newborn infant is the center of attention and is groomed frequently by other group members, although the mother does not allow them to hold or take the infant. In both the laboratory and the wild, the infant may attempt to get onto the mother's back after it is 2 weeks old, but is not allowed to regularly ride on her back until it is about a month old.

At 2 months of age, the infants are quite active. They often get off the mother for brief periods while she is resting, but remain in close proximity to her. They are always carried when the mother is moving, and have very little contact with animals other than the mother. In fact, for the first 2 or 3 months the infants are almost exclusively dependent on the mother. Unlike *L. catta* infants, *L. fulvus* infants are not shared during the first month, and infant sharing is very rare thereafter.

It was not until the infant reached 11 or 12 weeks old that it was seen to remain off

FIGURE 4. A *juvenile female* Lemur fulvus rufus *about one year of age.*

its mother for any length of time. This corresponds with observations (Klopfer 1974) on captive animals in which the average age that the infant was seen off its mother for at least 30% of the time was about 80 days. Infants of this age were seen playing alone or with the mother and other adults in the group. A mother was first seen weaning her infant when it was about 11 weeks old, and during this same week an infant was first encouraged (or actually forced) to move independently and make jumps between trees while the group traveled.

At 3 months of age, although the infants are still usually carried by the mothers, they are often made to travel and make jumps by themselves. This is sometimes done at the expense of the group's continued forward progression. The following example is taken from my field notes.

December 29:
8:27. A group leaves a kily tree in which it had been feeding, but an infant remains in the tree. A male goes back into the tree, up to the infant, and moves past it. The infant stays where it is. The other animals of the group have remained in the lower trees surrounding the kily. Two males go to the infant.
8:30. Now a female goes to the infant who attempts, but is not allowed, to get on her back. She slowly moves to the end of the branch and makes the jump to the smaller trees. One of the males follows her, stops at the end of the branch, marks, and then jumps. The second male has remained in the tree near the infant.
8:35. Another male goes to the infant and both adults go to the end of the branch. They stay there and ano-genital mark. The infant comes to the end of the branch. One of the adults jumps, and finally the infant follows. The other male stays in

the tree, marking, and then joins the others. The group then continues its forward progression into the small trees.

After 3 months of age, the amount of time the infants spend playing increases. However, since there are very few young animals in most *Lemur fulvus* groups, much of the play is done with adults. Two infants were seen playing together without adults in only 1 out of 8 play groups observed between December 23 and December 30 (when the infants were about 3–3½ months old). Play groups are not formed regularly as in *Lemur catta*, and no subgroups are noticeable.

By 3½ months, although some nursing continues, many of the infants are essentially independent of their mothers, feeding, traveling, and resting by themselves. However, other infants are still being carried, especially when the group travels. Between 3½ and 4 months of age, all of the infants become independent group members, although in the laboratory, nursing has been observed to continue occasionally for about 150 days.

In both *L. catta* and *L. fulvus*, 1-year-old juveniles are not yet adult size (Figure 4). By the age of two years the animals are adult size, but they do not breed until the following breeding season when they are 2½ years old.

■ SUMMARY AND CONCLUSIONS

Although these are only preliminary observations (quantitative and long-term investigations are needed), a number of differences between the socialization and development of *Lemur catta* and *Lemur fulvus* infants have been found. The most obvious of these are the relative precocity of infant *L. catta*, the prevalence of infant sharing in *L. catta* and the regular occurrence of large nursery groups made up of juveniles and infants in groups of *L. catta*.

The differences in the development of infants in these two species are summarized in table 2. These differences are most likely related to the habitat preferences of the two species.

Comparisons between cercopithecines show that among captive animals, infants of arboreal species become independent of their mothers at a later age than do infants of terrestrial species, such as vervets and baboons (Chalmers 1973). In a study focusing specifically on infant development in phylogenetically related species, Chalmers (1972) found *Cercopithecus aethiops* infants to be more precocious than more arboreally adapted species of cercopithecines (*Cercopithecus neglectus*, *C. mitis*, and *Cercocebus albigena*). The infants of the species reported to be the most arboreal (*C. neglectus* and *C. albigena*) spent the least amount of time away from their mothers. Vervet infants remained away from their mothers for longer periods of time and left their mothers at an earlier age than any of the other species. Chalmers' experiments indicate that the differences were not due to different degrees of restrictiveness on the part of the mother, nor to the rate of physical development of the infant. Infants of arboreal species simply did not leave their mothers as frequently as did those of terrestrial species.

TABLE 2

	INFANT REGULARLY RIDES ON MOTHER'S BACK	INFANT OFF MOTHER FOR 30% OF THE TIME	INFANT IS INDEPENDENT, E.G., TRAVELS AND FEEDS ALONE
Lemur catta	1 week	6–7 weeks	2½–3 months
Lemur fulvus	4 weeks	11–12 weeks	3½–4 months

Given the same rate of physical development, it is easy to see why it may be advantageous for infants of an arboreal species to leave their mothers less frequently and at a later age than those of terrestrial species. It is extremely difficult for a small and clumsy infant to move about in the trees. Even if a fall were not fatal, retrieval of a fallen infant could at times be extremely difficult or dangerous. At 3½–4 months old, *L. fulvus* infants were still hesitant and had to be coerced to make jumps that were necessary for the normal progression of the group. On the other hand, there may also be selective pressures on terrestrially adapted species for earlier locomotor independence. In species with long day ranges, the continuous carrying of infants, especially as they become larger and heavier, could be detrimental to the mother as well as the infant.

I would suspect that further studies comparing infant development in closely related terrestrial and arboreal forms (e.g., macaques or langurs) would reveal similar results. However, certain environmental factors may alter these relationships. For example, a very high predator pressure in certain terrestrial habitats may outweigh the selective advantages of having precocious infants. Furthermore, comparisons between species not phylogenetically closely related may not reveal the same relationships between terrestriality and precocity as those found between various species of *Cercopithecus* and *Lemur*. For example, there are too many factors which determine the rate of development to assume that terrestrial chimpanzees will have more precocious infants than arboreal lemurs, colobus monkeys, or orangutans.

Relationships between infant sharing and adult behavior are more obscure. Infant sharing has been postulated to have a number of different functions. For example, it has been related to general nonaggressiveness in Nilgiri langurs (Poirier 1970), closer binds between group members in sifakas (Jolly 1966), and the refinement of techniques of mothering abilities in subadult female vervet monkeys (Lancaster 1972). It is difficult to generalize these postulated functions in the above species to primates in general. Although infant sharing is quite rare in *L. fulvus* as compared with *L. catta*, the former is much less aggressive than the latter. Also, at least according to the data presently available, groups of *L. fulvus* are just as cohesive, and mothers of both species are equally skilled in caring for infants. Although subtle differences in these aspects of behavior might exist, they are yet to be demonstrated.

At this point it would be highly speculative to attempt to relate the differences in the interactions of infants with other animals of the group to general characteristics of the social structure of *L. catta* and *L. fulvus*. The juvenile-infant subgroups in *L. catta*, of course, reflect the general pattern of the relatively large group to divide into a number of subunits. After 3 months of age the *L. catta* infant interacts mainly with its peers, whereas the infant *L. fulvus* seldom interacts with animals of its own age. Further studies are needed, however, to determine the roles of specific adults and peers in the development of young, the importance of sex and individual differences, and the effect of genealogical relationship in the development of the species-specific patterns of social structure.

I have attempted to illustrate some of the relationships which exist between the development and socialization of *L. catta* and *L. fulvus* infants and the ecology and social structure of these species. To this purpose, however, I have only been able to give speculations concerning these relationships and, in most instances, to pose questions rather than find answers. Further long-term studies are needed to gain a better understanding of the correlates which exist between various aspects of behavior and ecology. At the present time, attempts are being made to establish protected reserves in Madagascar to ensure the continued survival of these endangered species. If the attempts are successful we may be able to learn more about the socialization, social structure, and ecology of the diurnal prosimians, and to make comparisons between their behavior and that of the anthropoids.

■ NOTES

1. Home ranges of *L. catta* may be as large as 23 ha in very arid environments with sparse vegetation (Budnitz and Dainis, 1975).

2. In fact all the Lemuriformes studied to date breed and give birth seasonally. However, stages of development and most aspects of socialization of the various species are extremely variable. Infant development and socialization have not been extensively studied within the suborder and no attempt will be made in this paper to compare *L. catta* and *L. fulvus* with other species of Lemuriformes.

3. *Lemur variegatus* is an exception in that the infant is left in a nest and is carried in the mother's mouth when moved.

■ ACKNOWLEDGMENTS

I would like to thank the following people for their comments and criticisms of earlier drafts of this paper: Linda Barnes, Marc Bekoff, Jeremy Dahl, Steve Easley, Alison Richard and Bill Sawyer. This study was supported in part by National Institute of Mental Health Research Fellowship MH 46268-01, and by Duke University and Washington University.

■ REFERENCES

Budnitz, N., and Dainis, K. 1975. *Lemur catta*: Ecology and behavior. In *Lemur biology*, eds. I. Tattersall and R. W. Sussman, pp. 219–235. New York: Plenum Press.

Buettner-Janusch, J. 1970. Genetic traits of lemurs. Paper read at the Conférence Internationale sur l'Utilisation Rationelle et la Conservation de la Nature, October, 1970, Tananarive, Madagascar.

Chalmers, N. R. 1972. Comparative aspects of early infant development in some captive cercopithecines. In *Primate socialization*, ed. F. E. Poirier, pp. 63–82. New York: Random House.

———. 1973. Differences in behavior between some arboreal and terrestrial species of African monkeys. In *Comparative ecology and behavior of primates*, eds. R. P. Michael and J. H. Crook, pp. 69–100. London: Academic Press.

Evans, C. S., and Goy, R. W. 1968. Social behavior and reproductive cycles in captive ring-tailed lemurs (*Lemur catta*). *J. Zool. London* 156:181–197.

Hinde, R. A. 1974. *Biological basis of human social behavior*. New York: McGraw-Hill.

Jolly, A. 1966. *Lemur behavior*. Chicago: University of Chicago Press.

———. 1972a. Troop continuity and troop spacing in *Propithecus verreauxi* and *Lemur catta* at Berenty (Madagascar). *Folia Primat.* 17:335–362.

———. 1972b. *The evolution of primate behavior*. New York: Macmillan.

———. n.d. A note on infant adoption by a male *Lemur catta*.

Klopfer, P. 1972. Patterns of maternal care in lemurs: II. Effects of group size and early separation. *Z. Tierpsychol.* 27:227–296.

———. 1974. Mother-young relations in lemurs. In *Prosimian biology*, eds. R. D. Martin, G. A. Doyle, and A. C. Walker, pp. 273–292. London: Duckworth.

Klopfer, P., and Klopfer, M. S. 1970. Patterns of maternal care in lemurs: I. Normative description. *Z. Tierpsychol.* 27:984–996.

Lancaster, J. B. 1972. Play-mothering: The relations between juvenile females and young infants among free-ranging vervet monkeys. In *Primate socialization*, ed. F. E. Poirier, pp. 83–104. New York: Random House.

Petter, J-.J. 1962. Recherches sur l'écologie et l'éthologie des lemuriens malgaches. *Mém. Museum Nat. Hist. Nat. (Paris), Ser.* A. 27:1–146.

Petter-Rousseaux, A. 1962. Recherches sur la biologie de la réproduction des primates inférieurs. *Mammalia*: (Suppl. 1) 26:1–88.

———. 1964. Reproductive physiology and behavior of the Lemuroidea. In *Evolutionary and genetic biology of primates*, vol. 2, ed. J. Buettner-Janusch, pp. 91–132. New York: Academic Press.

Poirier, F. E. 1970. The Nilgiri langur (*Presbytis johnii*) of South India. In *Primate behavior*, ed. L. Rosenblum, pp. 254–383. New York: Academic Press.

Sussman, R. W. 1972. An ecological study of two Madagascan primates: *Lemur fulvus rufus* Audebert and *Lemur catta* Linnaeus. Ph.D. dissertation, Duke University.

———. 1974. Ecological distinctions in sympatric species of *Lemur*. In *Prosimian biology*, eds. R. D. Martin, G. A. Doyle, and A. C. Walker, pp. 75–108. London: Duckworth.

Sussman, R. W., and Richard A. 1974. The role of aggression among diurnal prosimians. In *Primate aggression, territoriality, and xenophobia*, ed. R. L. Holloway, pp. 49–76. New York: Academic Press.

19 | A Comparison of Early Behavioral Development in Wild and Captive Chimpanzees

NANCY A. NICOLSON

- **INTRODUCTION**

The mother-infant relationship in nonhuman primates has always intrigued researchers in both the field and the laboratory. The greatest body of data on this topic has been collected for a few monkey species. Studies of mother-infant relations in the great apes are relatively few; observational conditions are often poor in natural habitats, and problems associated with the acquisition and care of large numbers of apes have put constraints on the amount of data that can be obtained in the laboratory. Although the slow rate of infant development and the long intervals between births allow provocative comparisons with human development, they can be inconvenient from the point of view of the researcher, who often has time limitations. In spite of these difficulties, some detailed observations of mother-infant relations in the great apes, especially the chimpanzee, have been made in both natural and captive environments. Goodall's (van Lawick-Goodall 1967) study of mother-offspring relations in free-ranging chimpanzees and the original account (Yerkes and Tomilin 1935) of chimpanzee mother-infant relations at the Yale Laboratories of Primate Biology are particularly noteworthy.

As Mason (1968) has pointed out, field and laboratory approaches to the study of primate behavior have developed independently, with many unfortunate consequences. Traditional differences in the two approaches have resulted in bodies of data which fail to complement each other. Although the literature on primate mother-infant relations is extensive, in only a few cases is direct comparison of the behavior of mother-infant pairs in natural habitats and captivity possible (i.e., brief accounts in Ransom and Rowell 1972; Itoigawa 1973).

This chapter investigates differences in the chimpanzee mother-infant relationship which probably reflect the kind of natural or captive environment in which the pairs live. Because a small number of pairs were observed, conclusions must be tentative. Nevertheless, the comparison will hopefully shed some light on the nature of environmental influences on chimpanzee social development, as well as pointing to problems that researchers might consider in the future.

■ SUBJECTS AND METHODS

Most of the data on which this chapter is based were obtained from detailed observations of three chimpanzee (*Pan troglodytes*) mother-infant pairs, each living in a different environment. Each pair was periodically observed from the time the infant was born, until it was 20 weeks old (see Tables 1 and 2).

The chimpanzee, Athena, gave birth on about June 22, 1973, to a female infant, Aphrodite. This mother-infant pair is a part of the free-ranging chimpanzee population living in the Gombe National Park, Tanzania, East Africa. The habitat and population structure have been described in greater detail elsewhere (van Lawick-Goodall 1968). During the study period, the mother and infant were always accompanied by Athena's juvenile son, Atlas, then almost 6 years old. Atlas and Aphrodite were Athena's only offspring; she had had a stillbirth in 1966 and a miscarriage in 1970. Athena and family often met with other subgroups of chimpanzees as they traveled within their home range. The family was found in groups comprised of individuals of all age and sex classes during the course of the study. Most members of the community to which Athena belongs are habituated to the presence of human observers and can be followed as they travel through the mountains. During the present study, it was possible to observe Athena, Atlas, and Aphrodite at close range with little apparent disturbance to their behavior. (Atlas occasionally chased observers or threw rocks at them in play.) Although over one-third of all observations of this family began in the vicinity of the research camp, data collected in this location represent only 5% of the total observation time. Athena and Aphrodite were observed for a total of 33.5 hours. Length of observation periods ranged from 30 minutes to 4 hours, with an average duration of 2.5 hours. Observation periods were distributed over the daylight hours (0700–1900). Most observations were made from a distance of 5–10 meters from the mother and infant. Binoculars (7 × 35) were used as needed, particularly at greater distances. Observations were dictated onto a tape recorder and were later transcribed onto checksheets with written summaries.

The other two mother-infant pairs lived at the Delta Regional Primate Research Center, Covington, Louisiana, in two very different environments. The wild-born, but captive-reared female, Gigi, gave birth on May 1, 1972, to her first infant, Delta. Mother and infant lived with six other wild-born chimpanzees (including Delta's father, Shadow) in a large seminatural enclosure (120m × 30m). The four females and three males arrived at the Center as infants and were raised together as a group or in subgroups. They had lived together in the outdoor enclosure since 1969. Estimated ages of individuals in the group at the time of this study ranged from 7–11 years. The oldest male, Rock, was removed from the enclosure on May 8, 1972; otherwise, group composition remained stable. The enclosure was located about 400m from the Center's main facilities and was surrounded by pine forest. The enclosure walls were of chain-link fencing with a top section of sheet metal, to a total height of about 5m.

TABLE 1 / BACKGROUND OF SUBJECTS

| INFANT | | MOTHER | | | | | STUDY CONDITIONS | |
Name	Sex	Name	Origin	Age at capture	Age during study period	Reproductive status	Environment	Other chimpanzees present
Aphrodite	f	Athena	Gombe National Park, Tanzania		19	multiparous	free-ranging	Atlas (male, 6 years) 1–18 others
Delta	f	Gigi	Liberia	3	10	primiparous	seminatural enclosure (120 × 30m)	Rock (male, 11 years) Shadow (male, 9 years) Bandit (male, 7 years) Polly (female, 9 years) Bido (female, 8 years) Belle (female, 8 years)
Buster	m	Kami	Sierra Leone	5	10	primiparous	cage (6 × 2.5 × 3.5m)	Max (adolescent male)

TABLE 2 / NUMBER AND LENGTH OF OBSERVATION PERIODS						
AGE OF INFANT (WEEKS)	APHRODITE (FREE-RANGING)		DELTA (ENCLOSURE)		BUSTER (CAGE)	
	Periods	Hours	Periods	Hours	Periods	Hours
1	1	3	13	42.5	5	4
2			15	27.5	10	9
3			14	18.5	11	10.5
4			9	11.5	10	9
5			10	10.5	6	3
6	1	3	6	4.5	4	2
7	1	3	9	7	2	1
8			10	6	4	2
9			12	7	4	2
10	2	5	11	6.5	4	2
11	1	3.5	6	3		
12	1	2.5	5	2.5	4	2
13	1	2	11	6	7	3.5
14			12	6.5	8	4
15			12	6	7	3.5
16	1	2.5	15	7.5	7	3.5
17	1	.5	11	5.5	5	2.5
18	2	4.5	14	7	3	1.5
19	1	4	8	4		
Total	13	33.5	203	189.5	101	65

The enclosure contained natural vegetation, three elevated platforms, many poles and tall stumps, and a variety of movable objects. The chimpanzees moved freely about the enclosure, except for an hour each day, when they were moved into an adjacent cage during cleaning and maintenance of the enclosure. Observations were made with 7 × 35 binoculars from an overhead observation deck. During the first month of Delta's life, intensive observations were recorded in the form of detailed notes; thereafter, observations were recorded directly onto checksheets with a brief written summary attached. A total of 189.5 hours of observation is included in this analysis. Observation periods during which checksheets were used generally lasted 30 minutes and never exceeded 1 hour. Observation periods were distributed over the daylight hours and the days of the week.

The chimpanzee female, Kami, was housed with an adolescent male, Max, in the cage complex for large primates at the Center's main facility. She had arrived at the Center in 1967; her country of origin was listed as Sierra Leone. Max arrived 1 month later, also from Sierra Leone. He had no contact with Kami, however, until 1972, when the two shared a cage for about 6 months. During this time, Kami gave birth to her first infant, Buster (May 9, 1972). (The father, Rock, lived in a different part of the building and had no contact with the infant.) Buster remained with his mother until 5 months of age; he was then removed from their cage and placed in the nursery. All observations of the mother-infant pair were made in the large cage they shared with Max. The cage consisted of an outdoor cage (6m × 2.5m × 3.5m) communicating with an indoor room (2.5m × 2m × 2m). A sliding door separated the two areas of the cage. Most observations were made while the animals were confined to the outdoor area where they were clearly visible to an observer seated immediately outside the cage.

Detailed notes were recorded during Buster's first three weeks; then observations were recorded on checksheets with written summaries. Kami and Buster were observed for a total of 65 hours. Observation periods lasted 30 minutes. Observations were usually made between 0630 and 0830, or between 1100 and 1200, so that regular maintenance and research activities in the building would not be disturbed. Observations were distributed over the days of the week.

The same methods of data collection were used in all three cases. The checksheet was divided into one-minute intervals and recorded the occurrence of the following kinds of behaviors if they occurred at any time during the minute:

1. The general activity of the infant (suckling, resting, etc.)
2. The position of the infant relative to the mother
3. Which member of the pair was responsible for each change in proximity
4. Mother-infant grooming, play, and other interactions
5. Interactions of the infant with chimpanzees other than its mother.

The mother's general activity was also recorded, on the minute interval only. Definitions of behavioral elements were based on Goodall's glossary of chimpanzee behavior patterns (unpublished manuscript), which is used as a standard reference for long-term studies at the Gombe Stream Research Centre. Due to the differences in observation conditions and schedules (see Table 2), behavioral measures are probably more accurate for the captive mother-infant pairs than for the free-ranging pair, particularly in the recording of the first appearances of patterns of behavior.

■ MOTHER-INFANT RELATIONS AND INFANT DEVELOPMENT

BEHAVIOR OF THE MOTHER WITH THE NEWBORN INFANT

Observation of all of the infants began on the first or second day postpartum. All mothers were carrying and handling their infants competently from the first time they were observed. They all rated at the top of the Rogers and Davenport (1970) 5-point scale of maternal proficiency, in which "good" mothers "carried the baby on the ventral surface, allowed the baby to grasp, and responded to the infant's vocalizations by readjusting, examining, or clasping it" (ibid.: 363). This level of maternal competence is typical of all wild mothers, primiparous or multiparous (van Lawick-Goodall 1967), but is not common in laboratory settings. Rogers and Davenport (ibid.) reported that only 1 out of 19 primiparous females who had spent less than 18 months with their own mothers received the top rating of maternal proficiency. Lower ratings reflected maternal responses ranging from incompetence or neglect to abuse. Females

FIGURE 1. *Gigi cradles her 2-week-old infant, Delta.* (Photo: C. E.
G. Tutin)

who had been with their own mothers more than 18 months performed better, with 6
out of 10 primiparous females receiving the top rating. Gigi had spent 2–3 years and
Kami, 5 years, in natural environments before capture and should, therefore, be
compared with the latter group.

During Delta's first 10 days, Gigi frequently placed her mouth over the infant's
face. Similar behavior in captive chimpanzee mothers has been reported by several
investigators (Yerkes 1943; Lemmon 1968; Rogers and Davenport 1970); the mothers
were apparently removing fluids from the neonate's nose and mouth. On Delta's
second day, Gigi responded to the infant's screams by making this kind of mouth-to-
face contact. Whatever its origin, this behavior was usually effective in quieting the
infant.

SUCKLING

Suckling is a behavior which has obvious adaptive value for the young infant. Perhaps
because regular suckling bouts are of such critical importance, there was little variability in bout frequency or duration among the infants. (A bout is defined as
continuous suckling activity, including any interruption less than 1 minute in duration.) During the first 4 months of life, Aphrodite, Delta, and Buster each suckled an
average of 3 times per hour. The mean length of suckling bouts decreased slightly
after the second month, from 3.9 to 2.9 minutes. Most suckling bouts lasted less than
5 minutes. Maximum bout length varied among the three infants, with Delta suckling as long as 10 minutes at a time, Buster 9 minutes, and Aphrodite 7 minutes.
Although Goodall (van Lawick-Goodall 1967) did not observe suckling bouts of

longer than 7 minutes after the infant's first month, the remainder of the data is in agreement with her findings.

Delta began suckling on the first day of life, but did not establish regular suckling until she was 5 days old. Aphrodite was suckling regularly by 2 days of age, as was Buster. All three infants were able to obtain the nipple without much difficulty. Athena cradled Aphrodite in such a position that the infant was able to reach either nipple with little effort. On Delta's first day, Gigi held her to the nipple and on one occasion removed Delta's mouth from the left nipple and repositioned her immediately on the right nipple. When Buster began to root, Kami moved him from a low cradling position to a position where his head was between her breasts and he was able to reach either nipple with ease. All of the mothers effectively assisted their infants in reaching the nipple. The captive mothers' proficiency is surprising in light of other descriptions of this interaction in primiparous mothers and their new infants:

> The nursing interaction is not automatic in the chimpanzee. Much learning is required for the development of skillful, regular, and efficient performance. Chimpanzees do not typically hold and carry the neonate at the breast area and primpiparous mothers do not reflexly place the hungry infant on the nipple. (Rogers and Davenport 1970:367)

Although suckling interactions were generally well coordinated for all mother-infant pairs, difficulties did arise. Inability to reach the nipple was the major cause of distress (as measured by frequency of whimpering) in all three infants during the first month. With the development of strength and muscle coordination, the infants were later able to pull themselves to the nipple without their mothers' aid. At first, infants suckled from only one breast in a single feed. Buster first switched from one breast to the other at 4 weeks, Delta at 10 weeks, and Aphrodite at 13 weeks.

Suckling bouts were usually ended by the infants. However, many bouts were interrupted or terminated when the mother moved, inadvertently jerking the nipple out of the infant's mouth. The two captive mothers sometimes ended a suckling bout by deliberately removing the infant from the nipple. When Delta was only 1 week old, Gigi pulled the infant off the nipple and successfully restrained her from further suckling, in spite of the infant's whimpers and screams. Kami restricted Buster in a similar manner when he was 6 weeks old. He was able to pull himself to the other nipple, however, and held onto it tenaciously. By the time Buster was 9 weeks old, Kami was hindering his attempts to suckle quite regularly. Although Kami thus prevented or delayed many suckling bouts, the percentage of time Buster spent in suckling (about 15% of the total observation time) did not differ from that spent by Delta or Aphrodite.

Although Athena did not deliberately interrupt suckling bouts once begun, she was observed to prevent suckling when Aphrodite was 13 weeks old:

> Aphrodite was very persistent in trying to suckle when apparently hindered by Athena; Athena either held Aphrodite away from the nipple or covered the nipple

with her hand. She also turned her back on Aphrodite a few times, leaving the infant sitting on the ground and struggling to move into ventral contact again. (Observation notes: September 20, 1973)

Aggressive maternal restriction of suckling has previously been reported in captive chimpanzees (Yerkes 1943), but is rare in wild chimpanzees. Earlier evidence (van Lawick-Goodall 1967) suggested that weaning in wild chimpanzees was mainly related to the drying up of the mother's milk. It now seems that weaning is a gradual process in which many mothers take some active, though not necessarily aggressive, part (Clark, this volume). The motivation underlying early prevention of suckling is unknown. In the three cases described, the mothers did not appear to be initiating early weaning; there was no significant reduction in the frequency or duration of suckling bouts. Maternal discomfort caused by suckling might be involved; Ransom (Ransom and Rowell 1972) found that primiparous baboon mothers, who have short nipples, are likely to experience some discomfort and may curtail suckling. It is also possible that mothers simply become irritable at persistent suckling attempts and try to limit them. This might be particularly true in restricted cage environments, as other studies have suggested. Spencer-Booth et al. (1965), for example, reported that a higher proportion of the suckling attempts of rhesus infants raised in cages with only their mothers were rejected than was the case for infants raised with their mothers in social groups.

FEEDING

Of the three infants studied, wild-born Aphrodite was the only one who began to eat solids before 5 months of age. She was chewing solid foods by 11 weeks. This is unusually early for a wild chimpanzee; chewing had not previously been observed in infants younger than 15 weeks (van Lawick-Goodall 1967). At 13 weeks, Aphrodite tried unsuccessfully to take fruit held by her mother. By 18 weeks, Aphrodite was able to obtain food from Athena by placing her lips next to her mother's mouth. Athena also allowed the infant to chew on banana skins she was holding.

Aphrodite was probably somewhat precocious in the development of feeding behaviors; the absence of such behaviors in Delta and Buster at the same age is quite normal. In addition, soft fruits and leaves are less abundant in captivity than in natural environments, and dry monkey chow is probably not very palatable to young infants.

TRAVEL

During travel, all mothers carried their infants in the normal ventro-ventral position throughout the first month. All were observed to give the infants additional support in the early weeks of life. During the first observation of Athena and Aphrodite, Athena walked tripedally, supporting the infant with one hand. She walked with the hunched gait characteristic of mothers with newborn infants. While brachiating, Athena held her knees up against her chest, cradling the infant. Once she was seen climbing in sloth position beneath a branch. By the time Aphrodite was 2 weeks old, she was able to cling unaided during travel. Athena continued to give the infant hand support now and then throughout the first 2 months. Gigi and Kami also adopted a tripedal, hunched gait during the infants' early days. Kami frequently hung in sloth position while moving along the top of the cage. By the time Buster was 2 weeks old, Kami hardly supported him at all; he was able to cling very well by himself.

Variations on the ventro-ventral position appeared in the second month. Dangling from the mother's ventrum by one or two hands was a position apparently initiated by the infant. Mothers often discouraged dangling by pressing the infants back into ventro-ventral contact. Buster began to dangle at 5 weeks, Aphrodite at 10 weeks, and Delta at 12 weeks. Dangling was common in all infants thereafter.

Aphrodite and Delta traveled in the ventro-ventral or ventral dangling positions exclusively during the study period. When Buster was a month old, Kami began to walk with him clinging to her arm or leg only. Soon she began to push him into a dorsal position. Consequently, ventral travel descreased from 100% of total travel

FIGURE 2. *Buster clings to his mother's back while she drinks.*
Photo: C. E. G. Tutin)

time during the first month, to 73% during the second, 37% during the third, and 27% during the fourth month. During the rest of the time spent in travel, Buster usually rode on Kami's back, although sometimes he could only manage to grab an arm or a leg before Kami began to move away.

Chimpanzees of the Gombe Stream population, with the encouragement of their mothers, begin to ride on their mothers' backs regularly from the time they are about 5–7 months of age (van Lawick-Goodall 1967). During her first 2 years, the infant Delta did not ride in the dorsal position at all. From 5 months of age until over 2 years, Delta lived with the rest of the original group on a small island in Lion Country Safari Park, Laguna Hills, California. When not clinging to her mother's ventrum, Delta usually walked next to her, maintaining contact with one hand. Distances covered were not great, due to the limited area of the island. Soon after her second birthday, Delta began to ride on her mother's back. After the group was moved to the larger Stanford Outdoor Primate Facility, Delta rode in this position regularly.

GROOMING

All three mothers began to groom their infants by the first or second day of life. Gigi and Kami groomed regularly from that time on. Athena was only seen four times during Aphrodite's first month, but she did groom the infant on three of these occasions. In all cases, grooming was performed attentively and at some length. Goodall (van Lawick-Goodall 1967) did not observe grooming by the mother of more than a few seconds' duration during the first few weeks of the infant's life, but more recent data indicate that some wild mothers do groom their newborn infants. Grooming of the neonate is common in captive chimpanzee mothers (Lemmon 1968; Rogers and Davenport 1970).

Quantitative data on mother-infant grooming were recorded after the first month for all infants. There was little difference among the three mothers on any measure of grooming employed: total percentage of minutes in which mother groomed infant, mean number of bouts per hour of observation, or mean bout length. (A bout was defined as continuous grooming activity, including any interruptions of less than 1 minute.) There was also no marked change in any of these measures over the 3 months for which data were obtained. The mothers groomed their infants in approximately 11% of the minutes observed. There was an average of 3.4 grooming bouts per hour, with a mean bout length of 2.2 minutes. Ninety-four percent of the mother-infant grooming bouts lasted less than 5 minutes; this figure is identical to that reported by Goodall (van Lawick-Goodall 1967) for mothers with infants under 1 year of age. However, in the present study only 50% of the bouts were under 1 minute, while in Goodall's sample over 80% of the bouts were that short. Maximum bout lengths were as follows: Athena/Aphrodite—12 minutes; Gigi/Delta—24 minutes; Kami/Buster—18 minutes. Grooming bouts over 10 minutes in length were extremely rare, being observed only 6 times in 143 hours of observation on all three mother-infant pairs.

Although little information is available concerning differences in mother-infant grooming interactions in natural and captive environments, some researchers have

FIGURE 3. *Athena grooms Aphrodite, then 1 month old.* (Photo: C. E. G. Tutin)

found differences in these interactions in mother-infant pairs living in social groups as compared with those living alone (i.e., rhesus monkeys, Spencer-Booth et al. 1965; baboons, Rowell 1968). In general, mothers groom their infants more when housed alone than in a group situation. Rowell (ibid.) found, however, that the addition of only one other baboon to the "isolated" environment resulted in a reduction of maternal grooming to a level similar to that observed in the group environment. Wolfheim et al. (1970), in comparing group and individually housed pigtail macaque mother-infant pairs, found no significant difference in grooming frequencies between the two environments.

Near the end of the study period, bald patches appeared over Delta's head and neck. Several members of this group sometimes pull out hairs while grooming themselves or others; although observers rarely saw this behavior directed toward Delta, overly zealous grooming presumably caused her baldness. It is quite possible that Gigi or others might have plucked Delta's head during the night. Yerkes and Tomilin (1935) reported that some of the captive mothers they observed habitually removed the hair from portions of their infants' bodies. Such behavior has not been observed in wild chimpanzees.

Aphrodite was the only infant observed to groom her mother during the study period; at 17 weeks, when this behavior first appeared, it was very short in duration and rudimentary in form. Delta made rough grooming movements at approximately 38 weeks. She was using more adult techniques by the middle of her second year (N. Merrick, personal communication).

Captivity may influence the development of infant grooming. In Goodall's study, infants did not groom before 26 weeks, but most infants developed adult techniques within the second year. In laboratory settings, grooming has rarely been observed in chimpanzees under 4 years of age (Yerkes 1933). Studies of captive and wild hamadryas baboons (Kummer and Kurt 1965) indicate a similar delay in the development of grooming in captive animals. Free-ranging 1-year-old infants engaged in social grooming during 15% of the minutes observed, while captive individuals did not groom each other regularly until 3 years of age. In the present study, grooming by infants was observed very infrequently. Hopefully, a comparison of older Gombe and Stanford infants will clarify what effect, if any, captivity in a seminatural environment might have on social grooming patterns and frequencies.

PLAY

All three mothers played with their infants, but play varied in frequency among the individual pairs. Athena played with Aphrodite much more frequently than either of the captive mothers. Kami played with Buster somewhat more frequently than Gigi played with Delta (see Figure 4). Play was usually distinguished from manipulative behaviors (see next section) on the basis of distinctive patterns, such as tickling or poking, seen only during play. In other cases, play was not clearly different in form from manipulation, but play patterns were more repetitive and were accompanied by a play-face.

Athena and Aphrodite were not systematically observed between the infant's first

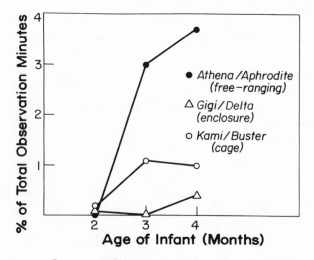

FIGURE 4. Percentage of minutes in which mother played with infant.

and sixth weeks. Mother-infant play was first seen in this pair when Aphrodite was 10 weeks old. On that occasion, Aphrodite actively participated in the play interaction and made a play-face. During the same week, Aphrodite was first heard to "laugh." ("Laughing" consists of soft panting sounds and characterizes social play [van Lawick-Goodall 1968].) Play was a regular feature of the mother-infant relationship from that point on, with Athena and Aphrodite participating equally. Play tended to increase as the infant grew older. Aphrodite's active participation in play was an early development relative to other Gombe infants. Goodall (van Lawick-Goodall 1967) first observed infant play-faces and "laughing" at 13–19 weeks, and the infants did not actively respond to social play until 14–19 weeks. Goodall found that the frequency of play was highly variable from one mother-infant pair to another.

Gigi and Kami began to play very occasionally with their infants when they were only a few weeks old. The infants did not respond, and sometimes they whimpered. Delta first made a play-face at 19 weeks of age, but was not an active participant in play. Mutual play between Kami and Buster was never observed.

MANIPULATION OF THE INFANT

Manipulation of the young infant (which, by our definition, does not include grooming or play) is a pattern of maternal behavior characteristic of captive chimpanzees which has not been observed in the Gombe population. This behavior has been interpreted as a form of maternal tuition, preparing the infant for growing independence (Yerkes 1943). Mason (1970: 269) has suggested that this form of behavior may occur in captivity, but not in natural habitats, because of "the combination of idleness and relief from the ordinary burdens of maternal care" that exists in captive environments. Both Kami and Gigi manipulated their infants in various ways from birth throughout the study period. Although Athena groomed and played with Aphrodite, she did not manipulate the infant in other contexts.

In the first month, the captive mothers spent considerable periods of time examining their infants. They lifted the infants up by the arms, breaking ventro-ventral contact to some extent. Gigi began lifting Delta in this way on the infant's first day, actually dangling Delta above her head. Mother and infant made soft vocalizations and Gigi "kissed" the infant on the mouth after lowering it into a normal cradling position. On other occasions, Gigi turned the infant around while lifting it, so that Delta's back faced Gigi's chest, or she swung Delta gently back and forth. Patting of the infant's head and back were also frequently observed. Delta sometimes whimpered during these manipulations, but usually she accepted Gigi's attentions rather passively.

By the time Delta was a month old, Gigi was manipulating her in a more purposeful manner, seemingly testing and encouraging the infant's developing motor skills. During one observation period, Gigi repeatedly placed Delta's foot against the enclosure fence, bending the infant's toes until they gripped the wire. When Delta was 6 weeks old, Gigi initiated "walking lessons." She would sit with Delta on the ground between her legs, the infant clinging to Gigi's body with her hands only; then Gigi would shuffle backwards, forcing Delta to take a few steps forward. This, too,

FIGURE 5. *Gigi lifts and examines Delta.* (Photo: C. E. G. Tutin)

caused Delta to whimper, but within a few weeks the lessons were tolerated, if not willingly engaged in. This form of tuition continued throughout the study period. Gigi later refined her technique, walking backwards bipedally while holding Delta's hands; this enabled Gigi to walk Delta somewhat farther and at a faster pace. When Gigi first began to break contact deliberately with Delta (the infant was then 16 weeks old), she would walk up to the enclosure fence, detach Delta, and place Delta's hands on the fencing. She would then back away a few meters, leaving the infant clinging to the fence, with her feet usually on the ground. Gigi watched Delta carefully during the separation and prevented all other animals from touching her. Sometimes Gigi turned and rubbed her perineum on Delta's back. A variation of the separation procedure was to put Delta in a shallow hole in the ground, again backing a few meters away and watching her.

Kami began to lift Buster up by his arms during his first week, though not as high as Gigi lifted Delta. At first, Buster was much less tolerant than Delta of breaks in ventro-ventral contact and protested loudly when Kami lifted him at all. By 3 weeks of age, however, Buster complained less, and Kami lifted him further away from her ventrum, but still did not dangle him as Gigi did her infant. Kami sometimes turned Buster upside down while examining him; she also swung him back and forth and stroked his head.

After Buster's first month, Kami's manipulations were mainly directed towards detaching him from her body entirely. When he was 9 weeks old, however, Kami did attempt to "walk" him, much as Gigi had done with Delta. She walked backwards bipedally, holding Buster's hands in her own. This behavior was not observed again.

In summary, whether or not the captive mothers were consciously trying to teach their infants anything, maternal manipulations probably encouraged the development of motor skills and independent behavior. The fact that captive mothers need devote less time to subsistence activities than free-ranging mothers does not explain why this particular form of manipulative behavior has developed in captivity. The adaptive significance of the mother's active role in promoting infant independence will be discussed in a later section.

MAINTENANCE OF PROXIMITY[1]

Quantitative data were obtained in the form of the percentages of observation minutes in which infants were in various positions relative to their mothers. In the first month, all infants remained in almost continuous ventral contact with their mothers. The percentage of time spent in ventral contact decreased during the next 3 months, with individual differences becoming more pronounced (see Figure 6). After the sixth week, Buster consistently spent less time in ventral contact with his mother than either Delta or Aphrodite. Delta was first observed out of ventral contact earlier than Aphrodite, but there was little difference in the overall percentage of time they spent in this position.

During the same period of time, the percentage of minutes the infants spent unsupported by their mothers, but still in contact, increased (see Figure 7). Interindividual differences paralleled those described above, with Buster spending increasingly more time in this position than the other two infants.

Mother-infant contact was first broken between Kami and Buster at 8 weeks, between Gigi and Delta at 11 weeks, and between Athena and Aphrodite at 18 weeks. (Goodall observed the first break in physical contact between Gombe mothers and infants when the infants were 16–22 weeks old.) No infant was recorded out of contact with its mother in over 1% of the minutes observed during the first 3 months. At 4 months of age, Aphrodite was hardly ever out of contact with her mother, Delta was out of contact in only 1% of the minutes observed, and Buster was out of contact in 30% of the minutes. These percentages represent occasions in which the infant spent any part of the minute out of contact; the percentage of minutes in which there was no contact at all never exceeded 0% for Aphrodite or Delta and reached only 4% for Buster during the fourth month. Sixty percent of the time Buster was out of contact, he was within arm's reach (about 1.5m in chimpanzees) of his mother, and 90% of the time he was less than 5m from his mother.

Measures of spatial proximity alone do not help to explain the roles of mother and infant in maintaining proximity. The relative contributions by mother and infant to the maintenance of proximity between them can be assessed by employing a function that takes into account the proportion of approaches or leavings over a given distance that are due to one partner or the other (Hinde and Atkinson 1970). Roles in the maintenance of ventro-ventral contact can also be assessed (Hinde and White 1974). The former function is expressed as the difference between the percentage of approaches due to the infant and the percentage of leavings due to the infant. This function will be positive if the infant is primarily responsible for maintaining proximity, and negative if the mother is. Similarly, roles in the maintenance of ventro-ventral contact are assessed by calculating the difference between the percentage of contacts

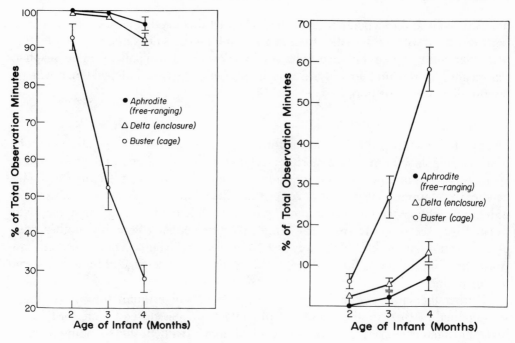

FIGURE 6. *Percentage of minutes in which infant was on mother's ventrum. Standard error is indicated where it is greater than 1%, representing day-to-day variability.*

FIGURE 7. *Percentage of minutes in which infant was off, but in contact with mother. Standard error indicated.*

initiated by the infant ("makes") and the percentage of contacts terminated by the infant ("breaks").

Van der Stoep (unpublished manuscript) examined data collected at Gombe for 8 months on six mother-infant pairs, with infants ranging in age from 6–36 months. The following general findings are relevant to the present study: in the Gombe population, the first breaks in contact are initiated by the infants, as they begin to explore the environment at a short distance from their mothers; until at least three years of age, infants seem to be relatively more responsible for breaking ventral contact than mothers; they are also more responsible for approaching the mother and for reestablishing ventral contact.

The only contact breaks observed between Athena and Aphrodite were initiated by the infant. Aphrodite also reestablished contact more frequently than her mother. Ten contact breaks were recorded between Gigi and Delta. Delta was responsible for none of the breaks and for 50% of the contact initiations. Thus she was relatively more responsible than Gigi for maintaining contact. The only large sample of "makes" and "breaks" (n = 307) and leavings and approaches (n = 108) was obtained from observations of Kami and Buster, during Buster's fourth month. Buster was responsible for 21% of the breaks in contact and for 71% of the contact initiations

(71% − 21% = 50%); he was responsible for only 9% of the leavings to beyond arm's reach and for 64% of the approaches (64% − 9% = 55%). In other words, Buster's mother frequently broke contact with him and moved beyond arm's reach, while he attempted to return to close proximity and contact. Buster played a greater role in maintaining proximity and contact, while his mother played a greater role in initiating separation.

The two captive mothers broke contact with their infants proportionately more than did Gombe mothers in Van der Stoep's (unpublished manuscript) sample. During the first deliberate breaks in contact, both Kami and Gigi cautiously detached their clinging infants and slowly moved a short distance away. When the infants whimpered or screamed, their mothers picked them up again. Gigi was particularly vigilant when other individuals were nearby. If they approached or reached toward the infant, Gigi often retrieved Delta immediately. Kami showed no increased protectiveness when the adolescent male, Max, was nearby, although she did threaten him away from Buster on occasion.

From the time Buster was about 13 weeks old, Kami began to leave him with increasing regularity. His vocalizations were no longer entirely successful in bringing about Kami's prompt return. The following is an excerpt from a typical 30-minute observation during Buster's fifteenth week:

Kami pushed Buster off her back by pinching his fingers, then backed away from him to a distance of 3m. After Buster screamed for 10 seconds, Kami approached. Buster suckled while Kami groomed him. Kami again broke contact, leaving

FIGURE 8. *Buster, 14 weeks old, clings to Kami's leg as cagemate Max looks on.* (Photo: C. E. G. Tutin)

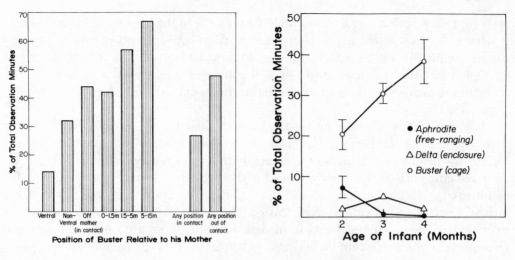

FIGURE 9. *Percentage of minutes in which the infant Buster whimpered, at different positions relative to his mother (months 2–4
combined).*

FIGURE 10. *Percentage of minutes in which infant whimpered.
Standard error indicated.*

Buster in the corner of the cage. After 8 seconds, Buster whimpered, and Kami
retrieved him. Two minutes later Kami broke contact and began to crawl backwards, slowly creeping off to the far corner of the cage. With pilo-erection, she
pant-hooted excitedly. Buster screamed, whimpered, and crawled a short distance
toward Kami. After Buster screamed almost continuously for 30 seconds, Kami
approached and pulled Buster into ventral contact. (Observation notes: August 14,
1972)

In general, Buster whimpered more frequently when out of contact with Kami than
when in contact, in any position (see Figure 9). Kami initiated most breaks in ventral
contact, and the more bodily contact was prevented, the more vigorously Buster
protested. Rate of whimpering was consistently higher for Buster than for either Delta
or Aphrodite (see Figure 10).

■ MOTHER'S GENERAL ACTIVITY

When not interacting with their infants, all mothers were engaged in maintenance
activities, social interactions with other chimpanzees, or rest. However, the proportion of time each mother devoted to the various activities differed markedly (see Table
3). Athena spent more time in feeding than any other single activity. This is a pattern
typical of free-ranging adult chimpanzees (Wrangham 1975). Kami spent very little
time feeding; she could obtain high-protein chow by walking into the indoor cage

TABLE 3 / PROPORTION OF TIME SPENT BY MOTHERS IN
VARIOUS ACTIVITIES

	FEEDS (%)	GROOMS SELF (%)	SOCIAL INTERACTION (%)	SITS IDLE OR RESTS (%)
Athena (free-ranging)	33	6	13	19
Gigi (enclosure)	11	2	11	53
Kami (cage)	2	6	1	64

area. (The 2% figure in table 3 may be biased by the fact that observations could not be distributed evenly over the hours of the day, and the indoor area was not always open. Nevertheless, it is very unlikely that Kami spent more time feeding than Gigi, who spent some time foraging for natural foods in the enclosure.) The proportion of time each mother spent in interactions with chimpanzees other than the infant is similar for Athena and Gigi. Most of Athena's interactions were with her son, Atlas. Gigi interacted with all five of her enclosure companions. Kami, on the other hand, spent very little time interacting with the adolescent male, Max, the only companion available. She spent the majority of the time simply sitting idle or resting.

The large percentages of time both captive mothers spent idle almost certainly reflect the nature of the environments rather than any intrinsic lethargy on the part of the chimpanzees themselves. In Kami's case, the social environment was as unchallenging as the physical environment. The mother's general response to characteristics of the environment could be expected, in turn, to affect the mother-infant relationship. As an example, the mother's general pattern of activity should influence the kinds of information the infant will acquire through observational learning. A chimpanzee infant's attention will surely be more frequently directed toward learning appropriate food items and feeding techniques if his mother spends 33% of each day feeding than if she spends only 11%.

■ INFANT'S RELATIONS WITH INDIVIDUALS OTHER THAN THE MOTHER

The three mother-infant pairs studied were almost never alone; each had at least one other chimpanzee nearby during all observation periods. Athena's juvenile son, Atlas (almost 6 years old), accompanied her throughout the study period. Forty-one percent of the time Athena and Aphrodite were observed, Atlas was the only other chimpanzee present. At other times, Athena and family were observed in groups of one to eighteen other individuals. Gigi and Delta had five other chimpanzees as constant companions in the enclosure: Shadow (male, 9 years), Bandit (male, 7 years), Polly (female, 9 years), Bido (female, 8 years) and Belle (female, 8 years). (A sixth individual, the 11-year-old male, Rock, was removed from the enclosure a week after Delta's birth.) Kami and Buster shared living quarters with the adolescent male, Max. Thus, all infants had some opportunity for interaction with chimpanzees other than their mothers.

Other chimpanzees typically showed interest in the newborn infants. The newborn, Aphrodite, was first seen with her mother in the vicinity of about ten other chimpanzees, most of whom were feeding. Some individuals directed greeting gestures to Athena, suggesting that this was Athena's first meeting with the group since the birth of the baby. Several adults, both male and female, and an infant approached to peer at Aphrodite. Athena seemed slightly upset by these attentions but tolerated most of them. None of the chimpanzees, other than Athena's son, tried to touch Aphrodite. Once Atlas reached to touch his sister's head, stroking it gently. Athena made no move to restrain him, but watched carefully.

Although Athena, Atlas, and Aphrodite were near other chimpanzees over half of the time they were observed, only Atlas interacted to any extent with the infant. This may have been due to greater interest on his part, greater permissiveness on Athena's part, or both. On two occasions, Athena distracted interested infants by playing with them. Two male infants and a juvenile male succeeded in touching Aphrodite a few times, but Atlas was the only individual other than Athena who groomed the infant. Atlas' interest in his sister increased as she grew older and more active. By the time she was 2 months old, Aphrodite began to respond actively to Atlas' attentions, reaching

FIGURE 11. *Group interest centers around Gigi and her newborn infant.* (Photo: C. E. G. Tutin)

out to him or chewing on his fingers. In these early interactions, Atlas was cautious. Athena was tolerant, but occasionally pushed his hand away or distracted him by grooming or playing with him. When Aphrodite was older, the three sometimes played together, with Athena and Atlas tickling each other and Aphrodite, and all "laughing" loudly.

In the Gombe population, infants are sometimes carried away from their mothers by other individuals, almost always older siblings (van Lawick-Goodall 1967). Observers saw Aphrodite "kidnapped" for the first time when she was 19 weeks old. On that occasion, Atlas carried Aphrodite away from her mother three times within five minutes. Athena's calm response suggested that Atlas might have started carrying his sister even earlier.

In the enclosure group, all of the chimpanzees took an active interest in the newborn infant, Delta. Gigi was very tolerant of the others, allowing them to touch and groom the new infant. All group members groomed Delta sometime within the first 2 days. Touching and manipulation of the infant's arms and legs became more frequent as Delta grew older. Grooming continued at a fairly constant rate (occurring during about 11% of observation minutes) over the first 4 months. Individual differences were relatively stable (see Figure 12). The females consistently groomed Delta more frequently than the males. (The females generally spent more time in social grooming than the males [McGrew and Tutin, personal communication].) For the most part, Gigi did not object to others touching or grooming Delta, but at times her patience was apparently exhausted. Bandit was remarkably persistent in his efforts to touch Delta, and when Gigi's grooming failed to distract him, she sometimes threatened him away. On one occasion, the frustrated Bandit finally threw a temper tantrum, but was still not permitted access to the baby. In most cases, Gigi

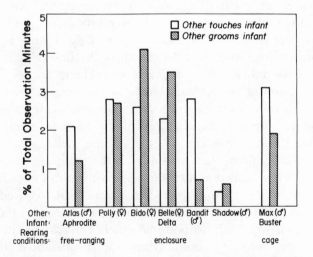

FIGURE 12. *Percentage of minutes in which individuals other than the mother touched or groomed the infant (months 2–4 combined).*

successfully curbed excessive interference by other individuals by changing her position slightly or by grooming the individual concerned.

Group members played with Delta infrequently, perhaps because she was so unresponsive. Belle and Polly tried to play with Delta at 11 and 19 weeks, respectively, but the infant remained passive on her mother's ventrum. Delta was never observed to engage in social play during the study period.

Members of the enclosure group tried to pull Delta away from her mother throughout the study period, but Gigi prevented them from actually taking the infant until she was 15 weeks old. At that time, Delta lost her grip on Gigi while Belle groomed her; Belle suddenly found herself holding the infant, quite by accident. She startled and squeaked with a fear-grimace. Gigi calmly took the infant back. On the next day, Bido deliberately pulled Delta away from Gigi; Gigi threatened Bido and retrieved the infant, in the midst of great group excitement. At 17 weeks, Delta crawled onto other chimpanzees (Belle, Polly, Bandit), who were sometimes able to hold her for a short time before Gigi took her back. A completely successful "kidnapping," with the abductor moving away with the infant, was not observed until Delta was 9 months old (M. Trudeau, unpublished manuscript).

In the small cage environment, the adolescent male, Max, was always close to Kami and her new infant. He showed a great deal of interest in the baby, approaching to peer at it closely. He touched Buster frequently, sometimes stroking the baby's head. Kami ignored Max or turned away from him. As Buster grew older, Kami usually permitted Max to touch and groom him as much as he cared, or dared to. (Kami easily dominated Max in any situation, and he was always cautious in interactions with her.) Buster crawled onto Max at 15 weeks of age. Max held him for over 3 minutes, until Kami approached and Buster crawled back to her. Max seemed eager to take Buster, but only after Kami had already left him. He never attempted to pull Buster away from Kami or carry the infant to another part of the cage.

In all three environments, interactions between the infants and individuals other than the mother were similar in form and frequency (see Figure 12). In the Gombe population, only siblings interact with infants with such a high frequency. The constant proximity of individuals in captive environments creates a situation analogous to the association of related individuals in a natural habitat. If frequency of association normally provides chimpanzees with any information about their probable degree of biological relatedness to other individuals, captivity could reduce individuals' ability to discriminate true kin. (It is interesting that Delta's father, Shadow, interacted less with the infant than any of the other group members.)

The nature of each individual's relationship with the mother influenced its interactions with the infant. Distinct partner preferences between certain individuals and dominance relations were clearly important, particularly in the captive groups. Both Gigi and Kami were able to dominate their companions in most situations, giving them the opportunity to choose which individuals could interact with their infants, and to what extent. Gigi did tolerate some individuals more than others, and as a consequence, some animals interacted with the infant more confidently than others.

■ SUMMARY

All three mothers cared for their newborn infants competently and showed affectionate responses to them. All mothers restricted suckling to some extent. The captive mothers were both observed to end some suckling bouts by removing the infant from the nipple. Prevention of suckling was most common for the caged mother, Kami. Nevertheless, all three infants spent about the same percentage of time suckling. Aphrodite was the only infant observed to eat solid food, respond to social play, or groom during the study period. Only Buster rode on his mother's back. All mothers groomed their infants during comparable percentages of time. Mothers differed in the frequency with which they played with their infants, with Athena (free-ranging) playing most and Gigi (enclosure) playing least. Both captive mothers manipulated their infants in a manner which encouraged independence, a pattern of behavior that has not been observed in the Gombe population. Gigi and Kami played a greater role in initiating and maintaining spatial separation from their infants than did the free-ranging mother, Athena. As a result, the three infants consistently differed on all measures of proximity to the mother, with Buster being most and Aphrodite least spatially independent of the mother. Buster showed the highest and Aphrodite the lowest level of distress, as measured by rate of whimpering. Mothers differed in their general pattern of activity, particularly in the amount of time each spent in subsistence activities or idle. Interactions with chimpanzees other than the mother were similar in form for the three infants; Delta had a higher overall rate of interaction with others because there were always more other individuals present in the enclosure.

■ DISCUSSION

VARIABLES THAT COULD CONTRIBUTE TO DIFFERENCES IN INFANT BEHAVIORAL DEVELOPMENT

Several variables could have contributed to the observed differences in measures of behavioral development and social interaction among the three mother-infant pairs. Factors that should be considered are the mother's dominance status and parity (which are often confounded with her age), the infant's sex and level of physical development, individual differences (broadly described as "personality"), and characteristics of the physical and social environment in which mother and infant live. Unfortunately, it was impossible to match all three subject pairs on any of these variables. The following section will attempt to evaluate the relative importance of each of the above factors to the development of differences in measures of infant independence.

Status of the Mother

Maternal dominance status may have important consequences for infant development, at least in some species. The mother's rank could affect the level of real or imagined danger to the infant from other group members. In a free-ranging population, Drickamer (1974) found that mortality was generally lower in the first year of life for infants of high-ranking rhesus mothers than for infants of low-ranking mothers. Although the mechanism is poorly understood, in many species high maternal rank can confer status advantages on offspring that sometimes persist into adulthood (i.e., Japanese macaques, Kawamura 1958; rhesus macaques, Koford 1963).

An analysis of status effects was not possible in the present study because no objective measurements of maternal dominance status, by any criterion, were available. As noted earlier, maternal status probably influenced the amount and kind of interaction occurring between the infants and other individuals, particularly in the captive groups. In the Gombe population, status effects were even less clear, because Athena and other free-ranging mothers spent a large percentage of time with only their dependent offspring nearby. It is my impression, however, that differences in dominance status among the three mothers in this study cannot account for the observed differences in infant development.

Parity of the Mother

At present, there is little information concerning the long-term effects of maternal age or parity on infant development. What information there is, mainly from laboratory studies, suggests two possible effects of parity: (1) primiparous mothers tend to be less competent than their multiparous counterparts in handling very young infants (i.e., Rogers and Davenport 1970), and (2) primiparous mothers tend to be more protective and less rejecting of their older infants than multiparous mothers (i.e., Seay 1966). There was no evidence of either of these effects in the present study.

Both captive females, although primiparous, had spent 2 or more years with their mothers in natural habitats before being captured. This exposure to normal mothering and social interaction was evidently adequate preparation for motherhood; neither of these mothers had had any experience with infants while in captivity. Nevertheless, they seemed equally as competent and relaxed in handling their young infants as the free-ranging multiparous female, Athena. Athena, in fact, was occasionally rather clumsy with her infant; at times during the baby's first 2 months, Athena bumped Aphrodite against branches while feeding in trees, causing the infant to whimper or scream.

Contrary to expectations based on parity, the primiparous mothers began to reject their infants earlier and more frequently than the multiparous mother. The primiparous mothers were certainly not over-protective of their infants. In summary, differences in infant development among the three pairs run counter to predictions of the effects of maternal parity; a more compatible explanation is needed.

Sex of the Infant

Sex differences in the development of infant independence have been reported for several primate species in laboratory groups. Mitchell (1968) and Jensen et al. (1968a) have suggested that mothers play an active role in promoting the earlier independence of male infants. Mitchell (1968) found that rhesus mothers restrained male infants less than female infants and withdrew from them more. Jensen et al. (1968a) reported similar results; they also found that pigtail macaque mothers punished their male infants more, thereby encouraging them to leave. Rosenblum (1974), on the other hand, claimed that male squirrel monkeys achieved earlier independence than females in spite of their mothers, who tried unsuccessfully to restrain them.

In the present study, the only male infant, Buster, became spatially independent of his mother relatively early in comparison to the two female infants. He spent increasingly less time in contact with his mother than either of the other infants (see Figures 6 and 7). Buster's mother was chiefly responsible for the initiation of independence. These findings fit the pattern of sex differences reported by Mitchell (1968) and Jensen et al. (1968a). There is reason to believe, however, that sex differences may not provide a complete explanation for the observed differences in mother-infant proximity. In the Gombe population there is no evidence at present that would indicate large sex differences in early infant development or maternal response. Few infants, male or female, break contact with their mothers earlier than Aphrodite did. Infants, not mothers, usually initiate the first ventures out of contact. No Gombe infants of either sex spend more than a small percentage of time out of ventral contact with their mothers during the first 4 months of life. It is possible, however, that subtle sex differences in the development of independence in chimpanzees do exist and may be amplified in some captive environments. Studies of rhesus macaques (Sackett 1972) and squirrel monkeys (Rosenblum 1974) have demonstrated that environmental characteristics may influence the development of independence to a different extent in male and female infants. Since early sex differences have not been studied in captive chimpanzees, there is presently no way of evaluating the possibility that there might be an interaction of the effects of sex and environment in this species.

Physical Development of the Infant

Human newborns, even those born at exactly the same gestational age, vary greatly in size and physical maturity (Tanner 1974). The same is undoubtedly true of nonhuman primate infants. As a result, differences in early behavioral development can be expected.

Infant weights could not be obtained and no other objective measurement of size was made during the present study. Nevertheless, it was clear that the three infants were not equally well-developed at birth. Aphrodite was a large infant by Gombe standards. Buster, though born a week after Delta, was larger than she throughout the study period. Aphrodite and Buster both had at least one tooth by 12 weeks of age; Delta's first tooth erupted at 16 weeks.[2] Delta was generally slower in the development of coordinated motor activity than the other two infants.

Because Aphrodite and Buster were similar in size and physical maturity, these factors could not have contributed much to differences in their behavior. But Delta's lag in physical development relative to the other infants very likely influenced several measures of behavioral development, especially measures of mother-infant proximity and play. If Delta had been as active and coordinated as Aphrodite and Buster, I suspect that mother-infant contact might have been broken earlier or more frequently. Also, Delta might have been more responsive to others' attempts to engage her in social activity, especially play.

Individual Differences

Individual differences, or "personality," cannot safely be ignored in studies of nonhuman primates. In rhesus mother-infant pairs raised under similar laboratory conditions, Hinde and Spencer-Booth (1971) found stable differences among the pairs when they were rank-ordered on a wide variety of measures of mother-infant interaction (i.e., time spent in contact, initiative in the maintenance of proximity, maternal rejections, etc.). Although some of the differences were related to the sex of the infant, or to the parity or dominance status of the mother, these factors did not appear to be of major importance. The authors concluded that subtle individual differences may have more significant long-term effects on infant development than more obvious status variables. Consistent individual differences are well documented in chimpanzee mother-infant pairs, both captive and free-ranging (Yerkes and Tomilin 1935; van Lawick-Goodall 1967, 1971).

In the present study, the effects of individual differences could not be isolated; they undoubtedly existed. Although it is impossible to prove that individual idiosyncracies did not produce all of the differences among the three mother-infant pairs, an evaluation of the possible effects of other variables is more useful in a search for general principles than an a priori assumption that the differences are essentially random. Because environmental effects on the mother-infant relationship have never been studied in chimpanzees, and because they are well-known in other species, this variable seems particularly worthy of consideration.

Environment

The literature concerned with environmental influences on behavioral development has expanded rapidly in recent years. Several studies have dealt specifically with the effects of various environmental conditions on mother-infant relations in nonhuman primates. Itoigawa (1973) compared mother-infant interactions in wild and captive Japanese macaques. Ransom and Rowell (1972) noted some differences in the behavior of wild and caged baboon mother-infant pairs. The majority of studies have compared the effects of different kinds of captive environments. In these experiments, mothers and infants of the following species have been observed: rhesus macaques (Spencer-Booth et al. 1965), pigtail macaques (Jensen et al. 1968b; Wolfheim et al. 1970; Castell and Wilson 1971), baboons (Rowell 1968), and squirrel monkeys (Kaplan 1972; Rosenblum 1974).

Interpretation of the results of these studies is complicated by the fact that environments may vary in so many respects. Even so-called natural environments actually range from entirely undisturbed habitats, to areas with limited human interference, to partially provisioned areas, and so on. Captive environments also vary—in size, physical complexity, social complexity, amount of human intervention, and other parameters. Simple comparisons are not possible.

The most consistently reported environmental effects on mother-infant relations involve the development of infant independence. Rosenblum (1971, 1974) has proposed a model to relate the level of environmental complexity or danger (combining physical and social variables) to the growth of independence.

In the first days of life, the degree of environmental complexity does not markedly influence the dyadic relationship; however, if an infant matures under conditions of low complexity a relatively slow and incomplete development of infant independence occurs. If maturation occurs under conditions of moderate complexity, there is a facilitation of independent functioning, but when very high levels of complexity are present there may well be a decrease or inhibition of the development of independence, perhaps mutually dependent on infant and maternal reactions. . . . With increasing age this U-shaped function relating complexity and attachment gradually would flatten with increased complexity, even at somewhat high levels, facilitating greater and greater levels of independent functioning when compared with more restricted circumstances. (Rosenblum 1974: 126–127)

(Rosenblum further suggests that this model should be qualified with respect to the sex of the infant.) Data from the majority of the studies mentioned above fit Rosenblum's model fairly well.

Data from the present study are also adequately described by the portion of the model that deals with early stages of development. Even the barren cage environment in which Kami and Buster lived was sufficiently rich in physical and social stimuli, by laboratory standards, to be classified as "moderately complex." During the first 20 weeks of life, Buster achieved greater independence (whether by his own initiative or not) than the other two infants, who lived in more complex environments. The nature of the interactions between Kami and Buster which probably facilitated independent behavior fits descriptions (Yerkes and Tomilin 1935) of other caged chimpanzee pairs. It seems that caged mothers actively encourage their infants to walk, ride in the dorsal position, and achieve some degree of spatial independence, resulting in earlier infant independence on average than has been reported for free-ranging pairs. Caged mothers firmly direct their infants' activities, using force, if necessary.

This behavior is in contrast to the gentle restrictiveness of free-ranging mothers, who rarely push their infants in any way toward early independence. No free-ranging chimpanzee mother has avoided contact with a young infant, or ignored its distress vocalizations, to the extent that Kami, the caged mother, did. Goodall (van Lawick-Goodall 1967) reported that all of the free-ranging mothers she observed restricted early attempts at independence for several weeks. For chimpanzees, then, the decrease or inhibition of the development of independence in highly complex environments, predicted by Rosenblum's model, is certainly related to maternal restrictiveness; differences in infant response to environments of various levels of complexity are less clear.

The seminatural enclosure in which Gigi and Delta lived falls somewhere on the continuum from moderately to highly complex environments. Accordingly, the growth of independence should have proceeded at a rate intermediate to that in cage and natural environments. This was in fact the case. With regard to the kinds of interaction that are probably related to infant independence (i.e., proximity, maternal restrictiveness, rejection), the relationship of Gigi and Delta showed patterns typical of both caged and free-ranging pairs. For example, Gigi and Delta were more similar to free-ranging than caged chimpanzees on measures of mother-infant proximity, but resembled caged individuals with respect to physical manipulation of the infant.

Although Rosenblum's model satisfactorily describes existing data, variations in mother-infant interaction resulting from the effects of environment can perhaps better be understood in terms of the selective forces which have favored the ability of mothers and infants to respond to new conditions. If the evolutionary logic of mother-infant relations can be expanded to account for at least some of the variations that appear in captivity, primate fieldworkers might find the results of laboratory studies more relevant to their own interests, and the findings of laboratory studies conducted under different conditions could be more easily compared.

ADAPTIVE SIGNIFICANCE OF VARIATIONS IN MOTHER-INFANT INTERACTION

It is a common opinion, especially prevalent among those who have observed animals in natural habitats, that behavior in captivity is, at worst, pathological and, at best, merely the vestige of behavior that would have been adaptive under more natural conditions. I will not argue that all behavior in captivity is adaptive (there is abundant evidence to the contrary), but that in many captive settings, animals simply do the best they can, adjusting their behavior to the circumstances in which they find themselves. The behavioral flexibility of primates is well known, and their ability to adapt to a new environment, even an unnatural one, should come as no surprise.

In natural populations of chimpanzees, the only stable group is the mother and her preadolescent offspring. It is crucial to the youngster's survival and eventual reproductive success that it remain with its mother for many years, even after nutritional weaning. Only the chimpanzee mother will provide adequate protection for her child in times of danger. And, in the sheltered environment provided by its mother, the young chimpanzee is able to learn important facts and skills: the geography of the home range, edible food items, the techniques of termite fishing, appropriate social responses, etc. This body of knowledge is a prerequisite for successful adulthood. Thus, it is adaptive for the chimpanzee mother to remain socially attractive to her child until it acquires sufficient physical strength and experience to venture out on its own safely. Most young chimpanzees do not leave their mother regularly until at least 9 or 10 years of age (van Lawick-Goodall 1973). Before this time, the mother may actively promote nutritional independence and encourage the juvenile to travel with her without being carried (Clark, this volume), but she apparently does not encourage spatial independence. It is the growing youngster who is most responsible for increasing the amount of time it spends at a distance from the mother.

The ways in which captive quarters differ from natural habitats alter the adaptive value of behaviors related to infant nutrition, safety, and learning. In captivity, food is abundant and of relatively high quality throughout the year; there are no predators or poisonous plants, and no risk of getting lost. Consequently, the young chimpanzee has less to learn about the environment. Under these conditions, earlier infant independence is in the evolutionary "best interests" of both mother and infant. If earlier independence does not conflict with the infant's needs for nourishment and protection, it means a higher probability of infant survival should the mother sicken, die, or disappear. It also means that the mother can begin to invest energy in the production and care of additional offspring somewhat earlier.

The faster rate of growth and development in captive chimpanzees relative to free-ranging individuals (Pusey and Riss, personal communication) indicates that captive standards of nutrition normally exceed those characteristic of the Gombe study area. In humans, better nutrition can shorten the length of lactational amenorrhea, a major determinant of birth interval in many populations (Short 1974; Frisch and McArthur 1974). The same is probably true of chimpanzees. Since weaning is only completed when the mother becomes pregnant again, better nourished captive mothers should be expected to wean their infants earlier than free-ranging mothers. Whether or not better nutrition increases the quality or amount of the mother's milk during the shortened period of lactation, the availability of high-protein supplementary food insures adequate infant nutrition during and after weaning. Yerkes and Tomilin (1935) reported that caged chimpanzee infants, if left with their mothers, were weaned between 2 and 3 years of age, or almost 2 years earlier than most free-ranging infants (Clark, this volume).

If, in captivity, earlier infant independence is evolutionarily advantageous for both mother and offspring, why does the offspring predictably object?[3] As Trivers (1974) points out, selection would have favored parental tendencies to mold offspring behavior more strongly than offspring tendencies to modify their own behavior under fluctuating environmental conditions:

> When circumstances change, altering the benefits and costs associated with some offspring behavior, both the parent and the offspring are selected to alter the offspring's behavior appropriately. That is, the parent is selected to mold the appropriate change in the offspring's behavior, and if parental molding is successful, it will strongly reduce the selection pressure on the offspring to change its behavior spontaneously. Since the parent is likely to discover the changing circumstances as a result of its own experience, one expects tendencies toward parental molding to appear, and spread, before the parallel tendencies appear in the offspring. (Trivers ibid.: 262)

In situations in which parental molding of offspring behavior actually occurs, parent-offspring conflict can be expected to arise and to continue until the parent brings about the desired offspring response, or the offspring changes its behavior voluntarily.

The behavioral repertoire of the newborn chimpanzee enables it to maintain close contact with its mother—it clings to her ventrum, and if contact is somehow broken, it whimpers or screams. For the first several months of life, behaviors related to suckling and the maintenance of contact make up the bulk of the infant's activities. There has not been selection for much flexibility in this pattern. Thus, the infant will resist maternal attempts to reduce the amount of physical contact.

The caged mother, Kami's, attempts to encourage independent behavior in Buster were met with strong opposition. He whimpered loudly and, if this failed to bring about his mother's return, he crawled doggedly after her. Unfortunately, observations of this pair could not be continued long enough for the outcome of the conflict to be determined. One testament to Kami's success in promoting Buster's independence is the fact that he survived the undoubtedly traumatic experience of permanent separation from her at the age of 5 months, an ordeal that has severely damaging effects on many wild youngsters, even those who are nutritionally self-sufficient (van Lawick-Goodall 1971, 1973).

A real measure of the adaptive value of variations in the mother-infant interaction depends on the reproductive success of the offspring. If, for example, the benefits of earlier independence were outweighed by long-term psychological effects of parent-offspring conflict that somehow reduced the reproductive abilities of the offspring, the patterns of interaction that gave rise to the conflict could not be considered adaptive. The solution to the problem of correlating offspring reproductive success with different styles of mother-infant interaction must await much more data from laboratory and field studies.

■ CONCLUSION

In light of the small number of chimpanzee mother-infant pairs who were observed, conclusions must be tentative. The three pairs, observed during the infants' first 20 weeks, were very similar with regard to the quality of maternal care of the neonate and measures of infant suckling. Because both of these aspects of the mother-infant interaction are directly related to infant survival, they are probably more resistant to environmental influences than those behaviors of less immediate biological consequence. The three pairs differed most markedly in measures of infant independence and maternal role in promoting independence. Differences in the sex and level of physical development of the infants may have been important contributing factors; more thorough studies of the effects of these variables on the chimpanzee mother-infant interaction would be worthwhile. Environmental effects were probably of major importance. Variations in patterns of mother-infant interaction can be understood in terms of the selective forces which have favored the ability of mothers and infants to respond to changing circumstances, when the benefits and costs associated with some patterns of interaction are altered. Data support the prediction that mothers will encourage infant independence earlier in less threatening environments and that this maternal strategy, although in the best interests of both mother and infant, may give rise to conflict between them.

■ ACKNOWLEDGMENTS

Dr. C. E. G. Tutin and Dr. W. C. McGrew began observation of both captive groups and provided advice and assistance during all stages of the research. I am especially grateful to Dr. Tutin for detailed notes and photographs. The following people also contributed data: C. Cochran, J. Crocker, P. Midgett, D. Riss, S. Simpson, and the field assistants of the Gombe Stream Research Centre. Thanks to Dr. H. Kraemer for advice during data analysis and to Drs. J. Goodall and D. A. Hamburg for their continuing support and encouragement. Support and facilities were provided by the following: the Grant Foundation; the Program in Human Biology and the Department of Psychiatry, Stanford University; the Delta Regional Primate Research Center (Dr. P. Gerone, Director); the Gombe Stream Research Centre (Dr. J. Goodall, Director), with the cooperation of the government of Tanzania.

■ NOTES

1. At present, there is no evidence from the Gombe data that individual differences on measures of mother-infant proximity are related to the sex of the infant.

2. Goodall (van Lawick-Goodall 1967) reports ages ranging from 12–16 weeks for the eruption of the first lower incisor in three Gombe infants.

3. This analysis does not imply that the responses of either parent or offspring must be *conscious*. Other mechanisms can be imagined, but are not of immediate concern here.

■ REFERENCES

Castell, R., and Wilson, C. 1971. Influence of spatial environment on development of mother-infant interaction in pigtail monkeys. *Behaviour* 39:202–211.

Drickamer, L. C. 1974. A ten-year summary of reproductive data for free-ranging *Macaca mulatta*. *Folia Primat.* 21(1):61–80.

Frisch, R. E., and McArthur, J. W. 1974. Menstrual cycles: Fatness as a determinant of minimum weight for height necessary for their maintenance or onset. *Science* 185:949–951.

Hinde, R. A., and Atkinson, S. 1970. Assessing the roles of social partners in maintaining mutual proximity, as exemplified by mother-infant relations in rhesus monkeys. *Anim. Behav.* 18:169–176.

Hinde, R. A., and Spencer-Booth, Y. 1967. The effect of social companions on mother-infant relations in rhesus monkeys. In *Primate ethology*, ed. D. Morris, pp. 343–364. Chicago: Aldine.

———. 1971. Towards understanding individual differences in rhesus mother-infant interaction. *Anim. Behav.* 19:165–173.

Hinde, R. A., and White, L. E. 1974. Dynamics of a relationship: rhesus mother-infant ventro-ventral contact. *J. Comp. Physiol. Psychol.* 86:8–23.

Itoigawa, N. 1973. Group organization of a natural troop of Japanese monkeys and mother-infant interactions. In *Behavioral regulators of behavior in primates*, ed. C. R. Carpenter, pp. 229–250. Lewisburg: Bucknell University Press.

Jensen, G. D.; Bobbitt, R. A.; and Gordon, B. N. 1968a. Sex differences in the development of independence of infant monkeys. *Behaviour* 30:1–14.

———. 1968b. Effects of environment on the relationship between mother and infant pigtailed monkeys (*Macaca nemestrina*). *J. Comp. Physiol. Psychol.* 66:259–263.

Kaplan, J. 1972. Differences in the mother-infant relations of squirrel monkeys housed in social and restricted environments. *Dev. Psychobiol.* 5:43–52.

Kawamura, S. 1958. The matriarchal social order in the Minoo-B group: A study on the rank system of Japanese macaque. *Primates* 1:149–156.

Koford, C. B. 1963. Rank of mothers and sons in bands of rhesus monkeys. *Science* 141:356–357.

Kummer, H., and Kurt, F. 1965. A comparison of social behavior in captive and wild hamadryas baboons. In *The baboon in medical research*, ed. H. Vagteborg, pp. 65–80. Austin: Texas University Press.

Lawick-Goodall, J. van. 1967. Mother-offspring relationships in free-ranging chimpanzees. In *Primate ethology*, ed. D. Morris, pp. 365–436. Chicago: Aldine.

———. 1968. The behaviour of free-living chimpanzees in the Gombe Stream Reserve. *Anim. Behav. Monogr.* 1:161–311.

———. 1971. Some aspects of mother-infant relationships in a group of wild chimpanzees. In *The origins of human social relations*, ed. H. R. Schaffer, pp. 115–128. New York: Academic Press.

———. 1973. The behavior of chimpanzees in their natural habitat. *Am. J. Psychiatry* 130:1–12.

Lemmon, W. B. 1968. Delivery and maternal behavior in captive reared chimpanzees, *Pan troglodytes*. Paper read at the A.A.A.S. meetings, Dallas, Texas.

Mason, W. A. 1968. Naturalistic and experimental investigations of the social behavior of monkeys and apes. In *Primates: Studies in adaptation and variability*, ed. P. C. Jay, pp. 398–419. New York: Holt, Rinehart, and Winston.

———. 1970. Chimpanzee social behavior. In *The chimpanzee*, vol. 2, ed. G. H. Bourne, pp. 265–288. Basel: S. Karger.

Mitchell, G. D. 1968. Attachment differences in male and female infant monkeys. *Child Dev.* 37:781–791.

Ransom, T. W., and Rowell, T. E. 1972. Early social development of feral baboons. In *Primate socialization*, ed. F. E. Poirier, pp. 105–144. New York: Random House.

Rogers, C. M., and Davenport, R. K. 1970. Chimpanzee maternal behavior. In *The chimpanzee*, vol. 3, ed. G. H. Bourne, pp. 361–368. Basel: S. Karger.

Rosenblum, L. A. 1971. The ontogeny of mother-infant relations in macaques. In *The ontogeny of vertebrate behavior*, ed. H. Moltz, pp. 315–367. New York: Academic Press.

———. 1974. Sex differences, environmental complexity, and mother-infant relations. *Arch. Sex. Behav.* 3:117–128.

Rowell, T. E. 1968. The effect of temporary separation from their group on the mother-infant relationship of baboons. *Folia Primat.* 9:114–122.

Sackett, G. P. 1972. Exploratory behavior of rhesus monkeys as a function of rearing experiences and sex. *Dev. Psychol.* 6:260–270.

Seay, B. 1966. Maternal behavior in primiparous and multiparous rhesus monkeys. *Folia Primat.* 4:146–168.

Short, R. V. 1974. Man, the changing animal. In *Physiology and genetics of reproduction*, eds. F. Fuchs and E. M. Coutinho. New York: Plenum Press.

Spencer-Booth, Y.; Hinde, R. A.; and Bruce, M. 1965. Social companions and the mother-infant relationship in rhesus monkeys. *Nature* 208:301.

Tanner, J. M. 1974. Variability of growth and maturity in newborn infants. In *The effect of the infant on its caregiver*, eds. M. Lewis and L. A. Rosenblum, pp. 77–103. New York: John Wiley & Sons.

Trivers, R. L. 1974. Parent-offspring conflict. *Amer. Zool.* 14:249–264.

Wolfheim, J. H.; Jensen, G. D.; and Bobbitt, R. A. 1970. Effects of group environment on the mother-infant relationship in pigtailed monkeys (*Macaca nemestrina*). *Primates* 11:119–124.

Wrangham, R. W. 1975. Behavioural ecology of chimpanzees in Gombe National Park, Tanzania. Ph.D. dissertation, Cambridge University.

Yerkes, R. M. 1933. Genetic aspects of grooming, a socially important primate behavior pattern. *J. Soc. Psychol.* 4:3–25.

———. 1943. *Chimpanzees: A laboratory colony*. New Haven: Yale University Press.

Yerkes, R. M., and Tomilin, M. I. 1935. Mother-infant relations in chimpanzees. *J. Comp. Psychol.* 20:321–359.

Part Four

EVOLUTIONARY PERSPECTIVE

20 Bio-Social Approach to Human Development

RICHARD C. SAVIN-WILLIAMS
DANIEL G. FREEDMAN

The appearance of E. O. Wilson's *Sociobiology: The New Synthesis* has given direction, depth and guidance to the field known as human ethology. It had previously consisted largely of "naturalistic" observations of preschool children in school settings (e.g., Blurton-Jones 1972; McGrew 1972) and unpublished studies (Freedman 1974a) and published speculation (Guthrie 1969; Wickler 1967) on the psychobiological meaning of the secondary sex characteristics. Wilson's book now opens the door between modern biology and the behavioral sciences, and his claim that a new era of sociobiological research is upon us seems entirely justified.

This report must be considered pre-Wilsonian, and it emphasizes studies done at the University of Chicago over the last decade by staff and students of the Committee on Human Development, working within an evolutionary framework.

These studies can be conveniently parceled into behavior-genetics studies, experimental studies of secondary sex characteristics, and studies on the development of status hierarchies. Our emphasis in this review will be on the third area, since it seems the most pertinent.

■ BEHAVIOR-GENETICS STUDIES

Our approach to behavior genetics has been strongly influenced by systems analysis and by its predecessor, holistic analysis. Basically, we are developmentalists who conceive of individual organisms as constructing their own reality on the basis of information exchange at all levels, including the genetic. Thus, two breeds of dog come to the same environment with different gene-mediated sets and therefore achieve different interactions with that environment; they consequently perform at different phenotypic levels. Freedman (1958) has proved the paradigm. Four breeds of dog were reared in two fashions by the investigator—"indulged" and "disciplined." Rearing began at three weeks of age, just as the pups began to hear, see, and move about. It was immediately clear that the passive E was a different stimulus for each breed: Shetland sheepdogs tentatively explored him, always mantaining a hesitance and gentleness. E's slightest movement caused them to start. Later, when trained to a

563

task, the slightest punishment had lasting effects, and it was at all times clear that E provoked both intense interest and vigilance in Shetland sheepdogs. Wire-haired terriers brought another approach to the same situation. The slightest suggestion of playfulness on E's part brought out a pup's play-aggression. Nipping and biting, within the playful context, started even as mobility was first achieved. By four and a half weeks, E had to wear gloves to prevent his skin being torn, and despite the fact that these pups never bit as hard as they obviously could, E could actually feel their jaws tremble as they tried to keep biting behavior "within bounds." Excitement and attraction, licking and leaping reached a point where the animals appeared beside themselves.

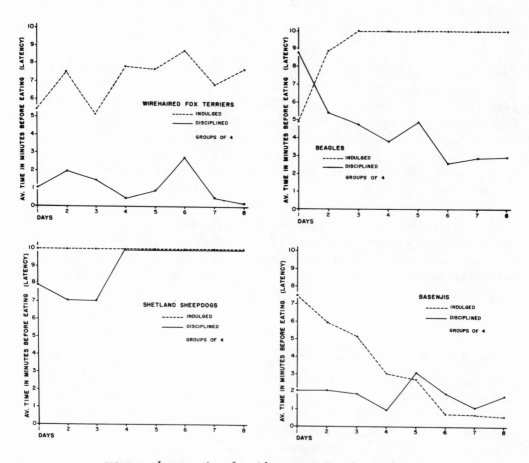

FIGURE 1. *Incorporation of punishment test. Four breeds of dog were raised under "indulged" or "disciplined" conditions from 3–8 weeks of age. They were then given a series of tests, including this one: Fresh ground meat was placed in the center of the room and the pup was punished for eating it. (All were attracted to it since they were on kibble rations.) E then left the room, and the time before eating was recorded through a one-way glass.*

While such super-excitement, biting and intense affection were the keys to wire-haired behavior in the indulged condition; in the disciplined condition, one had the sense that a spirit was being broken. Unlike the beagles, who were also greatly attracted to E, training of wire-haireds involved a thorough transformation of character—from a free, aggressive spirit to a somewhat depressed and compliant animal.

And so it went with the other two breeds (beagles and basenjis), each with a unique approach, each achieving unique interactions as a result, and each breed exhibiting unique phenotypic behavior over and above the usual individual differences (see Figure 1).

Our approach to human behavior genetics has been very similar. The first studies were of infant twins, and the relative development of identical and fraternal pairs was carefully followed over the first year and a half of life, with special emphasis on the earliest months. Insofar as each identical pair represented a single genotype, even as two members of a breed represent two highly similar genotypes, it was assumed that they would tend to similarly structure their shared environment (all twins were reared together). Similarly, fraternal pairs would tend to differentially negotiate their environment, thereby yielding substantially different phenotypes. And indeed this is what emerged from those studies (see Figures 2 and 3 for gross representations of the over-all findings). For example, if one identical had a tendency to smile as his mother whispered to him at three months, the other would tend also to smile in those circumstances. On the other hand, one fraternal might participate with his mother by

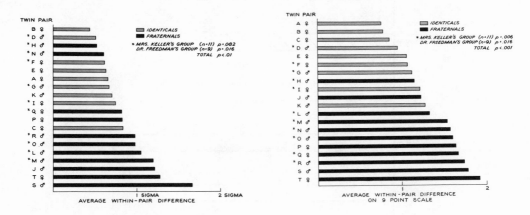

FIGURE 2. *Bayley Mental and Motor Scales. These are the summary results of monthly testing by two independent E's over the first year of life.*

FIGURE 3. *Bayley Infant Behavior Profile. Rating scales consisting of 21 items covering 12 categories of behavior. These are summary results of monthly testing by two independent E's over first year of life.*

staring intently yet not smiling, while his brother might smile readily but then immediately turn away or close his eyes (see Figure 4). That is, modes of participation with the caretaker were seen to vary with genotype.

The next step has involved looking at infants of inbred human groups. Do Navajo, for example, negotiate the early months of life in a unique manner compared, say, with Australian aboriginal infants? Is the total context, including its genotypic aspects, different in each cultural group, even as it was for our pups and our twins? Our initial results indicate that this is indeed so. Within the total contexts of differing cultures, differing genotypes and differing ecologies, it would appear that two such groups construct differing views of the world. The evidence is clear-cut that *at birth* such differing, ethnically bounded groups as Navajo, Australian aboriginal, or Nigerian

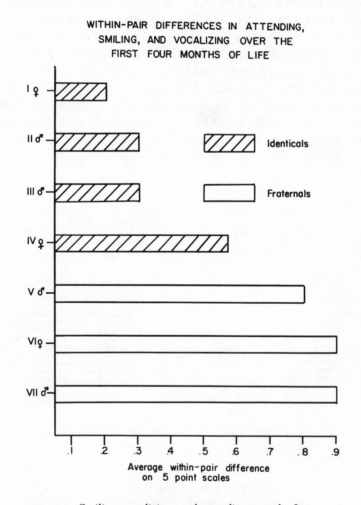

FIGURE 4. *Smiling, vocalizing, and attending over the first 4 months of life. Combined scale, based on weekly visits, starting at 2 weeks of age.*

Hausa differ considerably on a number of behavioral parameters (Freedman 1974). We present here our initial study (1969: 1227) of Chinese-American and European-American newborns:

> Twenty-four Chinese-American and twenty-four European-American newborns were examined while still in the nursery. The Orientals were largely of Cantonese background. All families were middle class and the bulk were members of a pre-paid health plan.
>
> The behavior scales (Brazelton and Freedman 1971) consisted of twenty-eight general behavioral items rated 1–9 and eighteen standard neurological signs, frequently used to screen for neural damage, rated 0–3. The twenty-five general items may be somewhat arbitrarily arranged into five categories as follows: (1) temperament—five items; (2) sensory development—three items; (3) autonomic and central nervous system maturity—eight items; (4) motor development—six items; (5) social interest and response—six items. As will be seen, any single item may overlap categories (e.g., lability of skin color can reflect temperament as well as autonomic maturity).
>
> All testing was done during September and October, 1968. Each test session lasted between 30 and 40 minutes. Testing was performed in the newborn nursery by the author's wife as the author watched, and scoring was done immediately afterwards in a room next to the nursery. Apart from a reliability sample which was marked independently, scoring depended on verbal agreement between the testers.
>
> A multivariate analysis of variance indicated that, on the basis of total performance, the two groups were decidedly different (p = .008). Further analysis indicated that the main loading came from the group of items measuring temperament, and those that seemed to tap excitability/imperturbability (p = .001). While the following discussion is based on mean ethnic difference on the distinguishing items, it should be emphasized that *there was substantial overlap in range on all scales between the Chinese and Caucasian infants.*
>
> The European-American infants reached a peak of excitement sooner (rapidity of build-up) and had a greater tendency to move back and forth between states of contentment and upset (lability of states). They showed more facial and bodily reddening probably as a consequence. The Chinese-American infants were scored on the calmer and steadier side of these items. In an item called *defensive movements*, the tester placed a cloth firmly over the supine baby's face for a few seconds. While the typical European-American infant immediately struggled to remove the cloth by swiping with his hands and turning his face, the typical Chinese-American infant lay impassively, exhibiting few overt motor responses. Similarly, when placed in the prone position, the Chinese infants frequently lay as placed, with face flat against the bedding, whereas the Caucasian infants either turned the face to one side or lifted the head. Inasmuch as there was no difference between the groups in the ability to hold the head steady in the upright, this maintenance of the face in the bedding is taken as a further example of relative

imperturbability, or ready accommodation to external changes. (Another possible explanation for these differences involves the obvious difference in nasal bone structure between Orientals and Caucasians, with the average Caucasian nose considerably larger.) In an apparently related item, rate of habituation, a pen light was repeatedly shone on the infant's eyes, and the number of blinks counted until the infant no longer reacted (shuts-off). The Chinese infants tended to habituate more readily.

There was no significant difference in amount of crying, and when picked up and consoled, both groups tended to stop crying. The Chinese infants were, however, often dramatically immediate in their cessation of crying when picked up and spoken to, and therefore drew extremely high ratings in consolability. The Chinese infants also tended to stop crying sooner without soothing (self-quieting ability).

To summarize, the majority of items which differentiated the two groups fell into the category of temperament. The Chinese-American newborns tended to be less changeable, less perturbable, tended to habituate more readily, and tended to calm themselves or to be consoled more readily when upset. In other areas (sensory development, central nervous system maturity, motor development, social responsibility), the two groups were essentially equal.

The next logical step involved a follow-up study so that we might trace the growing infants as they became members of their respective subgroups. The first such study was performed with ten Chinese-American and ten European-American mother-infant pairs from birth through four months of age (Kuchner 1976). The Chinese infants were at birth less excitable, more easily quieted when crying, and more adaptable to impositions of various kinds, just as in the study reported above. As a consequence, there was less need to interact with them and, indeed, over the course of the study, Chinese mothers were seen to pick up their babies far fewer times than did Caucasian mothers. The latter ended with far more interactions per unit time, even as the Caucasian infants ended with more vocalizations and more over-all movement. Indeed, while both sets of parents responded to an attitude questionnaire with similar ideas about child rearing, actual practices varied as a result of the totally different configurations in each set of households—differences due in part to those aspects of behavior seen to vary at birth.

That these differential starts in life result in later differences can as well be confirmed by nursery school studies. Nova Green (1969: 6) found substantially different patterns of behavior in Caucasian and Chinese nursery schools.

Although the majority of the Chinese-American children were in the "high arousal age," between 3 and 5, they showed little intense emotional behavior. They ran and hopped, laughed and called to one another, rode bikes and roller-skated just as the children did in the other nursery schools, but the noise level stayed remarkably low and the emotional atmosphere projected serenity instead of bedlam. The impassive facial expression certainly gave the children an air of dignity and self-possession, but this was only one element affecting the total impression. Physical movements seemed more coordinated, no tripping, falling, bumping or bruising was observed, no screams, crashes or wailing was heard, not even that common sound in other nurseries, voices raised in highly indignant moralistic dispute! No property disputes were observed and only the mildest version of "fighting behavior," some good natured wrestling among the older boys. The adults evidently had different expectations about hostile or impulsive behavior; this was the only nursery school where it was observed that children were trusted to duel with sticks. Personal distance spacing seemed to be situational rather than compulsive or patterned, and the children appeared to make no effort to avoid physical contact.

Doubtless, we are here witnessing the early ontogenetic form of substantial sub-cultural differences within the U.S., differences that have in substantial part been parlayed by gene pool differences. One could speculate further about East-West differences in philosophic and artistic tastes, but we will end our discussion here, saying only that such differences as do exist could not have arisen out of the blue. A contextual analysis merely asserts that a part of the observed ethnic variations must be attributable to the observed gene pool variations.

The question of prenatal causation versus genetic causation will be omitted here, although Freedman (1974) has discussed it elsewhere. Suffice it to say, that the evidence makes it unlikely that temperament is mediated by nongenetic prenatal events.

■ EXPERIMENTAL STUDIES OF THE SECONDARY SEX CHARACTERISTICS

Human secondary sexual characteristics are particularly open to experimental work within the evolutionary framework. For example, one can ask if such sex differences as breasts, beards, or the height and voice timbre differential have unique adaptive functions. One can then go about testing for differential psychological responsivity to these traits by modifying standard psychological techniques.

Our first such study involved putting beards on the men of certain cards of the Thematic Apperception Test to see if the elicited stories would thereby change (Figures 5 and 6). In administering the TAT, one is asked to "tell a story" about the scene, and the responses are then analyzed for underlying themes. Our hypothesis was that beards act as male-male intimidation devices (cf. Guthrie 1969), and that male respondents, in particular, would be differentially affected by the presence or absence of a beard. The results are shown in table 3. Whereas the male University of Chicago undergraduate respondents tended to see the unbearded young man in figure 5 as a youth receiving advice from a wiser, older man, the beard usually served to raise his status a notch or two, so that a response such as "two lawyers discussing a case" was not uncommon.

FIGURES 5a and 5b. *Card 7BM of Thematic Apperception Test, original and doctored version with beard. See Table 1.*

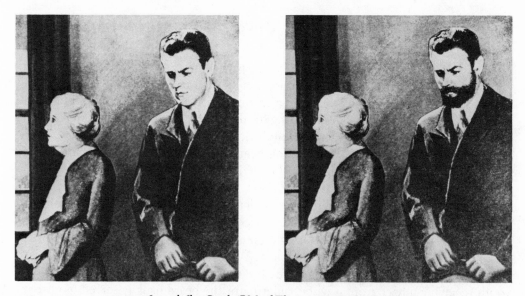

FIGURES 6a and 6b. *Card 6BM of Thematic Apperception Test, original and doctored version with beard. See Table 1.*

TABLE 1 / THEMES OF STORIES TOLD IN RESPONSE TO STANDARD
AND EXPERIMENTAL THEMATIC APPERCEPTION TEST CARDS.
HALF OF THE 12 MALE AND 12 FEMALE UNDERGRADUATE
SUBJECTS WERE ASSIGNED TO EXPERIMENTAL TAT SEQUENCES,
AND THE OTHER HALF WERE SHOWN THE CONTROLS. EACH S
WAS ASKED TO "TELL WHAT IS HAPPENING IN THE PICTURES."
STORIES WERE ANALYZED FOR CONSISTENT DIFFERENCES IN
CONTENT.

	TAT CARD 7 BM		EXPERIMENTAL CARD*	
	Females	Males	Females	Males
Submission of younger person	6	9	5	2
Equality or dominance of younger person	4	1	5	8
	TAT CARD 6 BM		EXPERIMENTAL CARD**	
	Females	Males	Females	Males
Both equally involved	8	8	2	3
Man acts, woman reacts	2	2	8	7

*beard added to younger man's face
**beard added to man's face

As can be seen, females did not make this distinction with anywhere near the consistency of male subjects, tending, thereby, to support the initial hypothesis. Assuming then that the male-male intimidation hypothesis is correct, beards might simultaneously serve to attract females, but we have not as yet performed the appropriate studies.

A second TAT card was used (Figure 6 and Table 1)—this time a young man with an older woman—and again the beard tended to raise the young man's status. Whereas without the beard he was most often in some sort of emotional debt and bond with the woman, with the beard he was usually seen as more independent. One story to the bearded version, "young Freud leaving home," seemed to sum it all up.

It can be objected that these are culture-specific responses, and the only answer is to repeat such studies in other cultures. As of now we have only supportive anecdotes obtained from anthropologists, and this is not the place to recount them. As an example of this sort of evidence, among beardless, warlike peoples, such as the Maori or Plains Indians of the last century, facial tattooing and painting were expressly performed to intimidate the enemy and to enhance the warrior's sense of power (Guthrie 1969). One can build almost endlessly on information like this, but clearly, experimental cross-cultural studies will be far more convincing.

We have performed TAT-type studies varying breast size—again with very consistent results only from the male respondents. The presence of even the barest outline of a breast (Figure 7) significantly increases the number of stories with courtship as a theme. Once again, speculation and educated observations abound on the sexual function of human breasts (cf. Wickler 1967), but let it suffice here to add that there is no correlation at all between breast size and milk production, and suckling cannot be the reason for the relatively large breasts of human females.

We have examined yet another hypothesis experimentally—that human smiling is, in part, homologous with the fear-grin of chimps and other primates. We have attacked this issue in several ways. In a study by Beekman (1970), a dominance hierarchy was ascertained in a mixed-sex class of eight year olds, by having all possible pairs of children rate one another's "toughness." Children were then paired in high and low dominance pairs, and in same-sex and mixed-sex pairs. Each pair was then given the task of drawing a picture together and the ensuing behavior was videotaped. It was found that low-dominance children gazed at and smiled at high-dominance children with significantly greater frequency with one important exception. Low-dominance boys did not gaze at or smile at the higher-dominance girls any more than the reverse was true (Table 2); we were struck that this follows the intuition that there

FIGURES 7a and 7b. TAT-like stimulus cards. The second card, differing only in outline of breast, received substantially more courtship stories from male respondents than did the first.

TABLE 2

I. MEAN PERCENT GAZING AT PEER

A. Same-sex, unequal-status pairs

		High-status child	Low-status child
First grade	Boys	7.22	19.52
	Girls	11.90	16.80
Third grade	Boys	3.65	8.50
	Girls	13.19	27.06

B. Cross-sex pairs

	Boys	Girls
First grade	8.82	17.35
Third grade	12.27	18.31

II. MEAN PERCENT SMILING

C. Same-sex, unequal-status pairs

		High-status child	Low-status child
First grade	Boys	51.25	42.54
	Girls	21.78	47.60
Third grade	Boys	20.50	25.83
	Girls	34.69	70.34

D. Cross-sex pairs

		Pairs with high-status girls	Pairs with low-status girls
First grade	Boys	21.21	28.11
	Girls	13.88	39.41
Third grade	Boys	27.91	26.95
	Girls	25.93	51.55

The 32 Ss were drawn from two first and third grade classes at the University of Chicago Lab School. These children had participated in a larger study for which each child ranked his classmates on a dominance hierarchy defined as "toughness." The Ss chosen for this study were the 8 members of each class whose peers had ranked them nearest the top or bottom of the classroom hierarchy. These high and low status boys and girls were paired in all possible combinations of status and sex, making a total of 6 cross-sex and six same-sex pairs for each class. The order in which the pairs interacted was randomized as much as possible. (Beekman 1970) is a tendency for girls to defer to boys, even when the former are clearly superior at a task. We thereupon set out on a further series of studies, including studies of newborns, aimed at elucidating these sex differences.

Observations in the newborn period have yielded few sex differences, but there is a major one now replicated in three different laboratories. Newborn females exhibit higher rates of reflexive smiling than do newborn males (see Freedman 1974). In a carefully controlled study by De Boer (in Freedman 1974a), the sex of the infant was unknown during experimental observations, boys had not yet been circumsized, and observations were made at standardized times and in the same place. Her overall results (Table 3) confirm that newborn females tend to smile more. Studies at subsequent ages through adulthood, in various settings, and across a number of cultures, consistently find females with lower thresholds of smiling (Freedman 1976). If human smiling is indeed an appeasement gesture—and it certainly may be described that way on many, if not most of the occasions one exhibits it—young females of our species are more specialized in the avoidance of aggression, and are particularly reticent to express it vis-à-vis males.

With this interpretation in mind, we performed the following study. Cronin (1976) trained eight-year-old boys and girls to play a game called "dodgeball" in which a circle of children try to hit a child in the center of the circle with the ball. When hit, the child returns to the circle and the hitter goes to the center. Cronin separated the children on the basis of ability, and four teams were developed—high- and low-ability

TABLE 3 / SEX DIFFERENCES IN INFANT SMILING			
	FREQUENCY OF SMILES		
SUBJECTS	FULL	ASYMMETRICAL	FLEETING
Female	27	25	169
(N=14)	X=1.9	X=1.8	X=12.0
Male	6	10	49
(N=9)	X=0.7	X=1.1	X=5.4
Difference	ns		ns	Significant
by sex		combined		at .05 level
		ns	

The 9 male and 14 female 2-day-old infants were all full-term, normal, black, and bottlefed. Each infant was delivered to the Experimenter by the nurse swaddled in its crib so that the sex was not apparent. None of the males had as yet been circumcised. The observations were done within 90 minutes of the infants' last feed, and all infants were predominantly in a light sleep or drowsy state during the 10-minute observation period. During the observation, the author recorded any indications of smiling into one of three categories: full smiling (at least one second), asymmetrical smiling, and fleeting smiles (short jerks of the corners of the mouth). The female infants were observed to exhibit more smiling of all types, but especially in the category of fleeting smiles.

boys and high- and low-ability girls; then, all teams were paired in a "tournament." Table 4 shows that in all cases but one, the team assessed as more skillful won their bout. However, as in the Beekman (1970) study, when high-skill girls were paired with low-skill boys, the girls completely collapsed, performing far below capacity. We can only conclude from these studies that girls, in the presence of boys, tend not to pursue their advantage, and instead, tend to give in. Guttman (1972), an experimental clinical psychologist, has documented similar behavior in five cultures (Navajo, Druze, Highland and Lowland Maya, American Anglo) and has developed a notion of the *parental imperative* to explain his results. Essentially, the parental imperative means that human females subdue their own egoistic drives in deference to that of their mates for the purpose of successfully rearing young; in this way, the female assures her mate's cooperation and "loyalty." Guttman gives further evidence that in all five cultures, after their period of fecundity, women tended toward greater expression of their own needs, and many males then had to defer to their mates for the first time.

In addition to the experimental corroboration in young females by Beekman (1970) and Cronin (1976) described above, it will be seen in the next section (Parker 1976) that both boys and girls behave differently when in same-sex or opposite-sex groups. Not only do these findings appear to reflect, generally, the nature of men and women in apparently universal roles vis-à-vis one another, but they have practical implications for education. We are now pursuing this in studies of females and males in same-sex and co-ed schools. Our initial results indicate a greater average degree of

TABLE 4 / SAME-SEX AND MIXED-SEX DODGEBALL GAMES

GAME	PLAYERS SKILL, SEX	PACE (RETRIEVALS PER HIT)	GROUP CENTRALITY (PERCENTAGE IN CENTER)
1	High girls Low girls	8.33	.63 .37
2	High boys Low boys	3.9	.55 .45
3	High girls Low boys	5.8	.27 .73
4	High girls High boys	4.2	.36 .64
5	Low girls Low boys	8.0	.44 .56
6	Low girls High boys	2.89	.00 1.00

Game characteristics. Note especially game number 3 in which girls previously ascertained as better players than the boys, nevertheless collapse in performance.

independence and egoism in girls from all-girl schools, and less competitiveness in boys from all-boy schools. In the same vein, the 1973 Carnegie Foundation Report on Higher Education has tabulated a far higher percentage of women from non co-ed schools in leadership positions.

We trust that this section has illustrated the various strategies that can be employed in pursuit of evolutionary hypotheses, and that it has also illustrated the varied and interesting topics such a pursuit can encounter.

▪ THE DEVELOPMENT OF STATUS HIERARCHIES

A primary empirical concern of human ethologists is the nature and function of dyadic dominance behavior occurring within a social group that systematically arranges group members into a hierarchical structure. This interest is not unique among human ethologists; comparative and social psychologists have traditionally concerned themselves with the formation and maintenance of group structure (see, for example, Schjelderup-Ebbe 1935 and Anderson 1937). The recent flurry of human studies with an evolutionary perspective is deeply indebted to these two disciplines for their hypotheses, methodologies, findings, and insights of naturally occurring social behavior.

However, several major provocative questions raised by comparative and social investigators remain unanswered:

1. To what particular patterns of behavior does "dominance" refer?
2. Do human groups manifest a dominance hierarchy as a result of intragroup dyadic behavior that structures or gives meaning to individual and group behavior?
3. How do the sexes differ in the above two questions?
4. What are the physical, mental, personality, and social characteristics of those who occupy varying status positions?
5. If a dominance construct exists, how stable is it over time and activities?
6. What are the individual benefits or prerogatives of various dominance positions?
7. Of what benefit is the dominance hierarchy to the group or species?

From the data of human ethological studies conducted in the last five years there is sufficient evidence to provide initial answers to these questions. After a short discussion of the contributions of comparative and social psychology, this section reviews the human ethological answers to the above seven questions.

COMPARATIVE PSYCHOLOGY

Common to most vertebrate species—and in some invertebrates as well—are behavioral interactions among group members that structure the group into a system of status differentiation necessary for group formation and maintenance. In fact, Alcock

(1975: 229) asserts that social animals which did not evolve a system of interindividual dominance relations became extinct because "their excess members lived longer during hard times, devoured the countryside, and caused the downfall of the entire group." Since the early 1900s, it has been reported that many species—including the cockroach, bumblebee, lobster, frog, fish, most birds, rat, cat, pig, and most primates—develop a dominance hierarchy—some in controlled laboratory situations where territoriality is thwarted, and others in their natural habitat (see recent reviews by van Kreveld 1970; Jolly 1972; Richards 1974; Syme 1974; Wilson 1975; Freedman et al., in preparation).

Initially, dominant-subordinate relationships were considered to be solely aggressive in nature, one animal controlling the behavior of another by force or fighting (Schneirla 1951; Scott 1953). Being dominant was not synonymous with being a leader; but group members did, via dominance behavior, arrange themselves into a hierarchical order, serving to reduce intraspecific fighting, to organize group members, and to adjust mutual relationships. Leader-follower relationships, on the other hand, were considered to be based on mutual attraction and the distribution of rewards in an independent-dependent manner (Scott 1953).

Schneirla (1951) asserted that the dominance hierarchy structure could not be unequivocally applied to all animal species or groups since such factors as ecology, physiology, sexual relations, food availability, population density, etc., often intervene to alter intragroup relations. Schneirla also questioned the adaptive function that a dominance hierarchy provides for a group—too much dominant behavior within a group prevents cohesion, stability, and unity by increasing interindividual distance. Unfortunately, however, Schneirla confuses dominance behavior with aggression.

Because of the prevalence of dominance hierarchies in the mammal kingdom, Scott (1953, 1969) sees little reason to doubt its existence in human groups, although it may be more subtle (less aggressive) in nonchildren groups. It is still empirically uncertain, though, if primatologist Zuckerman's (1932) early assertion—that one cannot adequately explain primate sociology without reference to the principle of dominance—is applicable to humans.

SOCIAL PSYCHOLOGY

The social psychological concept that is most analogous to the primatologist's dominance hierarchy is leadership. During the last fifty years social psychology has concerned itself with the nature of group leadership structures and with those who become leaders and followers. A leader is considered to be one who exerts social power: the ability to get others to do what one wants them to do, more than the reverse, or to influence others more than one is influenced by them. The leadership structure of a group may be based upon the emotional acceptance or the personal competence of particular individuals within the group (Glidewell et al. 1966). Most usually, only a few groups members are distinguished as leaders; the rest are considered to be followers, to varying degrees.

No attempt will be made here to review the massive leadership literature; adequate reviews are available elsewhere (Collins and Raven 1969; Gibb 1969). Important for this chapter is the doubtful conclusion that leadership and dominance can be equated, either in human or nonhuman primates (Gartlan 1968). In the former, leadership usually refers not to interpersonal or dyadic behavior, but to the ability of one or several individuals to exercise influence or control over the cognitions, attitudes, behaviors, or emotions of the group (Collins and Raven 1969). While leadership may be utilized in nonhuman primates to refer to the social role or function ascribed to it by social psychologists, more often it has its reference in action—to go before or to show the way, not necessarily equivalent to dominant behavior.

Social psychologists tend to consider dominance only in the context of behavior which verbally or physically threatens or aggresses (Sherif and Sherif 1964). For many, dominance has negative connotations:

> In dominative behavior, a person is rigid and inflexible, he has his mind made up; he does not reduce the conflicts of differences by finding a common purpose among differences; rather, he maintains or increases conflict or tension between himself and others, who differ from himself; he expends energy against or in opposition to others. In dominative behavior a person disregards the desires of others, he uses commands, threats, or force to gain his unyielding objectives; he attacks the status of others; he adds to the insecurity of others . . . it is the behavior of an insecure person (Anderson 1937: 403).

Such moralizing is not uncommon in social psychological writings; many distinguish between leadership and dominance by referring to dominant-type individuals who behave in the latter way as egotists (Whiting and Edwards 1974), headships (Gibb 1969), authoritarians (Adorno et al. 1950), and Machiavellians (Christie 1970) among others. Collins and Raven (1969) discriminate several types of power which may be exerted by individuals within a group: informational, reward, referent, expert, legitimate, coercive, and negative. The first five are attributes of leaders, the last two are attributes of those who exert dominant social power.

One may attempt to dominate others from various motivational desires: to service one's own interests, to gain desirable prerequisites, to counter-dominate, to humiliate or to aggress against another, or to assume leadership over the group (Maccoby and Jacklin 1974). One has a dominant personality if one desires power, prestige, and material gain, and if one manifests ascendance, assertiveness, and social boldness (Gibb 1969). Thus, one may or may not achieve leadership status by behaving in a dominant manner.

Question One: Defining Dominance

While species-specific modes of interindividual interactions will usually prevent a simple definition of dominance that is valid for all species (Callan 1970), there are, among closely related species, similarities in dominance behaviors which can be observed and delineated. These rely primarily upon nonverbal indices of dominance.

NONVERBAL INDICES OF DOMINANCE

Listed below are primatological indices of dominance that have beeen employed by observers of human groups.

1. Taking objects from another (Anderson 1937, 1939; Gellert 1961, 1962; Edelman 1968; Kaplan 1971; McGrew 1972; Thissen 1972; Sluckin 1975; Strayer and Strayer 1975; Savin-Williams 1976, in press, in preparation).
2. Winning physical encounters (Anderson 1937, 1939; Gellert 1961, 1962; Blurton-Jones 1967; McGrew 1972; Sluckin 1975; Savin-Williams 1976, in press, in preparation).
3. Supplantation, to displace another from a location (Kaplan 1971; McGrew 1972; Thissen 1972; Sluckin 1975; Strayer and Strayer 1975; Savin-Williams 1976, in press, in preparation).
4. Counter-dominance, resistance in complying with others (Anderson 1937, 1939; Gellert 1961, 1962; Edelman 1968; McGrew 1972; Savin-Williams 1976, in press, in preparation).
5. Physical assertiveness, e.g., hitting, jumping upon, chasing, wrestling, kicking, pulling, etc. (Gellert 1961, 1962; Ginsburg and Pollman 1975; Strayer and Strayer 1975; Savin-Williams 1976, in press, in preparation).
6. Focus of visual attention (Beekman 1970; Thissen 1972; Hold 1975).
7. Threat gestures, a movement to hit, kick, bite, etc. without making physical contact (Strayer and Strayer 1975; Savin-Williams 1976, in press, in preparation).
8. Spatial proximity (Wasserman and Stern 1975).

VERBAL INDICES OF DOMINANCE

While primatologists most often employ nonverbal indices of dominance, social psychologists are more apt to utilize verbal cues to connote dominance. Naturalistic ethological observers have been reticent to follow social psychological criteria; experimental investigators, i.e., those who pair individuals artificially, have been more inclined toward assessing relative dominance through the verbal indices listed below.

1. Verbal commands or suggestions which control the behavior of others (Anderson 1937, 1939; Sherif et al. 1961; Gellert 1961, 1962; Sherif and Sherif 1953, 1964; Kaplan 1971; Sluckin 1975; Savin-Williams 1976, in press, in preparation).
2. Verbally demanding attention (Anderson 1937, 1939; Gellert 1961, 1962; Thissen 1972).
3. Criticizing, reproving, or ridiculing others (Anderson 1937, 1939; Savin-Williams 1976, in press, in preparation).
4. Verbal counter-dominance (Gellert 1961, 1962; Savin-Williams 1976, in press, in preparation).

5. Winning verbal arguments (Savin-Williams 1976, in press, in preparation).
6. Verbal threats of physical harm (Savin-Williams 1976, in press, in preparation).
7. Verbal interruptions or success when two or more are talking at once (Kaplan 1971).

SOCIOMETRIC INDICES OF DOMINANCE

Several researchers have sought to assess group or interpersonal dominance relations via the cognitive judgments of the subjects or of those closely associated with the sample population. The exact question asked has varied from one study to another.

1. "Who is the toughest?" (Edelman 1968; Omark and Edelman, in press; Cronin 1975; Parker 1975; Sluckin 1975; Savin-Williams 1976, in press, in preparation).
2. "In a conflict would they win or get their own way?" (Gellert 1962; Edelman 1968; Sluckin 1975).
3. "Who uses force to secure material, to direct the behavior of others, and criticizes, blames, or reproves others?" (Anderson 1937, 1939; Gellert 1961, 1962; Cronin 1975).
4. "Who can control the group, suggest plans and activities and carry them out, and control the decision-making process?" (Sherif and Sherif 1953, 1964; Sherif et al. 1961; Gellert 1962).

INDICES OF SUBMISSION

Some time ago, Chapple (1940) noted that hierarchies depend upon the ability of one group member to evoke a proper response from another. Unfortunately, few of the ethological studies reviewed in this section have adequately recorded the various patterns of behavior which connote a subordinate status position. Several indices of submission, however, can be teased from these studies.

1. Withdraws or runs away (Gellert 1961, 1962; Blurton-Jones 1967; Strayer and Strayer 1975; Savin-Williams 1976, in press, in preparation).
2. Imitates or follows others (Gellert 1961, 1962; Blurton-Jones 1967; Kaplan 1971; Thissen 1972).
3. Assuming a submissive posture or giving an appeasement gesture (Ginsburg and Pollman 1975; Strayer and Strayer 1975; Savin-Williams 1976, in press, in preparation).
4. Giving up objects (Gellert 1961, 1962; Blurton-Jones 1967; Savin-Williams 1976, in press, in preparation).
5. Seeks orientation, permission, or help (Gellert 1961, 1962; Kaplan 1971; Strayer and Strayer 1975).
6. Verbally obeys or accedes (Gellert 1961, 1962; Kaplan 1971; Savin-Williams, 1976, in press, in preparation).

7. Losing physical encounters (Blurton-Jones 1967; Savin-Williams 1976, in press, in preparation).
8. Self-deprecation (Kaplan 1971; Savin-Williams 1976, in press, in preparation).

INTERCORRELATION OF DOMINANCE INDICES

While many of the studies reviewed above employed at least two methods of assessing the dominance rank-order—usually comparing a behaviorally derived rank-ordering with a dominance ranking made by informed others (staff, teachers, camp counselors)—far fewer report the intercorrelation of the various indices of behavior considered to reflect social dominance. For example, Gellert (1961) utilized six categories of behavior indicative of dominance, but she did not rank each dyadic member on the various indices and then compare each index with the others to test whether or not they produced a significantly similar dominance ranking. More systematic in this regard have been the camp studies of Savin-Williams (1976, in press, in preparation). In each cabin group, subjects ranked significantly the same (p < .01) on the eight behavioral indices of dominance. He thus aggregated these eight for the behavioral definition of dominance—a rank-order which correlated significantly with the campers' sociometric ranking of each other and with a second observer's assessment of relative dominance (r's ranging from .83 to 1.00; all p < .05).

Others have correlated the dominance hierarchy derived from observing behavior with that obtained from teachers' rankings of dominance. Anderson (1937) found an average correlation of .72 between teachers' ratings of who was expected to be dominant in a given situation and who actually was dominant, given the author's observations of the preschool children. Gellert (1962) asked teachers to rank-order a preschool classroom on the basis of where children stood in regard to directing the behavior of others, taking the lead in play, and pursuing one's own intentions; she discovered a significant (p < .01) tendency for relative position to predict the winner in experimental play situations. Others, however, have not recorded such high agreement between dominance based on behavioral criteria and on sociometric rankings. While Sluckin (1975) found a correlation of .73 in one group between the two methods of measuring dominance, in his other group the correlation was much lower (.49). Edelman (1968) suggested an age difference: teachers could not reliably predict among preschoolers the winner of staring contests by their ratings of who would win an argument, but could (p < .02) when the children were two to three years older.

Edelman (1968), Beekman (1970), Kaplan (1971), and Sluckin (1975) correlated subjects' rankings of relative toughness status with behavioral indices of dominance, with varying degrees of success. Sluckin and Edelman found young preschoolers could not agree on who was toughest; older children, however, could reliably judge toughness status. Still, in Sluckin's older children, the correlations between the behavioral and sociometric measures of dominance was low (.36 and .12); Edelman's correlation was significant (p < .03), but his children were 6 to 7 years old, 2 years older than Sluckin's 4-year-olds.

Beekman (1970) and Kaplan (1971) paired the extreme position individuals on Omark and Edelman's toughness hierarchy, giving them a "draw a picture together" task. Beekman found that the toughest dyad member as determined by the group sociometric ranking was significantly more likely than chance (p <.04) to be the recipient of higher behavioral dominance status—receiving the gazes of the other partner more than the reverse. Kaplan reported much the same trend (p <.005) between the group sociometric and behavioral indices of dominance (grab objects, gives suggestions, etc.).

Once the Sherifs' (1964) participant observers were able to reliably rank-order their naturalistic group of adolescents, an independent observer was asked to enter the group during an activity (e.g., as a baseball umpire) and to rank the group on the basis of social power. There was a significant agreement between these judges and the observer in all groups (r's ranging from .82 to .98; all p<.05). When the observer was able to elicit the judgment of the boys on the group's ranking of effective initiative, a significant intercorrelation (r's ranging from .88 to .90; all p <.01) was registered between the two rank-orders.

Question Two: A Group Dominance Hierarchy

While the previous section described the various methods of defining dyadic dominance relations, we now turn to research which purports to demonstrate that this dyadic behavior has meaning at a group level.

> Dyadic dominance describes the relative balance of social power between specific members in a social group, while dominance structures summarize the organization of such power relationships among all possible group members (Strayer and Strayer 1975: 1).

Two basic techniques, dyadic isolation and group observation, have been employed to study dominance structure within two age groups—preschool and elementary school children. Studies of adolescents are far fewer in number, except for the studies of social psychologists.

PRESCHOOL: DYADIC ISOLATION

One of the earliest studies on dominance (Hanfmann 1935) paired nine kindergarten boys, 5 years of age, in a play area for 30 minutes while two observers recorded everything both children said and did. After each session, the observers determined which pair member was the dominant one. A rank-order of dominance was obtained by counting the number of partners whom each boy dominated. While a linear "peck order" was evident for the lowest-ranking boys, the systematic arrangement of the top four was more complicated: while A dominated B, B dominated C and D, and C dominated D, both C and D dominated A. Hanfmann attributed this intransitivity to differing styles of dominant behavior: A was a violently disruptful individual, a style effective against a similarly oriented B but not against the social, controlling behavior of C and D. She considered A and B as dominant individuals and C and D as "leaders."

Another early investigator of dominance relations, Anderson (1937, 1939), studied 128 preschool (mean age, 51 months) and forty-five kindergarten (mean age, 63 months) children. Subjects were taken by pairs to a portable testing booth—a play area with a sand box and toys—where they were unobtrusively observed for five minutes by two adults. Each child was randomly paired with five other children of either sex from both their own and a strange play group. Because Anderson was not investigating consistency in dominance patterns, he did not construct a dominance rank-order of group members. Of interest are his methodology, sex comparisons, and correlations of dominance scores with other measures (to be discussed below).

Some 20 years later, Gellert (1961, 1962), essentially replicating Anderson's methodology, paired thirty-two male and female children, ranging in age from 46 to 62 months, for 20 minutes in a free-play setting while observers recorded from behind a one-way mirror. In the first study, same-sex individuals were paired for three sessions; a significant number (p < .01) of dyad members maintained the higher dominance score during the three play periods. In the 1962 study, children were observed in both same- and cross-sex dyads with pair partners varying with each matching. While some children remained remarkably consistent in dominance behavior regardless of whom they were paired with (the high- and low-ranking individuals), the majority varied in their frequency of dominance behavior as a function of the pair partner. She did not, however, construct an overall group hierarchy. There was also a tendency for individuals to display more dominance when they were paired with a low- versus a high-assertive individual—a finding consistent with Anderson, i.e., the amount of dominance behavior increases as the dominance status of the partner decreases.

A more cognitive approach has been taken by Omark and Edelman (in press; Edelman and Omark 1973). A photograph hierarchy test was given to two kindergarten classes. After it was ascertained that each child understood the meaning of "tough," they were asked to turn over the pictures, one at a time, of those classmates who were tougher than themselves. "Toughest" was thought to be the comparative term closest to the primatologists' concept of dominance (but see Maccoby and Jacklin 1974). While there was some agreement among the kindergarten children on "dyads of established dominance" (both X and Y independently agree that X is more dominant than Y), the percentage of agreement (40%) reflects no greater than random responses and is considerably below the percentage of agreement given by older children (see below). Still, toughest appeared to be more salient, i.e., having a higher percentage of agreement, than did other dimensions such as smartest, most friends, and nicest.

Finally, Wasserman and Stern (1975) paired 134 3- to 5-year-olds to see how close one dyad member would walk up to the other child when told by the adult to approach. It was assumed that a dominant child would come close to and fully face the goal child. Their three hypotheses were not supported by the data: that boys would be more dominant than girls; that 4-year-olds would be more dominant than 3-year-olds but less so than 5-year-olds; and that aggressive children (as reported by teachers) would be more dominant than less aggressive ones.

PRESCHOOL: GROUP OBSERVATION

More informative research has been conducted by unobtrusively observing pre-schoolers as they interact in a free-play setting. The belief, often explicitly stated, is that by observing the natural behavior of such naive subjects, one can better draw parallels between human and nonhuman primate groups.

Thissen (1972) sought to discover how early, developmentally, a dominance structure is apparent in children's play groups. He observed seven 2-year-olds, six boys and one girl, as they played at a day care center. Recording 549 dominance interactions in approximately 20 observation hours, Thissen found no consistent, *purely* linear dominance hierarchy across all types of dominance interactions. One boy, however, clearly dominated all others, and one boy was clearly subordinate to all others; the rest fluctuated in the middle range.

Other investigators, studying children ranging in age from 3 to 6 years, have noted a fairly well-defined and consistent dominance hierarchy during free-play activities. Sluckin (1975), observing a mixed-sex group of twelve 3- to 4-year-olds during a 4-week period, recorded 436 dominance interactions. He found no intransitive rela-tionships among tryads, i.e., if A dominated B and B dominated C more than the reverse, then A was also dominant over C. A linear dominance hierarchy was con-structed based on dyadic interactions; each child had at least an excess of two wins over the one immediately below him/her.

Thompson (1967), McGrew (1972), and Strayer and Strayer (1975) have all reported a stable, well-defined dominance structure among 3- to 5-year-olds. Strayer and Strayer's study is particularly interesting because it covers a 3-month period. During the last 6 weeks of observation, video samples were collected of naturally occurring conflict within the group between dyads of the seventeen children. Data were assessed on the degree to which observations of the 443 agonistic encounters corresponded to a linear dominance model—the percentage of observed dyadic dominance relations which follow the linear transitivity rule. While agonistic group interactions were not totally unidirectional, a fairly linear group dominance hierarchy was evident, especially whenever the dominant act led directly to a submissive re-sponse.

A cross-cultural verification of these findings has recently been conducted by Hold (1975). During free-play activities, she observed four groups totaling eighty-five West German children between the ages of 3½ and 6 years. Utilizing an instantaneous sampling technique, she focused on one child for a short time once every 5 minutes to see if they were looked at by other children and if they were involved in dominance behavior. In all four groups a dominance hierarchy based on attention structure (the number of times a child was looked at by three children simultaneously) was evident and especially linear at the extreme positions. This attention rank-order was then related to various categories of dominant and subordinate behavioral roles (see below).

These six studies do not confirm Blurton-Jones' (1967) earlier assertion—an assertion based on impressionistic observations rather than on empirical, quantifiable data—that the concept of a dominance hierarchy does not fit the organizational structure of preschool children in free-play groups. Interestingly, though, Blurton-Jones did note that some individuals regularly won fights while others consistently demonstrated submissive behavior. But the dominant individuals were not also the leaders or peace keepers or given priority access to objects.

ELEMENTARY-AGED CHILDREN

Much of the grade school research centers on the work of Omark and Edelman (in press; Edelman and Omark 1973) and an extension of their work by Parker (1975). Omark and Edelman gave first through fourth graders a cluster hierarchy test: individual class members' names were randomly grouped into clusters of six, including the word "me," and each subject was then asked to rank-order the six in terms of toughness; all class members' names were presented on paper, and each child was instructed to rank-order all fellow classmates in terms of toughness. In contrast to the kindergarten children, there was a general dyadic agreement on relative dominance status; average percentages ranged from 62% to 73% in the first and fourth grades, respectively. These percentages are significantly (p <.01) different from random guessing. Strayer and Strayer (1975) found much the same developmental trend when they observed dominance behavior in preschool and elementary school groups. Although a linear group dominance hierarchy and stable dyadic dominance relations were found at both age levels, dominance was a more "unitary social phenomenon" among the older children.

EARLY ADOLESCENTS

Beyond early and middle childhood there is little empirical evidence, especially from ethologists, that the formation of a dominance hierarchy is characteristic of human groups. Collins and Raven (1969) concluded that, whereas among groups of animals and human children a simple rank-ordering of power that generalizes across settings and activities has been found, there is little empirical data that such a phenomenon exists as humans become progressively older. Yet, social psychological studies of adolescent groups strongly suggest the existence of dominance differentiation operating to structure interpersonal behavior. Community and gang studies (Thrasher 1927; Whyte 1943: Suttles 1968) refer to a clearly enunciated and enforced system of power differentiation and patterned regularities within the adolescent reference group. In Whyte's street corner boys (1943), Doc was alpha because he was tough; his lieutenants, Danny and Mike, were next toughest.

More extensive observations have been made (Sherif and Sherif 1953, 1964; Sherif et al. 1961) in camp and city settings. Two dozen normal 10- to 12-year-old boys from the lower-middle class substratum spent 18 days together in a summer camp (Sherif and Sherif 1953; Sherif et al. 1961). After 3 days of camp, the boys were divided into two equal groups on the basis of camping and athletic skills, physical size, personality, intelligence, etc.; the goal was also to splinter initial interpersonal attractions. In all four groups, the staff, acting as participant-observers, was able to discern two well-defined, relatively hierarchical group structures on the basis of popularity and of social power (leadership or "effective initiative"). The two hierarchies were not the same, even though the most powerful individual was usually the best liked.

In 1964 Sherif directed his attention to more naturally occurring groups of male adolescents. Interested in the importance of reference groups, he sent graduate students into various socioeconomic settings to observe groups of six to twelve adolescents (15 to 18 years old). After several weeks the observers were able to gain an in-depth rapport with the group as coach, provider of transportation, etc. Instructed not to write in their presence, the observers recorded impressions and particular details of behavior for at least 100 observational hours per group. Over time, an observer was able to detect a patterned group structure based on power, control, and sanction of behavior. At first he was able only to reliably rank the extreme group members ("end anchoring"); in several more weeks with the group, he was able to decipher the power rank of the intermediate boys as well.

While it is doubtful that Sherif's power hierarchy is an actual dominance hierarchy as defined in ethological studies, it is valuable because of the age group studied, so seldom observed in its natural habitat by participant observation techniques. It also appears that Sherif's "captain" is closely analogous to an alpha individual in a dominance hierarchy.

Combining in-depth observations of specific dominance behaviors with various objective and sociometric measures in groups of six 12- to 14-year-old boys at a summer camp, Savin-Williams (1976, in press, in preparation), acting both as a cabin counselor and as an observer, found a stable, ordered dominance hierarchy in all three groups—formed perhaps within hours after group members met, and certainly by the third day of camp. For 5 weeks, Savin-Williams recorded dyadic verbal and physical dominance interactions among cabin members, utilizing the "all occurrences of some behavior" event sampling technique (Altmann 1974). Sampling was systematically dispersed throughout the daily schedule in five behavioral settings for an average of 3½ hours per day. The behavioral rank-order was derived by arranging the boys in each cabin group in order of the number of group members that they significantly dominated in pair-wise interactions. The sociometric dominance hierarchy was calculated for each group by averaging the individual rank-orderings of the campers when they were asked to list the campers in the cabin, including oneself, in order of toughness or dominance. As previously indicated, these two derivations of the dominance hierarchy were always significantly the same.

SUMMARY

Generally, then, one can conclude that in groups of young humans there are behaviors which, when performed in conjunction with others, become meaningful in structuring the group into a systematic arrangement of status differentiation based on social dominance. When this phenomenon has not been found, one can usually attribute it to one of two factors: a non-behavioral methodology or subjects who have not yet reached Piaget's concrete operations level of cognitive development.

Omark and Edelman (in press) and Parker (1975) failed to ascertain a consistent dominance hierarchy in young children, perhaps because they did not observe naturally occurring behavior or because, they speculate, the subjects had no cognitive ability to arrange objects or persons according to some quantified dimension on an ordinal scale (serialization). They believe that the logical operation of transitivity matures as the result of both social experience with other children and the maturation of cognitive structures, a necessary development for the formation of toughest hierarchies.

The ontogenetic trend is apparent only whenever subjects are asked verbally about the group structure, not when their behavior is observed. This implies that perhaps the methodology of ascertaining the dominance hierarchy may be more influential to whether the investigator will discover a group rank-order than the age of the sample which is studied. This is not to deny Strayer and Strayer's (1975) assertion that increased specific peer interactions *aid* the process by which dyadic dominance interactions "converge to produce a single group dominance hierarchy." It does appear, however, that as early as the age of 2 years, children begin to differentiate the extreme points of the dominance hierarchy, if not cognitively, then certainly behaviorally. With age, the group structure becomes more ordered and patterned and, by the first grade, children can not only behave as if there were a group hierarchy, but can also verbally report its existence. This conclusion speaks in part to the primatological debate about whether the group dominance structure exists in the mind of the animal or only in the mind of the observer (S. Altmann, personal communication). In studying children under age 6, one can be sure only that the group structure exists in the notebook of the adult observer; after that age it is also present in the cognitive awareness of the children themselves.

DOMINANCE: PREDICTORS AND INDICATORS

There is unfortunately little empirical evidence that reveals which behavioral indices of dominance best predict the overall dominance hierarchy. Thissen (1972) found toy-taking and spatial displacement yielded the most linear ordering of subjects. But Strayer and Strayer (1975) found physical attack and threat gestures which led to submissive acts elicited the most linear (98% of such behaviors corresponded to the linear transitivity rule) and rigid (94% of such behaviors corresponded to the established dominance relations hierarchy). Similarly, Savin-Williams (in preparation) reported verbal/physical threat behaviors to be the best predicator of the behavioral dominance hierarchy—only 8% of such behaviors transgressed the overall dominance rank-order. "Verbal ridicule," "verbal command," and "physical assertiveness with contact," in order, were the most prevalent dominance behaviors of early adolescent males. Thus, frequency of a particular kind of behavior does not necessarily imply its superiority in predicting relative status among group members.

The best example of this principle concerns aggressive behavior. While Hold

(1975) maintains that a dominant child is more likely to aggress because its power or authority derives from it as well as from some extra-group power, Sherif and Sherif (1964) have found that it is not necessary for alpha individuals to employ physical aggression to maintain power. Even though preschool teachers often misperceive dominance as merely aggression (Sluckin 1975), the amount of aggression or victories is less important than who wins or aggresses over whom (McGrew 1972). Beta- or gamma-ranked individuals may be far more involved in aggressive encounters than alphas or low-ranking group members (Savin-Williams, in preparation).

Question Three: Females and The Dominance Hierarchy

Callan (1970) notes that a general feature of primate groups is that dominance rela-tionships among females are relatively unorganized and unstructured when compared with intermale behavior. This does not imply, however, that females have relatively little social control or power within the heterosexual groups:

> I should like to suggest, roughly, that one structural feature common to a good deal of human and animal social life is that the males are the conspicuous participators, the upholders of the contours and corners of the social map, and that the position of the female is characteristically more subtle or even equivocal with respect to this map. (ibid.: 144)

While females may not be regularly engaged in dominance interactions among each other or with males, they are the "keepers-in-being" of the system as a system.

In a mixed-sex group, Omark and Edelman (in press) found that grade school boys occupied the top of the dominance hierarchy and girls the bottom with some overlap in the middle range—some girls were dominant over low-ranking boys. When heterosexual dyads were paired, there was almost unanimous agreement that the boy partner was the dominant one (p <.0002). Both sexes were equally perceptive of the overall group rank-order. Analyzing the sexes separately, Omark and Edelman (ibid.) found that, in general, the boys' hierarchy was more clearly delineated than that of the girls; while girls were aware of the boys' hierarchy, the reverse was not nearly so true.

Hold (1975), however, found a slightly different pattern: the majority of the female children were in the middle range of the dominance hierarchy, often imitating and seeking reassurance from high-ranking boys. Boys, on the other hand, tended to be either high or low in the dominance hierarchy—if the latter, then onlookers or outsiders. On an aggressive hierarchy, Sluckin (1975) reported the same pattern of sexual distribution. But when he arranged the individuals along the dominance dimension, in both groups a female occupied the beta position; other females were equally distributed throughout the rank-order.

But in other mixed-sex groups, contradictory results have been found. Strayer and Strayer (1975) did not obtain an extreme sexual stratification in dominance positions in either preschool or elementary school groups. The most dominant in one group was a female. In other studies, boys did not dominate girls on the "draw a picture together" task (Kaplan 1971); first girls and then boys had more dominant acts, not significantly so (Gellert 1961, 1962) and significantly so (Anderson 1937, 1939); and girls and boys were no different in distance stopped from the goal child or in angular placement of the body (Wasserman and Stern 1975). Others (Beekman 1970; McGrew 1972; Hold 1975) have reported, however, that girls are not as actively engaged in dominance behavior as are boys.

Kaplan's work (1971) suggests that the discrepancy in these studies may be due to the criteria employed to assess dominance. While toughness and physical aggression measures will usually place most boys at a higher rank than most girls, more feminine tasks, e.g., drawing pictures, personal closeness, and verbal behavior will elicit no sex differences in dominance. Or, as Maccoby and Jacklin (1974) write, perhaps status rank does not generalize to situations where aggression is not appropriate. The studies of Anderson (1937, 1939) and Strayer and Strayer (1975) remain exceptions, since their measures of dominance are physically defined in terms of aggression and physical threats. But when Anderson (1937) paired his preschoolers heterosexually, he found that the mean frequency of female dominance scores significantly declined as the boys significantly increased their dominance behavior. Now, boys expressed more dominance than did girls (approaches significance).

Cronin (1975) further substantiates Kaplan's hypothesis. Boys were usually nominated by peers as the toughest (84% of all nominations were boys); girls achieved social prominence via placing high on the "most grown-up" sociometric (82% of all nominations were girls). Opal, the girl who ranked highest on this dimension, dominated the female group whenever they played together and was most apt to participate in all-male activities.

Parker's (1975) extension of Omark and Edelman's (in press) research involved investigating the social interaction of same-sex classrooms. In such situations, neither group developed an agreed-upon dominance hierarchy based on the toughness sociometric. Girls overrated themselves as much as boys did, reflecting the increased emotional salience of toughness for girls whenever boys are removed from the group. In comparison to the mixed-sex groups of Omark, Parker found that girls maintained the same percentage of dyadic agreement on who is toughest; the boys, however, fell precipitously in agreement on toughness within the group.

> We speculate that in mixed-sex classes, the girls form an audience that elicits or encourages boys' toughness display behavior which in turn contributes to the working out of an agreed upon hierarchy. (ibid.: 67)

The future work of Cronin and Savin-Williams should clarify many of these research contradictions and uncertainties in regard to sex differences and similarities in dominance behaviors and group dominance structure. Cronin will be observing the naturalistic behavior of all-female groups between the ages of 8 and 14 years in playground situations, while Savin-Williams will replicate his summer camp research with all-female groups.

Question Four: Characteristics of Dominant and Subordinate Individuals

Many of the ethological studies reviewed in this chapter correlate various physical, mental, personality and social variables with the dominance hierarchy. The implicit goal is to derive factors that will adequately predict dominance position.

PHYSICAL CHARACTERISTICS

There is little consistent relationship between chronological age and dominance status in most groups investigated: Wasserman and Stern (1975) found no age correlate with dominance status; Anderson (1937, 1939) reported low correlations of .09 and .00 between the two measures; Thissen (1972) calculated a somewhat higher (.71) correlation between age and toy-taking; Hold's (1975) four groups ranged from −.20 to +.87 correlation between age and attention structure; and Savin-Williams (in preparation) found correlations between chronological age and dominance position to range from −.09 to +.60 in his groups of early adolescent boys. Hold (1975) suggests, at least with the younger preschool groups, that if the age difference is 2 years or greater, it is likely to have a significant effect on one's dominance position—but this could be due to other factors besides age: physical size, experience, intelligence, etc.

No investigator has yet reported a significant correlation between dominance status and height or weight (Anderson 1937, 1939; Thissen 1972; Hold 1975; Savin-Williams 1976, in press, in preparation). Savin-Williams did, however, discover relatively high correlations between adolescent dominance position and athletic ability (r's ranging from .71 to .94) and physical fitness (r's of .66). Pubertal status was significantly related (.89) in one group but not in the other two (.09, .31).

Thus, there appears to be little relationship between physical characteristics and the group dominance hierarchy, with the possible exception of athletic ability in adolescent groups.

MENTAL TRAITS

There is relatively little correlation between dominance status and mental age (Anderson 1937, 1939), mean length utterances (Thissen 1972), I.Q. (Savin-Williams, in preparation), creativity (Savin-Williams, in preparation), or smartness (Parker 1975).

PERSONALITY ATTRIBUTES

Few have attempted to relate personality traits or behavior with dominance status. While McGrew (1972) reports a significant positive correlation between teachers' ratings of aggression with the dominance hierarchy, others have found little consistent relationship between the two (Wasserman and Stern 1975; Hold 1975). Savin-Williams (in press, in preparation) found that in two of his groups there was a stronger tendency (.77, 1.00) for dominant individuals than for lower-ranking group members to perceive themselves as having many positive adjective and role characteristics; but in his other group (1976) the reverse occurred (−.54).

Omark and Edelman (in press) defined egocentrism as the tendency for overranking oneself on dominance in relation to the group consensus of where one ranks on the group hierarchy. While children from kindergarten through the third grade evidently become more accurate in their perception of self and others, at all age levels and at all status ranks some 70% of the children overrated themselves on toughness. Egocentrism was more prevalent in boys than in girls, perhaps, they suggest, because the competitive spirit is more of an issue for boys than it is for girls. By adolescence this proclivity declines (Savin-Williams 1976, in press, in preparation). Group members overrated themselves by at least one position in 40% of the forty sociometric self-ranks. By contrast, 10% underranked themselves, and 50% were accurate, i.e., the same as the group sociometric ranking. Thus, while accuracy of self-perception apparently increases from early childhood to adolescence, there is still a greater tendency for individuals throughout the dominance hierarchy to express egocentrism rather than self-abasement.

SOCIAL CHARACTERISTICS

McGrew (1972) reported a significant correlation between the dominance hierarchy and the teachers' ratings of the children on "sociability." Contact with teachers was not related to dominance status (Anderson 1937, 1939), while "nursery room experience" (the amount of time one has been in the same nursery classroom) was both significantly correlated (McGrew 1972; Hold 1975) and not related (Hold 1975) to dominance rank-ordering.

The relationship between peer popularity and dominance status is also confusing. In some groups, popularity and dominance rank were highly correlated, while in others they were not—r's ranging from −.33 to +.84 (Sherif and Sherif 1964; Parker 1975; Savin-Williams 1976, in preparation). There does appear to be a tendency for individuals to select as friends those who rank above rather than below oneself on the dominance hierarchy (Parker 1975; Savin-Williams 1976, in press, in preparation). Savin-Williams (in preparation) discovered that, generally, the higher status one achieves, the more likely peer group members are to attribute positive characteristics to him.

CONCLUSIONS

Thus, investigators have had little success in correlating various physical, mental, personality, and social attributes with the dominance hierarchy. Perhaps the problem lies in the attempt to predict all status positions by these attributes. More revealing have been studies which characterized not the entire dominance order but only the most dominant and subordinate positions. The alpha or toughest individual is, relative to the group, chronologically older (Thompson 1967; McGrew 1972; Savin-Williams, in preparation), physically taller and heavier (Thompson 1967; McGrew 1972), more athletic and physically fit (Savin-Williams, in preparation), and more popular (Sherif and Sherif 1953, 1964; Sherif et al. 1961; Parker 1975; Savin-Williams, in preparation). For example, Shaw, the Red Devil captain (Sherif and Sherif, 1953), was extremely daring, athletic, tough, attractive, intelligent, and popular with his peers. He was the captain who enforced his will by threats and by physical encounters. On his birthday, group members were surprised to discover that he was only 12 years old: "I thought you were at least 15!" exclaimed one boy.

Top-ranking individuals are often the focus of attention within the group. Omark and Edelman (in press) hypothesized, based on Chance's (1967) attention structure model, that high-ranking children would be more visible than would low-status children. A "ratio of agreement" was tested for the top two versus the bottom two ranking children of each age class and found to be statistically significant (p <.04), i.e., "there was more agreement among children on their perception of who is the child on the top than of the one at the bottom" (Omark and Edelman, in press: 26). Savin-Williams (1976, in press, in preparation) discovered the same trend in his adolescent groups. In three of the six sociometric dominance rankings (three groups), the alpha individual was unanimously chosen as number one; the other three times the alpha's average sociometric dominance rank ranged from 1.2 to 1.4. No other group member had such group consensus as to his relative dominance placement. Apparently, the group was most aware of Alpha's position, i.e., he was their center of attention.

Savin-Williams (in preparation) found omega-ranked boys to be poor athletes, physically unfit, pubertally immature, younger than other group members, followers, feminine, quiet, and religious. In other groups, low-ranking members were reported to be timid (Blurton-Jones 1967; Hold 1975) and unsocial (Hold 1975). Ginsburg (in preparation), defining the last third to fifth grade male child chosen for the school free-play kickball team as the omega individual, found that such children were significantly (p <.01) further from the group than other group members and, as a result, were seldom in physical contact with other children.

> . . . the omega children of this study invariably spent their time at the periphery of the play group. It often appeared that the stream of behavior altogether passed them by; when they did move into closer proximity with the group as a whole, they were frequently observed standing motionless while an entire entourage of children ran past. (ibid: 16)

Question Five: Stability of The Dominance Hierarchy

A group dominance hierarchy implies that individuals have a customary way of interacting with other group members that, though not absolutely static, remains relatively constant over time (Whyte 1943; Sherif and Sherif, 1953). Sherif and Sherif (1964) noted that when group activities change, there is often an alteration in leadership. But it is interesting to note that a former leader (or perhaps the overall leader, i.e., the one with the most social power) "lets" the other assume leadership. Many social psychologists substantiate the finding of the stability and resistance to change of a group hierarchy based on social power (a term closer to "dominance" than to "leadership") (Glidewell et al. 1966). While the leadership structure of a group may vary in accordance with the skills needed for a particular task or activity, the dominance hierarchy remains constant.

There is ethological evidence to support these claims. For over three months, Hold (1975) found that the average preschool child does not move more than three or four places up or down the dominance hierarchy in groups ranging in size from eight to forty-one members, regardless of the group activity. In all three groups of adolescent boys (Savin-Williams 1976, in press, in preparation), the dominance hierarchy was significantly the same (p <.01) regardless of the behavioral setting in which the boys were observed (rest hour, athletic games, discussions and meetings, mealtimes, and rising from and going to bed). Furthermore, this dominance hierarchy remained constant throughout the 5 weeks of camp with only minor shifting in position among low-ranking group members.

Question Six: Individual Benefits of Status Position

It is assumed by many human ethologists that a systematic arrangement of group members into a patterned dominance hierarchy is necessary for the continued existence of child and adolescent groups; thus, it is surprising that few of the studies reviewed in this chapter have systematically investigated the prerogatives that arise from either high or low dominance status within a social group. Yet, Alexander (1974: 326–327) maintains:

> Our basic statement must be that, in general, groups form and persist because all the individuals involved somehow gain genetically.

While it is difficult to speculate as to the genetic advantage of a particular status position given the human ethological studies conducted so far, it is possible to demonstrate "personal benefits" an individual may derive from a dominance position.

By virtue of their status, top-ranked individuals are able to acquire preferred toys or objects (Anderson 1937, 1939; Gellert 1961, 1962; Edelman 1968; Kaplan 1971; McGrew 1972; Thissen 1972; Sluckin 1975; Strayer and Strayer 1975; Savin-Williams 1976, in press, in preparation), spatial locations (Kaplan 1971; Thissen 1972; McGrew 1972; Sluckin 1975; Strayer and Strayer 1975; Savin-Williams 1976, in press, in preparation), and food (Savin-Williams, in press). Alpha is also the focus of attention within the group (by definition, see Hold 1975) and is able to control the actions of others (Savin-Williams 1976), both of which may be intrinsically rewarding (Washburn and Hamburg 1968). Others imitate him, seek his support, show friendly and anxious behavior towards him, and give him presents (Hold 1975).

A top-ranked individual does not need to propose games—he begins playing and the group follows his lead. He is able to move about more freely than lower-status individuals and is often allowed to innovate without inhibition from other group members (McGrew 1972; Hold 1975). If he is attacked by another, then it is likely to be through verbal means; lower-ranking members in the same situation face physical aggression as a means of correction (Hold 1975).

A review of studies on dominance provides little information concerning the benefits of a surbordinate rank. Sherif and Sherif (1964) suggest that all group members seek to be acceptable, to belong, to be approved by the peer group. To achieve this security some group members are willing to submit. Alexander (1974: 330) is more explicit:

> The subordinate also gains by his behavior: like the dominant he is informed by the interactions of the hierarchy when and how to display aggression, and when and how to withhold and appease and withdraw, so as to stay alive and remain in the group and be at least potentially reproductive for the longest period.

Question Seven: Group Benefits of Status Position

Chance (1967) and Eibl-Eibesfeldt (1975) ascribe to the dominance hierarchy the function of adding stability and expectancy to interindividual behavior, thus contributing cohesiveness to a group. Etkin (1964), emphasizing a slightly different function, asserts that the existence and maintenance of a stable group dominance hierarchy in most mammals ensures or at least serves to reduce intragroup aggression.

GROUP OBLIGATIONS AND ROLES

Because of their high dominance ranking, particular individuals have specialized group obligations and roles to perform expected of them by other group members. Failure to carry out these duties implies the loss of the prestige that has been bestowed upon the group (Whyte 1943). From a more egotistical point of view Alexander (1974: 327) notes:

> Whenever individuals derive benefits from group functions they may be expected to carry out activities that maintain the group, and thereby serve their own interests as well.

In preschool children, Hold (1975) noted that dominant children initiated games and then proceeded to organize them by forbidding, permitting, teaching, and distributing roles to other group members. When contact was made with another play group, the alpha individual was the one who posed and answered questions. He also served as a group arbitrator; in 13 of 23 observed instances of group struggles, the highest-ranking child physically interfered to protect the losing child. If a lower-ranking member were the arbitrator he would be more apt to utilize verbal means—a more indirect and inefficient technique.

An alpha in an early adolescent male group (Savin-Williams 1976) was expected by other group members to organize and direct athletic games, telling who to play where, and for how long. On hiking trips he walked near the middle of the single-file line except whenever the group walked over new terrain or became lost—then he would assume the lead position. During cabin meetings low-ranking members would often not vote until Alpha had expressed his opinion. Few ideas for group activities were passed without Alpha's recommendations (Savin-Williams, in preparation).

Doc, the captain of the street corner boys (Whyte 1943), determined and organized group activities, settled intragroup frictions and disputes, and defended the group against outsiders, much like Hold's dominant preschool children. When low-status members, Lou and Fred, dropped out of the Norton gang, there was little effect on the group as a whole; when, however, Doc and Danny (a lieutenant) left, the group disintegrated.

Similarly, Crane, the Bull Dog camp leader (Sherif and Sherif 1953) organized camping trips and softball games; when there was a debate over who should distribute a group prize, it was only Crane that all group members would allow to perform this task (cutting a watermelon). Crane was able to stop bickering without saying a word; he acted and no one complained. Thus, quite often a top-rank individual does not need to utilize physical force to enact his decrees.

The group thus benefits by having dominant individuals within the group: decisions are made, activities are organized, intragroup friction is avoided or reduced, and intergroup relationships are negotiated. And, as previously noted, dominant individuals receive benefits from the group for being dominant.

Research, or even speculation as to the obligations and roles of low-status group members, is practically nonexistent. Savin-Williams (1976) suggests, based on contact with his camp sample, that subordinate individuals are the workers who perform the tasks decided upon by their ranking superiors—a crucial function for the survival of any group.

REDUCTION OF INTRA-GROUP AGGRESSION

McGrew (1972) noted that the highest frequency of dominance behavior occurred during the first 7 days of group formation in his preschool children. Savin-Williams (1976, in preparation) reported that the number of dominance interactions occurring during the last 14 days was significantly ($p < .005$) less than the number of dominance interactions recorded during the first 14 days of camp. Temporary interruptions of this overall downward trend occurred when a new boy entered the group and when a low-ranking member was unexpectedly elected to a position of formal leadership by the camp administration. During these times, physical threats and assertive behavior increased sharply (Savin-Williams, in preparation).

Thus, the dominance hierarchy assigns roles and obligations to particular group members—which enhances a cohesive and well-functioning group—and regulates interindividual behavior in such a way so as to reduce the level of intraspecific aggression, which is beneficial to individuals and to the survival of the group.

Research Needs

As is true of most initial studies of a particular theoretical approach to an old problem, much still needs to be investigated. The major areas of research needs in regard to dominance and dominance hierarchies are briefly discussed below.

DEFINITION

If the concept of dominance is to be a meaningful term which can be reliably and validly applied to various species and intraspecific groups, then it is imperative to determine precisely which behaviors connote or define social dominance. A definition of dominance that is applicable to only one species or has reference to only one particular behavior pattern is doomed to obsolescence as a narrow and ungeneralizable tool. The studies reviewed above clearly indicate a multitude of dominance behavior patterns that have cross-species reference.

BEHAVIORAL INDICES

Future research should focus on the unobtrusive observation of nonartificial groups in natural habitats. Of primary interest should be the elucidation of the behavioral components of dominance, both species-specific indices and those characteristic of mammals in general. More emphasis should be placed on subtle indicators of dominance, e.g., facial expressions, postural orientations, speech patterns, gestures, stares and glances, and distancing. Not only the frequency of power-related acts, but also the intensity, duration, and latency of response should be investigated.

METHODOLOGY

While there is still a need for a diversified, methodological approach to conducting dominance research, a standard and acceptable means for constructing a group dominance hierarchy is sorely needed. Several methods have been utilized, making cross-study comparisons difficult: arranging group members on the basis of (1) the number of individuals (statistically significant or just "more than") a subject dominates; (2) the percentage of dyadic interactions in which one is the dominant member; (3) the frequency of dominant acts; (4) the percentage of observed dyadic dominance relations which correspond to the linear transitivity rule; and (5) sociometric averages of informed others, independent sources, and/or the subjects themselves.

RELATION BETWEEN MEASURES

The relationship between assessing a dominance hierarchy behaviorally versus sociometrically needs to be further explored. Questions which could be answered include: How reliable are shorthand methods such as sociometric exercises in assessing dominance? At what age can they be reliably utilized? At what cost can shorthand methods be utilized in terms of information lost that would have been attained via behavioral methods? Can non-group members reliably judge a group dominance structure? Etc.

It would also be interesting to ascertain the relationship between the personality score from tests which purport to measure dominance (e.g., the "Junior-Senior High School Personality Questionnaire") and the dominance behavior expressed by an individual.

PREDICTION OF STATUS

More attention should be paid to the prediction of dominance status—at least of the extreme positions—on the basis of physical, mental, personality, and social traits and attributes. Perhaps the predictive factors vary considerably by the age, sex, and culture of the subjects. This knowledge would elucidate a more fundamental issue: to what degree is one's dominance influenced by genetic, biological and environmental social factors?

SUBORDINATES

The studies reviewed in this chapter underscore the necessity of examining not only the intention and behavior of group members when they interact in a dominant manner, but also the response of those who are dominated—a point made by Chance (1967) some 10 years ago. Perhaps it is the behavior of subordinates which allows a group dominance hierarchy to form and be maintained. And as Hold (1975) asserts, it is not so much the kind, but the frequency of rank specific behavior that varies in accordance with the status of the individual.

LONGITUDINAL RESEARCH

Unfortunately, the reviewed research is not informative in regard to the exact processes by which a dominance hierarchy is formed and maintained over time, because most studies utilized preexisting groups. Exceptions to this trend are the early camp work of Sherif and Sherif (1953), Sherif et al. (1961) and Savin-Williams (in preparation). In the former there is little recording or precise measurement of actual dominance behavior—at least in the written reports. Savin-Williams has corrected this to a degree, but many of his campers knew fellow cabinmates from previous summers. His data do provide many important hypotheses in regard to the onset and development of group structure via discrete behavioral events—insights which require further research.

INDIVIDUAL AND GROUP BENEFITS

While there is a common-sensical appreciation of the benefits that individuals and a group derive from a dominance hierarchy, this needs to be grounded in empirical research. If dominance status gives priority of access to preferred and desirable goods, then this should be documented more fully than it presently is; if the dominance hierarchy reduces intragroup aggression, then this needs to be investigated, perhaps by observing groups in which the establishment of a dominance hierarchy is either artificially or naturally thwarted (for example, the entrance of a new, or the loss of an old group member). Perhaps more important, and more interesting, would be to examine why it is that some group members allow themselves to be dominated—what benefit do they receive by being subordinate?

GROUP DIFFERENCES

Collins and Raven (1969) assert that with an increase in age there is a reduced reliance upon aggression as a means for achieving dominance. This should be empirically verified, especially with individuals beyond the level of concrete operations in cognitive development. This suggests the possibility of real age differences in the formation and maintenance of dominance relationships and in the characteristics needed for, and the benefits derived from dominance. Various age, sexual, racial, and socioeconomic groups may differ not so much in kind but in the dependencies to which they rely upon various indices of dominance and in the different rates of expressed dominance.

PHYSIOLOGICAL AND NEUROLOGICAL CORRELATES

Because physiology and behavior are intimately interconnected, it behooves ethologists to become aware of the possible neuroendocrine and neurophysiological correlates of dominance and subordinance status (Kolata 1976). Already a major attempt to bridge the often wide gap between these two areas has been undertaken by Weisfeld (in preparation). The promise has yet to be realized.

A Counterproposal

This review of dominance hierarchies should not imply that there is general consensus that the primate form of dominance structure is applicable to human groups. Barkow (1974: 2–3), for example, believes that because of natural selection the primate tendency to form dominance hierarchies has been modified, such that humans form social hierarchies on the basis of "abstract principles" and "cognitive evaluations" of themselves and those they associate with:

> As one ascends the phylogenetic scale, a concept of social dominance purely in terms of threat and appeasement and preferential access to females and food become increasingly dubious.

The primate attention-structure or dominance hierarchy homologue in humans is prestige-striving:

> I am going to argue that natural selection has transformed our ancestors' general primate tendency to strive for high social rank into a more covert need to maintain self-esteem (ibid: 7).

For Barkow, self-esteem is achieved by evaluating the self as higher than others or, in primate terms, as nonhuman primates garner the attention of others, so humans desire the prestige afforded by others whether physically present or absent, real or imaginary, when they give their attention.

According to Barkow, early socialization experiences determine the rank at which one is willing to cease striving to achieve higher prestige or status. In Western culture, particular patterns of family interactions associated with early independence and mastery training may influence or even determine the need for social recognition (need achievement). Barkow believes that children and adolescents may differ from adults in the requirement that the givers of prestige be more physically present. Further, what any particular culture considers prestigious will be largely determined by the sociocultural evolution (history) of that culture. And if one destroys the prestige structure of such a human group, then one destroys the social organization and thus the group itself.

While Barkow's speculations are certainly interesting and plausible, the research reviewed in this chapter clearly indicates that the terrestrial primate tendency for arranging group members into a systematic and visible patterning of status positions is also characteristic of human children and adolescents. Perhaps Barkow's ideas and concepts are more applicable to human adult groups, and certainly more research on the prestige structure of adult groups is desirable.

■ REFERENCES

Adorno, T. W.; Frenkel-Brunswik, E.; Levinson, D.; and Stanford, R. N. 1950. *The authoritarian personality.* New York: Harper.

Alcock, J. 1975. *Animal behavior: An evolutionary approach.* Sunderland: Sinauer Association, Inc.

Alexander, R. D. 1974. The evolution of social behavior. *Annual Review of Ecology and Systematics* 5:325–383.

Altmann, J. 1974. Observational study of behavior: Sampling methods. *Behaviour* 49:227–267.

Anderson, H. H. 1937. Domination and integration in the social behavior of young children in an experimental play situation. *Genetic Psychology Monograph* 19:341–408.

———. 1939. Domination and social integration in the behavior of kindergarten children and teachers. *Genetic Psychology Monographs* 21:287–385.

Barkow, J. H. 1975. Social prestige and culture: A biosocial interpretation. *Contemporary Anthropology* 16:553–572.

Beekman, S. 1970. The relation of gazing and smiling behaviors to status and sex in interacting pairs of children. Master's thesis, University of Chicago.

Blurton-Jones, N. G. 1967. An ethological study of some aspects of social behaviour of children in nursery school. In *Primate ethology,* ed. D. Morris. London: Weidenfeld and Nicolson.

———. 1972. *Ethological studies of child behavior.* Cambridge: Cambridge University Press.

Callan, H. 1970. *Ethology and society: Towards an anthropological view.* Oxford: Clarendon Press.

Carnegie Commission on Higher Education. *Opportunities for women in higher education.* New York: McGraw-Hill.

Chance, M. R. A. 1967. Attention structure as the basis of primate rank orders. *Man* 2:503–518.

Chapple, E. D. 1940. Measuring human relations: An introduction to the study of the interaction of individuals. *Genetic Psychology Monographs* 22:1–147.

Christie, R. 1970. Scale construction. In *Studies in Machiavellianism,* eds. R. Christie and F. L. Geis, pp. 10–34. New York: Academic Press.

Collins, B. E., and Raven, B. E. 1969. Group structure: Attraction, coalitions, communication, and power. In *Handbook of social psychology,* vol. IV, eds. G. Lindzey and E. Aronson, pp. 102–204. Reading: Addison-Wesley.

Cronin, C. L. 1975. The place of girls in the social structure of children's groups. Unpublished paper, University of Chicago.

———. 1976. The performance of girls in same-sex and mixed-sex dodgeball games. Unpublished paper, Committee on Human Development, University of Chicago.

Edelman, M. S. 1968. An evolutionary approach to staring encounters and to the development of dominance hierarchies. Master's thesis, University of Chicago.

Edelman, M. S., and Omark, D. R. 1973. Dominance hierarchies in young children. *Social Science Information* 12:103–110.

Eibl-Eibesfeldt, I. 1975. *Ethology: The biology of behavior,* Revised edition. New York: Holt, Rinehart and Winston.

Freedman, D. G. 1958. Constitutional and environmental interactions in rearing four breeds of dogs. *Science* 127:585–586.

———. 1974a. Empirical studies in ethology. Mimeographed. Committee on Human Development, University of Chicago.

———. 1974b. *Human infancy: An evolutionary perspective*. New York: Halstead Press.

———. 1976. Infancy, culture, and biology. In *Developmental psychobiology: The significance of infancy*, ed. L. Lipsitt. New York: Halstead Press.

Freedman, D. G., and Freedman, N. A. 1969. Differences in behavior between Chinese-American and European-American newborns. *Nature* (London) 224:1227.

Freedman, D. G.; Omark, D. R.; and Strayer, F. F. in preparation. *Dominance relations: An ethological perspective on human conflict.*

Gartlan, J. S. 1968. Structure and function in primate society. *Folia Primatologica* 8:89–120.

Gellert, E. 1961. Stability and fluctuation in the power relationships of young children. *Journal of Abnormal and Social Psychology* 62:8–15.

———. 1962. The effect of changes in group composition on the dominant behaviour of young children. *British Journal of Social and Clinical Psychology* 1:168–181.

Gibb, C. A. 1969. Leadership. In *Handbook of social psychology*, vol. 4, eds. G. Lindzey and E. Aronson, pp. 205–282. Cambridge: Addison-Wesley.

Ginsburg, H. J. in preparation. Playground as laboratory: Studies of appeasement, altruism and omega children in a naturalistic setting. In *Dominance relations: An ethological perspective on human conflict*, eds. D. G. Freedman, D. R. Omark, and F. F. Strayer.

Ginsburg, H. J., and Pollman, V. A. 1975. An ethological analysis of non-verbal inhibitors of aggressive behavior in elementary school children. Paper read at the Biennial Meeting of the Society for Research in Child Development, Denver, Colorado.

Glidewell, J. C.; Kantor, M. B.; Smith, L. M.; and Stringer, L. A. 1966. Socialization and social structures in the classroom. In *Review of child development research*, eds. L. W. Hoffman and M. L. Hoffman, pp. 221–256. New York: Russell Sage Foundation.

Green, N. 1969. An exploratory study of aggression and spacing behavior in two preschool nurseries: Chinese-American and European-American. Master's thesis, University of Chicago.

Guthrie, R. D. 1969. Evolution of human threat display organs. *Evolutionary Biology* 4:257–302.

Gutmann, D. 1959. Men, women and the parental imperative. *Commentary*, December 1959, pp. 59–64.

Hanfmann, E. 1935. Social structure of a group of kindergarten children. *American Journal of Ortho-psychiatry* 5:407–410.

Hold, B. 1975. Attention structure and rankspecific behavior in preschool children. Paper read at the International Human Ethology Meeting, University of Sheffield, England.

Jolly, A. 1972. *The evolution of primate behavior*. New York: Macmillan.

Kaplan, A. 1971. The effects of sex and dominance on three dimensions of children's interactions. Unpublished manuscript, University of Chicago.

Kreveld, D. van. 1970. A selective review of dominance-subordination relations in animals. *Genetic Psychology Monographs* 81:143–173.

Kolata, G. B. 1976. Primate behavior: Sex and the dominant male. *Science* 191:55–56.

Kuchner, J. 1976. Chinese-American and European-American: A cross-cultural study of infant and mother. Ph.D. dissertation, University of Chicago.

Maccoby, E. E., and Jacklin, C. N. 1974. *The psychology of sex differences*. Stanford: Stanford University Press.

McGrew, W. C. 1972. *An ethological study of children's behavior*. New York: Academic Press.

Omark, D. R., and Edelman, M. S. in press. The development of attention structure in young children. In *Attention structures*, eds. M. R. A. Chance and R. Larsen. New York: Wiley.

Parker, R. 1976. Social hierarchies and same-sex peer groups Ph.D. dissertation, University of Chicago.

Richards, S. M. 1974. The concept of dominance and methods of assessment. *Animal Behaviour* 22:914–930.

Savin-Williams, R. C. 1976. An ethological study of dominance formation and maintenance in a group of human adolescents. *Child Development*.

―――. in press. Dominance in a human adolescent group. *Animal Behaviour*.

―――. in preparation. Dominance during early adolescence: Behaviors and hierarchies. In *Dominance relations: An ethological perspective on human conflict*, eds. D. G. Freedman, D. R. Omark, and F. F. Strayer.

Schjelderup-Ebbe, T. 1935. Social behavior in birds. In *Handbook of social psychology*, ed. C. Murchison, pp. 947–972. Worcester: Clark University Press.

Schneirla, T. C. 1951. The "levels" concept in the study of social organization in animals. In *Social psychology at the crossroads*, eds. J. H. Rohrer and M. Sherif, pp. 83–120. New York: Harper & Brothers.

Scott, J. P. 1953. Implications of infra-human social behavior for problems of human relations. In *Group relations at the crossroads*, eds. M. Sherif and M. O. Wilson, pp. 33–73. New York: Harper & Brothers.

―――. 1969. The social psychology of infrahuman animals. In *The handbook of social psychology*, vol. IV, eds. G. Lindzey and E. Aronson, pp. 611–642. Reading, Mass.: Addison-Wesley Publishing Co.

Sherif, M.; Harvey, O. J.; White, B. J.; Hood, W. R.; and Sherif, C. W. 1961. *Intergroup conflict & cooperation: The robbers cave experiment*. Norman: Institute of Group Relations.

Sherif, M., and Sherif, C. W. 1953. *Groups in harmony and tension*. New York: Harper.

―――. 1964. *Reference groups*. Chicago: Henry Regnery.

Sluckin, A. M. 1975. Three ways of measuring dominance in preschool children. Paper read at the Third International Human Ethology Workshop, Sheffield, England.

Strayer, F. F., and Strayer, J. 1975. An ethological analysis of dominance relations among young children. Paper read at the Biennial Meeting of the Society for Research in Child Development, Denver, Colorado.

Suttles, G. D. 1968. *The social order of the slum*. Chicago: University of Chicago Press.

Syme, G. J. 1974. Competitive orders as measures of social dominance. *Animal Behaviour* 22:931–940.

Thissen, D. M. 1972. Dominance structures in two-year-olds: Whether and if. Unpublished manuscript, University of Chicago.

Thompson, D. M. 1967. An ethological study of dominance hierarchies in preschool children. Unpublished manuscript, University of Wisconsin.

Thrasher, F. 1927. *The gang*. Chicago: University of Chicago Press.

Washburn, S. L., and Hamburg, D. A. 1968. Aggressive behavior in Old World monkeys and apes. In *Primates*, ed. P. C. Jay, pp. 458–478. New York: Holt, Rinehart and Winston.

Wasserman, G. A., and Stern, D. N. 1975. Approach behaviors between preschool children: An ethological study. Paper read at the Biennial Meeting of the Society for Research in Child Development, Denver, Colorado.

Weisfeld, G. E. in preparation. Physiological correlates of human dominance behavior. In *Dominance relations: An ethological perspective on human conflict*, eds. D. G. Freedman, D. R. Omark, and F. F. Strayer.

Whiting, B., and Edwards, C. P. 1973. A cross-cultural analysis of sex differences in the behavior of children aged three through eleven. *Journal of Social Psychology* 91:171–188.

Whyte, W. F. 1943. *Street corner society*. Chicago: University of Chicago Press.

Wickler, W. 1967. Socio-sexual signals and their intra-specific imitation among primates. In *Primate ethology*, ed. D. Morris, pp. 69–147. London: Weidenfeld & Nicholson.

Wilson, E. O. 1975. *Sociobiology: The new synthesis*. Cambridge: Harvard University Press.

Zuckerman, S. 1932. *The social life of monkeys and apes*. London: Kegan Paul.

21 Socialization in Mammals with an Emphasis on Nonprimates

MARC BEKOFF

Behaviour of a child must be adapted towards
survival as a child, as well as towards the
acquisition of information. A child is not
just a half-formed adult. (Blurton-Jones 1972a:9)

What factors during development determine the
special ways in which an individual animal
eventually will behave? What decides the specific
form and patterning of its behavior? What gives
a behavior pattern its unique character, making
it different from other behavior patterns? It
would be useless to pretend that the attempts to
answer these questions about the ontogeny of
behavior bring widespread agreement. Nor is there
harmonious consensus among those who study
behavior as to the ways these questions should
be answered or even about the nature of relevant
evidence. (Bateson 1976:1)

■ INTRODUCTION

In this chapter I shall discuss some general aspects of the development of social
behavior, specifically in nonprimate mammals.[1] One must keep in mind, however,
that primates are merely special kinds of mammals and that they follow many trends
and principles of behavior that have been uncovered in other mammals (Horwich
1972; Eisenberg 1973). All behavior presents us with a problem in development
(Manning 1971), and for a more complete understanding of adult behavior, ontogenetic (and phylogenetic) origins of carefully defined behaviors must be determined. Using the comparative-developmental method provides a powerful tool with
which to attack the problems at hand (Bekoff 1972).

For the purpose of this discussion, the term "socialization" will be used simply

and broadly to refer to the process(es) by which young organisms become "functional"
members of their respective species by somehow "acquiring" the variety of adaptive
skills that increase their individual reproductive fitness.

■ ROLE OF PARENTS AND OTHER ADULTS IN SOCIALIZATION[2]

There is a vast literature concerning many aspects of parental behavior of mammals,[3]
those factors important in the induction and/or maintenance of the "maternal" state,[4]
and the various modalities important for maintaining close proximity between the
mother and her young.[5]

Soon after birth, the young animal forms relationships with various other individuals (Hinde 1972; Richards 1974; Figure 1), the sex and age of which are dependent on the particular social organization of the species under study, its locale, and
the season in which the infant is born. For example, squirrel monkeys (*Saimiri
sciureus*) born early in the birth season have few playmates other than juveniles, while
infants born later in the season interact more with peers (Baldwin 1969). Variations in
group composition (age:sex ratios) in different habitats and consequent effects on
socialization require further study (Crook 1974).

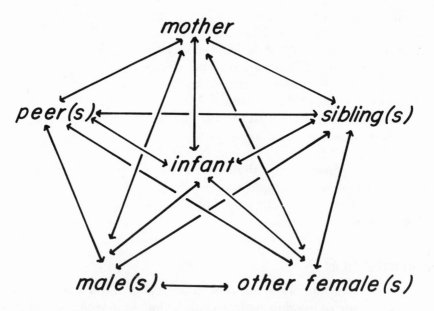

FIGURE 1. *A general schema of the possible social interactions
within a group in which there is an infant(s) (modified from Hinde
1972). The amount and direction of interaction depends on the
social organization of the species under study and may also vary
(intraspecifically) within different locales and ecological conditions.
"Males" and "other females" refer to individuals born at least one
litter prior to the infant.*

THE MOTHER

The mother-family unit is the basic unit in Mammalia (Eisenberg 1966). The mother is an important figure and is predictably the one female who is there at birth and shortly thereafter, although in some cases she may abandon her young. Besides the role that a mother plays in nursing her young, she may also be needed for protection, for orally stimulating their genitalia in order for them to urinate and defecate until they are able to eliminate on their own, for providing warmth until the young can thermoregulate, and/or for providing solid foods. She may also be instrumental in the initiation of independence or dispersal of the young (e.g., Jensen et al. 1968; Armitage 1974; Savidge 1974; Trivers 1974). Savidge (1974), studying social factors in the dispersal of deer mice (*Peromyscus maniculatus*), found that certain mothers advanced the rate of departure of their previous litter when a subsequent litter was born.

The length of time that young are dependent on their mother for providing different needs varies with the state of maturity of the newborn and the rapidity of development. For example, guinea pigs (*Cavia porcellus*) (Fullerton et al. 1974) and Alaska fur seals (*Callorhinus ursinus*) (Bartholomew 1959) are born in an advanced state of development (precocial) and there is a corresponding reduction of parental care. Consequently there is a very loose mother-infant bond. Altricial animals, those born in a relatively helpless condition, are usually dependent on the mother for some time after birth. This close association with the mother may play some role in the family group remaining together. For example, Rasa (1972) has suggested that late maturation of the dwarf mongoose (*Helogale undulata rufula*) is responsible for the group remaining together. In addition, the young are highest in rank below the mother and her consort and can, consequently, obtain food without competition from stronger adults. In kangaroos (*Macropodidae*), the young are dependent on the mother for a period of up to 1 year; however, there is very little interaction between them. Russell (1974) suggests that the low level of active caretaker tuition (particularly in the Red kangaroo, *Megaleia rufa*) is related to the high level of responsiveness and exploratory and manipulative behavior observed in the young. The length of dependency can also affect the productivity of a particular female—lengthened dependency resulting in diminished productivity. Martinka (1974) has suggested that in grizzly bears (*Ursus arctos*) in Montana, differences in productivity between years appeared to be related to variations in length of mutual relationships between mother and young. The presence of young can affect their caretakers' behavior (see note 2).

The mother may also affect her infant's social interaction with others. In various ungulates the young are reared in the virtual absence of peers (Altmann 1963; Rudge 1970). Rudge (1970), studying feral goats (*Capra hircus*), found that some young were isolated for periods of 1 month to 1 year, and that the mother interfered with the play of her own young and strange kids. Altmann (1963) compared elk (*Cervus canadensis nelsoni*) and moose (*Alces alces strassi*) and found that elk young play with other calves and occasionally yearlings and cows, while moose young play alone. At approximately 3 ½ weeks of age, elk young are integrated into the herd, while moose young achieve only limited integration. Likewise, unweaned harbor seal (*Phoca vitulina richardsonii*) young do not interact with one another or with adults other than their mother—social development taking place in nearly a social vacuum (Knudtson 1974). Adults interact only rarely outside of sexual encounters.

Another way in which a mother may affect the individuals with whom her infant(s) has contact is related to the mother's social rank. It has been suggested of various mammals that there is a correlation between a mother's rank and the social relationships that her offspring develop—offspring of dominant mothers themselves becoming dominant individuals (e.g., bison (*Bison bison*), Altmann 1963; wolves (*Canis lupus*), Rabb et al. 1967; new forest ponies (*Equus* spp.), Tyler 1972; various primates, Jolly 1972)—the caretaker providing a "model" for later assertive behavior of the young (Harper 1970). Loy and Loy (1974), studying a group of 33 juvenile rhesus monkeys (*Macaca mulatta*), found that dominance-ranking among juveniles was 95% predictable from prior knowledge of the mother's rank. In nonhuman primates there is also evidence that males play some parental role (Mitchell 1969) and that the mother's rank strongly influences the males with whom her infants will have contact (Ransom and Ransom 1971). The situation is not as clear in nonprimates.

OTHER FEMALES

In addition to the contribution of the mother, we must also consider the role of other females. Mammal studies indicate that females other than the mother may perform a number of "maternal" chores (e.g., Rowell 1961; Mech 1970; Spencer-Booth 1970; Hinde 1971; Moltz 1971; Jolly 1972). In mice (*Mus musculus*), Sayler and Salmon (1971) found that communally reared young showed an increase in growth rate when compared with young reared only with the mother, but that communal-rearing had little influence on later behavior. Schoenberger (1965) reports observing a virgin female wolf (*Canis lupus*) retrieving and transporting young pups. (An interesting observation is that of Frank [1952] who noted that young voles [*Microtus arvalis* Pall] resisted retrieval by females other than their own mother.) The effect that group attention from other females has on behavioral development is not clear, and after reviewing available literature, Spencer-Booth (1970: 158) concluded "it is difficult to make any useful summary of the behavior of females toward young other than their own, both because of the lack of good data on the subject and because very many species are completely unknown in this respect, and also because of the wide range of situations in which the young are reared."

MALES

The behavior of adult males toward young must also be considered. In a number of mammals, males perform various "maternal" tasks such as digging nests, feeding, retrieving, cleaning and grooming young. Care by males is common in rodents, fissiped carnivores, and primates (Mitchell 1969; Spencer-Booth 1970). The presence of an adult male can affect mating preferences in inbred mice (Yanai and McClearn 1972), as well as the physical development of female mice (Fullerton and Cowley 1971). Fullerton and Cowley observed that females reared in the presence of adult males (but not in physical contact) gained weight more rapidly, and showed earlier onset of eye and ear opening and eruption of lower incisors, than control animals. Mugford and Nowell (1972) found that mice reared with their father (although any male would probably do) until weaning were more aggressive than mice reared in the absence of a male. The effects of the differential rearing were still evident at 60 days of age. They believe young male mice become imprinted toward some odor that in later life elicits intermale aggression, and individuals reared in the absence of this odor consequently do not become imprinted and are less aggressive. Perhaps the best examples of the male's role in the rearing and socialization of young is provided by data collected on nonhuman primates (Mitchell 1969; Ransom and Ransom 1971), although extensive quantitative data are still absent. Ransom and Ransom (1971) point out that previous male-female ties in baboons (*Papio anubis*) influence the types of interactions in which the offspring will participate, and that adult male-infant interactions may influence later social relationships within the group. Spencer-Booth (1970), summarizing information on the behavior of males toward young, listed four points that should be considered: (1) How widespread is the phenomenon? (2) What is the extent to which males show the same behavior patterns as the mother and/or other females within the group? (3) What is the extent to which males show behaviors particular to their sex that contribute to the rearing of the young? and (4) What factors affect whether or not males will care for young if given the opportunity?

More data are needed concerning the effects of growing up in social groups differing in composition. Also, attention must be focussed on the effect of the young on adults. A nice example of how an infant may be used as a "tool" during interactions between Barbary macaques (*Macaca sylvanus*) is provided by Deag and Crook (1971). They observed that a male may use a baby "as a 'buffer' in a situation where an approach without that 'buffer' would lead to an increased likelihood of an aggressive response by the higher ranking animal" (ibid.: 196).

■ THE SOCIALIZATION OF YOUNG ANIMALS

THE PROBLEM

Mason (1971: 35) has written that "the central problem of psychosocial development can be stated quite simply: How is it that the newborn individual grows up to become a good member of his society? For 'good' you can substitute 'adequate,' 'effective,' 'normal,' 'socialized,' 'civilized,' 'decent,' or 'moral,' without altering the essential meaning." Loy (1974), summarizing behavior development in nonhuman primates, noted that there are two integrating phases of development: (1) the *developmental* phase, characterized by the appearance of new behavior patterns and development of new social roles and (2) the *social modification* phase, characterized by modification of existing behaviors and relationships. One of the reasons socialization is so difficult to study is the ease and rapidity with which it occurs. For example, Fuller (cited in Scott 1970) demonstrated that only two 20-minute periods per week of social contact for a puppy were adequate for the animal to establish reasonably normal social relationships. Other workers have also found that only minimal contact with con-

TABLE 1 / THE RELATIONSHIP BETWEEN THE DEVELOPMENT OF AGGRESSION AND SOCIAL ORGANIZATION IN CANIDS AND AGE AT DISPERSAL IN MARMOTS

CANID	FREQUENCY OF RANK-RELATED AGGRESSIVE INTER-ACTIONS DURING FIRST 6 WEEKS OF AGE*	FREQUENCY OF COHESIVE BEHAVIORS SUCH AS PLAY DURING FIRST 6 WEEKS OF AGE	SOCIALITY AS ADULTS	PROBABILITY OF INDIVIDUALS DISPERSING BEFORE OR AT SEXUAL MATURITY	EQUIVALENT TO [MARMOT**	AGE AT DISPERSAL
Red Fox	+++	+	Solitary	+++	Woodchuck	weaning
Coyote	++	++	Solitary-Semisolitary	++	Yellow-bellied Marmot	1 year
Wolf	−	++++	Highly social	−	Olympic Marmot	2 years

*"+" is a relative measure of frequency or probability (See Bekoff, in press)
**See Barash (1974)

specifics is adequate for normal social development. What *exactly* occurs during this early interaction remains a mystery. We are still unsure of the nature of the critical types of interaction during development, the prevention of which is responsible for the development of abnormal behavior (Beach 1968).

At birth and during early development, sex and individual differences become manifest. In addition, there may be a relationship between the ontogeny of behavior within a genus and later species-typical social organization observed among members of that genus (Table 1). For example, if one compares the social development of three members of the genus *Canis*—coyotes (*C. latrans*), wolves (*C. lupus*), and beagles (a domestic dog—*C. familiaris*)—there are significant differences in their ontogeny (Bekoff 1974, in press; Bekoff et al. 1975). Coyotes are more aggressive in early life and play less than either wolves or beagles, and coyote pups typically establish social relationships via severe, unritualized fights as early as 4 weeks of age. As adults, coyotes are less social than are wolves or beagles. How the early ontogeny of aggression is related to the later semisocial (or semisolitary) existence of coyotes is not precisely known; however, when the behavioral phenotypes of *individual* animals are studied, the possibility exists that individual differences in the ability to initiate social interaction and exploratory behavior may play some role. We have found that

TABLE 2 / THE SUCCESS (SUCCESSFUL PLAY-SOLICITS/TOTAL NUMBER OF PLAY SOLICITS) IN SOLICITING PLAY BY DOMINANT AND SUBORDINATE INDIVIDUAL COYOTES

		SUCCESS (%)
Coyote Pair 1*:	Dominant:	25
	Subordinate:	42
Coyote Pair 2*:	Dominant:	17
	Subordinate:	34

Intact coyote litter**

RANK	SUCCESS***
1	38
2	50
3	47
4	47
5	57
6	67

*from Bekoff (1974); hand-reared: differences are significant (p<.05).
**mother-reared; results are from approximately 100 hours of observations between 35–50 days of age; in this litter, severe dominance fights beginning on day 28 resulted in a linear hierarchy by day 35. (For further details see Bekoff, in press.)
***correlation between rank and success = −.79

higher-ranking individuals are the least successful at soliciting play from littermates (Table 2) and are also the most exploratory (Bekoff, in press). Although it remains to be tested in a larger number of coyote litters, it appears highly probable that animals that have had fewer social interactions with littermates would be more loosely tied to their group, and when this is combined with their exploratory nature, they would probably be the ones to leave the group earliest. The fate of other littermates is more difficult to predict and the role of adults in leading to dispersal of the young must also be considered. The fact that wolves have their "dominance" fights later in life, at approximately 9 months of age (Zimen 1975), may be important, since at the time of the fights they already have had significantly more interaction with one another than coyotes have had when they typically fight. Since it is assumed that wolf packs are formed from family groups (Mech 1970), this early social experience with littermates and other group members by young wolves would increase the probability of the group remaining together. The existence of large individual differences within the same species, and its implications for social ontogeny, social organization, and the evolution of behavior, require further study (Oliverio 1974; Bekoff, in press).

THE USE OF STAGES OR PERIODS OF DEVELOPMENT

A very common practice of researchers interested in social development is to divide the time-course of ontogeny into stages or periods (e.g., Scott and Marston 1950; Williams and Scott 1953; Thompson and Schaeffer 1961; Nice 1962; Scott 1962; Altmann 1963; King 1968; Stanley 1971; Horwich 1972, 1974; Scott et al. 1974; see also Schneirla and Rosenblatt 1963 and Denenberg 1964). Most workers follow the example of Scott and Marston (1950) of dividing ontogeny into four consecutive overlapping stages: (1) *neonatal*, (2) *transitional*, (3) *socialization*, and (4) *juvenile*, the duration of each depending on the species under study. (For a comparison of seven mammals see figure 5 in Scott and Marston [1950].) Hinde (1971) has reviewed the use of stages in studying behavioral development and concludes that they are generally not very helpful. He states that one major shortcoming of the application of stages to development is that their use tends to detract from the fact that development is a continuous, on-going process, and that what occurs in one stage is dependent on what occurred before, and will affect what occurs subsequently. In addition, the use of stages to provide a general picture of the social development of a species obscures individual differences. However, in order to make a general comparison between different species, a comparison of the age limits of certain stages may be useful, assuming the worker can be certain that the conditions under which the animals are being reared are as close to "normal" as possible for that species and that behavioral criteria are consistent.

CRITICAL PERIODS

The term "critical period" frequently is used in developmental studies. "Critical period" has been defined in a number of ways. Scott et al. (1974: 490), summarizing the applicability of critical periods to social ontogeny, wrote that "the theory of critical periods states that the organization of a system is most easily modified during the time in development when organization is proceeding most rapidly." If there is a uniform rate of development, then there is no critical period for the process under considera-tion. King (1968: 59) defines a critical period as "that period in development during which the probability of a behavior being emitted and the probability of it being reinforced by the environment are greatest... in most species the synchrony of emis-sion and reinforcement probability is well established." He also stresses that it is important to remember that some responses may develop simultaneously and interact and reinforce one another. The critical period has also been defined as the time in development when early experience has the greatest effect on later behavior (Fox 1966), and as the period of supramaximal sensitivity (Candland 1971). Candland (1971) is quite correct in writing that the way in which early stimulation affects later behavior is largely an unanswered question. Furthermore, we must always keep in mind that even if we say that *this* period is critical for the establishment of *this*, and *that* period critical for the establishment of *that*, we cannot ignore that what occurs in one period is not isolated from the development of the organism as a whole, and in the end, the critical period for "this and that" are interdependent. Development is a continuous process—the young organism passing through a series of overlapping, interdependent stages. One may only define *the* critical period for something in relative terms.

BEHAVIORAL DEVELOPMENT AND CENTRAL NERVOUS SYSTEM MATURATION

When studying the development of behavior, we must also consider the events going on under the skin—neural and otherwise (e.g., adrenal functioning [Candland and Leshner 1975]). During early life it appears that the central nervous system is in its most "plastic" state (Bekoff and Fox 1972; Devor 1975). The relationship between neural and behavioral ontogeny has been studied in many mammals (Tobach et al. 1971; Sterman et al. 1971; Altman et al. 1973 [see also Nadel et al. 1975]; Chevalier-Skolnikoff 1974; Sedláček 1974). Bronson (1965) has recently attempted to correlate behavioral data on the development of learning processes with neurological data on maturation of the central nervous system. He postulates three levels of CNS functioning: (1) the neocortex, (2) the subcortical forebrain (including thalamus, hypothalamus, and limbic system), and (3) the brain stem (including brain stem reticular system). Bronson (ibid.: 7) feels that "ontogenetically, the emergence of new behavioral capacities [may be seen] as a function of the sequential maturation of networks within the different levels." He also discusses the concept of critical periods relating the emergence of more complex learning capacities with CNS maturation. His discussion provides convincing evidence that "the defining variables (the signifi-cant stimuli, the behavioral effects, and the critical age range) of each critical period correspond closely with the perceptual capacity, the behavioral influence, and the maturational chronology of one of the three levels of the conceptual model" (ibid.: 20). Bronson's paper is an excellent example of an integrated approach to the prob-lems of behavioral ontogeny.

It is possible also to relate in detail CNS maturation to behavioral development (Table 3). At the onset of the "period of socialization" in dogs (approximately 3 weeks

TABLE 3 / GENERAL RELATIONSHIP OF PHYSICAL AND BEHAVIORAL DEVELOPMENT IN DOGS*

AGE IN WEEKS	STAGE OF PHYSICAL DEVELOPMENT	MOTOR REFLEXES AND SIMPLE CONDITIONING	RELATIONSHIP TO ENVIRONMENT	SOCIAL INTERACTION
Birth	Brain soft and jelly-like; sulci poorly formed	Random motor movements and twitches	Sensitive to: touch, taste, pain, cold and	Food-getting from mother
	Brain waves barely present:	Labial sucking reflex	hot stimuli	Grooming by mother
	activity	Positive thigmotaxic reflex	Underdeveloped systems: auditory,	Elimination reflexes stimulated by mother
	sporadic and asymmetrical	Positive thermotropism	olfactory, visual, temperature	Warmth from mother
	Cerebral cortex—two cellular layers distinguishable	Elimination reflexes Spontaneous head movement:	regulatory Searching and food-getting reflexes	Retrieving by mother (Beagles)
	Myelin coating on CNS nerve fibers under-	horizontal and vertical	present	Tactual contact with mother and litter mates
	developed except on trigeminal fibers and nonacoustic portion of auditory fibers	Crawl—slow and poorly coordinated; anterior portion of body better coordinated	Distress vocalization present	
	Retina undeveloped			
	Media of eye not transparent			
	Eyelids and auditory meatuses closed			
0				
	Olfactory stimuli affect EEG	Spontaneous muscle twitching when asleep	Limited olfactory sensitivity present but pup does not appear	General peak of mother pups
	Eyelid reflex present but weak	Rooting or "burrowing" reflex	to react to odor of dam	Mother seldom leaves pups
		Upward head reflex when pup sucking		
		Adult postural reflexes beginning to appear		General peak of non-protest vocalization
1				
		Gustatory appetitive and aversive conditioning obtained	Nociceptive withdrawal response appears	General peak of protest vocalization
		Backing movements appear		Pups begin to move after mother to nurse
	Rods and cones appear	Mobility increases perceptibly		

continued

		TABLE 3 / CONTINUED		
AGE IN WEEKS	STAGE OF PHYSICAL DEVELOPMENT	MOTOR REFLEXES AND SIMPLE CONDITIONING	RELATIONSHIP TO ENVIRONMENT	SOCIAL INTERACTION
2	Eyelids open (Beagles) Pupillary reflex present Ears open (Beagles) Teeth appear (Beagles)	Neonatal reflexes rapidly disappear Growls and barks elicited Rise up on legs Greater precision in sucking movements		Frequency of litter-dam separation increases
	Optic nerve myelinated	Stable conditioning patterns attained in several sensory motor areas	Startle response to sound appears (Beagles) Orientation toward visual and auditory stimuli	First notice human observer
3	EEG patterns differentiate for sleep and waking states	Straight form of propulsion appears Spontaneous twitching disappearing	Elimination outside the nest area	Critical period begins for primary social relationships Simple play interaction begins
	EEG affected by visual and auditory stimuli	Tail wagging develops (Beagles)	Temperature regulation system improved	Second peak of contact frequency between pups and dam
4	Cerebral cortex—six cellular layers distinguishable	Prancing appears Rapid maturation of motor patterns	Response to visual cliff, indicating depth perception	Tail wagging in response to social stimuli Litter moves as a group-allelomimetic behavior (Beagles)
	Myelin complete in ventrospinal nerve tracts	Pouncing on objects	Visual orientation toward distant stimuli	First indications of complete dominance of one pup by another (Beagles)
5	Retina fully developed	Hopping and running in leaps and bounds	Exploratory response to new objects Fear reactions appear (Beagles)	The "chase"—mother moves away while litter pursues her

continued

TABLE 3 / CONTINUED

AGE IN WEEKS	STAGE OF PHYSICAL DEVELOPMENT	MOTOR REFLEXES AND SIMPLE CONDITIONING	RELATIONSHIP TO ENVIRONMENT	SOCIAL INTERACTION
6	Oxygen consumption increases in cerebral cortex and sub-cortical areas, but decreases in medulla oblongata Brain adult-like in form, size, weight, and volume		Distress vocalization reaches a peak in strange pens (Beagles) Trial and error learning indicated	Pup is capable of strong attachments to a passive human Playful wrestling between pups (Beagles)
7	EEG appears adult-like Myelin distribution in white matter of cerebral cortex nears adult proportions	Walking and running movements still slightly wobbly	Fear reactions pre-dominate in un-socialized dogs Retrieving response appears Defecation in a definite place	Final weaning begins Pups exchange sharp nips in play Beginning of group attacks on one animal End of optimal period for forming primary social relationships

*Notations refer to dogs in general except where specified (Beagles). From Solarz 1970.

of age), EEG patterns differentiate for waking, resting, and sleeping states and become responsive to visual and auditory stimuli (Solarz 1970). Of the 33 tracts or regions in the CNS of the domestic dog for which Fox (1971: 143) followed the course of myelination, 28 show faint myelination by 21 days, and by 28 days all 33 show some myelination and 7 have already reached the adult level. Therefore, there is a temporal correlation between neural and social development. Between days 21 and 28, the CNS is undergoing rapid maturation and there are concomitant increases in percep-tual motor abilities and "active" interactions with littermates and other individuals.

As the organism ages, sensory as well as motor skills develop (Parker, this volume; Gibson, this volume). It is generally accepted that sensory systems develop according to the organism's needs (e.g., Anokhin's [1964] concept of heterochronous sys-temogenesis), and that the time-course of development of a particular system will depend on the "demands" being made on an infant at a given time. Gottlieb (1971) has compared the sequence of sensory development of various precocial and altricial mammals from independent evolutionary lines, and found that for six of these mam-mals (deermice [*Peromyscus* spp.], opossum [*Didelphys virginiana*], rat [*Rattus nor-vegicus*], domestic rabbit [probably *Oryctolagus cuniculus*], domestic cat [*Felis catus*], and humans [*Homo sapiens*]), the order of development of sensory systems is tactile, vestibular, auditory, and visual. The overall similarities in the sequence of development (homochrony) of sensory development appears to be species-general, while heterochronous development is species-typical. That is, across diverse species we observe the same sequence of sensory development, but within a given species the systems develop according to need; consequently there will be different time-courses of development when comparing one species to another. Gottlieb (ibid.) believes that the homochronous development of sensory systems is selectively neutral with respect to the adaptation of the newborn, since diverse species display the same developmental sequence. Heterochronous development, on the other hand, has been selected for, insuring species-typical adaptation of the newborn. There is really no reason for assuming selective neutrality of homochronous development (or of any other process). Strong, similar selective pressures operating on these animals could have resulted in this homochrony via parallel or convergent evolution, and this makes the phenomenon even more intriguing for future research.

It is obvious that the entire organism must be considered when discussing behavioral development, and that the development of social behavior is inseparably tied to CNS maturation and the ontogeny of sensory and motor abilities. Before discussing how the socialization process has been studied and how various skills are acquired, one more topic must be mentioned. The ways in which an organism changes externally is frequently as important to consider as are the covert changes occurring in the CNS.

PHYSICAL DEVELOPMENT AND SOCIAL BEHAVIOR

As an organism develops, it changes in a number of ways: size, shape, color, etc. An excellent example of the relationship between external appearance and social behavior is provided by Geist's work (1968a, 1968b, 1971) on mountain sheep (*Ovis* spp.). Mountain sheep show significant diversity in horn and body size among individuals (Figure 2) and strong sexual dimorphism. There are also differences in body coloration, particularly in *O. canadensis canadensis*, the small-horned sheep tending to be lighter than the large-horned ones. Males (rams) have a juvenile period lasting 5 to 6 years during which they show great changes in physical development. Between about 7 and 9 years of age the males reach full maturity. Females, on the other hand, mature only to the level of sexually mature yearling rams, and possess only two behavioral states. When they are in estrous, they act like young males (in accordance with their physical development) and when anestrous they act like sexually immature sheep. Geist's data clearly demonstrate that during behavioral development the ram uses the same behavior patterns progressively in different frequencies, and that "this plus the close correspondence between physical and behavioral maturation, indicates that the slight, progressive behavior changes from year to year are not likely due to learning" (1968a: 904).

In concluding this section, we see that overt and covert changes are important to consider when discussing behavioral development. Kuo's (1967) statement that all behavior is a functional by-product of the dynamic interrelationship of morphological factors, biophysical and biochemical factors, stimulating objects, developmental his-

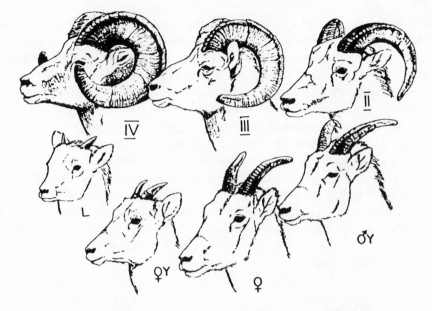

FIGURE 2. *Bighorn sheep can be classified into 8 sex-age classes,
seven of which are illustrated here. L = lamb, ♀Y = yearling
female, ♀ = mature female, ♂Y = yearling male, II, III, and IV
= different classes of rams. Note that females remain frozen in
development at the level of a sexually mature yearling ram. (See
text; after Geist 1968a, with kind permission of the author and of
the National Research Council of Canada, from the* Canadian
Journal of Zoology, *1968, 46:899–904).*

tory, and environmental context is well-taken, but unravelling the intertwining
threads of these and possibly other factors presents an exciting problem for future
research.

■ SOME METHODS OF STUDYING THE SOCIALIZATION PROCESS

THE DEPRIVATION EXPERIMENT

Since many mammals become socialized quite rapidly and the ways in which this
occurs remain a mystery, workers have used a number of techniques to deal with the
problem. In an attempt to "isolate" relevant variables, social isolation and deprivation
procedures have been used. Animals are reared under various conditions and the
effect of the treatment on later behavior is noted. For example, rhesus monkeys
(*Macaca mulatta*) reared in social isolation from other conspecifics show abnormal
behavior when allowed to interact subsequently with other rhesus monkeys, the effect
being somehow related to the environment in which the monkey is reared (Harlow
1969; Sackett and Ruppenthal 1973). The same effect can be achieved with dogs and
other mammals (Scott and Fuller 1965; Harper 1970; Fox 1971; Hinde 1974). Basically, severe restriction of early experience tends to produce behavioral anomalies,

but such studies do not actually tell us much about how animals become socialized—the variables or precise mechanisms that are involved. In addition, the effects that are obtained by using deprivation procedures may not only be due to the early environmental restrictions to which a young animal is subjected but, rather, may be due to later testing procedures and the stress that the animal experiences upon release from the impoverished environment (Fuller 1967).

Suffice it to say, social deprivation procedures have been criticized repeatedly because of the lack of control over what appear to be relevant variables (Lorenz 1965; Ewer 1968; Bateson 1976). For example, spatial and/or sensory restriction may be as important in producing aberrant behavior as is social isolation, and the effects of each are difficult to untangle. As Bronfenbrenner (1968: 702–703) has written in his exhaustive review of early deprivation in mammals, "there is need for systematic study in different species of the differential impact of social deprivation, at successive stages of development, with particular attention to the separate and joint contribution of mother-infant and infant-infant interaction as well as the role played by the physical environment... it will be a long time before research in this area suffers from unnecessary redundancy."[6]

CROSS-REARING

Another method of studying socialization and how the process may be altered is by use of the cross-fostering method—an animal of one species is reared with animals belonging to another species (e.g., Schutz 1964; Cairns and Johnson 1965; Cairns and Werboff 1967; Fox 1969; Beauchamp and Hess 1971; Denenberg 1971; Paschke et al. 1971; Immelmann 1972; Baker and Preston 1973; Bols and Wong 1973). Bols and Wong (1973) stress that cross-rearing is a gross treatment involving a wide spectrum of subvariables, and interpretation of results should be done with caution, especially as regards the effects of cross-rearing on later behavior.

In mice, the effects of cross-rearing may be strain-specific. For example, when C57BL/10J mice are reared with either a rat mother or rat aunt, there is a reduction in fighting; however, there is no reduction for similarly reared Swiss-albino mice (Denenberg 1971). In both strains, cross-rearing results in a reduction in the activity level both at weaning and in adulthood, and in Swiss-albino mice, there is also a marked reduction in adrenocortical reactivity to a novel situation. Denenberg and his colleagues have eliminated a number of possible causal factors such as biochemical differences in the milk supply between the mouse mother and the rat mother, as well as odors, and auditory and visual cues. Denenberg (1971: 92) concludes that "the differences obtained are caused by the mother's or aunt's behavioral interaction with the pups between birth and weaning. . . ." However, what the mother or aunt actually does that causes the behavioral and physiological modifications in cross-reared mice is still a matter of speculation. Hopefully, more careful analyses of pup-mother (aunt) interactions will clarify the situation.

Sex differences in cross-rearing experiments have also been found (e.g., Schutz cited by Immelmann 1972; Beauchamp and Hess 1971). Schutz found that sexual imprinting only occurs in males in a number of sexually dimorphic duck species. In females, early experience with another species has no effect on later sexual preference, and they react to conspecific males even if they have had no previous experience with them (see Immelmann 1972 for a review).

In an interesting study, Fox (1969) reared dogs with cats during the period of life corresponding to the socialization period of the dogs. He correctly accounted for differences in reflex development and CNS maturation (cat faster than dog) by using dogs that were 5–7 days older than the kittens. He used dogs (Chihuahuas) of approximately the same size as the cats. When the dogs were 16 weeks old a series of tests were conducted, one involving the reaction of the animals to their mirror-image. The cat-reared dogs were much less vocal, less active, and interacted fewer times and for a shorter duration with the mirror than did control animals. The experimental animals were then retested 2 weeks later after having 2 weeks of contact with other dogs, and Fox found large increases in the 4 measures used. Unfortunately, Fox did not retest his control group.

The results of one other report involving the effects of interspecies interaction on social development should be considered because of its theoretical implications (Baker and Preston 1973). Baker and Preston raised irus monkeys (*Macaca irus*) and patas monkeys (*Erythrocebus patas*) together. The youngsters differed in size, and the behavioral repertoires of these two species are also quite dissimilar. When the infants attempted to interact, they were unable to do so successfully because they were unable to "understand" what one another was doing because of the lack of similarity in behavioral repertoires. Consequently, neither of the two species functioned as the "proper receptor" for the behavioral activities of the other animal (see discussion below concerning the importance of reciprocal interaction for the development of communicatory skills), and the effect on socialization was reduced. Since the mothers of the infants were also in the cages, the effect of their presence must be considered. It would be interesting to study the effect of rearing two dissimilar animals together, each providing the *only* outlet for social interaction. If they did not interact because of large differences in their behavioral repertoires, then one might predict that they would behave similarly to "social isolates" when allowed subsequent interaction with conspecifics. The possibility that each might modify its repertoire to facilitate interaction, and the ways in which this might be accomplished, should be pursued. Cross-rearing techniques provide a good method for studying behavioral plasticity.

Although isolation and cross-rearing procedures have provided some gross data concerning social development, the precise ways in which these treatments affect behavior and the way in which young organisms acquire various abilities in "normal" circumstances are still not clear. The last section of this chapter will deal with the ways in which a number of adaptive skills are acquired or are thought to be acquired.

■ THE ACQUISITION OF ADAPTIVE SKILLS

In general, the more quickly the young can grow to the adult stage the more chance they have of reproducing before they get killed. Prolongation of youth can be selected

only if it provides some advantage which outweighs the increased chance of premature death. In man, this of course was provided by the learning that takes place during
childhood. Clearly, therefore, the prolongation of childhood could have begun only
after a certain degree of ability to learn from parental "instruction" had been established.[7] This ability to learn involves two components: not only the characteristics of
the nervous system of the learner, but also the parental behavior and social organization of the "instructors." We thus have a series: (1) Development of social behavior
favoring learning by the young which creates selective pressure for greater learning
ability; (2) Development of learning ability, particularly when young. This creates
selective pressure for a quantitative increase in the amount learned; (3) Development
of a longer learning period. This series will now become a circle with (3) making (1)
more "worth while" and (2) making (1) more easily changed, so that evolution along
these lines will be expected to proceed very rapidly until the point is reached where
the increased chance of premature death, acting as negative feedback, brings it to a
halt (Ewer 1960: 178).

GENERAL COMMENTS

During early life various skills are acquired. As the organism ages, it interacts increasingly with the animate and inanimate environment, and the information acquired
somehow enables the infant to "learn" what it can and cannot do and the significance
of its actions. There are a number of ways (not necessarily mutually exclusive) in
which a skill can be acquired. It may be: (1) innate or inborn, but not immune to
experiential variables, (2) acquired by some type of trial-and-error learning with
appropriate reinforcement, (3) acquired by observational learning and/or "imitation,"
(4) acquired by some form of "teaching," or (5) acquired in play, during which actions
from various contexts are performed.[8] When studying development, attention must
also be given to the ontogeny of motor and perceptual abilities (e.g., Held and Hein
1963; Hailman 1970; Hein et al. 1970; Hershenson 1971; Vestal 1973), particularly
the development of the ability to perceive stimuli, to differentiate aspects of a given
stimulus situation, and the way in which perceptual and motor systems become
integrated. The ways by which organisms generalize from one situation to another
should also be studied (Etienne 1973).

An organism's behavioral repertoire is somehow assembled during early life from a
reservoir of motor patterns, and Horwich (1972: 75) has noted the possibility of there
being "a correlation in mammals between the ontogenetic sequence of behavioral
component emergence and the temporal sequencing of the components in adult
behavior." The age at which various action patterns first appear, the change in
frequency and form over time, and the ways in which they are assembled into
temporal sequences require much greater analysis in order to test the generalizability
of Horwich's interesting suggestion.[9]

I would now like to consider the development of four types of behavior: (1)
predatory (food acquisition), (2) agonistic, (3) reproductive, and (4) communication,
scent-marking and the use of other modalities.[10] Other mammalian behaviors that
have also been studied developmentally include comfort and "toilet" behaviors (Ewer

1967; Fentress 1972; Horwich 1972), fear of predators (Müller-Schwarze 1972), and sniffing by rodents (Walker 1964).

PREDATORY BEHAVIOR—FOOD ACQUISITION

A young animal must somehow develop the ability to (1) recognize what is and what is not food and (2) successfully go out and get it. Indeed, malnutrition can affect behavioral development in many ways (Zimmerman et al. 1974). Although the following discussion is concerned primarily with meat eaters, herbivores also must acquire the ability to obtain a diet with the maximum amount of nutrients (Freeland and Jantzen 1974; Westoby 1974).

There have been a number of recent studies of a wide variety of mammals pertaining to the above two problems (e.g., Wüstehube 1960; Ewer 1968: 275–279, 1973; Galef and Clark 1972; Galef and Henderson 1972; Galef 1976; Eisenberg and Leyhausen 1972; Schaller 1972; Apfelbach 1973; Leyhausen 1973; Rasa 1973; Overmann 1976[11,12]).

In most carnivores, the young require solid food before they are able to obtain it on their own. In these cases, the mother and/or other adults may regurgitate partially digested food for the young (e.g., *Canidae*), may bring back a half-killed or fully-killed prey (Leyhausen 1973), or the mother may fetch the young to a recent kill (Schaller 1972). The experience the young have during these instances may help establish a predisposition for certain food types as well as provide "predatory practice" for the young (see below). Unfortunately, there is very little information available concerning what stimuli are used for prey identification and how the ability develops (Robinson 1969; Apfelbach 1973). Galef and Henderson (1972: 213) postulate that "Cues associated with diet eaten by a lactating female rat are transmitted to her young, probably via her milk, and are sufficient to markedly influence the food preferences of her young during weaning. For their first meals of solid food, the weanlings actively seek and preferentially ingest the diet their mother has been eating during the nursing period, even if that diet is relatively unpalatable." (See also Galef and Sherry 1973; Galef 1976.) Galef and Clark (1972) have demonstrated that rat pups' choice of diet is also affected by the physical presence of adults in the vicinity of a food site. Therefore, it can be concluded that mother's milk combined with the presence of adult rats at a feeding site serves to determine initial dietary selection by weanling rats. Apfelbach (1973) has shown that young polecats (*Putorius putorius* L.) learn to recognize the odor of future prey by smelling the food brought back by the mother, and social transmission of food sources by parents may also occur in pigeons (Neuringer and Neuringer 1974).

Assuming that somehow the animal learns what it can (should) and cannot eat, it is still faced with the problem of catching prey on its own or hunting as part of a coordinated group effort. For the sake of simplicity, I shall consider the case of a lone predator and a single prey. The ability to catch prey may be innate: a naive animal (an animal with no prior experience with prey) may successfully kill prey in a species-typical manner upon first exposure. This has been observed in polecats (Wüstehube 1960) and other mammals (Eisenberg and Leyhausen 1972) and a number of birds (loggerhead shrikes, *Lanius ludovicianus*, Smith 1973; American kestrels, *Falco sparverius*, Mueller 1974). However, some experience may be necessary to increase the "smoothness" and efficiency of carrying out the behavioral sequence (Eisenberg and Leyhausen 1972), or for learning the orientation of the killing bite. Rasa (1973) found that when young African dwarf mongooses (*Helogale undulate rufula*) leave their nest at approximately 24 days of age, practically all of the adult prey-killing repertoire is observed, except the killing bite, the orientation of which must be learned and which varies with different prey. Observational and/or "imitation" (see note 7) learning might also aid in the development of predatory abilities. Kuo (1930) found that kittens reared with a mother who killed rodents in their presence killed rodents earlier in life and more frequently than kittens reared alone or with a small rodent. Schaller (1972: 263) observed that young lions (*Panthera leo massaicus*) "have the opportunities to learn stalking techniques and killing methods by observing adults." Whether or not "teaching" occurs remains an open question. Kleiman and Eisenberg (1973: 645–646) wrote that "The female cat must cope with rearing offspring and teaching them to hunt without assistance of other adults." Ewer (1969) feels that the mother or other adults provide the situation in which young can experience prey killing without much danger from the prey, and that the young educate themselves. Kruuk and Turner (1967) observed a female cheetah (*Acinonyx jubatus*) carry a live Thomson gazelle (*Gazella thomsonii*) to her young, release it, and allow the young to chase it. Harper (1970), after reviewing relevant literature, has concluded that the role of the offspring in the development of their own behavior should be studied with less emphasis on the "educative" function of the caretaker(s).

The way in which predatory behavior develops at the level of the incorporation of individual actions into behavioral (temporal) sequences has been little studied. Such studies will certainly help in the application of Horwich's (1972: 75) suggestion (see above) about the relationship between the time (age) of emergence of various actions and the order in which they are performed during behavioral sequences by adults. In a recent study of the development of predatory behavior in young coyotes (Vincent and Bekoff in press), particular attention was paid to the ways in which 28 individual motor actions were linked together as individual coyotes had increasingly more experience with particular prey (mice). (The prey species must be considered in any analysis of predatory behavior; Eisenberg and Leyhausen 1972.) We also observed the age at which the actions first occurred outside of the "prey-killing" encounter, the order of appearance during encounters with a mouse, and the frequency with which they occurred during social play and agonistic interactions. The actions included (1) orienting towards the prey, (2) hesitant approach (often involving rapid withdrawals), (3) approach, (4) investigation of the prey (sniffing), (5) licking intentions directed towards the prey, (6) licking of the prey, (7) bite intentions, (8) nibbling, (9) biting, (10) pawing, (11) pinning, (12) pouncing, (13) lifting of the prey, (14) carrying, (15) shaking the prey from side to side, (16) tossing the prey with an exaggerated upward head movement, (17) circling with the prey in mouth (usually away from the observer(s) and/or other coyotes, (18) eating. Large individual differences were observed, and the sequencing of the above actions varied greatly. Furthermore, there was no relationship between what occurred during play (or agonistic interactions) and subsequent individual success in prey-killing (suggesting that the "practice theory" of play should be reconsidered in light of these and other data—for details see Bekoff 1976).

Although more data are also needed pertaining both to the acquisition of what is and what is not food and the ability to catch prey on one's own, the above discussion demonstrates that there are examples that fit into each of the categories listed above, and that species differences exist.

AGONISTIC BEHAVIOR

Agonistic behavior (after Scott and Fredericson 1951) refers to the cluster of behaviors including aggression (fighting and various types of threatening) and various forms of submission. I do not want to consider the reasons why young animals do fight or how the expression of agonistic behavior can be affected by early experience (see Eleftheriou and Scott 1971; Johnson 1972; Cairns 1972; Hinde 1974) but, rather, I would like to attend to the questions of how an animal develops the ability (1) to behave agonistically and (2) to control the expression of aggression in a conspecific. Intimately related to the acquisition of these abilities is the development of intraspecific communicatory skills.

In some mammals it appears that rudiments of "real" aggression first appear during "play-fighting" (Poole 1966; Ewer 1968, 1971; Cairns 1972). Poole (1966) found that of 13 patterns of behavior commonly observed during aggression in adult polecats, 9 appeared during aggressive play in the young by the time they were 8 weeks of age. At the time of their first appearance the patterns resemble those of the adult, so there does not seem to be any modification by experience as the animal grows older. That is, play does not provide practice for later "real" aggressive encounters. The 4 patterns that are absent from "play-fighting" are those that are used to "intimidate" the partner. In Australian fur seals (*Arctocephalus forsteri*), full neck displays (used by adults while threatening) first appear during "play fighting" between 3–4 weeks, and the open mouth threat display appears at approximately 2 weeks (Stirling 1970). However, since isolation-reared animals do fight, play-fighting experience cannot be considered a *necessary* step in the development of fighting skills (Cairns 1972). This also applies to socially reared coyotes who typically fight before they play.

Sex differences in fighting and threatening also appear during early life. For example, in the Northern elephant seal (*Mirounga angustirostris*), Rasa (1971: 89) observed that the weaned males perform "in almost 'slow motion' the behaviors associated with fights between adult bulls: throat pressing, bites to the neck, chest, flippers, and nose, rearing and slamming. The females perform the behaviors similar to those of the adult cows: loud vocalization delivered facing the partner with the mouth open and head nodding."

Canids provide a good group in which to compare the ontogeny of agonistic behavior. As mentioned above, the less social canids tend to fight earlier in life than do the more social canids (Bekoff 1974, in press). Furthermore, it is not uncommon to observe coyotes fighting before any play has occurred, while in wolves (and beagles), "play-fighting" almost invariably occurs before rank-associated agonistic encounters are observed. The first fights that are observed between infant coyotes are very severe and unritualized, and in a way, it would not be incorrect to refer to these as "presocialization" fights, for the animals have had very little social experience with one another at the time of fighting. It is very common to observe a highly aggressive animal continue to deliver uninhibited bites (often accompanied by vigorous shaking of the head from side to side) even after its opponent has assumed a submissive posture (rolled over on its side or back and emitting high-pitched vocalizations). Young coyotes also frequently "ignore" threats directed towards them. In a sense, we could say that these youngsters have not had enough social experience to acquire necessary communicatory skills, or perhaps their motor abilities may be more advanced than their sensory and association abilities. Fights that are observed in older coyotes and younger wolves and dogs (after they have engaged in "play-fighting") are generally less intense and more ritualized, that is, threat and submissive signals are "honored."

The response to an aggressive act is also important to consider. Eibl-Eibesfeldt (1967) and others have stressed that aggression cannot be studied apart from the response it evokes; however, the ontogeny of submissive signals has not been intensively studied. Submissive behavior usually results in terminating an actual or impending attack. In canids, two forms of submission occur, both of which can be traced to behaviors that were either directed toward the infant by the mother (stimulation to urinate or defecate) or performed by the infant during early food-begging (Schenkel 1967). Active submission, which appears to be derived from infantile food-begging, takes the form of muzzle-nudging and pushing, frequently accompanied by high-pitched whining. Passive submission is derived from the posture that the pup assumes when being stimulated to urinate or defecate by its mother; the animal lies on its back, frequently with one or both of its hindlegs spread, and remains immobile. High-pitched whining is frequently emitted and the lips are usually pulled back horizontally into a "submissive grin." Similarly, in primates, some submissive behaviors, for example, pucker lip expressions and lip smacking, appear to derive from infantile food-getting responses (Chevalier-Skolnikoff 1974).

REPRODUCTIVE BEHAVIOR

As with agonistic behavior, there is very little in the published literature, except for the plethora of studies showing that early experience, including hormone manipulations, can affect later reproductive performance (e.g., Harlow 1969; Beach 1968; Harper 1970: 85–86; Whalen 1971). One of the problems of studying the development of reproductive behaviors is that they do not appear to be the culmination of a continuous process of development (Whalen 1971). As Harper (1970: 85) has written: "Although early isolation may lead to disturbed reproductive behavior in mammals, *it is another matter to conclude that social interaction is necessary for the development of species-typical behavior patterns*" [italics mine]. He (Harper 1968) found that male guinea pigs, even after being isolated from birth until 80 days of age, made properly oriented mounts when first exposed to receptive females. Furthermore, Beach (1968) found that only minimal contact (15 minutes per day) was enough to override the effects of total isolation in male beagles. Socially isolated males were deficient due to abnormal orientation of mounts. Beach (ibid.: 236) poses an interesting question in concluding his study: "What are the critical types of interaction during development whose prevention is responsible for the unusual behavior shown by SI (semi-isolate) males?..." His statement also applies to the ontogeny of other types of behavior, and Scott's (1970) statement, that one of the difficulties of studying socialization is due to the rapidity (and subtlety) with which it occurs, must be recalled.

COMMUNICATION

Scent-Marking

Many mammals use various scents (urine, feces, glandular secretions, saliva) to communicate via olfactory channels with one another (Ewer 1968, 1973; Altmann 1969; Eisenberg and Kleiman 1972; Johnson 1973; for a review of the ontogeny of olfactory influences on the behavior of vertebrates see Cheal 1975). Berg (1944) studied the development of the micturition pattern in domestic dogs. In dogs, sexual dimorphism in urination patterns begins to appear at approximately 3–5 weeks of age, the females squatting and the males standing while urinating. The typical male leg-lift that usually appears between 19–40 weeks of age is under hormonal (androgen) control (Berg 1944; Martins and Valle 1948). Scent-marking with a midventral sebaceous gland by the Mongolian gerbil (*Meriones unguiculatus*) has been studied rather extensively by Thiessen and his colleagues. Gerbils of both sexes begin to mark at 5 weeks of age, and at 16 weeks of age, males begin to mark more than females (Lee and Estep 1971). In males, development of marking is related to the androgen status of the animal (Lindzey et al. 1968), castrates showing complete absence of both marking and the midventral gland. However, in females, gonadectomy has no effect on marking (Whitsett and Thiessen 1972). The response of gerbils to scents varies with the past experience of the animal (Thiessen, personal communication).

Müller-Schwarze and Müller-Schwarze (1972) studied the development of scent-marking in 3 male and 1 female pronghorn (*Antilocapra americana*). They found that 18 stages characterized the development of the male's response to urine, leading to the species-typical SPUD (*s*niff, *p*aw, *u*rination, *d*efecation) sequence. The young pronghorn are able to urinate without maternal assistance when they are 3–4 days of age. When they are 8 days old the young male will begin to go to the area where another male is urinating and urinate in the same area (distance attraction and social facilitation). At approximately 42 days of age *urination* and *defecation* (UD) are temporally correlated and at 6 months pawing started to precede urinating (PUD). Also at this age there is a difference in the response to male and female urine. The response to the female urine was SPSFUD (F = flehmen) and that to the male urine was SPSPFUD. At 6 ½ months, male urine on the ground releases SPUD in the proper sequence. As mentioned above, the analysis of sequences often provides insight into the development of a particular behavior pattern. This study does not support the hypothesis (Horwich 1972) concerning the correlation between the ontogenetic sequence of particular behaviors and the sequence in which they are later performed by adults.

One other aspect of olfactory communication that may well be of considerable importance to investigate concerns the question of whether or not there are changes in the composition of various scents with age. Doty and Kart (1972) found developmental changes in the histology of the midventral sebaceous gland in various *Peromyscus*. Might there also be changes in an individual's sensitivity to various scents as it matures? Wilson and Kleiman (1974) suggest that olfactory thresholds are lower for younger organisms when compared with adults, and relate this to olfactory input and play in various rodents.

Communication Via Other Modalities

The ways in which mammals acquire intraspecific social communication skills are little known, as are the ways in which developmental factors influence the use of signal movements (Hinde 1974). There are no studies that approach the elegance of recent work on bird-song (Thorpe 1961; Hinde 1969; Marler and Mundinger 1971; Marler 1976; Nottebohm 1971, 1972) or theories of language acquisition in human infants (Lenneberg 1967; Vetter and Howell 1971). The basic problem concerns "how messages are associated with signals and . . . how messages are associated with signals in contexts. . . ." (Beer 1971: 446), the communicative significance of an action requiring experimental verification (Marler 1965).

Hinde (1974) has written that the learning of signal movements by imitation is extremely rare. Mason (1971: 59) has suggested that it is "most likely the functions of social signals are learned according to the same principles that are involved in learning that the onset of a light will be followed by shock, or that a red placque covers a peanut." The development of the ability to communicate appears to involve a combination of (1) the performance of motor patterns that may in some cases be innate, (2) the pairing of these signals (stimuli) with particular eliciting stimuli, and (3) the production of a consistent response in other animals (Symons 1974). The way(s) in which reinforcement affects the frequency of occurrence of a signal movement still remains an open question (Hinde 1974).

Some signals may operate the first time they are used; however, it is usually assumed that as organisms accrue experience with conspecifics, they become better communicators (Figure 3). That is, they are better able to send the "appropriate" signal when called for, respond "properly" in those instances in which a signal is directed toward them, and predict the behavior of other individuals. Fedigan (1972) refers to this as "social perception"—"the ability to predict other individuals' behavior and to act appropriately. . . ." (p. 361). Whether or not play experience is necessary for the development of communicatory skills has only been studied (to my knowledge) in a nondeprivation setting by Symons (1974). He found that rhesus monkeys do *not* learn, refine, or practice agonistic signals during play-fighting.

More studies like Symons' are required. We all intuitively feel that during development young organisms acquire communicatory skills, and this belief is supported by the results of deprivation studies that clearly demonstrate that animals reared in the absence of other animals do show deficiences in social communication. Careful, patient observation, combined with well designed experiments, should help

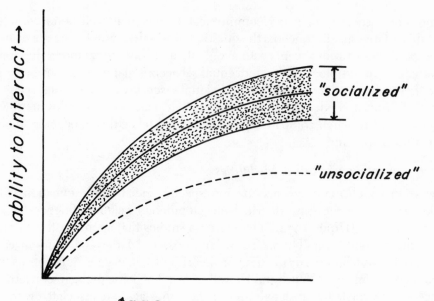

FIGURE 3. A *hypothetical relationship between the amount of social experience an organism has had and its ability to interact. "Interact" in this case refers to the animal's ability to send the "proper" signal when called for, to react "appropriately" to messages directed toward it, and to predict social behavior (after Bekoff 1976).*

shed some light on this extremely interesting and important aspect of social ontogeny. When studying the ontogeny of communication, Cullen (1972: 110) suggests that one must distinguish between "(1) the actual form of the social signals, (2) the individuals they are addressed to and (3) the responses they call forth in others." Analyses of temporal sequences of on-going behavioral interactions will undoubtedly prove helpful, by providing insight into stimulus (i)—response (reinforcement) contingencies.

■ CONCLUSION

I have attempted to conclude each section in place. I believe that very little is known about actual *mechanisms* of socialization in nonprimate mammals. It is hoped that this "devil's advocate" position will stimulate research that will answer questions about processes of species-typical social development in normally reared mammals. The results of deprivation and cross-rearing studies have provided some insight into the problems, but continued use of these procedures without an increase of careful, descriptive, nonexperimental and noninterventive studies both in the field and laboratory appears to be presently unwarranted. For example, we *still* do not know the role of play in the development of social behavior, and social deprivation procedures that remove the possibility for an individual to engage in play as well as other forms of social interaction are not providing additional insight. Another problem involves the definition of terms. The words "deprivation," "isolation," "restricted," "impoverished," "enriched," "critical period," and "play" still appear to mean different things to different people although frequently the differences are overlooked and unanimity of opinion is implied. As Georgescu-Roegen (1971: 211) has written: "Occasionally, the use of a term spreads through the scientific literature with amazing swiftness but without a valid birth certificate, that is, without having been defined in some precise manner. Actually, the swifter the spreading, the greater is everyone's confidence that the meaning of the term is perfectly clear and well understood by all." Until researchers agree on terminology and standardized procedures, progress will be impeded.

Marler and Hamilton (1966) have stressed that description is the most important phase in behavioral studies, and basic descriptive developmental studies are still needed. Both observable and unobservable (to the eye) events are important to consider in the process of the gradual unfolding of behavior during early life. The great importance of understanding social ontogeny in order to fully appreciate the complexities of adult behavior and mammalian societies in general makes developmental endeavours particularly exciting ventures.

■ ACKNOWLEDGMENTS

I would like to thank Jeremy Dahl, John David Henry, Peggy Hollyday, John A. King, and Frank Poirier for helpful comments on an earlier draft of this chapter and Ms. Kathleen Lorek for typing the manuscript.

■ NOTES

1. For those readers interested in behavioral development of birds see Allyn (1974), Bateson (1966, 1971, 1974), Salzen (1967), Sluckin (1972, 1973), Hess (1973), and Immelmann (1975).

2. Although I shall not discuss the ways in which the infant may affect its caretakers (e.g., Bell 1968; Harper 1971; Lewis and Rosenblum 1974), attention must be paid to the fact that adult-infant interactions are two-way affairs. Harestad and Fisher (1975) have observed differences in the social organization of nonpupping colonies of Steller sea lions (*Eumetopias jubata*) when compared with colonies in which pups were present.

3. Count (1958), Rheingold (1963), Eisenberg (1966), Mitchell (1969), Orians (1969), Harper (1970), Spencer-Booth (1970), Blurton-Jones (1972b), Goss-Custard, Dunbar, and Aldrich-Blake (1972), Trivers (1972, 1974).

4. Eisner (1960), Lehrman (1961), Rosenblatt (1967, 1970), Moltz (1971), Noirot (1972a), Terkel and Rosenblatt (1972), Lamb (1975).

5. *sounds:* Beer (1970), Okon (1970), Gould (1971), Espmark (1971), Noirot (1972b), Fraser (1973), Zippelius (1974), DeGhett (in press); *odors:* Klopfer (1971), Leon and Moltz (1971), Moltz (1971); contact through *tactile* channels such as caravan formation in *Crocidua:* Ewer (1968: 271), Zippelius (1972), Hellwing (1973). The role of visual cues is difficult to assess. It is usually assumed that there is some type of interrelationship between the various modalities (Herrenkohl and Sachs 1972).

6. Because of limited space I have refrained from getting into any discussion of the nature-nurture

controversy. Reviews may be read in Tinbergen (1951), Lorenz (1965), Lehrman (1953, 1970), Hinde (1970), R. J. Richards (1974), and Eibl-Eibesfeldt (1975). Beer (1975) presents an interesting discussion of the Lorenz "vs." Lehrman debates. Suffice it to say, behavior is neither entirely determined by "environment-expectant" nor "environment-dependent" factors, but rather is a result of an interplay between the two.

7. Interaction with peers must also be considered.

8. Sluckin (1972) and Davis (1973) have recently reviewed the field of observational and imitation learning, and this author (Bekoff 1976), Fagen (1976, in press), and Baldwin and Baldwin (1976 and this volume) have reviewed play and its relationship to social and physical development. Davis (1973) concludes that the word "imitation" is usually used as a label for ignorance. For an excellent and timely up-to-date view of the field of learning and its application to many behavioral problems see Hinde and Stevenson-Hinde (1973).

9. A way to think of the development of temporal sequences of actions would be as follows. Think of each action as a colored bead that is to be strung on a string. It would be possible to put these beads on the string in a variety of ways. If you had 5 beads—red, blue, green, orange, and yellow—and there were no restrictions placed on the use of the beads, you could string them in 3,125 ways! If during ontogeny they appeared in the order red, blue, green, orange and yellow, then according to Horwich, you would expect to see them performed in the same order when the animal performed the related behavior as an adult.

10. It is somewhat unrealistic to look at the development of one type of behavior without consider-ing how various behavioral systems interact. Various actions are used in a variety of contexts (at least superficially they look similar).

11. It would be interesting to have more data concerning the development of the use of "tools" in the acquisition of food (see Beck 1975 for a review of the literature) and the development of unique ways of obtaining food, such as the smashing of millipedes by mongooses (*Mungos mungo*) (Eisner and Davis 1967, McGrew, this volume).

12. For birds see Hogan (1973), Smith (1973), Mueller (1974); for reptiles see Burghardt (1968); for fish see Beukema (1968).

■ REFERENCES

Allyn, G. 1974. Mammalian socialization and the problem of imprinting. *La Terre et la Vie* 28:209–271.

Altman, J.; Brunner, R. L.; and Bayer, S. A. 1973. The hippocampus and behavioral maturation. *Behav. Biol.* 8:557–596.

Altmann, D. 1969. Harnen und Koten bei Saeugetieren. Wittenberg Lutherstadt: A. Ziemsen Verlag.

Altmann, M. 1963. Naturalistic studies of maternal care in moose and elk. In *Maternal behavior in mammals*, ed. H. L. Rheingold, pp. 233–253. New York: Wiley.

Anohkin, P. K. 1964. Systemogenesis as a general regulator of brain development. In *The developing brain, Progress in brain research*, vol. 9, eds. W. A. Himwich and H. F. Himwich, pp. 54–86. Amsterdam: Elsevier.

Apfelbach, R. 1973. Olfactory sign stimulus for prey selection in polecats (*Putorius putorius* L.).*Z. Tierpsychol.* 33:270–273.

Armitage, K. B., 1974. Male behavior and territoriality in the yellow-bellied marmot. *J. Zool. Lond.* 172:233–265.

Baker, R. P., and Preston, D. G. 1973. The effects of interspecies infants interaction upon social behavior of *Macaca irus* and *Erythrocebus patas*. *Primates* 14:383–392.

Baldwin, J. D. 1969. The ontogeny of social behavior of squirrel monkeys (*Saimiri sciureus*) in a semi-natural environment. *Folia Primat.* 11:35–79.

Baldwin, J. D., and Baldwin, J. I. 1976. Effects of food ecology on social play: A laboratory simulation. *Z. Tierpsychol.* 40:1–14.

Barash, D. P. 1974. The evolution of marmot societies: A general theory. *Science* 185:415–420.

Bartholomew, G. 1959. Mother-young relations and the maturation of pup behaviour in the Alaska fur seal. *Anim. Behav.* 7:163–171.

Bateson, P. P. G. 1966. The characteristics and context of imprinting. *Biol. Rev. Cam. Phil. Soc.* 41:177–220.

————. 1971. Imprinting. In *The ontogeny of vertebrate behavior*, ed. H. Moltz, pp. 369–387. New York: Academic Press.

————. 1974. Review of *Imprinting* by E. H. Hess. *Science* 183:740–741.

————. 1976. Specificity and the origins of behavior. *Adv. Study of Behavior* 6:1–20.

Beach, F. A. 1968. Coital behavior in dogs:III. Effects of early isolation on mating in males. *Behaviour* 30:217–238.

Beauchamp, G. K., and Hess, E. H. 1971. The effects of cross-species rearing on the social and sexual preferences of guinea pigs. *Z. Tierpsychol.* 28:69–76.

Beck, B. B. 1975. Primate tool behavior. In *Antecedents of man and after, III: Primate socio-ecology and psychology*, ed. R. Tuttle. pp. 413–447. The Hague: Mouton.

Beer, C. G. 1970. Individual recognition of voice in the social behavior of birds. *Adv. Study of Behav.* 3:27–74.

————. 1971. Diversity in the study of the development of social behavior. In *The biopsychology of development*, ed. E. Tobach, L. R. Aronson and E. Shaw, pp. 433–455. New York: Academic Press.

————. 1975. Was Professor Lehrman an ethologist? *Animal Behav.* 23:957–964.

Bekoff, M. 1972. The development of social interaction, play, and meta-communication in mammals: An ethological perspective. *Quart. Rev. Biol.* 47:412–434.

————. 1974. Social play and play-soliciting by infant canids. *Amer. Zool.* 14:323–340.

————. 1976. Animal play: Problems and perspectives. In *Perspectives in ethology*, vol. 2, ed. P. P. G. Bateson and P. H. Klopfer, pp. 165–188. New York: Plenum.

————. in press. Mammalian dispersal and the ontogeny of individual behavioral phenotypes. *Amer. Naturalist.*

Bekoff, M., and Fox, M. W. 1972. Postnatal neural ontogeny: Environment-dependent and/or environment-expectant? *Dev. Psychobiol.* 5:323–341.

Bekoff, M.; Hill, H. L.; and Mitton, J. B. 1975. Behavioral taxonomy in canids by discriminant function analyses. *Science* 190:1223–1225.

Bell, R. G. 1968. A reinterpretation of the direction of effects in studies of socialization. *Psychol. Rev.* 75:81–95.

Berg, I. A. 1944. Development of behavior: The micturition pattern in the dog. *J. Exper. Psychol.* 34:343–368.

Beukema, J. J. 1968. Predation by the three-spined stickleback (*Gasterosteus aculeatus* L.): The influence of hunger and experience. *Behaviour* 31:1–126.

Blurton-Jones, N. G. 1972a. Characteristics of ethological studies of human behaviour. In *Ethological studies of child behavior*, ed. N. G. Blurton-Jones, pp. 3–33. New York: Cambridge University Press.

————. 1972b. Comparative aspects of mother-child contact. In *Ethological studies of child behaviour*, ed. N. G. Blurton-Jones, pp. 305–328. New York: Cambridge University Press.

Bols, R. J., and Wong, R. 1973. Gerbils reared by rats: Effects on adult open-field and ventral marking activity. *Behav. Biol.* 9:741–748.

Bronfenbrenner, U. 1968. Early deprivation in mammals: A cross-species analysis. In *Early experience and behavior*, ed. G. Newton and S. Levine, pp. 627–764. Springfield: C.C. Thomas.

Bronson, G. 1965. The hierarchical organization of the ventral nervous system: Implications for learning processes and critical periods in early development. *Behav. Sci.* 10:7–25.

Burghardt, G. 1968. Comparative pre-attack studies in newborn snakes of the genus *Thamnophis*. *Behaviour* 33:77–114.

Cairns, R. B. 1972. Fighting and punishment from a developmental perspective. *Nebraska Symp. on Motiv.* 20:59–124.

Cairns, R. B., and Johnson, D. L. 1965. The development of inter-species social preferences. *Psychon. Sci.* 2:337–338.

Cairns, R. B., and Werboff, J. 1967. Behavior development in the dog: An interspecific analysis. *Science* 158:1070–1072.

Candland, D. K. 1971. The ontogeny of emotional behavior. In *The ontogeny of vertebrate behavior*, H. Moltz, pp. 95–169. New York: Academic Press.

Candland, D. K., and Leshner, A. I. 1975. Socialization and adrenal functioning in the squirrel monkey. *5th Int. Dongr. Primat*, 238–244.

Cheal, M. 1975. Social olfaction: A review of the ontogeny of olfactory influences on vertebrate behavior. *Behav. Biol.* 14:1–25.

Chevalier-Skolnikoff, S. 1974. The ontogeny of communication in the stumptailed Macaque, *Macaca speciosa. Contrib. Primat.*, vol. 2.

Count, E. W. 1958. The biogenesis of human sociality: An essay in comparative vertebrate sociology. *Homo* 9:129–146; *Homo* 10:1–35, 65–92.

Crook, J. H. 1974. Social organization and the developmental environment. In *Ethology and psychiatry*, ed. N. F. White, pp. 142–152. Toronto: University of Toronto Press.

Cullen, J. M. 1972. Some principles of animal communication. In *Non-verbal communication*, ed. R. A. Hinde, pp. 101–125. New York: Cambridge University Press.

Davis, J. M. 1973. Imitation: A review and critique. In *Perspectives in ethology*, vol. 1, eds. P. P. G. Bateson and P. H. Klopfer, pp. 43–72. New York: Plenum.

Deag, J. M., and Crook, J. H. 1971. Social behaviour and "agonistic buffering" in the wild Barbary macaque *Macaca sylvana. Folia Primat.* 15:183–200.

DeGhett, V. J. in press. The ontogeny of ultrasonic vocalization in *Microtus montanus. Behaviour.*

Denenberg, V. H. 1964. Critical periods, stimulus input, and emotional reactivity: A theory of infantile stimulation. *Psychol. Rev.* 71:335–351.

———. 1971. The mother as motivator. *Nebraska Symp. on Motiv.* 17:69–93.

Devor, M. 1975. Neuroplasticity in the sparing or deterioration of function after early olfactory tract lesions. *Science* 190:998–1000.

Doty, R.L., and Kart, R. 1972. A comparative and developmental analysis of the midventral sebaceous glands in 18 taxa of *Peromyscus*, with an examination of gonadal steroid influences in *Peromyscus maniculata bairdii. J. Mammal.* 53:83–99.

Eibl-Eibesfeldt, I. 1967. Ontogenetic and maturation studies of aggressive behavior. In *Aggression and defense* (Brain Function, vol. 5), eds. C. D. Lemente and R. B. Lindzey, pp. 57–94. Los Angeles: University of California Press.

———. 1975. *Ethology.* New York: Holt, Rinehart, and Winston.

Eisenberg, J. F. 1966. The social organization of mammals. *Handb. Zool.* (Berlin) 8(10):1–92.

———. 1973. Mammalian social systems: Are primate social systems unique? *4th Int. Congr. Primat.* 1:232–249.

Eisenberg, J. F., and Kleiman, D. G. 1972. Olfactory communication in mammals. *Ann. Rev. Ecol. System.* 3:1–32.

Eisenberg, J. F., and Leyhausen, P. 1972. The phylogenesis of predatory behavior in mammals. *Z. Tierpsychol.* 30:59–93.

Eisner, E. 1960. The relationship of hormones to the reproductive behavior of birds, referring especially to parental behaviour: A review. *Anim. Behav.* 8:155–179.

Eisner, T., and Davis, J. A. 1967. Mongoose throwing and smashing millipedes. *Science* 155:577–579.

Eleftheriou, B. E., and Scott, J. P., eds. 1971. *The physiology of aggression and defeat.* New York: Plenum.

Espmark, Y. 1971. Individual recognition by voice in reindeer mother-young relationship. Field observations and playback experiments. *Behaviour* 40:295–301.

Etienne, A. S. 1973. Searching behaviour towards a disappearing prey in the domestic chick as affected by preliminary experience. *Anim. Behav.* 21:749–761.

Ewer, R. F. 1960. Natural selection and neoteny. *Acta Biotheoretica* 13:161–184.

――――. 1967. The behaviour of the African giant rat (*Cricetomys gambianus* Waterhouse.) Z. *Tierpsychol.* 24:6–79.

――――. 1968. *Ethology of mammals.* New York: Plenum.

――――. 1969. The "instinct to teach." *Nature* 223:698.

――――. 1971. The biology and behaviour of a free-living population of black rats (*Rattus rattus*). *Anim. Behav. Mono.* 4:127–174.

――――. 1973. *The carnivores.* New York: Cornell University Press.

Fagen, R. 1976. Play, exercise, and training in animals. In *Perspectives in ethology*, vol. 2, eds. P. P. G. Bateson and P. H. Klopfer, pp. 189–219. New York: Plenum.

――――. in press. Selection for optimal age-dependent schedules of play behavior. *Amer. Naturalist.*

Fedigan, L. 1972. Social and solitary play in a colony of vervet monkeys (*Cercopithecus aethiops*). *Primates* 13:347–364.

Fentress, J. 1972. Development and patterning of movement sequences in inbred mice. In *The biology of behavior*, ed. J. A. Kiger, pp. 83–131. Eugene: Oregon State University Press.

Fox, M. W. 1966. Neuro-behavioral ontogeny: A synthesis of ethological and neurophysiological concepts. *Br. Res.* 2:3–20.

――――. 1969. Behavioral effects of rearing dogs with cats during the 'critical period of socialization.' *Behaviour* 35:273–280.

――――. 1971. *Integrative development of brain and behavior in the dog.* Chicago: University of Chicago Press.

Frank, F. 1952. Adoptionversuche bei feldmaeusen (*Microtus arvalis* Pall). Z. *Tierpsychol.* 9:415–423.

Fraser, D. 1973. The nursing and suckling behavior of pigs. I. The importance of stimulation of the anterior teats. *Brit. Vet. J.* 129:324–336.

Freeland, W. J., and Jantzen, D. H. 1974. Strategies of herbivory by mammals: The role of plant secondary compounds. *Amer. Nat.* 108:269–289.

Fuller, J. L. 1976. Experiential deprivation and later behavior. *Science* 158:1645–1652.

Fullerton, C. J.; Berryman, J. C.; and Porter, R. H. 1974. On the nature of mother-young interaction in the guinea pig (*Cavia porcellus*). *Behaviour* 48:145–156.

Fullerton, C. J., and Cowley, J. J. 1971. The differential effect of the presence of adult male and female mice on the growth and development of the young. *J. Genet. Psychol.* 119:89–98.

Galef, B. G. 1976. The social transmission of acquired behavior: A discussion of tradition and social learning in vertebrates. *Adv. Study Behav.* 6:77–100.

Galef, B. G., and Clark, M. M. 1972. Mother's milk and adult presence: Two factors determining initial dietary selection by weanling rats. *J. Comp. Physiol. Psychol.* 78:220–225.

Galef, B. G., and Henderson, P. W. 1972. Mother's milk: A determinant of the feeding preferences of weaning rat pups. *J. Comp. Physiol. Psychol.* 78:213–219.

Galef, B. G., and Sherry, D. F. 1973. Mother's milk: A medium for transmission of cues reflecting the flavor of mother's milk. *J. Comp. Physiol. Psychol.* 83:374–378.

Geist, V. 1968a. On delayed social and physical maturation in mountain sheep. *Canad. J. Zool.* 46:899–904.

――――. 1968b. On the interaction of external appearance, social behaviour and social structure of mountain sheep. Z. *Tierpsychol.* 25: 199–215.

――――. 1971. *Mountain sheep.* Chicago: University of Chicago Press.

Georgescu-Roegen, N. 1971. The entropy law and the economic process. Cambridge: Harvard University Press.

Goss-Custard, J. D.; Dunbar, R. I. M.; and Aldrich-Blake, F. P. G. 1972. Survival, mating and rearing strategies in the evolution of primate social structure. *Folia Primat.* 17:1–19.

Gottlieb, G. 1971. Ontogenesis of sensory function in birds and mammals. In *The biopsychology of development*, eds. E. Tobach, L. R. Aronson, and E. Shaw, pp. 67–128. New York: Academic Press.

Gould, E. 1971. Studies of maternal-infant communication and development of vocalizations in the bats *Myotis* and *Eptesicus*. *Comm. Behav. Biol.* 5:263–313.

Hailman, J. P. 1970. Comments on the coding of releasing stimuli. In *Development and evolution of behavior*, eds. L. R. Aronson, E. Tobach, D. S. Lehrman, and J. S. Rosenblatt, pp. 138–157. San Francisco: Freeman.

Harestad, A. S., and Fisher, H. D. 1975. Social behavior in a non-pupping colony of Steller sea lions (*Eumetopias jubata*). *Canad. J. Zool.* 53:1596–1613.

Harlow, H. F. 1969. Age-mate or peer affectional system. *Adv. Study Behav.* 2:333–383.

Harper, L. V. 1968. The effects of isolation from birth on the social behaviour of guinea pigs at adulthood. *Anim. Behav.* 16:58–64.

————. 1970. Ontogenetic and phylogenetic functions of the parent-offspring relationship in mammals. *Adv. Study Behav.* 3:75–117.

————. 1971. The young as a source of stimuli controlling caretaker development. *Dev. Psychol.* 4:73–88.

Hein, A.; Held, R.; and Gower, E. C. 1970. Development and segmentation of visually controlled movement by selective exposure during rearing. *J. Comp. Physiol. Psychol.* 73:181–187.

Held, R., and Hein, A. 1963. Movement-produced stimulation in the development of visually guided behavior. *J. Comp. Physiol. Psychol.* 56:872–876.

Hellwing, S. 1973. The postnatal development of the white-toothed shrew *Crocidura russula monacha* in captivity. *Z. Saeugetierkunde* 38:257–270.

Herrenkohl, L. R., and Sachs, B. D. 1972. Sensory regulation of maternal behavior in mammals. *Physiol. Behav.* 9:689–692.

Hershenson, M. 1971. The development of visual perceptual systems. In *The ontogeny of vertebrate behavior*, ed. H. Moltz, pp. 29–56. New York: Academic Press.

Hess, E. H. 1973. *Imprinting.* New York: Van Nostrand Reinhold Company.

Hinde, R. A., ed. 1969. *Bird vocalizations.* New York: Cambridge University Press.

————. 1970. *Animal behavior.* New York: McGraw-Hill.

————. 1971. Development of social behavior. In *Behavior of non-human primates*, vol. 3, eds. A. M. Schrier and F. Stollnitz, pp. 1–68. New York: Academic Press.

————. 1972. *Social behavior and its development in subhuman primates.* Eugene: University of Oregon Books.

————. 1974. *Biological bases of human social behaviour.* New York: McGraw-Hill.

Hinde, R. A., and Stevenson-Hinde, J., eds. 1973. *Constraints on learning.* New York: Academic Press.

Hogan, J. 1973. How young chicks learn to recognize food. In *Constraints on learning*, eds. R. A. Hinde and J. Stevenson-Hinde, pp. 119–139. New York: Academic Press.

Horwich, R. H. 1972. The ontogeny of social behavior in the gray squirrel (*Sciurus carolinensis*). *Z. Tierpsychol.* Suppl. 8.

————. 1974. Development of behaviors in a male spectacled langur (*Presbytis obscurus*). *Primates* 15:151–178.

Immelmann, K. 1972. Sexual and other long-term aspects of imprinting in birds and other species. *Adv. Study Behav.* 4:147–174.

————. 1975. Ecological significance of imprinting and early learning. *Ann. Rev. Ecol. Syst.* 6:15–37.

Jensen, G. C.; Bobbitt, R. A.; and Gordon, B. N. 1968. Sex differences in the development of independence of infant monkeys. *Behaviour* 30:1–14.

Johnson, R. N. 1972. *Aggression in man and animals.* Philadelphia: W. B. Saunders Company.

Johnson, R. P. 1973. Scent marking in mammals. *Anim. Behav.* 21:521–535.

Jolly, A. 1972. *The evolution of primate behavior.* New York: Macmillan.

King, J. A. 1968. Species specificity and early experience. In *Early experience and behavior*, eds. G. Newton and S. Levine, pp. 42–64. Springfield, Ill.: C.C. Thomas.

Kleiman, D. G., and Eisenberg, J. F. 1973. Comparisons of canid and felid social systems from an evolutionary perspective. *Anim. Behav.* 21:637–659.

Klopfer, P. H. 1971. Mother Love: What turns it on? *Amer. Sci.* 59:404–407.

Knudtson, P. 1974. Birth of a harbor seal. *Nat. Hist.* 88:30–37.

Kruuk, H., and Turner, M. 1967. Comparative notes on predation by lion, leopard, cheetah and wild dog in the Serengeti area, East Africa. *Mammalia* 31:1–27.

Kuo, Z. Y. 1930. The genesis of the cat's response toward the rat. *J. Comp. Psychol.* 15:1–35.

———. 1967. *The dynamics of behavioral development: An epigenetic view.* New York: Random House.

Lamb, M. E. 1975. Physiological mechanisms in the control of maternal behavior in rats: A review. *Psychol. Bull.* 82:104–119.

Lee, C. T., and Estep, D. 1971. The development aspect of marking and nesting behaviors in Mongolian gerbils (*Meriones unguiculatus*). *Psychon. Sci.* 22:312–313.

Lehrman, D. S. 1953. A critique of Konrad Lorenz's theory of instinctive behavior. *Quart. Rev. Biol.* 28:337–363.

———. 1961. Hormonal regulation of parental behavior in birds and infrahuman mammals. In *Sex and internal secretions*, ed. W. C. Young, pp. 1268–1382. Baltimore: Williams and Wilkins.

———. 1970. Semantic and conceptual issues in the nature-nurture problem. In *Development and evolution of behavior*, ed. L. R. Aronson, E. Tobach, D. S. Lehrman, and J. S. Rosenblatt. San Francisco: Freeman.

Lenneberg, E. H. 1967. *Biological foundations of language.* New York: John Wiley and Sons.

Leon, M., and Moltz, H. 1971. Maternal pheromone: Discrimination by pre-weanling albino rats. *Physiol. Behav.* 7:265–267.

Lewis, M., and Rosenblum, L. A., eds. 1974. *The effect of the infant on its caregiver.* New York: John Wiley and Sons.

Leyhausen, P. 1973. Verhaltensstudien an katzen. *Z. Tierpsychol.* Suppl. 2.

Lindzey, G.; Thiessen, D. D.; and Tucker, A. 1968. Development and hormonal control of territorial marking in the male mongolian gerbil (*Meriones unguiculatus*). *Dev. Psychobiol.* 1:97–99.

Lorenz, K. 1965. *Evolution and modification of behavior.* Chicago: University of Chicago Press.

Loy, J. 1974. Social development and modification among primates. *Amer. J. Phys. Anthro.* 40:144.

Loy, J., and Loy, K. 1974. Behavior of an all-juvenile group of rhesus monkeys. *Amer. J. Phys. Anthro.* 40:83–96.

Marler, P. 1965. Communication in monkeys and apes. In *Primate behavior*, ed. I. DeVore, pp. 544–583. New York: Holt, Rinehart, and Winston.

Marler, P. 1976. Sensory templates in species-specific behavior. In *Simpler networks: An approach to patterned behavior and its foundations*, ed. J. Fentress. Massachusetts: Sinauer.

Marler, P., and Hamilton, W. J. 1966. *Mechanisms of animal behavior.* New York: John Wiley and Sons.

Marler, P., and Mundinger, P. 1971. Vocal learning in birds. In *The ontogeny of vertebrate behavior*, ed. H. Moltz, pp. 389–450. New York: Academic Press.

Martinka, C.J. 1974. Population characteristics of grizzly bears in Glacier National Park, Montana. *J. Mammal.* 55:21–29.

Martins, T., and Valle, J. R. 1948. Hormonal regulation of the micturition behavior of the dog. *J. Comp. Physiol. Psychol.* 41:301–311.

Mason, W. A. 1971. Motivational factors in psychosocial development. *Nebraska Symp. on Motiv.* 18:35–67.

Mech, L. D. 1970. *The wolf.* New York: Natural History Press.

Mitchell, G. D. 1969. Paternalistic behavior in primates. *Psychol. Bull.* 71:399–417.

Moltz, H. 1971. The ontogeny of maternal behavior in some selected mammalian species. In *The ontogeny of vertebrate behavior*, ed. H. Moltz, pp. 263–313. New York: Academic Press.

Müller-Schwarze, D. 1972. Responses of young black-tailed deer to predator odors. *J. Mammal.* 53:393–394.

Müller-Schwarze, D., and Müller-Schwarze, C. 1972. Social scents in hand-reared pronghorn (*Antilocapra americana*). *Zool. Africana* 7:257–271.

Mueller, H. 1974. The development of prey recognition and predatory behavior in the American destrel, *Falco sparverius. Behaviour* 49:313–324.

Mugford, R. A., and Nowell, N. W. 1972. Paternal stimulation during infancy: Effects on aggression and open-field performance of mice. *J. Comp. Physiol. Psychol.* 79:30–36.

Nadel, L.; O'Keefe, J.; and Black, A. 1975. Slam on the brakes: A critique of Altman, Brunner, and Bayer's response-inhibition model of hippocampal function. *Behav. Biol.* 14:151–162.

Neuringer, A., and Neuringer, M. 1974. Learning by following a food source. *Science* 184:1005–1008.

Nice, M. M. 1962. *Studies in the life history of the song sparrow.* New York: Dover.

Noirot, E. 1972a. The onset of maternal behavior in rats, hamsters, and mice: A selected review. *Adv. Study Behav.* 4:107–145.

———. 1972b. Ultrasounds and maternal behavior in small rodents. *Dev. Psychobiol.* 5:371–387.

Nottebohm, F. 1971. Neural lateralization of vocal control in a passerine bird, I. Song. *J. Exp. Zool.* 177:229–262.

———. 1972. Neural lateralization of vocal control in a passerine bird. II. Subsong, calls, and a theory of vocal learning. *J. Exp. Zool.* 179:35–50.

Okon, E. E. 1970. The ultrasonic responses of albino mice to tactile stimulation. *J. Zool. Lond.* 162:485–492.

Oliverio, A. 1974. Evolutionary mechanisms in behaviour: An intraspecific genetic approach. *J. Human Evol.* 3:1–18.

Orians, G. 1969. On the evolution of mating systems in birds and mammals. *Amer. Nat.* 103:589–603.

Overmann, S. R. 1976. Dietary self-selection by animals. *Psychol. Bull.* 83:218–235.

Paschke, R. E.; Denenberg, V. H.; and Zarrow, M. X. 1971. Mice reared with rats: An interstrain comparison of mother and "aunt" effects. *Behaviour* 38:315–331.

Poole, T. B. 1966. Aggressive play in polecats. *Symp. Zool. Soc. Lond.* 18:23–44.

Rabb, B. G.; Woolpy, J. H.; and Ginsberg, B. E. 1967. Social relationships in a group of captive wolves. *Amer. Zool.* 7:305–311.

Ransom, T. W.; and Ransom, B. S. 1971. Adult male-infant relations among baboons (*Papio anubis*). *Folia Primat.* 16:179–195.

Rasa, O. A. E. 1971. Social interaction and object manipulation in weaned pups of the northern elephant seal, *Mirounga angustirostirs. Z. Tierpsychol.* 29:82–102.

———. 1972. Aspects of social organization in captive dwarf mongooses. *J. Mammal.* 53:181–185.

———. 1973. Prey Capture, feeding techniques, and their ontogeny in the African dwarf mongoose, *Helogale undulata rufula. Z. Tierpsychol.* 32:449–488.

Rheingold, H. L., ed. 1963. *Maternal behavior in mammals.* New York: John Wiley and Sons.

Richards, M. P. M. 1974. *The integration of a child into a social world.* New York: Cambridge University Press.

Richards, R. J. 1974. The innate and the learned: The evolution of Konrad Lorenz's theory of instinct. *Phil. Soc. Sci.* 4:111–133.

Robinson, M. H. 1969. Defenses against visually hunting predators. *Evol. Biol.* 3:225–259.

Rosenblatt, J. S. 1967. Nonhormonal basis of maternal behavior in the rat. *Science* 156:1512–1514.

———. 1970. Views on the onset and maintenance of maternal behavior in the rat. In *Development and evolution of behavior,* eds. L. R. Aronson, E. Tobach, D. S. Lehrman, and J. S. Rosenblatt, pp. 489–515. San Francisco: Freeman.

Rosenblatt, J. S.; Turkewitz, G.; and Schneirla, T. C. 1959. Early socialization in the domestic cat as based on feeding and other relationships between female and young. In *Determinants of infant behaviour,* vol. 1, ed. B. M. Foss, pp. 51–74. London: Methuen.

Rowell, T. E. 1961. Maternal behaviour in non-maternal golden hamsters (*Mesocricetus auratus*). *Anim. Behav.* 9:11–15.

Rudge, M. R. 1970. Mother and kid behaviour in feral goats. *Z. Tierpsychol.* 27:687–692.

Russell, E. M. 1974. Biology of kangaroos (*marsupalia-macropodidae*). *Mammal Rev.* 4:1–59.

Sackett, G. P., and Ruppenthal, G. C. 1973. Development of monkeys after varied experiences during infancy. In *Ethology and development,* ed. S. A. Barnett, pp. 52–87. London: Spastics International Medical Publication with Heinemann Medical Books Limited.

Salzen, E. A. 1967. Imprinting in birds and primates. *Behaviour* 28:232–254.

Savidge, I. R. 1974. Social factors in dispersal of deer mice (*Peromyscus maniculatus*) from their natal site. *Amer. Mid. Nat.* 91:395–405.

Sayler, A. and Salmon, M. 1971. An ethological analysis of communal nursing by the house mouse (*Mus musculus*). *Behaviour* 40:68–85.

Schaller, G. B. 1972. *The Serengeti lion*. Chicago: University of Chicago Press.

Schenkel, R. 1967. Submission: Its features and function in the wolf and dog. *Amer. Zool.* 7:319–329.

Schneirla, T. C., and Rosenblatt, J. S. 1963. Critical periods in the development of behaviour. *Science* 139:1110–1116.

Schoenberger, D. 1965. Beobachtungen zur Fortpflanzungsbiologie des Wolfes, *Canis lupus*. *Z. Saeugetierkunde* 30:171–178.

Scott, J. P. 1962. Critical periods in behavioral development. *Science* 138:949–958.

_____. 1970. Critical periods for the development of social behaviour in dogs. In *The post-natal development of phenotype*, eds. S. Kazda and J. H. Denenberg, pp. 21–32. London: Butterworths.

Scott, J. P., and Fredericson, E. 1951. The causes of fighting in mice and rats. *Physiol. Zool.* 24:273–309.

Scott, J. P., and Marston, M. 1950. Critical periods affecting the development of normal and mal-adjustive social behavior of puppies. *J. of Genet. Psychol.* 77:25–60.

Scott, J. P.; Stewart, J. M.; and DeGhett, V. J. 1974. Critical periods in the organization of systems. *Dev. Psychobiol.* 7:489–513.

Sedláček, J. 1974. The significance of the perinatal period in the neural and behavioral development of precocial mammals. In *Aspects of neurogenesis*, vol. 2, ed. G. Gottlieb, pp. 245–273. New York: Academic Press.

Sluckin, W. 1972. *Early learning in man and animal*. Massachusetts: Schenkman.

_____. 1973. *Imprinting and early learning*, 2d Edition. Chicago: Aldine.

Smith, S. M. 1973. A study of prey-attack behaviour in young Loggerhead shrikes,*Lanius ludovicianus* L. *Behaviour* 44:113–141.

Solarz, A. K. 1970. Behavior. In *The beagle as an experimental dog*, ed. A. C. Andersen, pp. 453–468. Iowa: Iowa State University Press.

Spencer-Booth, Y. 1970. The relationships between mammalian young and conspecifics other than mothers and peers: A review. *Adv. Study of Behav.* 3:119–194.

Stanley, M. 1971. An ethogram of the hopping mouse (*Notomys alexis*). *Z. Tierpsychol.* 29:225–258.

Sterman, M. D.; McGinty, D. J.; and Adinolfi, A. M., eds. 1971. *Brain development and behavior.* New York: Academic Press.

Stirling, I. 1970. Studies on the behaviour of the South Australian fur seal, *Arctocephalus forsteri* (Lesson). *Aust. J. Zool.* 19:267–273.

Symons, D. 1974. Aggressive play and communication in rhesus monkeys (*Macaca mulatta*). *Amer. Zool.* 14:317–322.

Terkel, J., and Rosenblatt, J. S. 1972. Humoral factors underlying maternal behavior at parturition: Cross transfusion between freely moving rats. *J. Comp. Physiol. Psychol.* 80:365–371.

Thorpe, W. H. 1961. *Bird song*. New York: Cambridge University Press.

Thompson, W. R., and Schaeffer, T. 1961. Early environmental stimulation. In *Functions of varied experience*, eds. D. W. Fiske and S. R. Maddi, pp. 81–105. Homewood: The Dorsey Press.

Tinbergen, N. 1951. *The study of instinct*. New York: Oxford University Press.

Tobach, E.; Aronson, L. R.; and Shaw, E., eds. 1971. *The biopsychology of development*. New York: Academic Press.

Trivers, R. L. 1972. Parental investment and sexual selection. In *Sexual selection and the descent of man*, ed. B. Campbell, pp. 136–179. Chicago: Aldine.

_____. 1974. Parent-offspring conflict. *Amer. Zool.* 14:249–264.

Tyler, S. J. 1972. The behaviour and social organization of the new forest ponies. *Anim. Behav. Mono.* 5: 87–196.

Vestal, B. M. 1973. Ontogeny of visual acuity in two species of deermice (*Peromyscus*). *Anim. Behav.* 21:711–719.

Vetter, H. J., and Howell, R. W. 1971. Theories of language acquisition. *J. Psycholing. Res.* 1:31–64.

Vincent, L., and Bekoff, M. in press. A quantitative analysis of the ontogeny of predatory behavior in coyotes, *Canis latrans. Anim. Behav.*

Welker, W. I. 1964. Analysis of sniffing of the albino rat. *Behaviour* 22:223–244.

Westoby, M. 1975. An analysis of diet selection by large generalist herbivores. *Amer.Natur.* 108:290–304.

Whalen, R. E. 1971. The ontogeny of sexuality. In *The ontogeny of vertebrate behavior*, ed. H. Moltz, pp. 229–261. New York: Academic Press.

Whitsett, J. M., and Thiessen, D. D. 1972. Sex difference in the control of scent-marking behavior in the Mongolian gerbil (*Meriones unguiculatus*). *J. Comp. Physiol. Psychol.* 78:381–385.

Williams, E., and Scott, J. P. 1953. The development of social behavior patterns in the mouse in relation to natural periods. *Behaviour* 6:35–64.

Wilson, S. C., and Kleiman, D. G. 1974. Eliciting and soliciting play: A comparative study. *Amer. Zool.* 14:341–370.

Wüstehube, C. 1960. Beitraege zur Kenntnis besonders des Spiel- und Beutefangverhaltens einheimischer Musteliden. *Z. Tierpsychol.* 17:579–613.

Yanai, J., and McClearn, G. E. 1972. Assortative mating in mice and the incest taboo. *Nature* 238: 281–282.

Zimen, E. 1975. Social dynamics of the wolf pack. In *The wild canids*, ed. M. W. Fox, pp. 336–362. New York: Van Nostrand Reinhold Company.

Zimmerman, R. R.; Geist, C. R.; and Wise, L. A. 1974. Behavioral development, environmental deprivation, and malnutrition. *Adv. Psychobiol.* 2:133–191.

Zippelius, H. M. 1972. Die karawanenbildung bei feld- und hausspitzmaus. *Z. Tierpsychol.*

———. 1974. Ultraschall-Laute nestjunger Maeuse. *Behaviour* 48:197–204.